ISLAND ECOSYSTEMS

US/IBP SYNTHESIS SERIES

This volume is a contribution to the International Biological Program. The United States effort was sponsored by the National Academy of Sciences through the National Committee for the IBP. The lead federal agency in providing support for IBP has been the National Science Foundation.

Views expressed in this volume do not necessarily represent those of the National Academy of Sciences or the National Science Foundation.

Volume

1 MAN IN THE ANDES: A Multidisciplinary Study of High-Altitude Quechua / *Paul T. Baker and Michael A. Little*

2 CHILE-CALIFORNIA MEDITERRANEAN SCRUB ATLAS: A Comparative Analysis / *Norman J. W. Thrower and David E. Bradbury*

3 CONVERGENT EVOLUTION IN WARM DESERTS: An Examination of Strategies and Patterns in Deserts of Argentina and the United States / *Gordon H. Orians and Otto T. Solbrig*

4 MESQUITE: Its Biology in Two Desert Scrub Ecosystems / *B. B. Simpson*

5 CONVERGENT EVOLUTION IN CHILE AND CALIFORNIA: Mediterranean Climate Ecosystems / *Harold A. Mooney*

6 CREOSOTE BUSH: Biology and Chemistry of *Larrea* in New World Deserts / *T. J. Mabry, J. H. Hunziker, and D. R. DiFeo, Jr.*

7 BIG BIOLOGY: The US/IBP / *W. Frank Blair*

8 ESKIMOS OF NORTHWESTERN ALASKA: A Biological Perspective / *Paul L. Jamison, Stephen L. Zegura, and Frederick A. Milan*

9 NITROGEN IN DESERT ECOSYSTEMS / *N. E. West and John Skujins*

10 AEROBIOLOGY: The Ecological Systems Approach / *Robert L. Edmonds*

11 WATER IN DESERT ECOSYSTEMS / *Daniel D. Evans and John L. Thames*

12 AN ARCTIC ECOSYSTEM: The Coastal Tundra at Barrow, Alaska / *Jerry Brown, Philip C. Miller, Larry L. Tieszen, and Fred L. Bunnell*

13 LIMNOLOGY OF TUNDRA PONDS: Barrow Alaska / *John E. Hobbie*

14 ANALYSIS OF CONIFEROUS FOREST ECOSYSTEMS IN THE WESTERN UNITED STATES / *Robert L. Edmonds*

15 ISLAND ECOSYSTEMS: Biological Organization in Selected Hawaiian Communities / *Dieter Mueller-Dombois, Kent W. Bridges, and Hampton L. Carson*

US/IBP Synthesis Series | 15

ISLAND ECOSYSTEMS

Biological Organization in Selected Hawaiian Communities

Edited by

Dieter Mueller-Dombois
University of Hawaii

Kent W. Bridges
University of Hawaii

Hampton Carson
University of Hawaii

ISLAND ECOSYSTEMS
Biological Organization in Selected Hawaiian Communities

ISBN-10: 1-932846-27-1
ISBN-13: 978-1-932846-27-0

Library of Congress Control Number: 2012950531

THE BLACKBURN PRESS
P. O. Box 287
Caldwell, New Jersey 07006 U.S.A.
973-228-7077
www.BlackburnPress.com

CONTENTS

FOREWORD

This book is one of a series of volumes reporting results of research by U.S. scientists participating in the International Biological Program (IBP). As one of the fifty-eight nations taking part in the IBP during the period of July 1967 to June 1974, the United States organized a number of large, multidisciplinary studies pertinent to the central IBP theme of "the biological basis of productivity and human welfare."

These multidisciplinary studies (Integrated Research Programs) directed toward an understanding of the structure and function of major ecological or human systems have been a distinctive feature of the U.S. participation in the IBP. Many of the detailed investigations that represent individual contributions to the overall objectives of each Integrated Research Program have been published in the journal literature. The main purpose of this series of books is to accomplish a synthesis of the many contributions for each principal program and thus answer the larger questions pertinent to the structure and function of the major systems that have been studied.

Publications Committee: US/IBP
Gabriel Lasker
Robert B. Platt
Frederick E. Smith
W. Frank Blair, Chairman

PREFACE to the 2012 Printing US/IBP Synthesis 15

The books of the US/IBP Synthesis Series, a **U**nited **S**tates Contribution to the **I**nternational **B**iological **P**rogram, went out of print after Hutchinson Ross Publishing Company closed its business in the late 1980s.

This book, *ISLAND ECOSYSTEMS: Biological Organization in Selected Hawaiian Communities*, (first published 1981, republished 2012) remained in strong demand over the following years. It still serves as a basic background document for future research. It was the first book that focused on biodiversity assessment of ecosystems in a remote archipelago of the Pacific, a major task accomplished through a coordinated multi-disciplinary team of 35 senior and 25 junior field researchers. The senior researchers are listed on the frontal pages xvii-xx. Most of them were taxonomists and evolutionary biologists, who were encouraged to sample their specific organism groups along ecological transects at predetermined focal sites. The minimum requirement for each participating specialist was to identify his (her) species assembly at that transect location and to quantify each species in a form optimal for the group. Sample plot sizes were adapted to species behavior. For example bird surveys needed larger sample areas than soil arthropods or plant community assemblages.

Seventy-seven Technical Reports were published during the active field research phase from 1970 through 1976. They are listed on pages 550-554 and are available in the Hamilton Library of the University of Hawai`i. A similar number of articles was published in journals appropriate for each group of species. The synthesis of results was accomplished through use of ecological methods explained in the book, *Aims and Methods of Vegetation Ecology* (Dieter Mueller-Dombois and Heinz Ellenberg, John Wiley & Sons 1974, republished by The Blackburn Press 2002). These methods relate primarily to the Synthesis Tabulation Technique and to establishing Life-form or Guild Spectra. The latter allowed to specify aspects of species function and their roles in the community.

Since its first publication in 1981, *Island Ecosystems* has spawned other books with the writers' involvement, *Vegetation of the Tropical Pacific Islands* (with F. Ray Fosberg, Springer-Verlag 1998); *Hawai`i, The Fires of Life* (with Garrett A. Smathers, Mutual Publishing, Honolulu 2007); *Biodiversity Assessment of Tropical Island Ecosystems: PABITRA Manual for Interactive Ecology and Management* (with K.W. Bridges and Curtis Daehler, Bishop Museum Press 2008); and *`Ōhi`a Lehua Rainforest, Born Among Hawaiian Volcanoes, Evolved in Isolation. The Story of a Dynamic Ecosystem with Relevance to Forests Worldwide* (with James D. Jacobi, Jonathan P. Price, and Hans Juergen Boehmer, available Winter 2012 from Amazon.com).

Dieter Mueller-Dombois, PhD, Dr. h.c.
Kailua, HI
August 15, 2012

PREFACE

This book is a synthesis of the methods and results of thirty-five investigators who used a team approach to study the biological organization in selected natural ecosystems of the Hawaiian Islands. The research was done between 1971 and 1976 as a part of the U.S. contribution to the International Biological Program (IBP). The Hawaii IBP was one of three subprograms under the Origin and Structure of Ecosystems Program, which formed a complement to the five US-IBP biome programs. The Hawaii subprogram was financed almost entirely by the National Science Foundation. A small supplementary grant was received from the Bishop Estate Corporation of Hawaii. Both grants are herewith gratefully acknowledged.

During the research phase, progress was reported in a series of technical reports and in the open literature. Our technical reports were known as US/IBP Island Ecosystems Integrated Research Program (IRP) Technical Reports. They were distributed to other IBP groups and to a number of libraries in mimeographed form by the Department of Botany, University of Hawaii (at Manoa) in Honolulu. Seventy-seven reports were produced; these are now permanently available on microfiche. Throughout this book reference is made to individual technical reports. These are cited as TR (Technical Report) numbers (for example, TR 16), occasionally with author names (for example, Howarth, TR 16) in place of dates. Open literature citations are given additionally for reports that were subsequently published, wherever this information was made available during the preparation of this synthesis volume. A list of running numbers with topics of the technical reports is appended to this volume. To obtain a technical report, write to: Reprography Department, Hamilton Library, 2550 The Mall, University of Hawaii at Manoa, Honolulu, Hawaii 96822 U.S.A. A microfiche sheet holds up to sixty report pages. The charge per microfiche sheet is sixty cents including air postage in the United States. International mailing requires additional postage. Prepayment is requested according to University of Hawaii fiscal regulations.

This synthesis is a radical reduction of the data and results that were accumulated by the program participants during the five-year research phase. By necessity, many details were omitted, even though some of them—for example the discovery of new island species—merit special summaries and synoptic discussions.

Part I outlines the scope of the book, gives some of the background, and presents the framework for our integrated research project. The information presented here is based to some extent on our 1970 research proposal and on the many workshops that were held, particularly during the first three years of the project. The research proposal was an initial multidisciplinary team effort with major contributions provided by A. J. Berger, J. L. Gressitt, D. Mueller-Dombois, G. C. Ashton, J. W. Beardsley, D. E. Hardy, and C. H. Lamoureux. Sheila Conant, E. Alison Kay, and F. J. Radovsky contributed further information to the introductory survey.

The main thrust of our IBP subprogram was directed to discovering principles relating to the organization of island organisms in various natural communities and ecosystems. As such we divided our results into four main parts.

Part II relates to the spatial organization of thirteen organism groups (plants, birds, rodents, canopy arthropods, soil arthropods, soil microfungi, and so on) along a high island mountain gradient that traverses from alpine scrub into montane tropical rain forest. Because of too many disturbed situations and sampling difficulties, it was not possible to extend this mountain transect down to sea level. Part III deals with the community structure and general niche differentiation of organisms in two representative island ecosystems, a montane rain forest and the newly discovered but widely occurring Hawaiian lava tube ecosystem. Part IV is concerned with the short temporal behavior of island organisms in a variety of ecosystem types. Only five organism groups are discussed because of insufficient time-sequence data available for other groups. Some aspects of longer-term temporal relationships are dealt with in part III under ecosystem maintenance trends and in chapter 16. Part V reports on the structure of the genetic systems and microevolutionary developments in representative island species, with particular emphasis on drosophilid flies.

Each of the four main parts is divided into three components; an introductory chapter, one or two result chapters, and a higher-level synthesis chapter. Readers who are interested primarily in the principles developed from this research can obtain this information by reading the four higher-level synthesis chapters for each result part. These are chapter 4 (for part II), chapter 8 (for part III), chapter 11 (for part IV), and chapter 14 (for part V).

A major challenge in the synthesis procedure was to develop a unified data-processing and display technique for the multidisciplinary research components, which are reported in parts II, III, and IV. We found such techniques by using well-established models developed in the field of vegetation analysis and synthesis. For part II, the spatial gradient analysis, we used the European synthesis table technique (Mueller-Dombois and Ellenberg, 1974) and converted the tabulated data into a diagrammatic display model borrowed in part from Whittaker's (1967) bell-shaped species curves. A departure from the Whittaker model in our display format is the separate arrangement of species curves on each species distribution diagram. For part III, the community structure and niche differentiation analysis, we used the data display method developed for plant life-form spectra. This method works equally well for the quantitative display of animal guilds. With this display model, taxonomic diversity of structural and functional species groups can be shown numerically and diagrammatically in any ecosystem context. For part IV, the annual-cyclic behavior in organism groups, we used a simple modification of the species distribution diagrams. Instead of showing quantitative changes in species distribution over a spatial gradient, quantitative changes in species distribution and other phenological phenomena, such as vegetative flushing and flowering, were displayed diagrammatically over a time period.

These three models for uniform data display, which were used for the different organism groups in the result portions of parts II, III and IV, became a major facilitation for writing the higher-level synthesis chapters.

Finally part VI gives a general summary and the major conclusions with regard to the fundamental question: What is unique about the ecology of island ecosystems? Both the ecological and microevolutionary approaches, which involve very different research methods, came to essentially similar conclusions. These, quite unexpectedly, represent considerable departures from some of our initial hypotheses. The last chapter, which addresses the question of island forest maintenance under natural conditions, is particularly directed to the more practical concerns of land managers and policy makers.

The general conclusion derived from our IBP studies is that biological conservation of island ecosystems, in spite of other pressures, is a realizable goal. Indigenous island biota are not so easily displaced from their habitat as has often been assumed. Humans play by far the most important role. Because of this added responsibility, our efforts to understand the dynamics of island ecosystems should be increased through continued research and information feedback with land managers.

Throughout the text we have used the Hawaiian spelling for Hawaiian place names, certain geological terms, and for common names of plants and birds. This spelling includes two diacritical marks, the glottal stop and the macron. The correct spelling is given in the latest Hawaiian Dictionary (Pukui and Elbert, 1975), which we followed. The glottal stop replaces a consonant by separating the vowel sounds, and the macron puts an accent on the syllable. For example, the Hawaiian word, 'a'ā, which we have used several times, refers to the coarse, clinker-type lava. The word has several meanings depending on where the diacritical marks are placed. 'A'a means to dare, to take a risk or accept a challenge. 'Ā'ā means an inability to speak intelligently, to stutter or stammer. It also refers to a small person or dwarf. A'a means a small root, vein, artery, nerve, or muscle.

The use of Hawaiian spelling has become more and more important in recent years with advancing liguistic research of the Polynesian languages. Its special way of spelling can, therefore, no longer be ignored in scientific publications, particularly if these address an international readership, such as found in the different nations of the Pacific basin and elsewhere.

During its preparation, the manuscript was reviewed in successive increments by F. R. Fosberg, who provided many constructive criticisms. T. Nakata prepared our line drawings. We are grateful to both of these individuals for their important contributions. Our sincere appreciation also goes to Barbara Carr and Harriet Matsumoto for their untiring and excellent services in typing the manuscript.

Finally, we wish to acknowledge two grants received for the production of a camera-ready book manuscript, one of $3500 from the University of Hawaii Foundation and another of $1200, which was provided by a memorial fund for Setsuko Nakata, respected staff member of the Department of Entomology, B. P. Bishop Museum.

Dieter Mueller-Dombois
Kent W. Bridges
Hampton L. Carson

LIST OF CONTRIBUTORS

G. E. Baker
Department of Botany, University of Hawaii at Manoa, Honolulu, Hawaii 96822

J. W. Beardsley
Department of Entomology, University of Hawaii at Manoa, Honolulu, Hawaii 96822

A. J. Berger
Department of Zoology, University of Hawaii at Manoa, Honolulu, Hawaii 96822

B. M. Brennan
Department of Agricultural Biochemistry, University of Hawaii at Manoa, Honolulu, Hawaii 96822

K. W. Bridges
Department of Botany, University of Hawaii at Manoa, Honolulu, Hawaii 96822

H. L. Carson
Department of Genetics, University of Hawaii at Manoa, Honolulu, Hawaii 96822

S. Conant
Department of General Sciences, University of Hawaii at Manoa, Honolulu, Hawaii 96822

R. G. Cooray
Department of Botany, University of Hawaii at Manoa, Honolulu, Hawaii 96822

E. M. Craddock
Department of Genetics, University of Hawaii at Manoa, Honolulu, Hawaii 96822

M. D. Delfinado
Department of Entomology, University of Hawaii at Manoa, Honolulu, Hawaii 96822

M. S. Doty
Department of Botany, University of Hawaii at Manoa, Honolulu, Hawaii 96822

D. Fujii
Department of Entomology, University of Hawaii at Manoa, Honolulu, Hawaii 96822

W. Gagné
Department of Entomology, B. P. Bishop Museum, Honolulu, Hawaii 96819

J. L. Gressitt
Department of Entomology, B. P. Bishop Museum, Honolulu, Hawaii 96819

D. E. Hardy
Department of Entomology, University of Hawaii at Manoa, Honolulu, Hawaii 96822

F. G. Howarth
Department of Entomology, B. P. Bishop Museum, Honolulu, Hawaii 96819

W. E. Johnson
Department of Genetics, University of Hawaii at Manoa, Honolulu, Hawaii 96822

C. H. Lamoureux
Department of Botany, University of Hawaii at Manoa, Honolulu, Hawaii 96822

J. R. Leeper
Department of Entomology, University of Hawaii at Manoa, Honolulu, Hawaii 96822

J. E. Maka
Department of Botany, University of Hawaii at Manoa, Honolulu, Hawaii 96822

W. C. Mitchell
Department of Entomology, University of Hawaii at Manoa, Honolulu, Hawaii 96822

D. Mueller-Dombois
Department of Botany, University of Hawaii at Manoa, Honolulu, Hawaii 96822

L. J. Newman
Department of Biology, Portland State University, Portland, Oregon 97207

Y. K. Paik
Department of Genetics, University of Hawaii at Manoa, Honolulu, Hawaii 96822

F. J. Radovsky
Department of Entomology, B. P. Bishop Museum, Honolulu, Hawaii 96819

G. A. Samuelson
Department of Entomology, B. P. Bishop Museum, Honolulu, Hawaii 96819

G. Spatz
Institut für Grünlandlehre der Technischen Universität, München, West Germany

D. Springer
Department of Entomology, University of Hawaii at Manoa, Honolulu, Hawaii 96822

W. A. Steffan
Department of Entomology, B. P. Bishop Museum, Honolulu, Hawaii 96819

W. W. M. Steiner
Department of Genetics, University of Hawaii at Manoa, Honolulu, Hawaii 96822

M. F. Stoner
Department of Biological Sciences, California State Polytechnic University, 3801 W. Temple Avenue, Pomona, California 91678

K. C. Sung
Department of Entomology, University of Hawaii at Manoa, Honolulu, Hawaii 96822

J. M. Tenorio
Department of Entomology, B. P. Bishop Museum, Honolulu, Hawaii 96819

P. Q. Tomich
Research Unit Hawaii State Department of Health, Honokaa, Hawaii 96727

L. J. Watson
Department of Botany, University of Hawaii at Manoa, Honolulu, Hawaii 96822

I

Introductory Survey

1

Some Bioenvironmental Conditions and the General Design of IBP Research in Hawai‘i

D. Mueller-Dombois

SCOPE AND OBJECTIVES

The broad research aims of the Hawai‘i IBP were twofold: to concentrate on aspects that are unique and different in island ecosystems as compared to continental ecosystems and to assist in solving regional problems relating primarily to wildland management and conservation of biological resources. On this basis, four general objectives were defined:

1. To determine why some organisms in Hawai‘i have undergone speciation, while some of the most successful have not.
2. To determine why some ecosystems in Hawai‘i are stable and some fragile.
3. To develop models relating to the variables that contribute to stability and diversity in Hawaiian ecosystems.
4. To determine the rates of evolution in different organism groups in Hawai‘i and the factors affecting these rates.

These four objectives required research approaches in two different disciplines: studies in evolutionary biology, microevolution, and ecological genetics (objectives 1 and 4) and studies in ecosystem and general ecology (objectives 2 and 3). One of our challenges was to integrate these two basically different areas in a combined study approach.

For each objective we established initial working hypotheses.

Objective 1

That speciation is the result of biological hardship.

That speciation occurs only if the population develops isolating mechanisms.

That speciation occurs only if the population develops in areas of geographic isolation.

That successful nonspeciated populations have larger gene pools.

Objective 2

That species diversity is positively related to stability.

That life-form diversity may be more important than species diversity to ecosystem stability.

That climatic factors have significant effects on stability.

That native ecosystem stability is directly related to the vigor of the two major native tree species, *Metrosideros collina* subsp. *polymorpha* and *Acacia koa*.

Objective 3

That Hawaiian ecosystems may be modeled for predicting the outcome of alterations.

That Hawaiian ecosystems may be modeled for comparison of their structures with those of continental ecosystems and other island ecosystems.

Objective 4

That speciation is a function of time.

It was obvious from the start that these objectives and hypotheses could not receive a complete treatment within the five-year time frame given for our project. Instead they must be viewed as longer-term objectives. We restricted our research to a few case examples and to a subset of the four longer-term objectives, which we expressed in four synthesis themes: part II, “Spatial Distribution of Island Biota”; part III, “Community Structure and Niche Differentiation”; part IV, “Temporal Relationships of Island Biota”; and part V, “Genetic Variation within Island Species.”

GENERAL CHARACTERISTICS OF ISLAND ECOSYSTEMS

Island biota have received considerable attention from evolutionary biologists since Darwin (1859) developed the theory of speciation from his observations of the Galapagos finches. Carlquist (1965, 1970, 1974) presented comprehensive reviews together with theories that may explain the often bizarre life forms developed on isolated islands. MacArthur and Wilson (1967) considered island ecosystems ideal laboratories for the development of evolutionary theory. Fosberg (1963, 1966, 1967), Dorst (1972), and others (for example, UNESCO Expert Panel on MAB Project 7, 1973) have written conceptual papers dealing with the precarious balance and fragility of island ecosystems. Three important ecological conditions render island ecosystems different in degree from continental ecosystems: geographic isolation, small size, and recent time.

Geographic Isolation

This condition is responsible for a screening effect on the number and kinds of organisms that can reach an island. A small percentage of these becomes established as populations, which may then radiate into the various habitats available on an island. In time, a successfully radiating population may become adapted by specializing. The original population may thus become differentiated or fragmented into a number of subpopulations. The result is the often-made observation on high islands that related taxa have an unusually wide ecological amplitude. The screening effect itself, however, results in biotic disharmony in the sense of missing taxa. This means that certain taxa and life forms that one may expect to find in environmentally equivalent habitats on other islands or on neighboring continents are absent altogether.

This leads to the question of community saturation and the resultant stability-fragility relationships. The two phenomena work against each other: *disharmony* suggests nonsaturated conditions, *radiation* suggests a filling effect with respect to available habitats and niches. The stability-fragility relationships in island ecosystems are not so easily predicted (as are often attempted on a more casual basis) because community saturation depends not only on taxomic diversity but on the ecological characteristics of the participating taxa as well. Study of the ecological roles of such taxa requires detailed and intensive research.

Small Size

As a rule, islands are thought of as small land masses surrounded by water. Different opinions exist with regard to defining Australia as an island or a continent, but for the purpose of studying differences between island ecology and continental ecology, islands should be defined as small land masses surrounded by water. For island research the UNESCO Expert Panel on MAB 7 (1973) suggested an approximate upper size limit of 10,000 km^2, a limit that includes the island of Hawai‘i, the largest island in the Hawaiian archipelago.

The biological implication of small land size is that population sizes of perennial organisms also tend to be small. Even in the mountainous or high islands, which usually have much greater land masses than do low coral islands or atolls, the recurrence of similar habitats is rather limited in comparison to most continental mountain ecosystems. Small areas restrict the size as well as the gene flow and composition of populations developing there. Greater genetic homogeneity may imply increased specialization. Whether this leads to a greater fragility of island populations cannot be answered as generally as is usually done because population stability depends on both the ecological properties of the population and the nature of the regional perturbations.

Recent Age

One biological consequence of recent geological age is that tropical island rain forests on high volcanic islands are much younger than most tropical continental forests. In certain areas the forests may have undergone more or less uninterrupted succession from giant equisetum-lycopod and seed fern forests to primitive gymnosperm and angiosperm forests. In contrast, the origin of most existing forests on high volcanic islands is within the modern angiosperm era. For example, the oldest parts of the high Hawaiian Islands are estimated to be merely six million years old (Macdonald and Abbott, 1970). Fosberg (1948) estimated that only one arrival form was required to become successfully established every twenty thousand to thirty thousand years to account for today's native angiosperm flora of a little over seventeen hundred taxa. The shorter geological time available for community development may account in part for a lower diversity in tropical island as compared to tropical mainland communities. In that respect tropical island communities are similar to temperate mainland communities, which can be considered species depauperate compared to tropical continental communities (Doty et al., 1969).

Unique Evolution

It can be argued that the three ecological conditions of geographic isolation, small size, and recent age are not unique to islands, but their biological consequences are most clearly developed on islands. The main point is that these conditions have resulted in a different evolution in island ecosystems. Thus on islands, more so perhaps than in biogeographically different continental ecosystems, the effect of a unique evolution becomes evident in many ecological relationships. The ecological relationships involve primarily the interactions among native species and among native and exotic species in Hawai'i. Many of the exotic species were introduced by humans within the last two hundred years. Humans thus effectively broke the natural isolation barrier of the Hawaiian Islands.

Community Structure

Because of unique evolutions and species assemblages on islands, the structure of island communities is expected to be unique also. However, this is true only at the level of species composition and quantitative distribution. At the level of dominant plant life form, islands are not unusual. Nearly all world plant-formation types can be found on high islands. There are grasslands, bogs, alpine tundras, savannas, closed evergreen rain forests, open seasonal forests, scrub formations, and deserts, to name a few of the more common world formations or biome types. These are conditioned by the peculiarities of climates and soils just as they are on continents. This gross structural similarity combined with the unique evolution of island biota and communities establishes the scientific and practical relevance of deriving general principles from the study of island ecosystems.

SOME BIOLOGICAL AND ENVIRONMENTAL CHARACTERISTICS OF THE HAWAIIAN ISLANDS

Biogeography

The native Hawaiian biota is the product of several million years of evolution on an isolated chain (figure 1-1) of volcanic islands extending 2,400 km (1,500 mi) northwest to southeast, from 28°N to 18°N latitude, in the Pacific (figure 1–2). Surrounded by deep water (over 5,000 m deep), the islands' nearest neighbors are Johnston Island, 720 km (450 mi) southwest; the Line Islands, 1,600 km (1,000 mi) south, and Wake Island, 1,930 km (1,200 mi) to the east. This extreme isolation of the island chain has undoubtedly been the most influential of all the factors that have contributed to the development of a terrestrial flora and fauna unique as to disharmony and speciation.

The native terrestrial biota, now estimated at between ten thousand and fifteen thousand species, probably descended from fewer than one thousand introductions, which must have colonized the islands against very great obstacles. Many groups of plants and, especially, animals dominant or abundant on other Pacific islands and on the continents bordering the Pacific are absent from the Hawaiian chain. Other groups occur as strays or relicts. Still others that are rare elsewhere are dominant in Hawai'i. Representation apparently reflects, in part, superior vagility (propensity for dispersal) over long distances by wind, water, or long-ranging sea birds. This thesis is supported by the noticeable lack of native nonflying mammals, land reptiles, amphibians, and freshwater animals in general, by the presence of plants, many of which originally must have had small seeds adapted to wind and bird dispersal (Carlquist, 1966a, 1966b), and by the capture of certain insects and spiders in air traps over the ocean (Gressitt and Yoshimoto, 1964).

Speciation rates during the course of evolution of the native biota have differed among the various endemic groups of flora and fauna. In the case of the single bat *(Lasiurus cinereus semotus)* found in the islands, there has been no local speciation. Among the birds, where seventy kinds have apparently evolved from fifteen ancestors, speciation has been moderate but nevertheless remarkable. In other instances large numbers of species have presumably evolved from single ancestors. The insect family Drosophilidae, with 600 to 700 species (Carson et al., 1970; Hardy, 1974), and the plant genera *Dubautia, Argyroxiphium,* and *Wilkesia* are examples.

Many factors have influenced speciation. These include such isolating mechanisms as different times of arrival, variations in gene pools (through

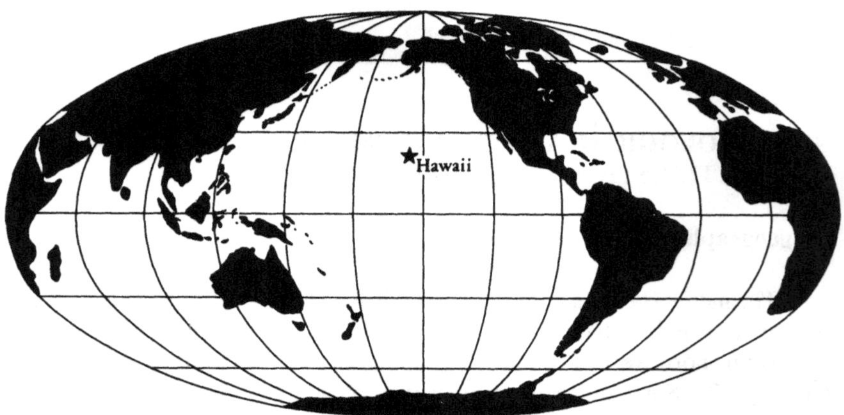

FIGURE 1-1 *Location of Hawai'i in a global perspective.*

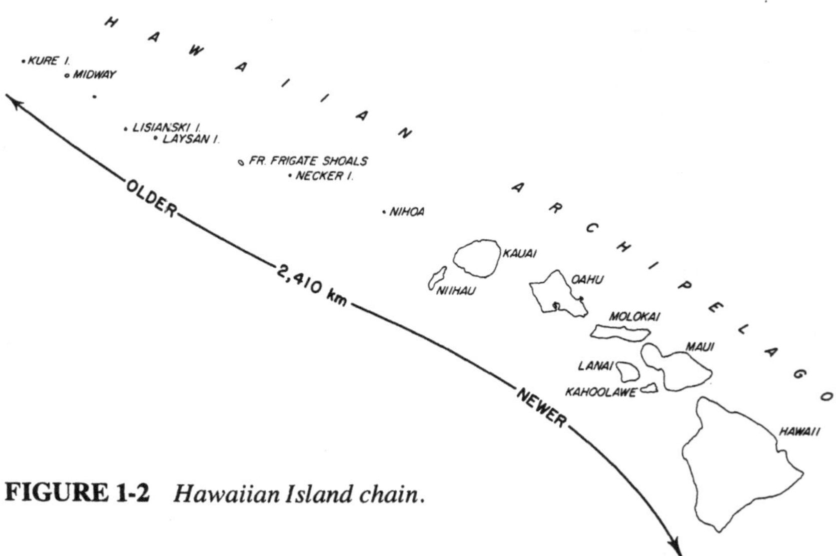

FIGURE 1-2 *Hawaiian Island chain.*

perhaps a single gravid female, egg batch, or seed), differing success in the new environment, and different situations with respect to genetic trends, population levels, adaptation, competition, and parasitism. Speciation has also been influenced by age differences among the islands. The distinctive chronological gradation so noticeable in the topography (from northwest to southeast within the archipelago) is reflected in the biota of the islands. Many other features of the archipelago serve as isolating mechanisms. These include small offshore islands, kīpukas (older vegetation surrounded by recent lava flows), lava tubes, and narrow valleys on the older high islands (Kaua'i, O'ahu, Moloka'i, and Maui).

Since the advent of humans in Hawai'i, profound ecological changes have taken place. Large numbers of weed or pest species have been introduced, both intentionally and accidentally. Many plants that are common garden ornamentals or hedges elsewhere have become serious weeds, blanketing large areas cleared of forest for grazing and invading native forests.

Similarly various mammals and birds have been introduced as escaped pets. Others have been released by organizations or local government agencies for hunting purposes. Cattle, horses, goats, sheep, deer, and pigs have devastated thousands of acres of natural vegetation and continue to threaten some of the remaining stands. Mongooses and rats have helped reduce native birds. Many species of insects and other invertebrates have also ravaged the native environment. An average of sixteen insect species is accidentally introduced to the islands each year (Beardsley, 1962).

The transformation of the biota from native to introduced is not restricted to the lowlands but extends to mountain ridges, valleys, and summits where hiking trails, sightseeing roads, radar, and other human-associated phenomena occur. Many temperate plants such as *Rubus* have become established in these areas. Native plants have disappeared and been replaced by the invaders in practically all areas where the original vegetation was cleared to make way for towns, roads, agriculture, grazing, gardening, and other developments. Exceptions occur where special precautions such as fences (effective for only a few of the invaders), weed-control, or other special measures have been taken.

Associated with the recession of the native vegetation from the lowlands and parts of the mountains is the decrease in the number of native birds, insects, and snails and the extinction of some of these species. It has become obvious that native animals cannot survive when the vegetation of their natural habitat is seriously disturbed.

Volcanism

The Hawaiian chain is estimated as twenty-five million years old in the northwest (Midway Island) and is progressively younger toward the southeast (figure 1-2). Ages of the older sections of the high islands (estimated by the potassium-argon method) are as follows: Kaua'i 5.6, O'ahu 3.3, Moloka'i 1.8, Maui 1.3, and Hawai'i 0.8 million years old (Macdonald and Abbott, 1970). At the southeast end, on the island of Hawai'i, lava still bubbles or fountains forth and runs down toward the sea. Thus at one end of the island chain are mountains with freshly cooled basalt lavas covering much of the land surface; and at the other end are only sandy, largely nonigneous, calcareous atolls or very low islands. All stages are to be found in between. Kaua'i, the oldest major high island, has old soil types and relatively species-rich communities; Hawai'i, the youngest major island, appears to have the simplest biotic communities, with some surfaces as yet unpopulated.

As lava flows from the tops of mountains, it leaves a chemically uniform belt transect of new volcanic surface stretched downward through a great range in climate. Often alongside such lava flows are mature communities that may serve as standards toward which the sere on the new lava flow can be expected to progress. This enables the study of community development on the same substratum through a wide variety of climatic conditions. Such uniform surfaces range in age from prehistoric to historic (historic refers to the last two hundred years). Thus one can study the results of seral development in a given climate on substrates at different ages over the past few hundred years.

Another interesting phenomenon associated with lava flows is the formation of kīpukas, islands of vegetation. Molten lava flows like water toward the lowest elevation. Consequently when lava flows around hills, islands of older lava remain, which may be heavily forested. Successive eruptions from different vents in a rift zone may result in kīpukas bound by lava flows of different ages and at different stages of revegetation. Kīpukas may also occur at the leading edge of a particular lava flow that stopped because the eruption came to an end. Other types of kīpukas may be formed by the general vagaries of lava flows. In addition to providing sites of different stages of succession under similar climatic conditions, kīpukas are prime areas for finding endemic plants and animals.

The Hawaiian lavas are highly vesicular basalts that differ in gas and heat content. The hotter, more gaseous lava type is called *pāhoehoe;* it flows like a river during an eruption and chills with a smooth or ropy surface. The cooler, less gaseous lava type is called *'a'ā*. It moves more sluggishly than pāhoehoe and has a highly fractured clinker-like or rock-rubble surface.

When it flows, pāhoehoe lava usually roofs over with a cooling crust, a process that insulates the flowing lava from atmospheric cooling and thus enables the lava to flow great distances from the vent. These roofed-over crusts often form channels that drain partially, creating a system of lava tubes or caverns within the flows. Older lava tubes develop into discrete ecosystems with many species of highly specialized obligatory cavernicoles related to endemic surface species. These lava tube ecosystems offer a rich area for evolutionary and ecosystem research, including comparisons with continental cave ecosystems.

Climate

Hawai'i is the only state in the United States that lies within the tropics. It is also the only state composed of relatively small islands completely surrounded by ocean. These facts contribute to its unique climate.

Descriptions of the macroclimates of Hawai'i have been provided by Blumenstock (1961), Britten (1962), Price (1966), and Blumenstock and Price (1967). They describe the main Hawaiian Islands as the summits of volcanic mountains that arose from the bottom of the sea (figure 1-3). The mountainous nature of Hawai'i is indicated by the fact that 50 percent of the state land area lies above an elevation of 600 m (2,000 ft), and 10 percent lies above 2,100 m (7,000 ft). The maximum elevations (in meters) of the six major islands are: Hawai'i, 4,206; Maui, 3055; Kaua'i, 1,597; Moloka'i, 1,515; O'ahu, 1,231; and Lāna'i, 1,207.

Almost half of the land in the state lies within 8 km (5 mi) of the coast. Only about 5 percent, all on the Island of Hawai'i, is more than 33 km inland. Thus the marine influence on the climate is pronounced.

During most of the year the northeast trade winds account for the dominant air movements over the state, and rainfall distribution is influenced primarily by the trades and the terrain. From May through September the trades are prevalent 80 to 95 percent of the time. From October through April the trades are prevalent only 50 to 80 percent of the time. On the average, major storms, associated with cold fronts, lows, and upper air lows and troughs, occur from two to seven times per year, usually between October and March. During such winter storms the dry leeward lowlands receive most of their annual rainfall. In fact, most areas of the state, except for the Kona coast of Hawai'i, have higher rainfalls in the winter than in the summer, although most areas with high rainfall remain relatively wet all year. Hurricanes occasionally pass close to Hawai'i, but between 1904 and 1967 only four came close enough to affect the islands

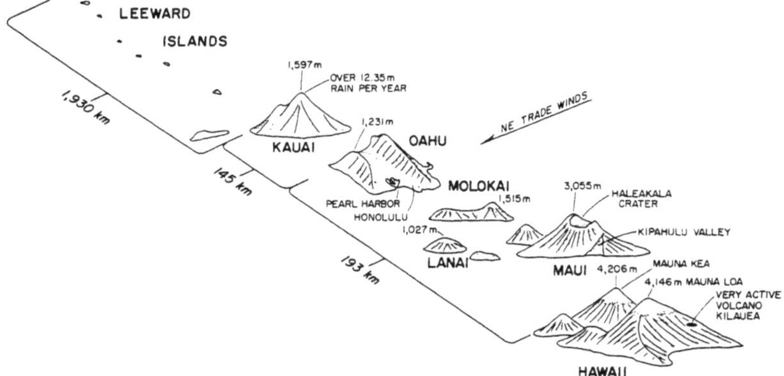

FIGURE 1-3. *Generalized topographic profile of Hawaiian Islands*

and only one actually passed through the islands. Tornadoes also occur at infrequent intervals.

Day length in Hawai'i is relatively uniform throughout the year. In Honolulu the longest day (including twilight) is fourteen hours, ten minutes; the shortest is eleven hours, forty minutes. The uniform day length and the small annual variation in the altitude of the sun above the horizon result in relatively small variations in the amount of incoming solar energy. This and the nearly constant flow of fresh ocean air of relatively uniform temperature over the islands are the major factors that contribute to the very slight seasonal changes in air temperature in the islands.

The overall pattern in Hawai'i is one of equable temperature conditions. Below 1,520 m (5,000 ft) the difference in mean monthly temperatures between the warmest and coolest months does not exceed 5°C (41°F), and the average daily range in temperature is between 4° and 9°C (39°–48°F) on the windward side. On the leeward side, the average daily range is greater, between 8° and 11°C (46°–52°F). Above 1,520 m altitude, the mean daily range in temperature is still greater, between 9°C (48°F) and 14°C (57°F) (Blumenstock, 1961). The highest temperature on record is 38°C (100°F), but temperatures above 35°C (95°F) are extraordinarily rare even in the dry leeward lowlands, and outside such areas temperatures of 32°C (90°F) and above are quite uncommon. The lowest temperature on record is −10°C (14°F), recorded at 3,055 m (10,020 ft) on Haleakalā, Maui. When long-term records from the summits of Mauna Kea and Mauna Loa, above 4,115 m, are available, it is possible that temperatures as low as −15°C (5°F) or less may be recorded.

Under trade wind conditions a temperature inversion is typically present between about 1,520 and 2,130 m (figure 1-4). This inversion layer is correlated with a moisture discontinuity and has an effect on both relative humidity (RH) and rainfall. Below the inversion, the RH commonly averages 70 to 80 percent in windward areas and 60 to 70 percent in leeward areas. Above the inversion, the relative humidity is generally below 40 percent and often as low as 5 to 10 percent.

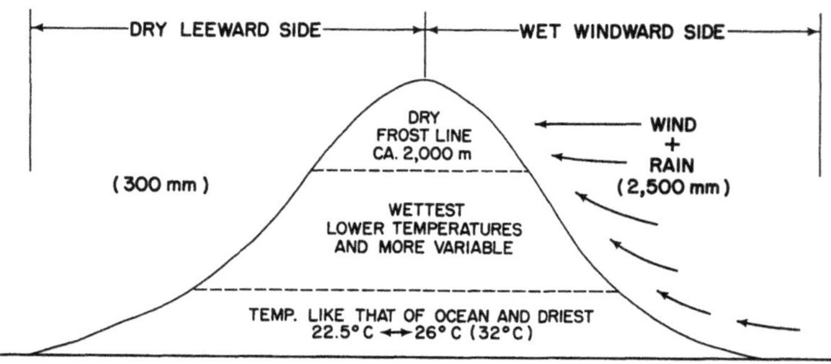

FIGURE 1-4. *Generalized climatic cross-section through higher Hawaiian mountains.*

While the average annual rainfall of the state as a whole is about 1,880 mm (70 in), variation from place to place is so great as to make this figure meaningless. At Kawaihae on the leeward coast of Hawai'i, the average annual rainfall is less than 170 mm; at Mount Wai'ale'ale on Kaua'i it is 12,350 mm (and has exceeded 15.2 m). Rainfall gradients are extremely steep, exceeding 1,000 mm per km in many places. Along the 4 km line from Hanalei Tunnel to Mount Wai'ale'ale on Kaua'i, the gradient is 1,800 mm per km. In general, leeward lowlands and mountain peaks above 2,500 m elevation are the driest areas, while maximum rainfalls are recorded at or near the crests of mountains less then 1,800 m high, and at 610 to 1,220 m on the windward slopes of the higher mountains.

Soils

The soils in the Hawaiian Islands have developed from volcanic ash and basaltic lava under a wide range of climatic conditions. Another soil area in some coastal sections and in the leeward islands has developed from marine deposits, such as coral limestone and sand. Cline (1955) gives an excellent description of the Hawaiian soils according to the earlier system of great soil groups. This system is now superseded by a new comprehensive system of soil taxonomy (U.S. Department of Agriculture, 1975). Of course, the general relationships are unaffected by the new system. The Hawaiian soils are at least as varied as the climatic zones in Hawai'i, and they are not unique but can be related closely to similar soil types elsewhere in the tropics.

Soils developed in humid tropical lowlands to mid-altitudes on the windward sides of the older high islands are mostly Oxisols and Ultisols. These are characteristically reddish-brown (because of *in situ* ferric iron precipitation), deeply weathered soils, low in silicates and cations, and of weakly acid to strongly acid reaction. Soils developed in dry climates (mostly on the leeward sides) are often lighter in color, less strongly leached of cations, even partly calcified, and of less acid to circumneutral reaction. Most of the soils in high altitudes are geomorphologically recent and poorly or incipiently developed from parent materials. On steep slopes, soils are also poorly developed because erosion keeps pace with weathering, constantly removing soil particles and layers developed from the parent materials. Soils on the geologically young Island of Hawai'i are mostly Inceptisols and Histosols. Lithosols are also of wide occurrence there and on east Maui.

The Hawaiian soils were remapped according to the new comprehensive system of soil classification (Sato et al., 1973, for Hawai'i Island; Foote et al., 1972, for all other high islands). A general summary was provided by Uehara (1973). Ten soil orders are recognized for the Hawaiian Islands following the new system. These may be grouped into three broader environmental categories as shown in table 1-1.

TABLE 1-1. *Soil Orders in Hawaiian Islands*

Soil Order	Description	Occurrence
Well-developed soils in humid climates		
Oxisols	Strongly weathered oxide soils; formerly low humic latosols and humic ferruginous latosols	Relatively flat land at low elevations
Ultisols	Ultimately leached soils; with argillic horizons resulting from downward translocation of fine clays; formerly humic latosols	Steeper slopes and less stable landscapes at somewhat higher elevations than oxisols; montane tropical rain forests on older higher islands
Spodosols	Spodos = woodash = podzols	Bog and swamp soils in high rainfall zones, e.g., Alaka'i Bog soil on Kaua'i
Alfisols	Much like ultisols but less leached of nutrients; Al-Fe soils or formerly pedalfer soils	NW O'ahu in subhumid climate under forest
Well-developed soils in seasonal or dry climates		
Aridisols	Formerly red desert soils	Very dry zones, mostly confined to the northern leeward shore of Hawai'i Island; annual rainfall less than 200 mm
Vertisols	Vert = inverted; soils that swell and shrink with wetting and drying; formerly dark magnesium clays or grumusols	Several places in lowlands on leeward O'ahu, especially on talus slopes and floors of deeply dissected amphitheater-headed valleys
Mollisols	Moll = soft; soils with deep, crumb-structured A_1 horizons; rather rich in nutrients; in part formerly latosolic brown forest soils	Moderately dry climates (e.g.; soils of Kīpuka Kī and Kīpuka Puaulu on transect 1 in Hawai'i Volcanoes National Park)
Little-developed or geomorphically recent soils		
Histosols	"Organic tissue" soils; shallow soils of predominantly organic materials (5–20 cm deep)	Overlying relatively recent lava; under rain forests; mostly on Island of Hawai'i
Inceptisols	Incipient = young soils, but not very recent, from volcanic ash with weakly developed horizons	Cover extensive areas on Hawai'i on slopes of Mauna Kea and in Hawai'i Volcanoes National Park; also on East Maui
Entisols	Ent = recent; lacking horizons; formerly regosols	On volcanic ash near tree line on Mauna Kea and Mauna Loa; on old beach sand (N. O'ahu); on recent alluvial deposits

TABLE 1-1. (Continued)

Soil Order	Description	Occurrence
Lithosols*	Lithos = rock; essentially rock outcrop, includes all recent lava flows (ʻaʻā and pāhoehoe) and older rock bluffs and stony substrates.	Dominant surface on Mauna Loa and Kīlauea volcanoes; older rock outcrops occur locally in the mountains of all high islands

*This name is not used officially among soil orders but seems to fit the category "miscellaneous land types" which in the Hawaiian islands includes a variety of widely distributed rocky substrates.

Flora

The terrestrial native flora of the Hawaiian Islands is remarkable. It contains about 1,440 species of angiosperms (St. John, 1973), 170 species of pteridophytes (Fosberg, 1948), 160 species of liverworts (Miller, 1956), 240 species of mosses (Bartram, 1933, 1939; Hoe, 1974), 700 species of lichens (Magnusson, 1954), and an unknown number of algae and fungi. Endemism is extremely high, especially in the vascular flora. Over 96 percent of the native species of angiosperms are endemic, as are about 65 percent of the pteridophytes, 75 percent of the liverworts, 65 percent of the mosses, and 38 percent of the lichens. Although the fungi and the terrestrial and freshwater algae have not been thoroughly studied, it appears that they exhibit little, if any, endemism. No endemism seems to be present in the blue-green algae recorded from any habitats in Hawaiʻi.

The vascular flora apparently has been derived from a relatively small number of species that reached the Hawaiian Islands by long-distance dispersal. An estimated 272 original successful immigrants could have given rise to the native angiosperm flora, while 135 immigrants could account for the pteridophyte flora (Fosberg, 1948). In Hawaiʻi today one finds expressed in the flora all stages of evolutionary development: widespread genera with some species only slightly or not at all different from representatives elsewhere in the world, widespread genera with many distinct and well-defined endemic Hawaiian species, and endemic genera.

The affinities of Hawaiʻi's flora seem to be primarily with areas west and south of the Hawaiian Islands, with surprisingly little affinity with the nearest continental area, North America. Fosberg (1948) estimates that of the original immigrants of the angiosperms, 40.1 percent show affinities with the Indo-Pacific region, 16.5 percent show affinities with the Austral region, 13.3 percent show affinities with the American region, and 12.5 percent show pantropic and cosmopolitan affinities. Among original pteridophyte immigrants, 48.1 percent show Indo-Pacific affinities, 20.8 percent show pantropic and cosmopolitan affinities, and 11.9 percent show American affinities. While additional studies will contribute information that may change these percentages, the overall pattern of affinities exhibited by the flora seems fairly clear and is unlike that of any other area in the world. In terms of largest numbers of well-recognized species and varieties, some of the best-represented families of angiosperms are tabulated in table 1-2.

The native flora is a disharmonic one, and several plant groups that are widespread in the tropics and might be expected to be native in Hawai'i are not. Fosberg (1948) considered the following groups to be significantly absent from the native flora: *Ficus, Piper,* Cunoniaceae, Bignoniaceae, Araceae, Gymnospermae, and mangroves. All of these groups except the Cunoniaceae have been introduced to Hawai'i in some numbers in recent years. Mangroves have been successful, probably because there were no other plants occupying their particular ecological niche. In less than fifty years mangrove swamps have developed on Moloka'i, O'ahu, and Lāna'i. A few species of *Ficus* (Moraceae) and *Spathodea campanulata* (Bignoniaceae) have become naturalized to some extent but usually only in otherwise disturbed areas. They are not yet causing serious problems. *Piper methysticum* has persisted in the wild since its cultivation by the Polynesians. Other than these, the plants in the groups significantly absent from the native flora have not seemed to spread beyond cultivation.

The most successful species of the introduced flora are those that have become thoroughly naturalized, have extensive distribution ranges, and most often seem to be in competition with the native flora. They include *Prosopis pallida, Leucaena latisiliqua, Acacia farnesiana* (Leguminosae); *Lantana camara, Stachytarpheta* spp. (Verbenaceae); *Rubus rosaefolius, R. penetrans*

TABLE 1-2. *Best-Represented Families and Genera in Native Angiosperm Flora of the Hawaiian Islands*

Family	Approximate number of species and varieties	Genera with largest numbers of taxa
Compositae	200+	*Bidens, Dubautia, Lipochaeta*
Gesneriaceae	200+	*Cyrtandra*
Campanulaceae		
subfamily Lobelioideae	150+	*Cyanea, Clermontia, Lobelia*
Rubiaceae	100+	*Hedyotis, Gouldia, Coprosma, Psychotria*
Labiatae	100+	*Phyllostegia, Stenogyne*
Rutaceae	100±	*Pelea, Zanthoxylum*
Loganiaceae	75±	*Labordia*
Euphorbiaceae	75±	*Euphorbia* (subgen. *Chamaesyce*)
Gramineae	65+	*Panicum, Eragrostis*
Cyperaceae	50+	*Cyperus*
Pittosporaceae	50±	*Pittosporum*
Piperaceae	50±	*Peperomia*
Caryophyllaceae	45±	*Schiedea*

(Rosaceae); *Psidium guajava, P. cattleianum, Rhodomyrtus tomentosa, Eugenia cumini* (Myrtaceae); *Myrica faya* (Myricaceae); *Melastoma malabathricum, Tibouchina semidecandra, Clidemia hirta* (Melastomataceae); *Opuntia megacantha* (Cactaceae); *Bidens pilosa, Pluchea* spp., and many other genera (Compositae); *Commelina diffusa* (Commelinaceae); *Andropogon glomeratus, A. virginicus, Panicum maximum, Paspalum conjugatum, Pennisetum clandestinum, P. setaceum,* and many other genera (Gramineae).

Although many monographic treatments of genera and families of Hawaiian plants are available, the only comprehensive manual of the Hawaiian flora was published by Hillebrand in 1888 (the 1965 facsimile edition is widely available). This manual is badly out of date and treats only about 1,000 species. Rock's (1913) flora of the native trees is limited to selected species but is still very useful. Neal's (1965) garden flora provides brief descriptions and some keys for many of the cultivated and introduced weed species. Degener's *Flora Hawaiiensis* (1932–present) is not yet completed. It treats somewhat over half of the native species only. A comprehensive checklist of the flowering and seed plants in Hawai'i has recently been produced by St. John (1973). According to his estimate, the number of seed plants in the Hawaiian Islands is 2,744 native species and varieties and 4,987 introduced species and varieties. Of the exotic taxa, 1 percent are now growing in wildland habitats.

Vegetation

The wild vegetation in the leeward lowlands consists mostly of introduced plants. In this warm, tropical winter-rainfall and summer-drought climate, one can distinguish several structural formations. Most typical is the perennial grass savanna. The more dominant grasses are *Heteropogon contortus* and *Rhynchelytrum (Tricholaena) repens.* The woody plants may be scattered trees or shrubs, or shrubs variously clumped, often forming thickets. Among trees, members of the Leguminosae are important, particularly *Prosopis pallida* and *Leucaena latisiliqua.*

Open seasonal evergreen scrub and forest occur above this lowland vegetation on the leeward side. Where well preserved (a few places only), this is taxonomically quite a rich tree vegetation, with at least twenty native tree species. Among the more common ones are *Sapindus oahuensis, Diospyros sandwicensis, Erythrina sandwicensis,* and *Canthium odoratum.* However, most of this subhumid, summer-dry environment is converted to introduced vegetation showing variously open grass-shrub, shrub, and tree vegetations. Among the quantitatively more important exotic trees are *Psidium guajava, Schinus terebinthifolius,* and *Eugenia cumini.*

In the montane environment, evergreen rain forest predominates, with Myrtaceae (*Metrosideros,* also the introduced taxa *Eugenia* spp. and *Psidium* spp.) and native tree ferns (*Cibotium* spp.). The native legume tree *Acacia koa* prevails locally in patchy stands in the submontane part of the rain forest zone. The wettest parts of this zone contain bogs on gently sloping or level terrain. One example is the Alaka'i Bog on Kaua'i with dwarf forms of the *Metrosideros* tree and *Cheirodendron* spp.

The rain forest vegetation reaches to the highest elevations on Kaua'i (1,590 m), O'ahu (1,230 m), and Moloka'i (1,500 m). On the two higher-mountain islands, Maui (3,055 m) and Hawai'i (4,206 m), the rain forest vegetation is replaced upward (at about 1,700 m) by mountain parkland or savanna. There, native legume trees *(Acacia koa* and *Sophora chrysophylla)* are again important. Above this occurs subalpine open forest and scrub. The scrub shows heath, that is, shrubby members of Ericaceae (*Vaccinium* spp.) and those of a closely related family, Epacridaceae *(Styphelia)*. In the alpine scrub and near the tree line (on Mauna Kea near 2,800 m) are individuals of the peculiar herbaceous phanerophytes, the silverswords (*Argyroxiphium* spp.). This genus is found only in Hawai'i. Homologs in the tropical Andes are the giant espeletias and on Kilimanjaro, the treelike senecios. These three taxa are all from the family Compositae. Above the alpine scrub is the alpine tundra or stone desert. On Mauna Loa the alpine tundra extends from 3,000 m to the summit. On Mauna Kea it extends from 4,000 m to the summit. The alpine tundra is characterized by a sparse moss *(Rhacomitrium lanuginosum),* by crustose lichens (such as *Lecanora* spp.), or by both.

This general sequence of altitudinal zonation is quite similar to that on several continental tropical mountains (Walter, 1971). However, the floristic communities are very different. Hawaiian vegetation zones have been described by several authors, including W. Hillebrand (1888), Rock (1913), Egler (1939), Ripperton and Hosaka (1942), Krajina (1963), and Knapp (1965). A summary comparison of three of these zonation schemes is shown in table 1-3.

Floristic communities have been identified by Fosberg (1961, 1972) on a very general level. He recognizes about thirty major ecosystems for Hawai'i. On a more detailed (larger-scale) level, Hatheway (1952) and Wirawan (TR 34) described floristic communities for the native dry forest remnant on O'ahu. Various other vegetation segments have been studied floristically for their community formation and ecology. These segments include the coastal communities on O'ahu (Richmond and Mueller-Dombois, 1972), the dry-grass communities on O'ahu (Kartawinata and Mueller-Dombois, 1972), the dry-zone vegetation on O'ahu (Egler, 1947; Mueller-Dombois and Spatz, 1975a), the east flank vegetations on Mauna Loa and Mauna Kea (Mueller-Dombois and Krajina, 1968), the vegetation in Hawai'i Volcanoes National Park (Mueller-Dombois, 1966; Newell, 1968; Mueller-Dombois and Fosberg, 1974), and the vegetation on recent volcanic materials (Atkinson, 1969, 1970; Egler, 1971; Smathers and Mueller-Dombois, TR 10).

TABLE 1-3. *Ecological Zones in Hawaiian Islands: Summary Comparison of Three Zonation Schemes*

F. Egler (1939)	R. Knapp (1965)[a]	V.J. Krajina (1963)
Xerotropical	I. Hot dry zone	Important key plants
Makai zone (seaward)	4. Very dry	1. *Prosopis* (kiawe) zone
Middle zone Mauka zone (inland toward mountains)	3. Moderately dry	2. *Leucaena* (koa haole) 3. *Psidium* (guava)—*Lantana*
	5. Zones 3 and 4 mixed; Ni'ihau and Kahō'olawe	
Pluviotropical	II. Rain forest zone	
Guava zone Koa zone	1. Submontane	4. *Acacia koa—Psidium—Styphelia* (koa—guava—pukiawe) 5. *Acacia koa—Nephrolepis—Paspalum* (koa—Boston fern—Basket grass)
'Ōhi'a zone Cloud zone	2. Montane	6. *Metrosideros—Cibotium* ('ōhi'a—tree fern) 7. *Cheirodendron trigynum—Cibotium* (olapa-tree fern) 8. *Cheirodendron dominii—Clermontia—Gunnera* (lapalapa—oha—wai—ape'ape) 9. *Oreobolus*—bog zone (Alaka'i swamp and others)
Only on Maui & Hawai'i	III. Cool dry-forest and heath zone	
	6. Upper montane or subalpine	10. *Metrosideros—Sadleria* ('ōhi'a—amaumau fern) 11. Koa—Boston fern—*Dryopteris paleacea* 12. Koa—*Sophora* (mamane)—*Styphelia* (pukiawe) 13. *Sophora—Styphelia—Vaccinium* (ohelo)
	IV. Alpine	
	7. Cold desert, from about 3000 m up	14. Lichens, mosses, (*Argyroxiphium*)

[a]Number sequence relates to vegetation zone map in Knapp (1965:320).

Note: In addition a strand zone is present on all islands. It covers most of the leeward islands in their entirety.

Land Arthropods

With the relative paucity of species and individuals of vertebrate animals in the native fauna of Hawai'i, the arthropods assume great importance in the terrestrial biota from the standpoint of dominance in numbers of species, numbers of individuals, and ecosystem function. More than five thousand endemic species of land arthropods have been described or sorted for studies in progress. The total number will be considerably larger, although many species are facing extinction with the current rapid change in the environment.

The insect fauna of an area is closely associated with the native vegetation. This is truer for Hawai'i than for continental areas, for several reasons. Because the Hawaiian archipelago is extremely isolated, natural immigration to the islands has been highly selective. The subsequent naturalization probably results in an even greater selectivity. Primary colonizers must have been plants, but the pattern of later successful colonization by plants was influenced by the arrival of pollinators, including arthropods. These first plants probably were followed by insects of several types, largely feeders on living, freshly decaying, or dead plants. Because of the few vertebrate animals colonizing, those groups parasitic on vertebrates, and living in their nests, cadavers, and droppings, are few in number. Insects parasitic and predaceous on other insects must have come secondarily or evolved from already established species. Insects with narrow host requirements were less likely to establish themselves. Such species are better known from intentional introductions for biological control of pests.

In the course of their evolution since reaching these islands, the native insects to a great extent have become fully dependent upon endemic plants, often in a very narrow host relationship. With the advent of more and more introduced plants, there has been rather little evidence for the attachment of endemic insects to introduced plants. Thus the native fauna becomes restricted in occurrence or extinct as its plant associates are affected by humans and introduced biota.

Table 1-4 is a summary of the currently known endemic Hawaiian arthropods. Many of the endemic species belong to the principal phytophagous groups (Heteroptera, Homoptera, Lepidoptera, and Coleoptera). Many others belong to groups feeding in decomposing plants (especially Diptera; many Coleoptera). Hardy (personal communication, 1974) estimates that when all species are known, the actual number of endemic insect species (except for the Diptera) will be much larger than the figures given in table 1-4. Hardy estimates that there are eight thousand to ten thousand species of endemic insects.

The disharmonic nature of the endemic arthropod fauna left vacant ecological niches, which were often filled by adaptive shifts in the groups present (for example, a shift from phytophagous to predaceous forms). More recently arthropods introduced by humans have filled these niches. By adaptive shifts, these exotic arthropods often displace endemic species. Ants, mosquitos, ter-

TABLE 1-4. *Tentative Summary of Endemic Hawaiian Terrestrial Arthropods*

Order	Families with endemic spp.	Genera with endemic spp.	Endemic genera	Endemic species
Pseudoscorpionida	?	?	?	?
Acarina	sev.	?	?	100+
Araneida	10	28	8	100+
Amphipoda	1	4+	2+	30+
Isopoda	1	1	—	1+
Symphyla	1	2	—	2
Diplopoda/Chilopoda	1	1	—	15
Palpigrada	1?	1?	?	?
Tardigrada	1?	1?	—	?
Collembola	1	1	—	1+
Thysanura	1	1	—	2
Orthoptera	2	7	6	50+
Dermaptera	1	?	?	2+
Psocoptera	3?	3	2	200
Mallophaga	2	3	—	5
Odonata	3	3	2	29
Thysanoptera	3	6	3	30
Heteroptera	11	37+	25+	650+
Homoptera	5	32	21	352+
Neuroptera	3	6	5	54
Lepidoptera	20	93	47	863+
Coleoptera	19	102	71	1,500
Diptera	26	84	12	1,500+
Hymenoptera	15	44	30	710+
Totals[a]	126	458	234	6,196

Source: Based on published volumes of *Insects of Hawaii* by Zimmerman (1948a–e, 1957, 1958a–b) and Hardy (1960, 1964, 1965), with estimates of additions based on existing collections and on recent discoveries by the US-IBP Island Ecosystems Integrated Research Program.

[a]Totals do not include figures with question marks.

mites, and cockroaches are lacking in the endemic fauna, but several species of each are now well established and have had great impact on both the human economy and the endemic biota.

Among the insects proper, thirty-three orders are recognized. Only fifteen of these are included in the native fauna. Thirteen additional orders have been introduced by humans. Humans have introduced about fifteen hundred species of insects, as well as some two hundred kinds of mites and other arthropods.

Land Snails

Since the first decade of this century the Hawaiian land snails, especially tree snails of the family Achatinellidae, have been cited as classic examples of the results of evolutionary phenomena. Of the two best-known families, the Achatinellidae and Amastridae, the latter is the more primitive and probably the older of the two in the islands. The amastrids are predominantly ground snails, feeding on decaying vegetation and fungi. They are both oviparous and viviparous. Most are dull and inconspicuous in color, but in shell form they exhibit remarkable diversification. The largest members of the family Amastridae, *Carelia,* may reach 85 mm in length. *Carelia* are now confined to Kaua'i, but fossils have been found on Ni'ihau. Smaller, more elaborately sculptured amastrids occur on O'ahu, Maui, Moloka'i, Lāna'i and Hawai'i.

The nine families of native Hawaiian land snails are represented by 37 genera and some 1,000 species and subspecies. Of these, 4 genera and 215 species and subspecies belong to the family Achatinellidae. The Achatinellidae are, for the most part, highly colored and viviparous. They feed on fungi and on decaying vegetation on the leaves and trunks of trees. No achatinellids are known from Kaua'i. On O'ahu, Maui, and Moloka'i, they have differentiated into several genera and hundreds of specific entities, some apparently confined to particular ridges and valleys (Baldwin, 1887; Hyatt, 1912–1914).

Introduced land snails and slugs are primarily of concern as agricultural pests in most regions of the world. In Hawai'i introduced snails are not only agricultural pests, but they also decimate the native land snail fauna and act as intermediate hosts for a number of parasites. Disruption of the native land snail fauna stems from the introduction of the giant African snail, *Achatina fulica,* on Maui and O'ahu in 1936. Efforts to eradicate the snail failed, and in 1957 carnivorous snails from a variety of countries were introduced in an attempt at biological control. Of twenty-three species representing fourteen genera introduced at that time, at least two have become widely established: *Gonaxis quadrilateralis* and *Euglandina rosea*. *Euglandina,* which in its native habitat in Florida preys on tree snails, has moved into Hawai'i's native forests, where it is a serious threat to the tree-dwelling Achatinellidae.

Other mollusks introduced either accidentally or on purpose are of less significance to the native biota: *Bradybaena similaris* and the slugs *Dendroceras laeve* and *Veronicella alte* are common garden pests, and *Bradybaena, Opeas javanicum, Subulina octona, Dendroceras,* and *Veronicella* are hosts for the rat lungworm.

Land Birds

Hawai'i now has two groups of land birds: endemic and exotic. Seven families of world birds have endemic genera, species, or subspecies in Hawai'i, and an eighth family (Drepanididae, Hawaiian honeycreepers) is endemic to the Hawaiian Islands.

The honeycreeper family is one of the most spectacular examples of adaptive radiation among animals. Plumage colors in the subfamily Psittirostrinae range from dull olive-drab to bright yellows and oranges. Bill shapes range from heavy parrot-like and finch-like to short, slender warbler-like to extremely long and decurved types. In the subfamily Drepanidinae plumages are pure or some combination of white, black, scarlet, and vermilion. Bill shapes, always decurved, vary in length. Although definitive studies remain to be done, it is clear that niche differentiation, especially with respect to feeding, is equally as spectacular as morphology. In Hawai'i the Drepanidids alone have filled ecological niches that in continental ecosystems are not filled within a span of several orders of birds. For example, there are Drepanidids with feeding strategies similar to those of parrots, finches, warblers, woodpeckers, nectar feeders, and creepers.

From a presumed single ancestral colonizing species, twenty-three species and twenty-four subspecies of honeycreepers evolved. Of these, seven species and seven subspecies of honeycreepers are thought to be extinct. An additional twenty-nine species or subspecies are now classified as rare and endangered. The other endemic birds are listed in table 1-5. Of these, seven species or subspecies are presumed to be extinct, and six others are listed as rare and endangered.

Very little is known of the life history and ecology of any of the extinct birds. In recent years the nest and eggs of several endemic species have been described (Berger et al., 1969; Berger, 1969a, 1969b, 1970), and studies of the breeding biologies of the O'ahu 'elepaio (*Chasiempis sandwichensis gayi;* Conant, 1977), the Hawai'i 'amakihi (*Loxops virens virens;* Berger, 1969a), and four honeycreepers on Kaua'i (Eddinger, 1970) have been completed. An intensive study of the palila (*Psittirostra bailleui;* van Riper, 1978) has also been completed recently. However, most of the endemic birds have not been thoroughly studied, and their habitats and populations continue to decline.

TABLE 1-5. *Endemic Hawaiian Land Birds Excluding Honeycreepers* (Drepanididae)

Family	Scientific name	Common name
Accipitridae	*Buteo solitarius*	Hawaiian hawk or 'io
Strigidae	*Asio flammeus sandwichensis*	Hawaiian owl or pueo
Corvidae	*Corvus tropicus*	Hawaiian crow or 'alalā
Turdidae	*Phaeornis palmeri*	Small Kaua'i thrush
	Phaeornis obscurus	Hawaiian thrush
Sylviidae	*Acrocephalus f. familiaris*	Laysan millerbird
	Acrocephalus familiaris kingi	Nihoa millerbird
Muscicapidae	*Chasiempis sandwichensis*	'Elepaio
Meliphagidae	*Moho braccatus*	Kaua'i O'ō
	Moho apicalis	O'ahu 'O'ō
	Moho bishopi	Moloka'i 'O'ō
	Moho nobilis	Hawai'i 'O'ō
	Chaetoptila angustipluma	Kioea

Conant (TR 74) has begun a study of population dynamics in an attempt to define critical habitat on the basis of maximum density, frequency, and diversity of native bird species.

The introduction of parasites and diseases new to the Hawaiian avifauna is one of the most serious consequences of exotic bird introductions. The role of bird malaria, as well as avian pox and other parasitic infections, in the decline of the Hawaiian avifauna is unknown, but available evidence suggests that it is serious (Warner, 1968).

The exotic linnet *(Carpodacus mexicanus)* and ricebird *(Lonchura punctulata)* are known to be important agricultural pests, yet new potential pests, such as the recently established warbling silverbill *(Lonchura cantans)* on Hawai'i (Berger, 1975), continue to be intentionally or accidentally introduced.

At present there is little evidence to demonstrate that exotic bird species compete seriously with native birds for ecological resources. However, the Japanese white-eye *(Zosterops japonica)* coexists with native birds where they are found, and it exploits many of the same food resources (Guest, TR 29). The Japanese white-eye is probably the most abundant and widespread land bird in Hawai'i. Studies of this species and other potentially competitive species are being conducted in habitats where they coexist with native birds to determine whether they compete successfully with native birds and what the consequences of such competition are for the native avifauna.

Effects of Humans on the Biota

The native Hawaiian biota, which had evolved in the absence of humans and of large grazing or carnivorous mammals, seems to have been in a delicate balance within the island ecosystems before the arrival of humans. About thirteen hundred years ago when the Polynesians first arrived, they brought with them some twenty-five species of plants for food, medicine, fiber, and other purposes. These included coconut, taro, banana, breadfruit, candlenut, paper mulberry, ti, sweet potato, and various yams. They also brought rats, dogs, pigs, and jungle fowl. The Polynesians and the organisms that they brought with them certainly influenced the native biota, but one can only speculate on the magnitude of their effects. Undoubtedly the impact of the Polynesians was greatest in areas where the Hawaiians lived and grew their crops. On the other hand, Hawaiians considered land, plants, and animals as the property of, and to be held in trust for, the gods, an outlook that resulted in a form of practical conservation. Fish and shellfish were collected only in season. Thus although the initial effects of Polynesian colonization on the native Hawaiian biota may have been great, they were certainly less drastic than the effects brought about by sustained contact with European cultures.

It is likely that new ecological equilibria were established at some time after the extensive colonization of the islands by the Polynesians and that these equilibria were operative in 1778 when Captain Cook arrived.

Cook released the goat and a second breed of pig in 1778. In 1793 Captain Vancouver released cattle and sheep. Horses were introduced in 1803. Cats and commensal rats and mice probably arrived on some of the earliest ships. The last feral horses and cattle are still to be found in some areas of the Kona coast of Hawai'i today. Sheep, goats, and pigs today endanger the native forests and other ecosystems on most of the main islands.

While the large-hoofed animals were rapidly increasing in numbers, they degraded the vegetation over wide areas by feeding on the seedlings of native woody plants, by local overgrazing, and by trampling. Many plants were introduced, intentionally and otherwise, and these occupied the areas newly opened up by animals. Since early records are so incomplete, we cannot estimate the number of species of the native biota that became extinct in the century following Cook's visit. For the last hundred years, better records are available. As extensive areas of land were cleared for agricultural purposes, the native biota disappeared almost completely from these areas. The delicate balances existing in the native biota were easily upset. As the plants disappeared, the birds, insects, and terrestrial mollusks associated with them also disappeared.

Humans have intentionally and unwittingly introduced to the islands fauna that have adversely affected the native biota. Introduced species of land snails and slugs continue to threaten endemic land mollusks. Introduced birds, and the parasites and diseases carried by them, may have seriously affected the native birds (Warner, 1968). Certain introduced insects are important in the spread of these bird diseases and parasites. Additional species of insects and other arthropods are accidentally introduced each year (Beardsley, 1962). Exotic arthropods affect not only the endemic biota but also the human economy. These human-caused biological perturbations present a constantly growing and changing problem.

Even after the exceptional biotic resources of Hawai'i came to be appreciated by the scientific community, the desire for "progress" and economic development on the part of the general public was, and still is, so great that policies to protect and manage Hawai'i's unique and most precious resource—its native biota—are only slowly being put into effect.

ORGANIZATION OF INTEGRATED RESEARCH

Organism Groups Studied in the Program

A major interest of the project participants was the study of well-preserved Hawaiian ecosystems with a dominance of native species. We believed that the opportunity to study these species was decreasing, and we thought that studies of well-preserved ecosystems would provide basic information against which we could compare ecosystems variously disturbed by the invasion of exotic species.

Plants, arthropods, and birds received the greater emphasis because most of the native species are found in these categories. A few species from each category were selected for studies in microevolution and genecology. The more cosmopolitan groups, the fungi and terrestrial algae, were also studied. Among insects and other arthropods, the groups emphasized were Diptera (particularly Drosophilidae and Sciaridae), Coleoptera, Heteroptera, Homoptera, Lepidoptera, and Acarina. Other arthropods were included according to their ecological role, such as pollinators, seed feeders, and ectoparasites. Notable gaps that could not be filled because of limited funds were, among botanical organisms, the bryophytes and lichens and, among animals, the snails.

A general aim was to study the ecological roles of these organism groups in selected ecosystems. This included a consideration of interactions between native and human-introduced species. Therefore exotic species were also studied in each category, and special emphasis was given to introduced mammals, particularly to the impact of rodents (rats, mice, and mongooses) and feral ungulates (goats and pigs) on the vegetation.

The time constraints of the project precluded a uniform advance of knowledge on the functional role of each organism group. Some of the groups required a considerable amount of taxonomic background work. However, our synthesis aim was to develop a unified basis. We accomplished this by concentrating our efforts for this project primarily on the structural (rather than the functional) aspects of ecosystem analysis. The main synthesis areas covered by this approach relate to distribution and quantitative composition of species along spatial gradients, population fluctuations in relation to phenological events and annual variations in climate, community structure and niche differentiation in selected ecosystems, and genetic structure and variation of selected populations. The full elucidation of the ecological role of each species and species group can be considered a further step for which the groundwork has been laid.

Conceptual Modeling

The problem of bringing studies on organism groups together into an integrated approach is not as easy to solve as one may expect. The US-IBP Biome studies on the American continent had focused their research on ecosystem metabolism. For such functional studies there was a relatively simple model: the primary production-consumer-decomposer-mineral cycling processes of an ecosystem. These component processes provided a reasonable conceptual vehicle for organizing ecosystem research in most IBP teams.

In contrast, the focus of our program was on studies of species interactions in an evolutionary and ecosystem context. We set the particular goal of finding out what, if anything, makes island ecosystems different from continental ecosystems. We expected few differences in the gross functions of primary production, decomposition, and mineral cycling. For example, the measurement of rates of photosynthesis and respiration of an island evergreen tropical montane rain forest was not expected to be greatly different from a continental evergreen temperate montane rain forest, except for the climatic variables involved. Of course, a major difference was to be expected in the consumer component of the island forest, largely because this ecosystem component evolved without the mammalian herbivores.

The study focus on species interactions had to be brought into a framework of ecosystem structure that was useful to everyone on the program. We selected two ecological models for this purpose: environmental gradient analysis and structural community analysis.

Environmental gradient analysis along a major mountain slope provided us with a model to study the spatial variation of species and species groups in relation to one another and to known variations in habitat factors. The model was useful for investigating the quantitative relationships and spatial positions of native and introduced biota, for contrasting wide- and narrow-ranging species, and for defining degrees of adaptation to specific segments along the gradient. Time variations include such phenomena as phenological events and cyclic variations in the biota.

Structural community analysis in a large, homogeneous, relatively undisturbed montane rain forest provided a model to study the population structure of important organism groups and, in part, their niche differentiation in the forest. This gave us an index for predicting the dynamics of this important native island ecosystem.

Site Coordination

We achieved integration by coordinating our research in two sites: the mountain slope and the rain forest. Each participant was required to sample according to a specified layout in these sites. The layout involved fourteen focal sites along the mountain slope and four specified transects in the rain forest. Once each person had contributed to the IBP site studies, that person was free to investigate other sites deemed important for the study of his or her particular organism group. But the requirement for participation in our team research was to contribute to the two primary IBP sites.

We developed a hierarchical sampling scheme that was related to the size and general environmental sensitivity or behavior of our respective organism groups. On this basis, we recognized three sampling groups: (1) The soil and litter sampling group included the participants concerned with the soil and litter fungi, the soil algae, the soil arthropods, and the litter-inhabiting Diptera. (2) The tree sampling group included the participants concerned with phytophagous insects in general, *Metrosideros* psyllids, *Acacia koa* psyllids, and the cerambycid bark and stem beetles. And (3) the total site sampling group included the participants concerned with vascular plants, rodents, birds, and general habitat insects. Because our sampling levels were hierarchical in size, sampling group 1 was to be included within sampling group 2, and group 2 within group 3. The basis for this was to attain reasonable homogeneity across the three sampling levels so that group 1 sampling was meaningful in the framework of group 2 sampling, and group 2 sampling was meaningful in the framework of group 3 sampling.

Sampling and Data Processing

Data analysis and processing was built into the sampling scheme. For example, the mountain slope sampling data always involved the collection of three parameters: the sampling location and date, the species recorded, and the quantity of each species recorded. These three parameters lent themselves to computer processing by the two-way table technique and dendrograph methods. Similarly time-phase data were adapted to processing by the two-way table technique.

Four strategically placed climatic stations were operated throughout the duration of the research. The climatic data on rainfall, throughfall, temperature, and relative humidity (from hygrothermographs) were processed weekly and transferred to the computer. Various tabulations, useful for interpreting behavior of all organism groups studied, were printed periodically as individual sheets and technical reports. These were distributed to all participants.

Communication and Administration

During the first two years we held quarterly workshops involving all program participants. These workshops were filled out primarily with conceptual modeling. In the second two years, we held workshops twice a year and an annual review symposium once a year (in November). In addition to presentations and discussions of methods and results, this meeting served to clarify the content for the annual progress report and budget renewal proposal. Between meetings we communicated by telephone and memoranda. There was no need to establish a newsletter, since nearly all program participants were from the University of Hawai'i or the Bernice Pauahi Bishop Museum, both in Honolulu.

A major communication item was our Technical Report series, which was established in December 1970. The first results appeared in February 1972, one and one-half years after we launched the program. After that, Technical Reports were produced more or less continuously, one or two every month. We continued the series during the synthesis phase until the end of the final funding year, August 31, 1975.

The Technical Report series served to communicate preprints or manuscripts, individual progress, and data sets important to members of our team. The inside cover of each report contains a notice stating that the information is of a preliminary nature and was prepared primarily for internal use in the US-IBP Island Ecosystems Program. Thus a Technical Report was purposely not considered a publication. This was important because it allowed authors to submit manuscripts to journals of their choice at their convenience. The customary time lag of one to two years from the time a manuscript is submitted to a journal to the time it is published would have defeated our efforts to communicate our results rapidly within the group. Moreover, production of a Technical Report guaranteed the author's title to a particular piece of research, an important consideration in a program of interfacing research segments involving a number of investigators.

We produced 250 copies of each report, an operation that involved one typist full time. Yet the operation was small enough to be flexible, and it did not require a major expense. This number of reports, moreover, allowed for distribution to interested individuals and institutions (such as the Institute of Pacific Islands Forestry, the Hawaii State Department of Land and Natural Resources, the National Park Service, all other US-IBP programs, and a few international programs). All Technical Reports are now on microfiche and are available from the University of Hawaii Library. A list of titles of all Technical Reports is appended to this volume.

The program was initiated by three codirectors: J. Linsley Gressitt, Andrew J. Berger, and Dieter Mueller-Dombois. Mueller-Dombois was asked to be the scientific coordinator. In the day-to-day administration of the program, he was helped by an assistant director, Kent W. Bridges, and an administrative assistant, Lynnette Araki. Additional office staff included a typist, Barbara Carr; a computer programmer, G. Virginia Carey; and in the final year, a technical editor, Winifred Yamashiro. A draftsman, Tamotsu Nakata, assisted the program participants in their needs for technical drawings. At the research site, a site manager, James D. Jacobi for most of the time, cared for the four vehicles (two with four-wheel drive) and for the maintenance of the two small houses rented from the Park Service. He also serviced the four climatic stations, maintained the day-to-day liaison with the National Park Service, and gave general assistance to program participants. Communication to the research site was by telephone.

RESEARCH AREA

Geographic Location

The IBP study sites were located on the Island of Hawai'i in and near Hawai'i Volcanoes National Park (figure 1-5). The Island of Hawai'i, locally called the "Big Island," is approximately 100 km by 100 km in size (exactly 10,480 km^2). The IBP field quarters consisted of two small houses located in the service quarters area of the National Park Service, near the east end of transect 1 on figure 1-5, at 1,200 m (3,950 ft) altitude. The field quarters could be reached by car from Hilo airport within forty minutes.

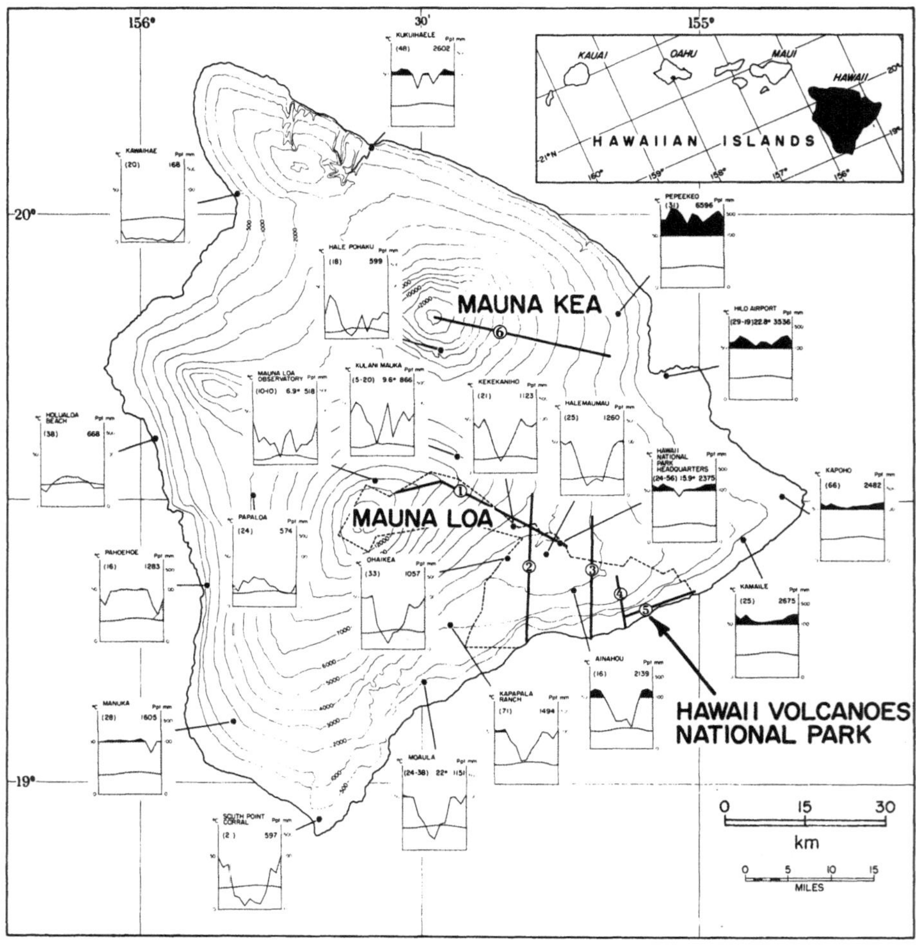

FIGURE 1-5. *Orientation map of Hawai'i showing location of IBP transects 1–6. Transect 1 on the east flank of Mauna Loa was chosen as the test site for the integrated environmental gradient analysis. The 80 ha Kīlauea rain forest study site is at the north end of transect 2. Dashed lines show outline of Hawai'i Volcanoes National Park. Contour lines are given in feet. IBP field quarters were near the east end of transect 1. The diagrams show mean monthly rainfall (mm) and temperature (°C) curves plus mean annual rainfall of twenty-one weather stations.*

Among the six transects shown on figure 1-5, transect 1 (on the east flank of Mauna Loa) was chosen as the test site for the integrated environmental gradient analysis. Fourteen focal sites were established along the Mauna Loa transect. These sites ranged from 3,050 m (10,000 ft) in the very sparse alpine scrub down to 1,190 m (3,900 ft) in the montane rain forest. The other five transects served as validation sites or for extension of individual studies. A second site for integrated sampling was established as an 80 ha (200 acre) plot at the north end of transect 2, near 1,520 m (5,000 ft) altitude. This site is outside the park in a well-preserved native rain forest in the Kīlauea Forest Reserve. The site was reached by a ten-mile (16 km) jeep trail leading through a cattle ranch.

Reasons for Choice of Area

A major reason for choosing this area was the availability of a site-preparation study, "Atlas for Bioecology Studies in Hawai'i Volcanoes National Park" (Doty and Mueller-Dombois, 1966). This 507-page document includes a checklist of vascular plants, a vegetation map with vegetation profiles, a description of the air-photographic coverage, the geology and volcanic events, the soils, the weather and climate, and a summary review of the biological studies done so far in the area, together with a bibliography.

Other reasons included the availability of relatively undisturbed sites still in original vegetation, the improved protection from vandalism of field instrumentation inside the park, and the expressed interest of the National Park Service in having the IBP study done within its territory and of the private landholding Bishop Estate Corporation in having a study done in the Kīlauea Forest Reserve. Finally, there were important logistical advantages to choosing Hawai'i Volcanoes National Park. We were offered field living quarters, laboratory, library, and herbarium facilities.

Characteristics of the Area

The area is geologically special in that it is within the range of active volcanism. The soil substrates are all still relatively recent volcanic surfaces. The last lava flow from Mauna Loa occurred in 1950. The Kīlauea Volcano (at 1,200 m) in the center of Hawai'i Volcanoes National Park is almost continuously active. Thus all results from our studies have to be interpreted with respect to a geologically very young mountain system. This, however, lends itself to interesting comparisons to older volcanic mountainous islands occurring in similar climates. In figure 1-5, the climate of the study area is shown in relation to the rest of the island by twenty-one climate diagrams prepared according to the method of Walter (1971).

On the diagrams, the mean monthly air temperature is plotted with reference to the left-hand ordinate, resulting in the nearly horizontal curve on each diagram. The temperature intervals on the left ordinate are 10°C, starting with 0°C at the abscissa. The abscissa indicates the months of the year, with January at both ends and July in the center. Mean monthly rainfall is plotted with reference to the right-hand ordinate. The intervals are 20 mm from 0 mm at the abscissa upward to 100 mm. From 100 mm on upward, each interval represents 200 mm. The amount of rain in excess of 100 mm is indicated by a black field (see eastern part of the island, figure 1-5). Wherever the rainfall curve (which is the more irregular curve on each diagram) undercuts the temperature curve at the selected scale (of rainfall:temperature = 2:1), a drought period is indicated (see diagrams on south and northwest sides of island). The mean annual rainfall in mm is written at the right-hand top of each diagram. The mean annual temperature appears as a figure to the left only for stations with temperature records.

The mean annual temperature near the top of transect 1 is 6.9°C. At this elevation (near 3,350 m or 11,000 ft), ground frost occurs each night throughout the year. At the east end of transect 1 (at 1,200 m or 3,950 ft) the mean annual temperature is 15.9°C. At sea level (Hilo Airport) the temperature is 22.8°C.

The month-to-month temperature curve indicates typical insular tropical climates at all elevations because of the small difference (4°–6°C) between summer and winter temperatures. The mean annual temperature at the north end of transect 2 in the Kīlauea Forest study site can be extrapolated as approximately 14°C. Rainfall in the Kīlauea Forest site is similar to, though somewhat less than, rainfall at the park headquarters on transect 1. At the Kīlauea Forest site, rainfall is between 1,800 and 2,000 mm per year, and there is usually a short dry season in June, when the monthly rainfall can be less than 100 mm. Along the Mauna Loa transect (transect 1), the rainfall decreases from about 2,400 mm at its east end in the rain forest to about 500 mm at its upper end (near 3,660 m or 12,000 ft) in the alpine moss *(Rhacomitrium)* desert. In addition to the overall decrease upslope, rainfall is particularly low during June. At the tree line (near 2,440 m or 8,000 ft) rainfall is also markedly low in September.

The climate along the Mauna Loa transect compares well with that along the Mauna Kea transect (Mueller-Dombois and Krajina, 1968). The climate diagrams show wet conditions on the east side of the island. Summer drought climates occur in the south. The reverse—winter drought climate—occurs on the west side at Holualoa beach. A true desert climate occurs on the northwest side of the Big Island at Kawaihae (168mm).

The vegetation of the study area has been described in the *Atlas for Bioecology Studies in Hawai'i Volcanoes National Park* (Doty and Mueller-Dombois, 1966).

II
Spatial Distribution of Island Biota

2

Introduction

D. Mueller-Dombois and K.W. Bridges

REASONS FOR CHOICE OF TRANSECT APPROACH

Distribution studies of biota and ecosystem components can be done at all geographic scales. Most distributional studies of island biota have been concerned with comparisons between islands and island groups, such as those of MacArthur and Wilson (1967). On large, oceanic islands, it is also useful to study biotic distribution patterns within the island itself. In this approach, emphasis is placed on understanding how the generally depauperate biota adapt, compete, and evolve. The former approach is generally given to a study of dispersal, establishment, and, to some extent, evolution.

In this investigation the within-island approach was chosen for a number of reasons. Primarily it allows a study oriented toward examining the island ecosystems rather than separate organism-oriented studies. Such an investigation also could be carried out within the time limits given to the IBP studies, and the results would be expected to allow comparisons with the other ecosystem-oriented IBP projects. The area selected for the spatial distribution analysis was Hawai'i Volcanoes National Park. Here we concentrated our analysis along an existing mountain transect (transect 1 on figure 1-5).

A distribution study along a major mountain slope allows for comparisons with similar studies in continental mountain ecosystems. The approach ties into a general area of ecological theory that has its roots in the ecological series concept of Russian authors (for example, Sukachev, 1928), in the ecological species group derivation techniques of Ellenberg (1956; Mueller-Dombois and Ellenberg, 1974), and in the environmental gradient approach of Whittaker (1967) that was developed in temperate, continental mountain systems.

Environmental gradient analysis is essentially a field-experimental approach. It utilizes the experimental principle of changing the environmental factor or factor complex to which one wishes to test a biological response while attempting to hold all other factors constant. The latter is extremely difficult or next to impossible in field environments, and more often than not one finds that several factors change concomitantly, randomly, or even erratically. The goal is to select sampling sites in such a way that the gradient inconsistencies are minimized. Whittaker (1956, 1960) achieved this by comparing only sites of similar physiographic position at different elevations.

In an altitudinal gradient analysis, the changing factor (or factor complex) is generally the climate. A number of other factors may be kept constant so that the climatic influence may be observed. The particular selection of constant factors depends largely on the organisms being studied. For example, a well-drained substrate of similar geochemical composition or similar parent material may be important for vegetation studies, and the availability of the same tree species may be important for studies of host-adapted insects.

By choosing an altitudinal gradient on an island, it is possible to make comparisons of the spatial distribution of native versus exotic species and their deviations from continental gradients. Such an approach also provides an opportunity to examine other trends critically. The temporal changes in the abundance of organisms along a climatic gradient is one example. The study of the genetic structure of the organisms along the gradient should help reveal the pattern of their microevolution. For these sorts of reasons, the altitudinal gradient analysis approach was considered ideally suited for providing answers to several of the initial working hypotheses.

In Hawai'i quantitative studies of species distributions have more than an academic interest. Because of the high proportion of rare and endangered species native to the islands, particular attention is given to their preservation. This is expressed in the establishment of natural areas and by various efforts to expand the present range of native species.

MAUNA LOA TRANSECT AS A CASE EXAMPLE OF A MAJOR ENVIRONMENTAL GRADIENT

Extensive mountain slopes, suitable for environmental gradient analysis, may be found in nearly all climatic regions of the world. They occur in continental as well as insular settings. They usually display a reasonably self-evident altitudinal zonation of vegetation belts. For example, the altitudinal vegetation zones in Hawai'i, on Mauna Loa and Mauna Kea, are closely comparable to those of other tropical mountains of the world (Troll, 1959; Walter, 1971).

Geologically Mauna Loa is a very young mountain. Its biota and community formation can thus be compared with the biota and community formation of older mountains with similar elevations and climatic zonations. As a young volcanic mountain, Mauna Loa is typically shield-shaped, meaning that it has a very gradual smooth slope descending radially in all directions from the summit. There are no dissected valleys, streams, or tributaries, and topographic variations within the same altitudinal segment are so minor that they have no significant effect on the vegetation. Thus the problem of eliminating the effect of a "topographic moisture gradient" (Whittaker, 1956, 1960) in altitudinal gradient analysis does not exist for such a mountain. All sites within a given elevational segment are well drained and of uniform topographic position, a characteristic that makes for little complication in site selection. The only major discontinuities are caused by a few intercepting lava flows of recent historical origin, issued within the last two hundred years. These flows are narrow enough so that they were avoided in sampling. The sampling sites are located on the older, prevailing surfaces and are comparable in the sense that they are from volcanic materials of similar geochemical composition (Macdonald and Abbott, 1970). Any substrate differences that may occur can be interpreted in terms of physical factors.

Important differences in the vegetation formations along the transect slope can be attributed to climate. Evidence of this is given by the close similarity of vegetation formations along a parallel east flank transect on Mauna Kea (Mueller-Dombois and Krajina, 1968), a mountain of much greater age. Both temperature and rainfall decrease upslope in a similar pattern on both mountains. Rainfall on these mountains is controlled mostly by the trade winds so that a given slope direction is exposed to a spatially rather stable atmospheric moisture pattern. The east flank of Mauna Loa, which was chosen for our transect analysis, is exposed to the northeast trade winds. It thus is representative of the wet side of a major trade-wind-intercepting mountain.

MAUNA LOA TRANSECT IN RELATION TO OTHER PARK TRANSECTS

The geographic location of the Mauna Loa transect is shown as transect 1 on figure 1-5. It extends from 12,000 ft (3,660 m) near the summit of Mauna Loa, downslope on its east flank in the park to 3,920 feet (1,190 m) near the summit of the Kīlauea crater. Transect 1 spans a distance of 22 mi (35 km). The transect cuts through four of the six environmental sections recognized for the park (Mueller-Dombois and Fosberg, 1974). The four sections are (1) the alpine section, from the top of Mauna Loa to 8,500 ft (2,590 m); (2) the subalpine section, downslope to 6,700 ft (2,040 m); (3) the montane seasonal section, downslope to 3,800 ft (1,160 m) and south of transect 1; and (4) the montane rain forest section, in part covered by the east end of transect 1 (see figure 2-1). The other two environmental sections in the park, not traversed by the Mauna Loa transect, are (5) the submontane seasonal section and (6) the coastal lowland section; both are on the south slope of Kīlauea Volcano. These sections are traversed by transects 2 to 5 on figure 1-5. These transects were not included in the integrated analysis. A description of all six park transects is given by Mueller-Dombois (1966: 396–437).

PROFILE DIAGRAM AND PHOTOGRAPHS OF ECOSYSTEMS

The profile diagram on figure 2-2 indicates the general relationships of climate, vegetation, and substrate along the Mauna Loa transect. Twelve seg-

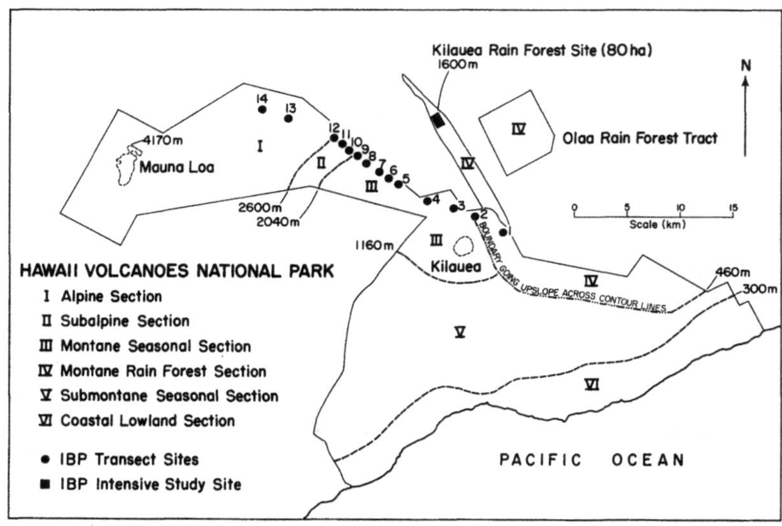

FIGURE 2-1. *Hawai'i Volcanoes National Park. The map shows the six environmental sections, the fourteen Mauna Loa transect sampling sites, and the Kīlauea Rain Forest site.*

ments are shown, defined on the basis of mappable differences in vegetation structure and dominant species along the slope (Mueller-Dombois, 1966, 1967). The combination of the vegetation types with their environmental components (climate, topographic position, and soil substrate) allows them to be recognized as ecosystem types. The twelve segments are:

Segment 1	Alpine stone desert.
Segment 2	*Rhacomitrium* moss desert.
Segment 3	Sparse alpine scrub.
Segment 4	Alpine aggregate scrub.
Segment 5	*Metrosideros* tree line ecosystem.
Segment 6	Open subalpine *Metrosideros* scrub forest.
Segment 7	Mountain parkland (formed by *Acacia koa* tree communities, *Styphelia* shrub communities, and grass communities).
Segment 8	*Acacia koa-Sapindus* savanna.
Segment 9	Closed kīpuka forest (segment 8 interdigitates locally with segment 9).
Segment 10	Open *Metrosideros* dry forest.
Segment 11	Open *Metrosideros* rain forest.
Segment 12	Closed *Metrosideros* rain forest.

Figures 2-3 through 2-6 and their legends provide a brief description of the twelve segments or ecosystem types.

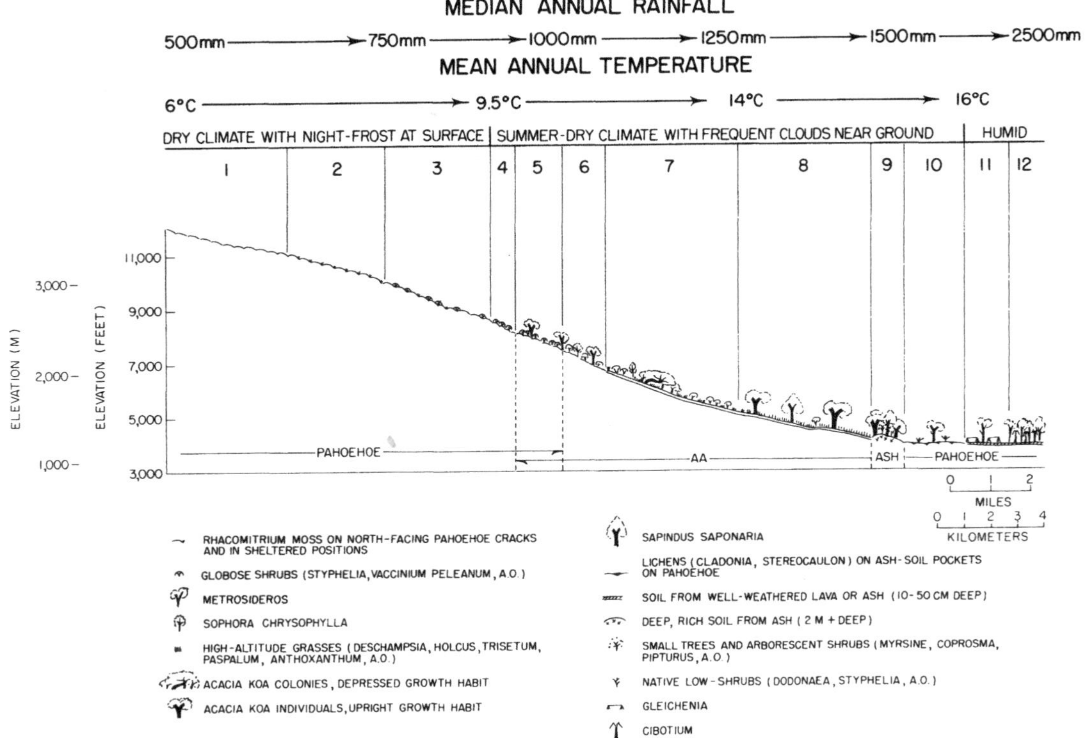

FIGURE 2-2. *Profile diagram of Mauna Loa transect*. For geographic location see transect 1, figure 1-5.

1. *View of Mauna Loa summit (13,680 ft, 4,170 m) and its upper east flank forming the skyline to the left downslope to about 10,000 ft (3,050 m). Photograph taken from summit area of Mauna Kea, 35 km away. The light- and dark-colored areas are lava flows of different kinds and ages. The light-colored fields are pāhoehoe flows (mostly smooth pavement type or ropy rock plates with cracks); the dark-colored fields are mostly ʻaʻā flows (rock rubble type lava). The older flows are oxidized and light brown in color. They show up grayish on the photograph.*

2. Segment 1: *Vegetationless stone desert extending from summit to about 11,000 ft (3,350 m). Here substrate is composed of ropy pāhoehoe lava.*

3. Segment 2: Rhacomitrium *moss desert, from 11,000 to 10,000 ft (3,050 m). The moss* R. lanuginosum *var.* pruinosum *is found on the older, oxidized lava. Here it occurs only in very scattered small colonies in the trade-wind-facing (northward-oriented) crevices. The white hyaline tips give the moss a snowlike appearance.*

4. Segment 3: Vaccinium-Styphelia *low-scrub desert, or sparse alpine scrub, from 10,000 to 8,550 ft (2,600 m). The two native shrubs* Vaccinium peleanum *and* Styphelia douglasii *grow here in widely scattered formation. Additional species are the ferns,* Pellaea ternifolia *and* Asplenium trichomanes, *and two native grasses,* Agrostis sandwicensis *and* Trisetum glomeratum, *which are extremely sparse. The dominant surface feature is still the oxidized pāhoehoe lava.*

FIGURE 2-3 *Alpine section (Mauna Loa transect). Dry, cool climate: 500–750 mm rainfall per year; 6–9.5°C mean air temperature; nocturnal ground frost each night of the year. Profile segments 1–3 on figure 2-2.*

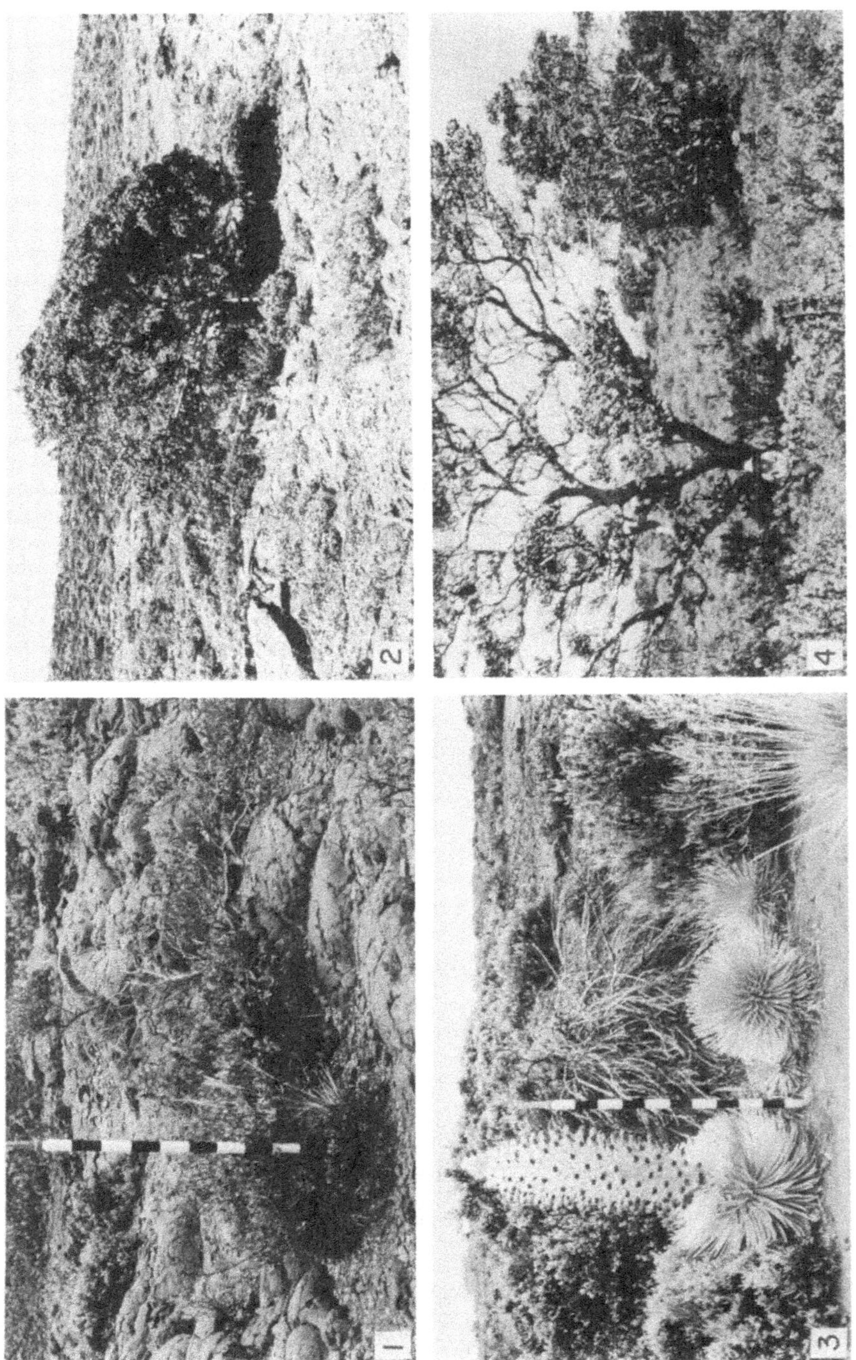

1. Segment 4: *Scattered globose aggregate-scrub, from 8,550 to 8,200 ft (2,500 m). Additional native shrub species occur:* Dodonaea viscosa *and* Coprosma ernodeoides, *which often grow aggregated in clumps with* Vaccinium *and* Styphelia. *The native grass* Trisetum glomeratum *grows sparsely on the fine soil dust collected around the shrub clumps (on the photograph).*

2. Segment 5: Metrosideros *tree line ecosystem, from 8,200 ft (2,500 m) to 7,500 ft (2,290 m). Widely scattered* Metrosideros collina *subsp.* polymorpha *var.* incana *trees grow 4–5 m tall, with native shrub species aggregated beneath.*

3. Segment 5: *The Hawaiian silversword* (Argyroxiphium sandwicense) *thrives in the tree line ecosystem, but here it was planted by the National Park Service. The native bunchgrass,* Deschampsia australis, *is in the foreground. The shrubs are mostly* Styphelia douglasii *(on the photograph).*

4. Segment 6: *Open* Metrosideros *scrub forest, from approximately 7,500 ft (2,290 m) to 6,700 ft (2,040 m). The change from segment 5 to 6 is very gradual.* Metrosideros *trees grow less widely scattered than in the tree line ecosystem, and the shrubs become denser downslope. At approximately 7,500 ft (2,290 m) elevation, the lava substrate is no longer the dominant surface feature. Instead shrubs cover most of the surface. The native legume tree* Sophora chrysophylla *occurs scattered in the lower part of this segment (not on the photograph).*

FIGURE 2-4. *Subalpine section (Mauna Loa Transect). Summer-dry climate: ~1,000 mm rainfall per year; 9.5–12°C mean air temperature; frequent clouds near ground. Profile segments 4–6 on figure 2-2.*

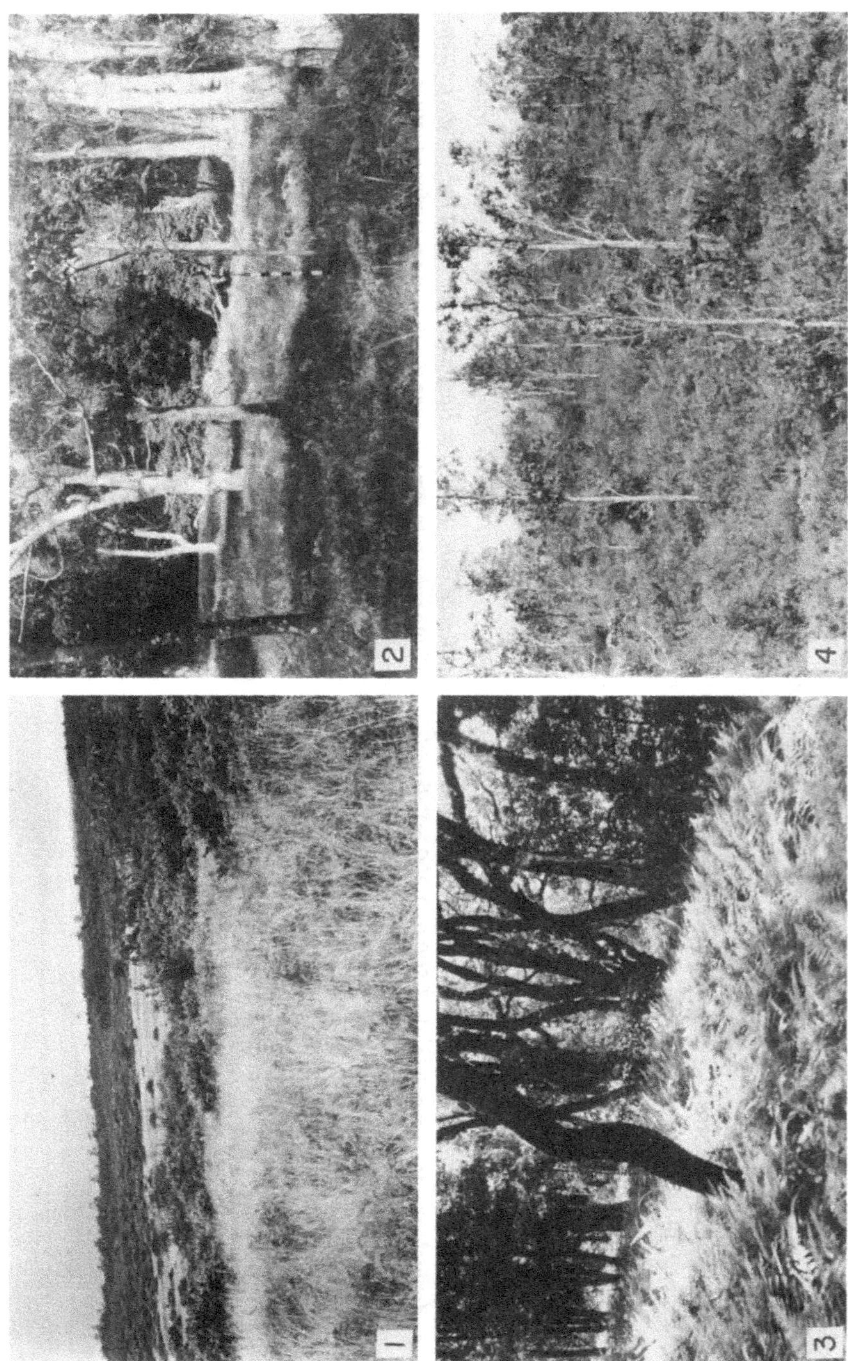

1. Segment 7: *Mountain parkland ecosystem, from 6,700 ft (2,040 m, at end of Mauna Loa strip road) to approximately 4,500 ft (1,370 m). This ecosystem contains three distinct communities:* Acacia koa *tree colonies, tall-shrub communities of mixed* Styphelia douglasii *and* Dodonaea viscosa *bushes, and grassland. The grassland, which forms the matrix surrounding the two types of woody plant communities, is dominated locally by the native* Deschampsia australis, *but where disturbed by pigs, it is dominated by the European weedgrass* Holcus lanatus.

2. Segment 8: Acacia koa-Sapindus *savanna, from 4,500 ft (1,370 m) to 4,000 ft (1,220 m). The major difference from the mountain parkland ecosystem is the absence of the* Styphelia-Dodonaea *tall-shrub communities. Also several other introduced grass species become dominant, particularly* Paspalum dilatatum. *The dominant tree is* Acacia koa, *which is occasionally associated with* Sapindus saponaria *(not on photograph).*

3. Segment 9: *Closed kīpuka forest. The Hawaiian word* kīpuka *stands for island of vegetation surrounded by more recent lava flows. The taller trees (up to 25 m) are the same as in the savanna. A well-developed lower tree layer is comprised of several species, including* Myrsine lessertiana, Coprosma rhynchocarpa, Psychotria hawaiiensis, Osmanthus sandwicensis, Sophora chrysophylla, Pipturus albidus, *and others.* Microlepia setosa *is a locally dominant fern in the undergrowth (on photograph). A detailed description is given by Mueller-Dombois and Lamoureux (1967). This closed mesic forest interdigitates locally with segment 8, between 4,250 and 4,000 f (1,300–1,220 m).*

4. Segment 10: *Open* Metrosideros-*lichen forest with native low shrubs, at approximately 3,950 ft (1,200 m) elevation.* Metrosideros collina *subsp.* polymorpha *var.* incana *grows in the old pāhoehoe cracks that appear to control the density of stocking. The trees are rather short (3–10 m). The depressions between outcropping pāhoehoe blocks are covered with pockets of finer soil and ash, which support a poor growth of lichens* (Cladonia *spp. and* Stereocaulon), *grasses and herbs* (Andropogon virginicus, Bulbostylis capillaris, Pteridium aquilinum). *Most of the surface is covered by native shrubs,* Dodonaea viscosa, Styphelia tameiameiae, Raillardia scabra, Vaccinium reticulatum, *and* Coprosma ernodeoides.

FIGURE 2-5. *Montane seasonal section (Mauna Loa transect). Summer-dry climate: 1,100–1,500 mm rainfall per year; 12–16°C mean air temperature; frequent clouds near ground. Profile segments 7–10 on figure 2-2.*

1. Segment 11: *Open* Metrosideros-Gleichenia *forest, at approximately 3,950 ft (1,200 m) elevation. Tree growth is similar to preceding type, but undergrowth is dominated by the matted fern,* Gleichenia (= Dicranopteris) emarginata. *In places not dominated by the fern, there are individuals of the herbs* Dianella sandwicensis, Hedyotis centranthoides, Lycopodium cernuum, *and* Andropogon virginicus. *The woody fern,* Sadleria cyatheoides, *is frequent also.*

2. Segment 12: *Closed* Metrosideros-Cibotium *forest, at 3,920 ft (1,190 m). This is the prevailing montane rain forest type in the park. It is characterized by a stand of relatively even-sized* Metrosideros *trees with diameters at breast height of 40 to 65 cm and uniform crown-canopy heights from 14 to 20 m. Tree boles are rather naked and rarely covered with epiphytes. A distinct lower stratum is formed by the tree ferns (*Cibotium *spp.). Associated smaller trees include* Myrsine lessertiana, Ilex anomala, Coprosma ochracea, *and* Cheirodendron trigynum. *Common shrubs are* Vaccinium calycinum, Cyrtandra platyphylla, Broussaisia arguta, Pipturus albidus. *Herbs frequently found include* Gahnia gahniaeformis, Briza minor, Isachne distichophylla, Hedyotis centranthoides, Lycopodium cernuum, *and* Peperomia *spp.*

3. Segment 12: *Inside view of the rain forest near Thurston lava tube. Tree fern* (Cibotium glaucum) *behind the meter stake,* Sadleria cyatheoides *in foreground, left.*

4. Segment 12: *View to Kīlauea Iki crater floor in rain forest section near end of transect 1.*

FIGURE 2-6. *Montane rain forest section (Mauna Loa Transect). Year-round humid climate: 2,000–2,500 mm rainfall per year; ~16°C mean air temperature. Profile segments 11 and 12 on figure 2-2.*

ALTITUDINAL DISTRIBUTION OF MAJOR SITE FACTORS

From a transect-wide viewpoint the major site factors can be grouped into three environmental components: climate, soil substrate, and mechanical influences.

Climate

Two important climatic factors form continuous gradients along the transect: median annual rainfall and mean annual air temperature. The relationship of both factors is indicated at various points along the transect on figure 2-2. The general trend is for the median annual rainfall to increase from 500 mm in the alpine section downslope along the transect to 2,500 mm in the montane rain forest section. The mean annual air temperature (as recorded in meterological shelters 1.5 m above the ground) increased downslope from 6°C in the alpine section to 16°C in the montane sections of the transect.

The mean monthly variation of these two parameters can be gleaned from the climate diagrams on figure 1-5. In the montane rain forest section, the average monthly rainfall is above 100 mm except in June, when the rainfall may drop to 80 mm (figure 1-5, Hawai'i National Park headquarters station). In the mountain parkland (segment 7 on figure 2-2), there is a pronounced dry season in June when the rainfall curve touches or even drops below the mean monthly temperature curve (figure 1-5, Kekekaniho). Near the top end of the transect in the alpine section, a second dry season occurs during September (Mauna Loa observatory).

The variability of the rainfall from year to year can be seen both in changes in the annual amounts and in changes in the monthly distribution patterns. Considerable variation in the annual rainfall totals was seen during the three years of intensive sampling (1972–1974); the mean standard deviation of the annual rainfall sums for three IBP sites (4, 6, and 9) was 214 mm (Bridges and Carey, TR 22, TR 38, TR 59).

The mean monthly temperature curve forms almost a straight line at each station, indicating tropical insular climates at all elevations (figure 1-5). The warmest months are August and September and the coolest January and February, but the mean monthly temperature range between summer and winter is generally within 4°C. The daily temperature range is much greater and varies from about 18°C at high elevations to 5°C near sea level. At Hawaii National Park headquarters (near the low end of the transect) the mean daily temperature range is about 15°C. In the alpine section, freezing air temperatures occur from November through March, as indicated by the fine dotted line below the mean monthly temperature curve on the Mauna Loa observatory diagram (figure 1-5). Nocturnal ground frost temperatures were recorded during the warmest month of the year (August 1966; Mueller-Dombois and Krajina, 1968) on the east flank of Mauna Loa down to 8,600 ft (2,600 m).

Several other climate parameters are also of interest as they vary along the transect. Fog drip under trees is a common phenomenon along the transect from the tree line ecosystem at 8,200 ft (2,500 m) downslope through the mountain parkland; this includes the entire climatic zone defined on figure 2-2 as summer-dry climate with frequent clouds near ground. Fog drip was measured by Juvik and Perreira (TR 32) along the transect with louvered screen cylinders of 2,691 cm^2 surface area mounted 3 m above ground level. They found that fog drip added 65 percent to the rainfall at the tree line at 8,200 ft (2,500 m) on Mauna Loa. The contribution of fog drip decreased above and below this point. In the savanna (figure 2-2, segment 8) it was 49 percent. In terms of absolute fog interception, the maximum water yield was recorded in the mountain parkland at 5,200 ft (1,580 m) elevation, where 638 mm of fog drip water was collected over a period of seven months (October–April 1972). Thus one can conclude that the trees and the plants growing under them in the tree line ecosystem receive at least 1.65 times the annual rainfall amount shown on figure 2-2; in the mountain parkland they may receive about 1.5 times the amount. The exact amounts vary, of course, with the height and surface area of the tree, its branching and foliage arrangement, and other factors affecting condensation.

Pan evaporation rates for three transect sites have recently been determined by Clark, Austring, and Juvik (1975). Their data show a generally increasing rate of evaporation from 2.60 mm per day in segment 5 (the open *Metrosideros* scrub forest in the subalpine section) to 3.18 mm per day in segment 10 (the open *Metrosideros*-native scrub lichen forest in the lower end of the montane seasonal section). These data are based on ninety-eight days of recording between July 1974 and May 1975.

Soil Substrate

The major substrate types are indicated on figure 2-2 as pāhoehoe, ʻaʻā, and ash.

Pāhoehoe is the prevailing substrate in the alpine section along the transect from segments 1 through 5. Pāhoehoe is the type of lava that is particularly hot (1,000°C) and liquid during the time of extrusion, when it flows like a slow-moving river before it cools down *in situ*. The cooling results in an uneven pavement-like surface. Several variations of pāhoehoe have been described by Jones (1943). The main plant substrate is made up of the many cracks, folds, and fissures occurring between the bedrock slabs of the pāhoehoe (figure 2-3, photographs 3 and 4). Only the oxidized, reddish-brown or buff-colored lava, which is the oldest and prevailing substrate, was included in the transect sampling. A few recent lava flows that occur here and there along the transect (such as shown in photograph 2 on figure 2-3) were excluded from the analysis. The cracks and fissures in the pāhoehoe are so numerous in this upper area of the transect that the scattered occurrence of shrubs (in segment 3, figure 2-2) cannot be explained as a lack of suitable substrates for plant growth. Instead the sparse growth appears to be a function of climate (low rainfall and low temperature) and geological recency of the substrate. Over long periods of time an increase in the density of the vegetation can be expected. This conclusion is based on comparisons of the vegetation cover at similar altitudes and in similar climates in other tropical mountains (Mueller-Dombois and Krajina, 1968; Walter, 1971).

In segment 5 (tree line ecosystem) the profile diagram (figure 2-2) shows an overlap of the two major types of lava: pāhoehoe and ʻaʻā. ʻAʻā lava consists of individual rock chunks of various sizes, ranging mostly between 5 and 30 cm in diameter. These rock chunks are loosely stacked up in the form of a sheet, usually less than 1 m deep, above a solid pāhoehoe-type core (Macdonald, 1945). The fissures between the rock chunks are filled with fines (mostly from volcanic ash) along the Mauna Loa transect. On first observation, it seemed that the upper limit of tree growth was correlated with the ʻaʻā lava. However, subsequent observations revealed that the first trees forming the tree line on Mauna Loa occur also on pāhoehoe. Such a situation is shown on photograph 2 of figure 2-4. ʻAʻā lava prevails throughout the middle section of the transect from the subalpine open scrub forest through the savanna (segments 6–8, figure 2-2).

In the same direction downslope, there is a gradual increase of fines from volcanic ash. In the tree line ecosystem, most lava cracks are already filled, and there are a few scattered pockets of fine, yellow dustlike ash on the pāhoehoe surface (such as shown in the foreground on photograph 3, figure 2-4). Such ash pockets occur throughout segment 6, but the general surface is strewn with ʻaʻā rocks, and the spaces between the rock chunks are filled almost everywhere with fine soil from ash. The volume of ash increases further downslope, where it forms initially a thinly overlying soil blanket (10–25 cm deep) in the upper part of the mountain parkland. This soil blanket is frequently interrupted by lava rock outcrops. Farther downslope, rock outcrops become less frequent until they disappear entirely in the savanna. The ash blanket in the savanna is generally over 70 cm deep. The kīpuka forests in segment 8 have very deep soils (over 5 m deep in places) that developed from ash dunes (Mueller-Dombois and Lamoureux, 1967). The same applies locally to the savanna, which interdigitates with the closed kīpuka forest in segment 9. More detailed soil variations from the alpine section to the kīpukas are described by Mueller-Dombois (1966), but the general trend is one of an increasing number of soil pockets from the alpine through the subalpine section downslope to an area of a shallow discontinuous ash blanket (in the upper mountain parkland) becoming a more continuous ash blanket (in the lower part) that increases in depth to over 70 cm in the savanna.

East of the deep-soil savanna and kīpuka, the substrate changes abruptly at the border to the open *Metrosideros* dry forest (segment 10, figure 2-2). Here again, the prevailing substrate is pāhoehoe with a shallow, highly discontinuous ash blanket. The substrate is physically not unlike that of the subalpine open scrub forest (segment 6) were it not for the difference in the outcropping lava type. Farther eastward, the ash blanket again becomes continuous in the open rain forest (segment 11), and it reaches depths greater than 50 cm in the closed rain forest (segment 12).

The original chemical composition of the lava types and ash were all very similar since they originated from the same source of magma. However, the pāhoehoe and ʻaʻā along the slope originated from the Mauna Loa volcano, while the ash blanket through the mountain parkland and savanna and the pāhoehoe at the east end of the transect originated from the Kīlauea volcano.

The parent rock material is basaltic, showing relatively high amounts of calcium (about 10 percent) and magnesium (6–10 percent). The original silicon content is very high (near 50–70 percent). It drops in the course of weathering and becomes very low in older rain forest soils. But all soils along the Mauna Loa transect are geomorphologically very recent, and there is as yet little secondary clay formation. According to the soil-order classification given in chapter 1, the transect soils represent four of the eleven soil orders found in the Hawaiian Islands. The four are Lithosols (throughout the alpine and subalpine section), Entisols (the mountain parkland soil), Mollisols (the savanna and kīpuka soils), and Histosols (in the montane rain forest section).

The organic content in the surface soil contains about 12–15 percent carbon in the rain forest soils, and the pH is between 4 and 5. Throughout the savanna and mountain parkland, the organic content is similarly high, ranging from 10 to 15 percent, but the pH is less acid, ranging generally from 5.5 to 6.5 (Mueller-Dombois, 1966).

Soil analyses on the mineral concentrations are shown for the transect profile in figure 2-7. The relative abundances have been plotted in such a way as to allow their comparison with the normal ranges of mineral concentrations established for optimal plant growth. The original values are given in ppm by Stoner (TR75:85).

Mechanical Factors

The more important mechanical factors can be summarized in three groups: herbivory, ground disturbance, and fire.

Herbivory in the form of cattle grazing was a long-standing influence in the mountain parkland and savanna (Fagerlund, 1947) before the area was incorporated into the national park system in 1927. However, grazing rights were still upheld during World War II and up to 1948, when finally all cattle were removed (Apple, 1954). This undoubtedly had the pronounced effect of reducing *Acacia koa* tree communities to only scattered survivors of old-growth trees, as can be observed today on the adjacent ranchland (Cuddihy,

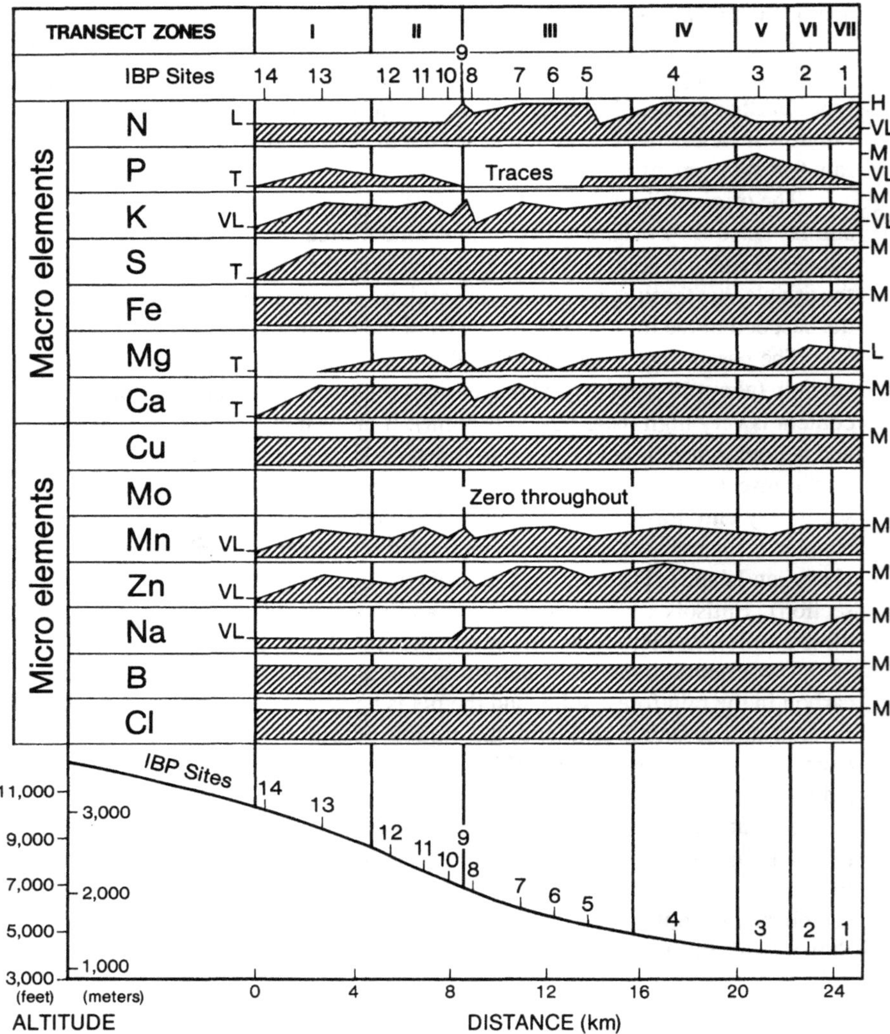

FIGURE 2-7. *Mineral concentrations in the surface-soil along Mauna Loa transect plotted as values relative to optimal plant growth (as determined for agricultural crops and considered generally available to plant growth). H = high, M = medium, L = low, VL = very low, T = trace.*

1978). Baldwin and Fagerlund (1943) considered cattle grazing the only significant herbivore influence at that time. However, feral goats (*Capra hircus* L.) roamed the same territory probably a few decades after their introduction by Captain Cook in 1778. Their range of distribution then was undoubtedly wider than that of cattle, since they have been seen throughout the transect from the tree line ecosystem (segment 5 on figure 2-2) to open *Metrosideros* dry forest (segment 10). They stay away only from the rain forest. Their concentration center along the transect is in the mountain parkland (segment 7). However, the control measures exercised by the park service are causing a changing pattern of distribution throughout the park. Their greatest abundance and disturbance effects have been in the coastal lowland section (TR 13). Their number in the montane seasonal section was estimated to be between one thousand and two thousand in 1966 (Mueller-Dombois, 1967). In this area they move about in small herds of about ten to thirty individuals where they feed particularly on the sucker reproduction of *Acacia koa*. The park service stepped up its goat-control program in 1960 (Gerdes, 1964), when from one thousand to five thousand goats were killed in the entire park. The effect of this stepped-up goat-control program became evident in a comparison of aerial photographs taken in 1954 and in 1965. According to these photographs, the vegetatively reproducing koa colonies expanded by a rate of 0.5 to 2.5 m per year into the surrounding grass matrix (Mueller-Dombois, 1967). The effect of goats on koa reproduction was subsequently studied in detail in an exclosure at 6,200 ft (1,890 m) and throughout the mountain parkland at various places (Spatz and Mueller-Dombois, 1973). It was found that the goat population did not prevent sucker reproduction, but their browsing effect seemed to produce an abnormally dense stocking of suckers in a haphazard pattern. Where these suckers had a chance to grow in height beyond their browsing reach (at about 1.7 m), the suckers tended to form denser koa stands than one would expect without the mechanical damage caused by goats during the sucker initiation stage. Since 1972 the goat-control program has been intensified so koa suckers are now sprouting in the grassland around nearly all koa colonies. However, it is doubtful that the grassland matrix in the parkland will ever disappear under koa because of other mechanical influences.

Ground disturbance from feral pigs (*Sus scrofa* L.) is a very common mechanical influence throughout the mountain parkland and particularly in the savanna (segment 8). Their effect on the grass-cover composition has been studied experimentally (Spatz and Mueller-Dombois, TR 15). It was found that pig rooting favors decidedly the replacement of the dominant native bunchgrass *Deschampsia australis* by the introduced European weedgrass *Holcus lanatus*.

According to Park Service records, no major fire had occurred in the mountain parkland and savanna areas for at least six decades. However, at least infrequent fires must have occurred in the past as evidenced by the charcoal bits found in the surface ash soil throughout the area (Mueller-Dombois, 1967; Vogl, 1969). Charcoal was found at 70 cm soil depth under the kīpuka forest (segment 9) and dated as 2,170 ± 200 years old (Mueller-Dombois and Lamoureux, 1967). This indicates that fire has been a long-interval natural perturbation in this transect area, which probably includes segments 7 through 9. Natural fires can be caused by intervening lava flows during volcanic activity.

In spite of the dynamic situation resulting from these mechanical influences, the central area of the transect can be considered relatively stable and in balance with the prevailing climate and substrate. This can be said for two reasons: a generally similar vegetation (savanna and parkland, although much more disturbed) occurs on the east flank of Mauna Kea (Mueller-Dombois and Krajina, 1968), and the spatial scale of the ecosystems as defined on the profile is wide enough to allow for small-area dynamic changes without affecting their overall structure.

HYPOTHESES OF SPECIES DISTRIBUTION

We had several initial working hypotheses that involved special consideration of species distributions. One of these was that endemic island birds, insects, and other subsidiary life forms evolved primarily in adaptation to the community structure-forming dominant native plant species. To test the degree of spatial association of native biota, we sampled along the Mauna Loa transect, which cuts through the four environmental sections.

Another initial working hypothesis was that climatic factors have significant effects on ecosystem stability. We surmised from general observations that rain forest ecosystems would be more stable than seasonal ones. As an initial index of native ecosystem stability, we intended to use the proportions of native and exotic population sizes within organism or life form groups of an ecosystem, realizing, however, that stability involves persistence over time in the presence of certain disrupting forces or perturbations. Since both rain forest and seasonal environments occur along the Mauna Loa transect, data from our studies should allow us to test this general hypothesis of ecosystem stability.

Investigations of species-distribution problems similar to ours had been conducted in continental ecosystems prior to our study. In order to compare our results with those of the other studies, we adopted the hypotheses proposed by these investigators, with suitable modifications and elaborations.

According to Whittaker (1970, 1975), there are currently four hypotheses on species distribution patterns:

1. Competing species, including dominant plants, exclude one another along sharp boundaries. Other species evolve towards close association with the dominants and toward adaptation for living with one another. There thus develop distinct zones along the gradient, each zone having its own assemblage of species adapted to one another, and giving way at a sharp boundary to another assemblage of species adapted to one another.

2. Competing species exclude one another along sharp boundaries, but do not become organized into groups with parallel distributions.
3. Competition does not, for the most part, result in sharp boundaries between species populations. Evolution of species toward adaptation to one another will, however, result in the appearance of groups of species with similar distributions.
4. Competition does not usually produce sharp boundaries between species populations, and evolution of species in relation to one another does not produce well-defined groups of species with similar distributions. Centers and boundaries of species populations are scattered along the environmental gradient.

All of the four spatial distribution patterns seem possible, although few tests have been made on the hypotheses. Whittaker's studies in continental temperate mountain ecosystems support the last hypothesis. However, Daubenmire (1966) suggests an altitudinal zonation of dominant tree species in eastern Washington and northern Idaho that supports the first hypothesis.

A slightly different approach has been taken by Terborgh (1971), who studied the altitudinal distribution of bird species on an Andean mountain gradient in Peru through four ecological zones. He suggests three models of altitudinal distribution. His first model implies that species distributions overlap broadly and vary continuously along an altitudinal gradient in parallel with continuously varying environmental and biological factors. By biological factors he means, for example, tree species or canopy characteristics that provide the food source and shelter of birds. This pattern is the exact equivalent of the one proposed in Whittaker's fourth hypothesis; the implied causes are also the same. Terborgh's second model hypothesizes that distribution limits of species are controlled by competitive exclusion. As the resulting pattern, he suggests sharp discontinuities between cogeneric species along the gradient but broadly overlapping ranges of taxonomically unrelated species. This model is similar to Whittaker's second hypothesis. Terborgh's third model suggests that distributional limits are determined by site-factor discontinuities along a gradient. In this case the resulting distribution pattern would be one of zonation of communities with reasonably sharp boundaries. Terborgh uses the term *ecotone* for such boundaries. Except for the underlying reasons, Terborgh's third pattern would be the same as the one proposed in Whittaker's first hypothesis. Whittaker's hypotheses do not consider such a situation, although along any complex gradient, one can expect steep environmental subgradients to occur as a normal phenomenon.

For the reasons suggested in our initial working hypothesis, we hypothesized that the distribution pattern of the native biota along the Mauna Loa transect may follow the pattern proposed in the first and/or third hypotheses of Whittaker: evolution toward adaptation to one another resulting in spatially associated species groups. In contrast, we further hypothesized that the introduced biota would follow the distribution pattern suggested in Whittaker's fourth hypothesis.

The fundamental question we are raising has been very clearly stated by Robert H. MacArthur (1972:161):

> A critical question remains: Do different plant species change synchronously, or does each have independent distribution? If they change synchronously, vegetation types are more than a mere convenience; they are real and hence necessary as a subject of study. Whittaker (1969) has spent much of his life investigating this and has shown fairly convincingly for mountains in the United States that plants appear and disappear independently as we go up a mountain. Holdridge might dispute this, for the tropics at least, where he believes plants change synchronously. However, no one has carried out in the tropics a study like those of Whittaker and we must await such a study before we can pass final judgment on whether life zones are real in nature or whether they are the scientist's convenient but arbitrary classifications.

Our study adds two further dimensions to MacArthur's statement: we are concerned not only with plant species but a number of complementary biota, and our gradient is in a tropical island situation and on a geologically young mountain.

The gradient study has further implications for the relationship of species diversity and community integration. This relationship is usually interpreted for homogeneous or uniformly heterogeneous habitats (for example, Poore, 1964). In the context of uniform habitats, species diversity has been called alpha diversity by Whittaker (1970). He distinguishes this from beta diversity, the species diversity along habitat gradients. Whittaker holds that the highest degree of integration is accomplished by a high beta diversity. High beta diversity implies accommodation of a large number of species with restricted distributions along a given environmental gradient as opposed to a few wide-ranging species on the same gradient. Low beta diversity appears to be a characteristic along altitudinal gradients on oceanic islands, at least in Hawai'i. In the tropics, this is a peculiarity only for islands. Whether this also means poor integration, needs further examination. An increase of beta diversity through exotic species invasions appears to show the opposite, a decrease in integration.

FIELD SAMPLING DESIGN

The sampling for different organism groups required techniques that are more or less specific for each major group. Some of the sampling techniques required more effort than others. Therefore it was expected that the number of sampling locations would vary among investigators. For this reason, some degree of stratification became necessary that would yield a relatively high sampling efficiency. This was done by utilizing the transect segments as a reference framework for the establishment of a certain minimum number of focal or coordinated sampling sites and by stratifying the vegetation within focal sites into life form communities.

During an IBP field meeting in August 1972, we decided as a team that each transect investigator would sample his or her organism group in a minimum of fourteen identical locations along the Mauna Loa transect. The common locations were those where a majority of the investigators had already sampled prior to this decision. The fourteen focal IBP transect sites are briefly described in table 2-1. Figure 2-8 portrays the locations of these focal sites with their altitudinal limits on a profile diagram that corresponds in scale to the one shown on figure 2-2.

The originally contemplated sampling layout was to have a focal site rather systematically distributed at every 1,000-foot level (~300 m) of altitude. This resulted in the following seven stations:

Elevational level (ft)	Type of vegetation	IBP site number
4,000 (1,220 m)	Closed rain forest	1
5,000	Mountain parkland	5
6,000	Mountain parkland	8
7,000	Subalpine scrub forest	10
8,000	Near tree line	12
9,000	Sparse alpine scrub	13
10,000 (3,050 m)	Upper limit of sparse alpine scrub	14

Seven additional sites were established to intensify the sampling network and to include a few more site variations. Three additional sites were established at the 4,000-foot level to cover the major site variations between the closed rain forest and the mountain parkland. These are sites 2 (open *Metrosideros* rain forest), 3 (open *Metrosideros* dry forest), and 4 (savanna). Four other sites were established at the middle section of the transect: two additional ones in the mountain parkland (sites 6 and 7), one at the border of the mountain parkland with the subalpine scrub forest (site 9), and one at the border between the latter and the tree line ecosystem (site 11). Three sites are ecotonal (sites 9, 11, and 14) in relation to the previously defined transect segments or ecosystem types. It was felt that this would add interest for the subsequent analysis of species distributions along this mountain gradient. A few investigators also sampled in between the focal sites, particularly for the vascular plant analysis, which was more or less continuous with at least three samples for every 500-foot (~150 m) interval.

TABLE 2-1. *Focal IBP Sites on Mauna Loa Transect*

IBP site number	Location and elevation	Vegetation	Transect Segment
1	Thurston lava tube, 3,920 ft (1,195 m)	Closed *Metrosideros-Cibotium* (ʻōhiʻa tree fern) forest	12
2	Sulphur bank, 4,000 ft (1,220 m)	Open *Metrosideros-Gleichenia* (ʻōhiʻa matted fern) forest	11
3	Tree molds area, 4,000 ft (1,220 m)	Open *Metrosideros*-native shrub-lichen forest	10
4	Kīpuka Kī near climatic station, 4,200 ft (1,280 m)	*Acacia koa-Sapindus* savanna	8
5	Power line trail, 4,920 ft (1,500 m)	Mt. parkland ecosystem, *Acacia koa* colony	7
6	IBP climatic station, 5,250 ft (1,600 m)	Mt. parkland ecosystem, *Acacia koa* colony	7
7	Keamoku flow, just above 5,650 ft (1,720 m)	Mt. parkland ecosystem, *Acacia koa* colony	7
8	Above goat exclosure, 6,200 ft (1,890 m)	Mt. parkland ecosystem, *Acacia koa* colony	7
9	End of Strip Road, 6,700 ft (2,040 m)	Mt. parkland ecosystem, *Acacia koa* colony	7 to 6 transition
10	7,000-foot level, (2,130 m)	Open *Metrosideros* scrub forest (scrub with scattered trees)	6
11	7,500-foot level, (2,290 m)	Open *Metrosideros* scrub forest (scrub with scattered trees)	6 to 5 transition
12	8,000-foot level, (2,440 m)	*Metrosideros* tree line ecosystem (open scrub with scattered trees)	5
13	9,000-foot level, (2,745 m)	*Vaccinium-Styphelia* low scrub desert (very sparse scrub)	3
14	10,000-foot level, (3,050 m) Puu Ulaula area	*Vaccinium-Styphelia* low scrub desert (very sparse scrub)	3 to 2 transition

A further stratification within focal sites was considered desirable because of the different behavioral scale of our organism groups. This relates to an easily recognizable quantitative classification of plant communities on a finer level of structure. Three life form communities were defined for this purpose:

1. Tree community: An aggregation of woody plants that form a closed canopy (at least 60 percent ground cover) covering at least 60 m^2 ground surface (two or more trees), and that have their maximum crown biomass at 5 m height or taller.
2. Shrub community: A grouping of woody plants having their maximum crown biomass between 0.2 and 5 m height. Over 50 percent of the shoot biomass must be of the shrub life form (woody plants). No ground-cover limit was assigned. Therefore shrubs can be very scattered (for example, one per 100 m^2), but they must always be present in greater quantity than herbaceous plants.

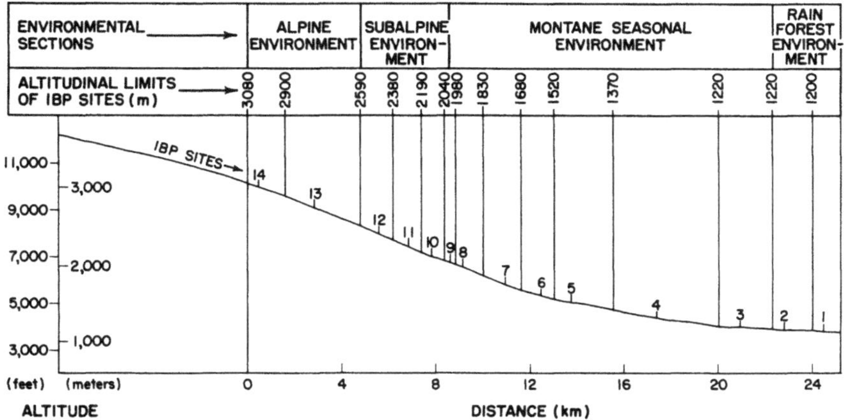

FIGURE 2-8. *Profile Diagram of Mauna Loa Transect. Locations and altitudinal limits of the fourteen IBP focal sites are shown in relation to the four environmental sections.*

3. Grass community: A grouping of herbaceous plants among which more than 50 percent of the shoot biomass must be of the grass life form. No ground-cover limit was assigned. Therefore grass communities can also be very sparse or desertlike, but herbaceous life forms must dominate.

The three quantitative structural definitions match closely what most field biologists would intuitively classify as tree, shrub, and grass communities, respectively.

The site comparison table (table 2-2) presents a summary of IBP focal sites of the sample locations and numbers as used by individual investigators.

DATA ANALYSIS

When investigating species distributions, it is necessary to have an analytical methodology to demonstrate the distribution patterns objectively. Whittaker (1967) has employed two techniques. He used a sample stand ordination that he called the double weighted-average technique, and he performed species ordinations by drawing plots of species distributions in the form of his well-known bell-shaped curves along the gradients that he investigated. We adopted a similar approach here, although it is based on different analytical techniques. We believe that the techniques we adopted would allow us to make more discriminatory interpretations of our data.

Several criteria were employed in our selection of analysis techniques. It was most important that the techniques be able to handle the type and quantity of data collected in the investigations. This includes both quantitative and

TABLE 2-2. *General Site Comparisons*

IBP focal site	Site name	Elevation feet (m)	Investigators and Organism Groups															
			Baker: Leaf and litter fungi	Beardsley: Sap sucking insects	Conant: Birds	Doty: Soil algae	Gagné: Canopy insects	Hardy: Litter insects	Lamoureux: Tree phenology	Mitchell: Blossom-feeding insects	Mueller-Dombois, Spatz: Vascular plants	Nishida: Tree insects	Paik: *Drosophila*	Radovsky: Soil arthropods, ectoparasites	Samuelson: Wood and bark beetles	Steffan: Sciaridae	Stoner: Soil fungi	Tomich: Rodents
1	Thurston	3,920(1,190)			1	18	1	3	7		1	2		18	√	1	2	6
2	Sulphur Bank	4,000(1,220)					2	4	6		2	3	1		√		3	16
3	Tree Molds	4,000(1,220)		3	3	11,12	3	5	5	3	3		2	10,11,12	√	3	4	5
4	Kīpuka Kī	4,200(1,280)	4		4	8,9,16	4	6,7			4		3	8,9,16	√	4	5,6	4,15
5	Powerline	4,920(1,500)			5	20		8			5			20	√	5	7,8	3
6	Climate Sta	5,250(1,600)	3			6	6	9	3		6		4	6,7	√	6	9	
7	Keamoku	5,650(1,720)		7	7	25				7	7		5		√	7	10	
8	Goat Excl.	6,200(1,890)						10	2		8		6	5	√	8	11	2
9	End Strip Rd	6,650(2,030)	5	9		4	9	11	1	9	9		7	4	√	9	12	
10	7,000′	7,000(2,130)	2	10	10	3	10	12		10	10	4	8	3	√	10	13	1
11	7,500′	7,500(2,290)				29		13			11		9	2	√	11	14	
12	8,000′	8,000(2,440)	1	12	12	1	12	14,15		12	12		10	1	√	12	15	10,13
13	9,000′	9,000(2,740)				31		16,17			13				√		16	11,12
14	10,000′	10,000(3,050)				33		18			14				√		17	14

Note: The numbers for each investigator refer to his or her site sampling scheme. Sampling sites falling on the transect are aggregated at the nearest IBP focal sites. Specific qualifications are given in the chapters discussing each investigator's results.

qualitative data. The proven usefulness of the techniques was considered to be important; the introduction of new types of analysis was avoided because it could compound the interpretation problems. The volume of the data to be analyzed and the need for uniformity in the analysis procedures required that we utilize computer analysis techniques. Where possible, already-operational programs were obtained. In selecting these programs, the ease of data preparation, the program capacity, and the ease of installation on the local computer were evaluated. When necessary, additional programs were written to supplement the series of analysis programs.

The series of analytical programs has proven to be quite flexible and easy to use. The data are all stored in a common machine-readable format. These data are then processed with a program that selects the relevés* to be analyzed and, based on the input of a few control parameters, produces a data set formatted for the various analysis programs.

Sample Ordination by Dendrograph Technique

The analysis objective for sample ordination was to determine the similarity patterns among the various sample locations along the transect by subjecting the total species content of each spatial sample to a similarity test. Whittaker's (1967) double-weighted-average technique resulted in scatter diagrams of sample stands not unlike those produced by the Wisconsin ordination technique (Bray and Curtis, 1957; Beals, 1960; Newsome and Dix, 1968). The two techniques differ very much in detail, but their objective is the same: to demonstrate sample similarities geometrically. The geometric distance among samples then allows one to recognize the presence or absence of groups or clusters of similar samples.

During the past decade dendrograms have become more widely applied in multivariate analyses of this sort (Frenkel and Harrison, 1974; Orloci, 1975, 1978). Like the Wisconsin ordination technique, dendrograms are a diagrammatic tool for displaying the content of a similarity matrix (Mueller-Dombois and Ellenberg, 1974). The dendrogram technique generates clusters of samples at various degrees of similarity. However, the recognition of ecologically meaningful clusters presents problems of interpretation similar to those in geometric methods.

For our purposes we used the dendrograph technique of McCammon (1968; McCammon and Wenniger, 1970) because of its flexibility and tested usefulness. A dendrograph differs from the usual dendrogram in that the between-group distances or similarities are also calculated and shown diagrammatically by scaling the distances along the x-axis. However, the main feature

---

**Relevé* is the French word for "abstract." We use this word throughout this text as the equivalent of community sample (a sample containing usually more than one species).

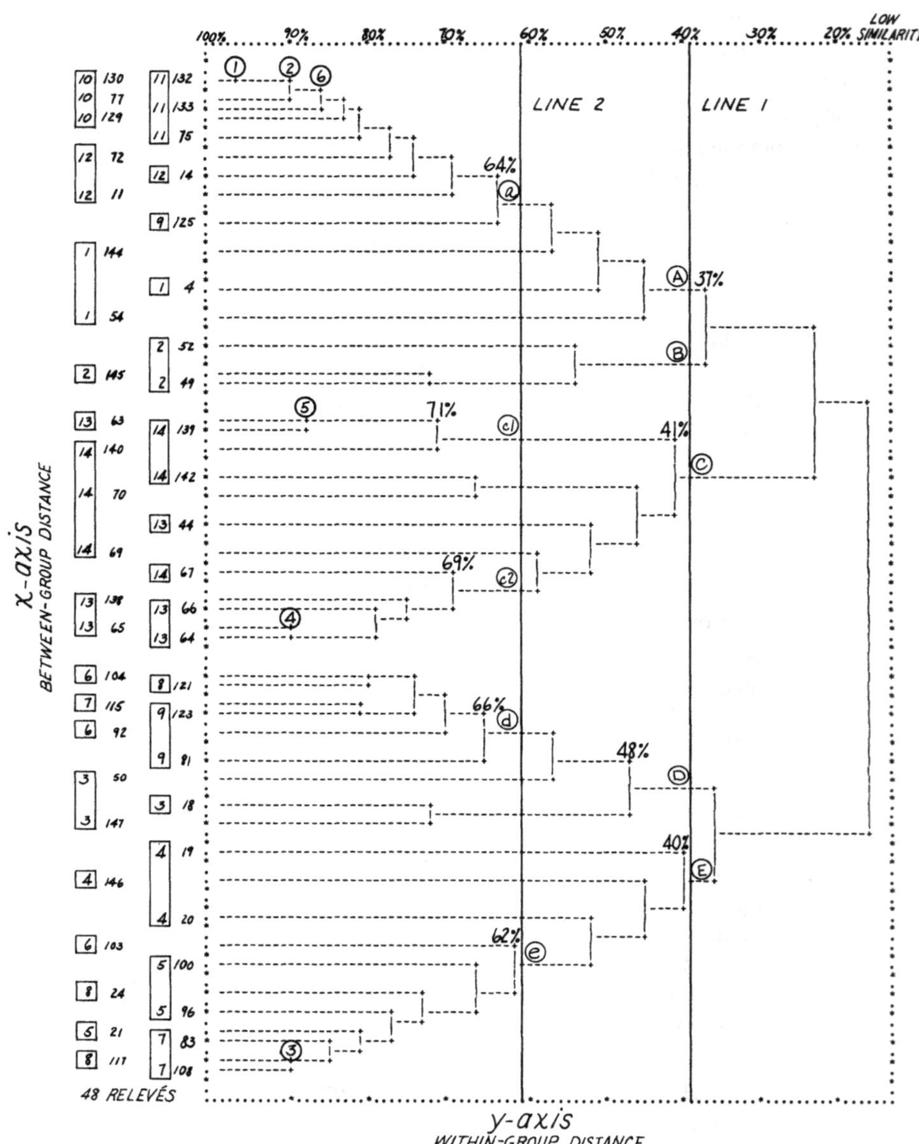

FIGURE 2-9. *Dendrograph based on forty-eight vegetation samples compared with quantitative Sørensen index of similarity. Blocked-out numbers are IBP site numbers 1–14. Numbers not blocked out are relevé numbers.*

of any dendrogram or dendrograph is the within-group distances or similarities, which are shown along the y-axis of each dendrogram or dendrograph.

Figure 2-9 is a dendrograph comprised of forty-eight community samples or relevés (thirty-six tree and twelve shrub relevés) that describe the fourteen IBP focal sites. Individual relevé numbers are shown on the left-hand side (the two rows of numbers that are not blocked out) along the x-axis of the dendrograph. Next to the individual relevé numbers (along the x-axis) are the corresponding IBP focal site numbers, which are boxed for clarification. The y-axis (top and bottom on the diagram) gives a similarity scale from high (100 percent) at left to low at right. The asterisks along the y-axis occur at 10 percent similarity intervals.

The relevé groups or clusters are indicated by horizontal lines connecting the various relevés. The distance of this horizontal line from the y-origin (at 100 percent similarity) shows the within-group similarities. Since the clusters are hierarchically arranged, it is important to evaluate both the number of relevés that have been grouped and the similarity of the group. Ecologically meaningful groupings are not established automatically. They have to be identified. But before we proceed with the ecological interpretation, we will briefly describe how the dendrograph was constructed.

The general procedure employed in determining the relevé similarities is illustrated below. A description of the other similarity indexes is given in Mueller-Dombois and Ellenberg (1974). First, a matrix of similarity indexes or community coefficients was calculated using Motyka's quantitative modification of Sørensen's index. This index reads as follows:

$$IS = \frac{2w}{A + B} \times 100$$

where
IS = index of similarity
A = sum of quantitative values of all species in relevé A
B = sum of quantitative values of all species in relevé B
w = sum of the smaller of the two quantitative values of the species that are common to relevés A and B.

The number of comparisons involved are $n(n\text{-}1)/2$, which for forty-eight relevés comes to 1,128 IS values. Table 2-3 shows part of the computer-derived similarity matrix. (The remaining part was omitted to save space and because the procedure can be explained from this restricted section of the matrix.)

The second major step in the analysis is the construction of the dendrograph. The dendrograph program of McCammon (1968) is based entirely on the similarity matrix, so that the dendrograph is merely a mathematically derived diagrammatic display of the content of a similarity matrix. Among various cluster-analysis techniques available (Sneath and Sokal, 1973; Orloci, 1975, 1978), McCammon's program can be defined as an agglomerative, centroid, unweighted pair-group method. The procedure is explained briefly in five steps:

TABLE 2-3. *Partial Similarity Matrix*

Relevé	117	121	123	125	129	130	132	133	138	139	140	142	144	145	146	147
77	3.1	2.7	14.2	33.0	78.9	84.2	86.8	74.4	11.7	26.8	41.6	5.8	29.7	32.7	0.2	46.9
81	30.5	71.7	50.4	23.8	7.7	8.3	5.8	8.2	5.6	4.9	4.8	5.3	0.0	0.0	3.8	34.1
83	83.9	53.6	55.9	26.6	6.5	6.9	4.7	6.9	0.3	0.0	0.3	0.0	0.0	0.0	10.0	21.2
92	57.1	65.3	68.0	28.8	5.3	5.6	3.0	5.6	0.4	0.0	0.3	0.0	0.1	0.0	11.3	13.9
96	74.2	40.6	43.0	25.1	4.7	5.0	2.6	2.8	0.0	0.0	0.0	0.0	0.2	0.0	18.1	19.4
100	40.7	41.6	44.1	29.0	4.8	5.1	2.6	5.0	0.4	0.0	0.3	0.0	0.3	0.0	17.7	21.0
103	38.0	39.2	48.7	31.4	4.6	4.9	2.5	2.5	0.0	0.0	0.0	0.0	0.2	0.0	23.1	18.9
104	56.4	80.7	59.6	20.6	14.6	15.6	13.3	15.8	0.0	0.0	0.0	0.0	0.0	0.0	11.1	48.9
108	90.0	64.4	64.7	24.1	2.8	3.0	2.7	5.0	0.5	0.2	0.4	0.0	0.0	0.0	8.9	16.8
115	65.3	81.1	81.5	28.7	2.9	3.1	3.0	5.7	0.4	0.0	0.3	0.0	0.0	0.0	9.4	20.2
117		61.7	62.1	23.4	2.9	3.2	3.0	5.3	0.6	0.2	0.6	0.0	0.0	0.0	7.5	17.8
121			77.3	24.9	2.5	2.7	2.6	5.1	0.0	0.0	0.0	0.0	0.0	0.0	9.2	36.6
123				45.0	14.4	15.5	13.2	15.3	5.0	4.4	15.9	0.0	0.0	0.0	16.2	19.3
125					35.4	37.9	36.2	28.3	10.0	22.3	31.6	4.7	0.0	0.0	21.7	21.3
129						81.7	81.3	70.7	8.8	20.4	32.6	4.4	26.1	28.4	1.9	41.2
130							97.1	85.6	10.3	23.4	36.7	5.4	27.4	30.0	2.0	43.4
132								85.3	10.6	24.2	37.5	5.0	28.0	30.7	0.1	42.1
133									10.2	9.7	24.9	9.9	27.8	30.5	0.1	41.8
138										50.2	43.2	38.5	0.0	0.0	0.0	0.3
139											63.5	29.2	0.0	0.0	0.0	0.1
140												17.3	0.0	0.0	0.0	3.3

Notes: Relevé numbers are shown on left side and at top; the matrix values are similarity indexes in percentage based on the quantitative modification of Sørensen's index.

The complete matrix contains $\frac{n \times (n-1)}{2} = \frac{48 \times (47)}{2} = 1{,}128$ values. The complete set of similarity indexes is not shown here because the omitted values are not needed for the explanation of the dendrograph technique.

1. The computer is programmed to convert all percentage *IS* values into metric distance values by the arc cosine transformation. This transformation is done so that the similarity values can be averaged correctly in subsequent computations. This avoids the problem of averaging percentage values. The arc cosine transformation changes high-similarity values into low-distance values and vice-versa (for example, *IS* 90% = arcos 0.451 and *IS* 10% = arcos 1.471).
2. The computer then searches the matrix for the lowest arc cosine value (which corresponds to the highest *IS* value). By detecting this value it has located the two relevés that are most similar in their quantitative species content. In our example this pair is comprised of relevés 130 and 132. The two relevés share an *IS* value of 97.1 percent (see table 2-3). This value is plotted as output cluster ① on the dendrograph (figure 2-9).
3. The computer then averages all distance values of the two relevés (130 and 132). This implies a reduction of the two columns in the similarity matrix (under relevés 130 and 132) to one and a reduction of $n-1 = 47$ values (for a second matrix not shown here).
4. The computer then searches the entire matrix again for the lowest-distance value. This value may be found among the previously calculated ones, or it may be among the forty-seven new average values. In our example it was among the new average values. This is indicated by output cluster ② on the dendrograph (figure 2-9), which joins relevé 77 at 90.3 percent within-group similarity to the already established first cluster formed by relevés 130 and 132.
5. The computer then calculates a new set of average values by combining relevé 77 with the former average values of relevés 130 and 132. As a result the matrix is reduced once more. In our example, the third highest *IS* value (or the third shortest-distance value) was located among the originally calculated values on table 2-3. It is the *IS* value of 90 percent similarity shared by relevés 117 and 108. This relevé pair was plotted as output cluster ③ on the dendrograph.

The computation cycles are repeated in this way until all relevés are clustered. The dendrograph is not printed until the entire network sequence of output clusters is established. All numerical values for the output clusters and their sequence are shown in table 2-4. However, the cosine values for within- and between-group distances were not reconverted into percentage similarity values. The percentage values have been added subsequently for the first six output clusters. Their sequence can be found on the dendrograph by moving a straightedge from left (high within-group similarity) to right (low similarity).

For an ecological interpretation of the dendrograph, it is important to locate the most meaningful clusters. If ecologically meaningful clusters or relevé groupings occur at all, they can be expected to be in the mid-range of similarity. This is because at high within-group similarities, all relevés or

TABLE 2-4. *Computer-Derived Output Clusters*

Relevé	Output Cluster		Final Results Within Group	Final Results Between Group	
130	①	97.1%	0.2410	0.2410	97.1%
132	②	90.3%	0.4435	0.5448	85.5%
77	⑥	86.5%	0.5256	0.6078	82.1%
133	8		0.5837	0.6707	
129	10		0.6286	0.7185	
75	15		0.6888	0.8394	
72	17		0.7320	0.8617	
14	24		0.8030	1.0513	
11	29		0.8880	1.2281	
125	32		0.9712	1.3454	
144	37		1.0361	1.3609	
4	41		1.1087	1.5080	
54	44		1.1879	1.3603	
52	34		1.0041	1.1246	
145	21		0.7630	0.7630	
49	46		1.3358	1.4936	
63	⑤	87.9%	0.4972	0.4972	87.9%
139	22		0.7744	0.9129	
140	42		1.1456	1.2598	
142	26		0.8410	0.8410	
70	39		1.0909	1.1975	
44	36		1.0318	1.2627	
69	31		0.9395	1.2008	
67	25		0.8089	0.9391	

Relevé	Output Cluster		Final Results Within Group	Final Results Between Group	
138	16		0.7220	0.7960	
66	13		0.6480	0.7465	
65	④	90.0%	0.4510	0.4510	90.0%
64	47		1.4079	1.5250	
104	12		0.6321	0.6321	
121	18		0.7359	0.7913	
115	9		0.6181	0.6181	
123	23		0.7858	0.8606	
92	28		0.8600	1.0084	
81	33		0.9734	1.2567	
50	38		1.0857	1.2773	
18	20		0.7628	0.7628	
147	45		1.2066	1.2750	
19	43		1.1625	1.4224	
146	40		1.1048	1.4212	
20	35		1.0257	1.4270	
103	30		0.9110	1.1012	
100	27		0.8476	1.1139	
24	19		0.7411	0.8512	
96	14		0.6861	0.7718	
21	11		0.6290	0.7058	
83	7		0.5521	0.6029	
117	③	90.0%	0.4506	0.4506	90.0%
108					

Note: Includes corresponding within- and between-group arc cosine differences. These values, converted into percentages (carried through only for output cluster 1 to 5), are the basic values used for plotting the dendrograph (figure 2-9).

community samples are unique, while at low similarities they are all joined into one large all-inclusive group or cluster. On the dendrograph, the all-inclusive cluster occurs at 16 percent within-group similarity. In our example this cluster puts together all four environmental sections of the Mauna Loa transect: the alpine, subalpine, montane seasonal, and montane rain forest.

Rules for Identifying Major Dendrograph Clusters

At this point one can detect ecologically meaningful clusters in two ways: deductively by identifying those clusters that best define various predetermined transect segments or ecosystem types, or inductively, by establishing certain arbitrary rules that automatically isolate a number of clusters in the mid-range of within-group similarity.

Since we were aiming at objectively defining transect zones by a uniform procedure applicable to several different organism groups, we chose the second approach. For this purpose we established arbitrary cutoff lines (which can be used as rules) that permit objective identification of dendrograph clusters in the mid-range of within-group similarity. A major cluster should be defined from the number of replicate samples. For example, in this case where a minimum of three replicate samples was used to describe each of the fourteen IBP sites, a major cluster was defined as a grouping of at least three relevés.

The two cutoff lines shown on the dendrograph are defined as follows. Line 1 isolates the maximum number of major clusters by minimizing the number of single clusters. In case of ambiguity the number of major clusters takes precedence over the single clusters. Line 2 isolates any number of major clusters where the number of single clusters is just under 25 percent. In our example line 1 identifies five major clusters (A, B, C, D, and E) at a level of within-group similarity ranging from 37 percent to 48 percent as shown on figure 2-9. Line 2 identifies five major clusters (a, c1, c2, d, and e), which here can be called higher-similarity subclusters occurring at a within-group similarity range of 62 percent to 71 percent. These latter subclusters, which are portions of the major clusters, are relatively more homogeneous.

Transfer of Dendrograph Clusters to Zonation Diagram

The ten dendrograph clusters that were identified by the two cutoff lines must now be interpreted ecologically. For this purpose their major information content was transferred into what may be called a zonation diagram (figure 2-10). It shows that cluster A combines IBP site 1 (the closed rain forest) with IBP sites 9 through 12 (extending from the upper limit of the mountain parkland through the subalpine scrub forest to, and including, the tree line ecosystem). This line 1 cluster, which combines thirteen relevés at 37 percent within-group similarity, is thus putting very unlike units into one group. At line 2 the rain forest is split off, and the subcluster "a" represents a more homogeneous segment that combines only the tree line ecosystem and subalpine forest with the upper limit of the mountain parkland at the higher within-group similarity of 64 percent. The meaning of each cluster is easily read from figure 2-10. For example, cluster B defines the open montane rain forest, cluster C defines the alpine section (with IBP sites 13 and 14), and cluster D defines a rather heterogeneous combination of upper mountain parkland and open seasonal *Metrosideros* forest (IBP site 3). Again at the second cutoff line, subcluster "d" combines only the upper mountain parkland relevés, while the montane seasonal forest is split off. Cluster E combines the mountain parkland with the savanna.

The clusters as defined by the two cutoff lines therefore indicate homogeneous zones at different levels of within-group similarity. The question then arises as to what zonal limits are most strongly indicated by the samples. Here

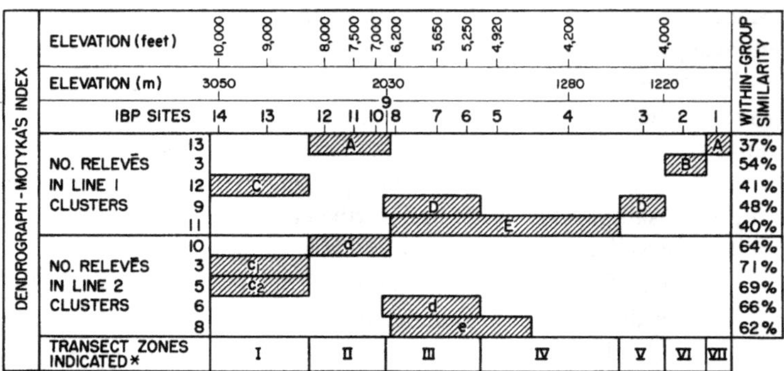

*Based on at least two cluster limits

FIGURE 2-10. *Zonation Diagram. Altitudinal distribution of dendrograph clusters formed from forty-eight vegetation samples (thirty-six tree relevés plus twelve alpine shrub relevés; 169 species) by quantitative Sørensen index. I = sparse alpine scrub; II = open subalpine scrub forest; III = mountain parkland, excluding site 5; IV = savanna, including site 5 of mountain parkland; V =* open Metrosideros *dry forest; VI = open* Metrosideros *rain forest; VII = closed* Metrosideros *rain forest.*

again a general ruling has been adopted that can be applied to other sample sets as well. We draw a boundary where at least two cluster limits occur on the zonation diagram. This we feel provides a minimum standard for a strong zonal boundary. The resulting transect zones are shown and defined at the bottom of the zonation diagram. The boundary of the savanna has been moved up to about 5,000 ft (1,520 m) by the sample data and the rules imposed on them. The predetermined boundary was set at 4,500 ft (1,370 m). However, in earlier publications (Mueller-Dombois, 1966; Mueller-Dombois and Krajina, 1968), the same boundary was also set at 1,520 m, which indicates that the mountain parkland-savanna boundary is a very gradual or transitory one. The clusters can be investigated further as to the species that are particularly responsible for the sample clusters.

Species Ordination by Two-way Synthesis Table Technique

After determining the community-sample pattern, the next analysis objective (as in Whittaker's gradient analysis model) was to extract the species-distribution trends from our sample sets. To do this we adopted the Ceska-Roemer (1971) program because of its unique flexibilities. This program is a close computer simulation of the Braun-Blanquet synthesis table technique (Mueller-Dombois and Ellenberg, 1974:chap. 9).

Both the dendrograph and the two-way synthesis table start their analysis from the same raw data table. In this table each row represents a species, and each column represents a relevé or sample site. Within the table, the occurrence of a species in a relevé is entered as its cover, frequency, abundance, or simply presence (depending on the organism being studied). An empty table cell indicates the absence of a species from a relevé.

In the dendrograph technique, the columns of the raw data table were compared, producing a similarity matrix. Such an analysis has been referred to as Q-technique in numerical taxonomy (Cattell, 1952, Sneath and Sokal, 1973).

In our second objective, we were particularly interested in comparisons between the species, that is, in an analysis of the between-row similarities and differences. This type of analysis is known as R-technique. The Ceska-Roemer program (similar to the original Braun-Blanquet method) is unique in that it performs both a Q- and an R-type analysis on the data set. For this reason, the program can be called a two-way table technique (Greig-Smith, 1964).

The primary emphasis in this type of analysis is the identification of groups of species with similar distribution ranges. The distribution of these groups can then be used to identify transect zones. Since this is an independent form of analysis (with quite different assumptions), zones established in this analysis may be compared to those from the dendrograph analysis. In addition, the Ceska-Roemer program isolates the species responsible for any emerging relevé group or cluster. Moreover, it performs a distribution analysis of the species throughout a given set of samples or relevés with regard to the species spatial limits and spatial parallels (association of occurrences).

Analysis of Species with Restricted Distributions

In contrast to the dendrograph technique, the two-way table technique is programmed to reject at the beginning those species that are present in all or nearly all relevés because these omnipresent species do not contribute any information to the differentiation of the zones. The species rejected at the start of the analysis are all those that are present in 66 percent or more of the relevés under comparison. This is an arbitrarily set limit that conforms to a general norm used in phytosociological work. The remaining species have limited distribution ranges.

Definition of Associated Species Groups

The analysis procedure required the identification of groups of species that are similar because of the pattern of their distributions. The primary difficulty comes in providing a precise definition of similarity in species distributions.

The general concept of species distributions is quite clear. We also know that, even in restricted areas, species rarely coincide exactly in their distributions. Two species, however, may both occur at a majority of the sites sampled along their range. At a few sites one of these species may occur in the absence of the other. If these two species occur at the same sites most of the time, these species have a coincident range (over the common sites) and are thereby associated.

The next step is to search for more species that may be considered associated. Some new species may have distributions nearly coincident with the distributions of the original species in the group, but the new species may not be present at a few of the sites common to the first two species. These new species may also occur at a few sites other than the common sites. Thus, as species that characterize the associated group of species are added, two types of distribution exceptions may be found. One of these is the occurrence of a species at a site outside the group of sites. The other is the absence of a species from sites common to the group. The precise thresholds that we established for these two types of exceptions control the way species were associated in the group selection process.

In the Ceska-Roemer program, the use of threshold values for species selections is controlled in rule I: a species is recognized as associated with a group if it occurs in at least X percent of the sites or relevés of the group and is not present in more than Y percent of the sites outside this group. The actual percentage values used for X and Y can be changed as an option in the program. When X is 50 percent and Y is 10 percent, associated species are all those that occur together in 50 to 100 percent of the sites of the group and in not more than 10 percent of the sites outside this group. This is known as the 50/10 option, which has been applied empirically in many previous phytosociological tables. The other four options are 50/20, 66/10, 66/20, and 66/33. The most discriminating option is 66/10, which is followed by 50/10, 66/20, 50/20, and 66/33, in that order. Thus the first two are the stronger and better criteria for species associations. Application of rule I in the program requires a comparison of all species occurrences in the program by whatever option is stipulated. Since this is a comparison between rows, rule I performs an R-type analysis. Associated species cannot be fewer than two species. If a species cannot be grouped with at least a second species according to rule I, it is rejected as a single species. In this case it appears in the final table as an ungrouped species, together with those initially rejected omnipresent ones.

Definition of Relevé or Site Groups.

If the sites in which certain associated species occur were known in advance, it would be relatively simple to search the raw data table for the species that conform to rule I. This is the technique adopted in a program by Spatz and Siegmund (1973). It would be possible also to develop such a species extraction program in conjunction with the dendrograph technique, which allows one to define groups of similar sites. However, using the approach of Ceska and Roemer, relevé groups were defined by the associated species themselves during the operation of the program.

While forming associated species groups in rule I, the *X* percent part of the option allows species to be absent in some of the sites that are generally common to the group. As the group gains new species, however, it is possible that many of the species could be absent at the same site. To ensure that the absences are spread equitably, Ceska and Roemer apply rule II, which performs a Q-type analysis. This rule stipulates that a site is accepted in the species group if it contains at least *X* percent of the species associated with that group. For reasons of homogeneity, the value of *X* is the same as in rule I. If, for example, the 50/10 option is used, a relevé must contain at least half of the species of an associated species group in order to be part of this group of relevés. Rule II therefore imposes a restriction on the method in species-poor areas where sometimes single species are important for community differentiation. However, this restriction is not serious in our analysis because such single species appear in the final table anyway. Such ungrouped species can be used in the species-distribution analysis along the transect just as well as the grouped ones.

Procedure and Sorting Result

The mathematics of the program is quite complex, but the procedure and final outcome can be explained without the mathematical background:

1. The computer selects an initiating species whose relevés form a first tentative relevé group. This initiating species is chosen arbitrarily from the species with high presence throughout the sample set (that is, from those that are present in just under 66 percent of the relevés).
2. All remaining species are then compared (by R-type analysis) as to their matching occurrences with the first group-initiating species according to one of the options of rule I. Species that fit the association criterion are temporarily retained; those that do not are temporarily rejected. In this first search process, a number of species may have been found that fit this criterion.
3. The computer then performs a Q-type analysis by scanning the relevés in which these species occur, and any relevé that contains less than 50 percent or less than 66 percent of the species (depending on which rule I option is applied) is rejected. This is the application of rule II.
4. The computer then compares the species rows again for their matching occurrences, because by the rejection of some relevés by rule II, some species may now fall outside the specific rule I option. Thus the same option of rule I is applied in a second computation cycle (a repetition of

step 2), and this is followed again in the same cycle by rule II (a repetition of step 3).

5. Additional computation cycles by an alternation of rules I and II are performed until all spatially coinciding species are isolated from the set.

Frequently an initiating species is rejected in subsequent computation cycles as single species, which then forms a one-species group. These single species are listed at the end of the printout table that gives the final results. This table also includes the originally rejected omnipresent species and the rare species that occurred only once or twice throughout the entire sample set.

In our example, 169 species were contained in the set of 48 relevés. For this reason, only an extract of the whole table, which lists 19 species, is here reproduced as table 2-5. This table extract is sufficient for the purpose of explaining the remainder of our analysis procedure. The table shows 4 of the 9 groups that were generated by the 50/10 option of rule I.

The relevé numbers are listed along the top of the table; the IBP focal site numbers have been added for orientation. The relevés that have been included in the groups are shown in combination with the species that are characteristic of each group. The species are listed (by abbreviated name code) along the left side of the table and are separated into the associated species groups. The relevés that make up each group are shown by bracketing the top and bottom of the associated species group with asterisks in the relevé columns. For example, the group at the top of the table consists of relevés 117, 129, 130, 132, 133, 11, 14, 72, 44, 63, 64, 65, 66, 138, 67, 139, and 140 and is characterized by four species (*Pellaea ternifolia, Asplenium adiantum-nigrum, Asplenium trichomanes,* and *Agrostis sandwicensis*). This has been numbered group 2, based on the order in which it was identified in the computation cycles. The other groups have also been labeled in the output table. The numbers and symbols appearing in each species row are quantitative values (here Braun-Blanquet cover-abundance values) that were recorded for the species at that sample location in the field.

The "dictate" option of the program has been applied in the run that produced the table. This option allows one to arrange the sequence of samples in any order. This is a very desirable feature, and it facilitated our data interpretation in relation to the known environmental gradient. As shown at the head of the table, the IBP sites are arranged in order from 1 to 14—that is, from the low to the high end of the Mauna Loa transect. The program may also be used to sort the relevés into a new order forming coherent blocks, a feature useful for other purposes.

Extraction of Species Groups into Zonation Diagram

It is now possible to extract the species groups generated by the program into a zonation diagram similar to the way it was done for the clusters of the

TABLE 2-5. *Extract of a Final Two-Way Table*

IBP Focal site no.	1			2			3			4			5			6			7			8			9			10			11			12			13						14					
	0	0	1	0	0	1	0	0	1	0	0	1	0	0	1	0	1	1	0	1	1	0	1	1	0	1	1	0	1	1	0	1	1	0	0	0	0	0	0	0	0	1	0	0	0	1	1	1
Relevé no.	0	5	4	4	5	4	1	5	4	1	2	4	2	9	0	9	0	0	8	0	1	2	1	2	8	2	2	7	2	3	7	3	3	1	1	7	4	6	6	6	6	3	6	6	7	3	4	4
	4	4	4	9	2	5	8	0	7	9	0	6	1	6	0	2	3	4	3	8	5	4	7	1	1	3	5	7	9	0	5	2	3	1	4	2	4	3	4	5	6	8	7	9	0	9	0	2
Species group 2																							*						*	*		*	*	*	*	*	*		*	*	*	*	*			*	*	
PEL TER																											+			+		+	+	1	1	1	1	+	1	1	1	1	+			+	1	
ASP ADI																				R		+	R						+	+		+		1	1	1	R			R		R				+	R	
ASP TRI																							R		+	R				R		+		1	1	+	+		R	R	R						R	
AGR SAN							R																						R	R			1			+	1		1		+		+		+		+	1
																							*						*	*		*	*	*	*	*	*		*	*	*	*	*			*	*	
Species group 1					*		*	*																				*	*	*	*	*	*	*	*	*												
MAC GAH					+		1	1	1																			1	1	1	1	1	1	1	+	1										R	1	
COP ERN					3		+	+	1																		2	1	2	2	1	2	2	2	2	+												
LUZ HAW					1		1	2					+													+	1	1	1	1	+	1		R		+												
RAI CIL					R			1																				+	+	1		+	2	1	1	R												
COP MON																												1	1	2	1	2	2	2	2	2												
PLE THU								+		R																			+	+	1	+	+	1	+													
POL PEL				+	+		1	2																				+	+		1	+																
					*		*	*																				*	*	*	*	*	*	*	*	*												
Species group 3							*	*					*	*	*	*	*	*	*	*	*	*	*	*	*	*	*																					
HOL LAN					+		+	R		3	2	2	2	1	1	1	2	1	1	1	1			1	1	2	2																					
CAR WAH				+			R	+					2	+	1	1		1	1	2	2	2	1	1	1	1	1						1	+														
ACA KOA									2		2	2	5	4	3	3	3	3	4	4	3	5	4	3		3	2																					
STY TAM					2		3	3	3				2			1		3	2	2	2	1	2	3	3	2																						
PAN TEN													1	1	1	2	1	1	1	2	2	2	1	2	2	2																						
							*	*					*	*	*	*	*	*	*	*	*	*	*	*	*	*	*																					
Species group 4					*		*	*	*				*	*	*	*	*	*	*	*	*	*	*	*	*	*	*	*	*	*	*	*	*	*	*	*												
DES AUS					1		+	2					2	1	1	3	1	1	3	2	2	4	2	2	2	2	2	1	1	1	1	1	1	1	1	1												
PTE AQU					1		2		2	2	3	2	2	2	2	1	2	1	1	+		1	+	+	1	1	2		1	1	1																	
DOD SAN					+		1	2	2							+		2	1	R	+							2	2	2		2	2	2	1	1												
					*		*	*	*				*	*	*	*	*	*	*	*	*	*	*	*	*	*	*	*	*	*	*	*	*	*	*	*												

Note: Rules used: 50 percent inside, 10 percent outside.

dendrograph. For this purpose we may ignore the few species occurrences outside the relevé groups. However, they are not ignored in the final analysis. Figure 2-11 shows the completed extraction of the nine groups generated from the forty-eight relevés. They are here arranged in decreasing altitude from the upper to the lower end of the Mauna Loa transect. The computer-generated group number is written into each block graph. The number of relevés combined by the group are recorded on the left side. On the right side appear the numbers of species belonging to each group.

We now can apply the same criterion for the definition of transect zones as was applied in the zonation diagram giving the dendrograph results; that is, we can draw a boundary where at least two group (or cluster) limits are indicated by the data. In this case the result is almost identical to that generated by the dendrograph technique except that the boundary between zones III and IV now coincides with the predetermined boundary. Therefore it seems valid to accept that boundary in favor of the one generated by the dendrograph analysis.

Ordination of Species Distributions along the Mauna Loa Transect

By use of the established transect zones, species groups, and single species, it is now relatively simple to demonstrate diagrammatically the important species-distribution trends in a form comparable to Whittaker's bell-shaped curves of species distribution along gradients. This was done by averaging the quantitative value for each species by IBP focal site and by plotting this value

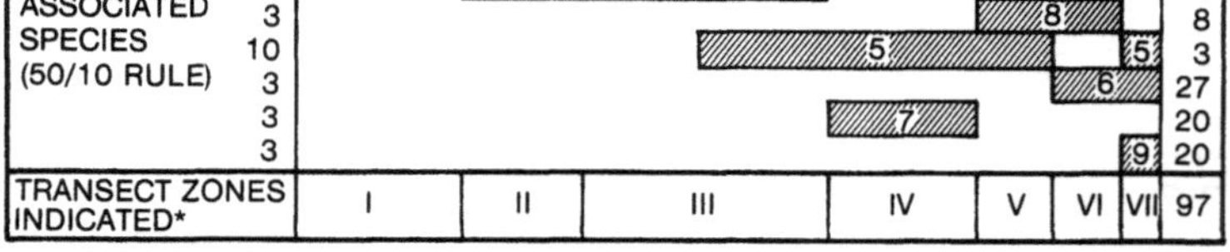

ELEVATION (feet): 10,000 9,000 8,000 7,500 7,000 6,200 5,650 5,250 4,920 4,200 4,000

ELEVATION (m): 3050 2030 1280 1220

IBP SITES: 14 13 12 11 10 9 8 7 6 5 4 3 2 1

NO. OF RELEVÉS GROUPED BY ASSOCIATED SPECIES (50/10 RULE)	Group	NO. SPECIES IN GROUP
16	2	4
12	1	7
17	3	5
28	4	3
3	8	8
10	5	3
3	6	27
3	7	20
3	9	20

TRANSECT ZONES INDICATED*: I II III IV V VI VII — 97

*By at least two group distribution limits.

FIGURE 2-11. *Zonation diagram. Altitudinal distribution of associated species groups derived by the two-way table technique. I = sparse alpine heath scrub; II = open subalpine* Metrosideros *scrub forest; III = mountain parkland,* Acacia koa *tree colonies; IV =* Sapindus-Acacia koa *savanna; V = open* Metrosideros *dry forest; VI = open* Metrosideros *rain forest; VII = closed* Metrosideros *rain forest.*

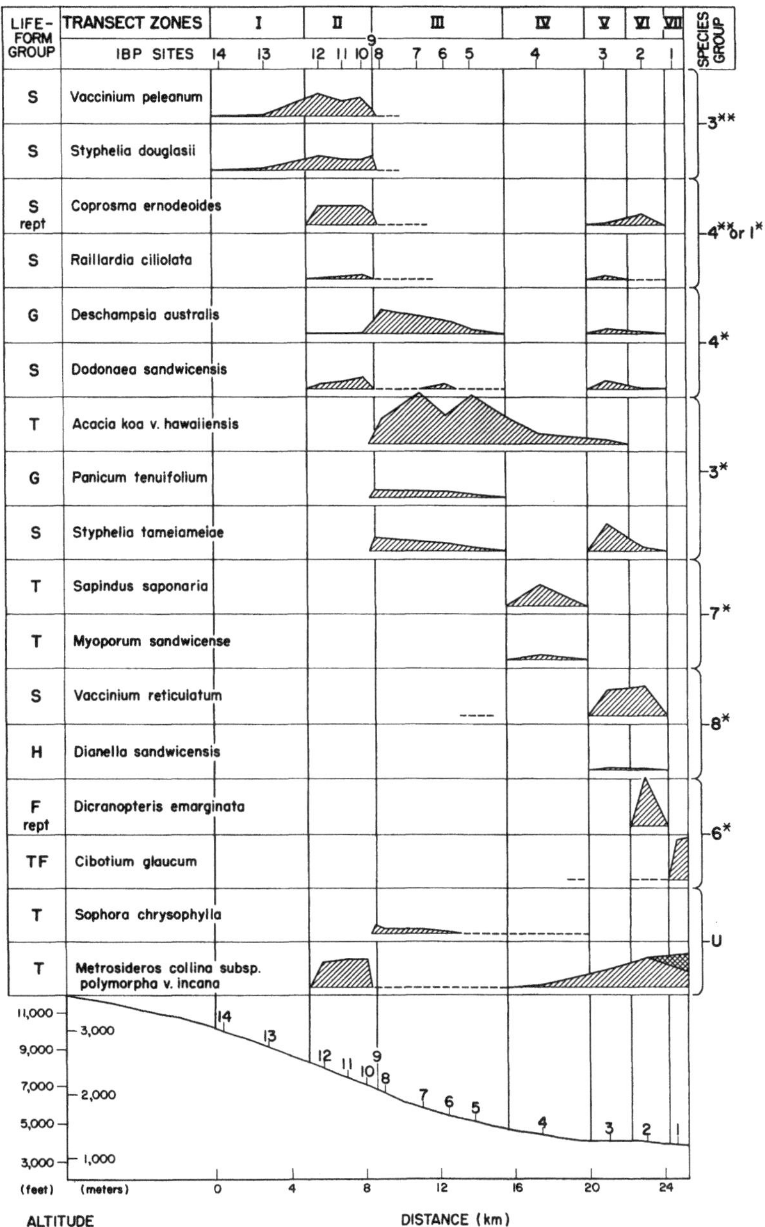

LIFE-FORM GROUP
TRANSECT ZONES
I
II
III
IV
V
VI
VII
IBP SITES
14 13 12 11 10 9 8 7 6 5 4 3 2 1
SPECIES GROUP
S Vaccinium peleanum
S Styphelia douglasii
3**
S rept Coprosma ernodeoides
S Raillardia ciliolata
4** or 1*
G Deschampsia australis
S Dodonaea sandwicensis
4*
T Acacia koa v. hawaiiensis
G Panicum tenuifolium
S Styphelia tameiameiae
3*
T Sapindus saponaria
T Myoporum sandwicense
7*
S Vaccinium reticulatum
H Dianella sandwicensis
8*
F rept Dicranopteris emarginata
TF Cibotium glaucum
6*
T Sophora chrysophylla
T Metrosideros collina subsp. polymorpha v. incana
U
11,000
9,000
7,000
5,000
3,000
3,000
2,000
1,000
(feet)
(meters)
0 4 8 12 16 20 24
ALTITUDE
DISTANCE (km)

over each of these sites along the mountain gradient. Figure 2-12 shows the result for selected native plant species as they occur along the Mauna Loa transect.

In contrast to Whittaker's form of presentation, species population curves were not smoothed into bell-shaped curves. Instead population quantities within species were connected by straight lines, a minor difference. However, our diagram incorporates three major departures from Whittaker's diagrams:

1. Individual species population curves are shown separately. This provides for a clearer overview than a plot of several species on the same base line and y-axis segment.
2. Species population curves are ordinated by amplitude restrictions from the left side of the graph (high altitude species, group 3) to the right side of the graph (rain forest species, group 6).
3. All quantitatively important plant life forms are admitted to the test; the test thus is not restricted to woody plants as in Whittaker's diagrams.

The group numbers on the right side of the diagram are the computer-derived numbers from the two-way tables. Here groups were utilized from two rule I options: the 66/10 and 50/10 options. In addition, two ungrouped or single species are shown at the bottom of the diagram. In contrast to the zonation diagram, species ranges extending beyond the group limit are shown in the distribution diagram exactly as they were recorded in the field. Dashed lines indicate rare occurrences either in the relevés or as observed outside. The greater population quantities of most species occur inside the group limits. This is of interest insofar as the two-way table analysis generates a quantitative sorting in spite of being operated by only presence and absence criteria in this program.

FIGURE 2-12. *Distribution diagram of selected native plant species along Mauna Loa transect. Transect zones I–VII derived from two-way table analysis. Members of species groups (*) derived from 50/10 option, group members (**) derived from 66/10 option, u = ungrouped species. Dashed lines mean present but not recorded in relevés or rare in relevés. Cross-hatch-*Metrosideros collina *subsp.* polymorpha *var.* macrophylla. *Life-form symbols: S = shrub, rept = reptant or creeping, G = grass, T = tree, H = herbaceous plant other than grass, F = fern, TF = tree fern.*

3

Altitudinal Distribution of Organisms along an Island Mountain Transect

D. Mueller-Dombois, G. Spatz, S. Conant, P.Q. Tomich, F.J. Radovsky, J.M. Tenorio, W. Gagné, B.M. Brennan, W.C. Mitchell, D. Springer, G.A. Samuelson, J.L. Gressitt, W.A. Steffan, Y.K. Paik, K.C. Sung, D.E. Hardy, M.D. Delfinado, D. Fujii, M.S. Doty, L.J. Watson, M.F. Stoner, and G.E. Baker.

VASCULAR PLANTS*

Sampling Layout

For the vegetation analysis, samples (relevés) were laid out more or less continuously along the Mauna Loa transect from the beginning of the slope at 1,190 m (3,920 ft) to the upper limit of the alpine scrub at 3,050 m (10,000 ft). The sampling objective for continuity was to place at least three relevés into each of the twelve 150 m (500 ft) altitudinal intervals. Samples were stratified for three reasons. The first concerns intervening lava flows. The only environmental stratification considered necessary was to avoid placing samples on recent lava flows that traverse the Mauna Loa transect at a few places. Since there are no important topographic variations on the east flank of Mauna Loa other than the major mountain gradient itself, there was no need to stratify samples by a topographic moisture gradient (*sensu* Whittaker, 1956, 1960, 1970). The second reason was community structure. Because of an irregular distribution pattern along the Mauna Loa transect of the major community-structure-forming life forms, vegetation samples were stratified into tree, shrub, and grass community samples. Finally, the previously defined ecosystem boundaries were ignored in the sampling layout along the slope from the savanna (ecosystem type 8) to the alpine scrub (type 3, figure 2-2). In ecosystem types 10 to 12, which occur at the 1,200 m (4,000 ft) level of the transect, samples were placed more or less centrally into the types. This was

*By D. Mueller-Dombois and G. Spatz

TABLE 3-1 *Grouping of Relevés by IBP Sites*

IBP site	Altitude Meters (feet)	Altitudinal Range m (ft)	Altitudinal Interval m (ft)	Relevé Numbers: Tree relevés	Relevé Numbers: Shrub relevés	Relevé Numbers: Grass relevés
1	1,190	None		4	None	None
	(3,920)			54		
				144		
2	1,220	None		49	None	None
	(4,000)			52		
				145		
3	1,220	None		18	None	None
	(4,000)			50		
				147		
4	1,280	1,220–1,370	150	19	None	40
	(4,200)	(4,000–4,500)	(500)	20		90
				146		91
5	1,500	1,370–1,520	150	21	87 101	85 98
	(4,920)	(4,500–5,000)	(500)	96	89	88 102
				100	99	97 150
6	1,600	1,520–1,680	160	92	23 106	22 107
	(5,250)	(5,000–5,500)	(500)	103	86	95
				104	94	105
7	1,720	1,680–1,830	150	83	84	82 112
	(5,650)	(5,500–6,000)	(500)	108	110	109 113
				115	114	111 116
8	1,890	1,830–1,980	150	24	25	26 120
	(6,200)	(6,000–6,500)	(500)	117	119	27
				121	122	118
9	2,030	1,980–2,040	60	81	78	79 126
	(6,650)	(6,500–6,700)	(200)	123		80 127
				125		124 135
10	2,130	2,040–2,190	150	77	76	None
	(7,000)	(6,700–7,200)	(500)	129	128	
				130		
11	2,290	2,190–2,380	190	75	73 134	None
	(7,500)	(7,200–7,800)	(600)	132	74	
				133	131	
12	2,400	2,380–2,590	210	11	12 71	None
	(8,000)	(7,800–8,500)	(700)	14	13 137	
				72	43	
13	2,740	2,590–2,900	310	None	44 65	None
	(9,000)	(8,500–9,500)	(1,000)		63 66	
					64 138	

TABLE 3-1 *(Continued)*

IBP site	Altitude Meters (feet)	Altitudinal Range m (ft)	Altitudinal Interval m (ft)	Tree relevés	Relevé Numbers Shrub relevés		Grass relevés
14	3,050	2,900–3,080	180	None	67	139	68
	(10,000)	(9,500–10,100)	(600)		69	140	141
					70	142	143
Total relevés				36	38		32

done to increase the sampling efficiency per relevé in this 7 km long horizontal transect section.

The relevé sizes were based on the minimal area concept (Mueller-Dombois and Ellenberg, 1974) for sampling at least 95 percent of the vascular plant species in each type of life-form community. The sizes adopted for tree communities were 400 m^2, for shrub communities 200 m^2, and for grass communities 100 m^2. In each relevé, all vascular plant species were listed, and their quantities were estimated according to the Braun-Blanquet cover-abundance scale. A total of 106 relevés were sampled, 32 in grass communities, 38 in shrub communities, and 36 in tree communities. Subsequently all relevés were grouped by altitudinal interval and location to correspond to the fourteen IBP sites. This grouping is shown on table 3-1. Each of the fourteen IBP sites is described by at least three relevés of a given life-form community, wherever these were available along the transect interval. Table 3-1 indicates that tree communities were available upslope to IBP site 12. For the highest two sites (13 and 14), sparse alpine scrub is the prevailing vegetation, and shrub community samples are used to describe the focal IBP sites in this altitudinal range. Shrub communities occur only above the savanna, from where they are continuously present to the upper limit of the alpine scrub. Table 3-1 also indicates that grass cover is continuous from the savanna (site 4) at 1,280 m to the upper limit of the mountain parkland (site 9) at 2,030 m. The grass relevés at IBP site 14 relate to a sparse alpine grass cover that occurs as a local variant on cinder, not on lava rock substrates.

Tree Community Zonation

Figure 3-1 affords a direct comparison of the clustering patterns and transect zones that were derived from the same data set by the two independent analysis methods. The data set is comprised of forty-eight woody plant community samples (thirty-six tree community relevés from 1,190 m elevation to the tree line at 2,590 m and twelve shrub community relevés above the tree line) that were used to describe the fourteen IBP sites with their altitudinal limits as indicated at the bottom of the figure. The two independent analysis methods are the dendrograph technique (with Motyka's index) and the two-way table technique (with the 50/10 rule).

The relevé clusters of the dendrograph analysis show little coincidence with the species relevé groups of the two-way table analysis. This is not

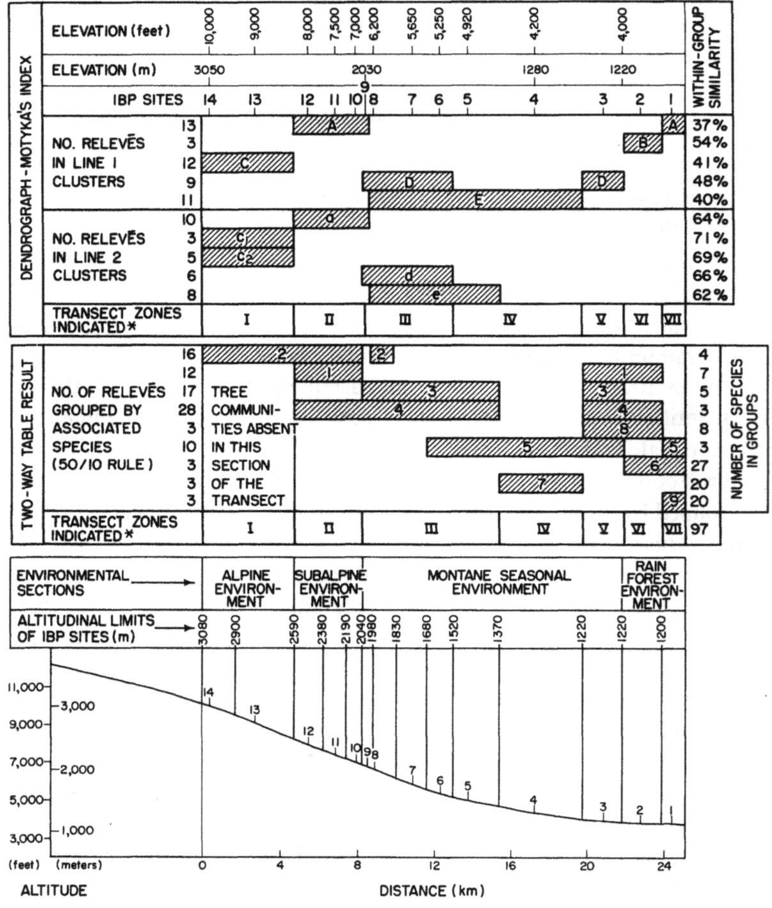

FIGURE 3-1. *Altitudinal transect zones derived from dendrograph and two-way table techniques of forty-eight plant community samples. Transect zones resulting from two-way table analysis are: I, sparse alpine heath scrub, II, subalpine* Metrosideros *scrub forest; III, mountain parkland,* Acacia koa *tree colonies; IV,* Sapindus-Acacia koa *savanna; V, open* Metrosideros *dry forest; VI, open* Metrosideros *rain forest; VII, closed* Metrosideros *rain forest.*

surprising since the clustering criteria of the two methods are so different. In the dendrograph technique, the entire species content of each relevé is compared against every other relevé of the set for its quantitative-floristic similarity; that is, each species is considered of equal importance.

In the two-way table technique, species are not considered of equal importance for establishing relevé similarities. Instead only those species with similar distributions throughout certain subsets of the relevé set are considered the key to relevé clustering or group formation. In this case 97 of the 169 species showed such grouping tendencies.

In spite of the different clustering patterns of the same data set, the resulting transect zones show remarkable similarity. Both methods yielded seven zones according to the criteria, which were to draw a zonal boundary where at least two cluster or group limits occurred. The zonal boundaries are alike except for one—that between zones III and IV. The dendrograph analysis moved the savanna boundary upslope between sites 5 and 6 near 1,520 m (5,000 ft), whereas the two-way table technique placed the boundary near 1,370 m (4,500 ft) elevation, where it was placed subjectively prior to this analytical procedure (see figure 2-2). This boundary also caused difficulty during the mapping process of the park's vegetation (Mueller-Dombois and Fosberg, 1974) because of its more transitory character.

Figure 3-2 shows another variation of the same two methods of zonation or community analysis for the same data set. The dendrograph analysis is based on a different similarity index, the index of Spatz (1970).* Spatz's index is still more discriminatory than the widely applied index of Motyka used for figure 3-1. Spatz's index, a quantitative modification of Jaccard's (1901) qualitative index, incorporates both qualitative and quantitative species values in a very balanced manner (see evaluation in Mueller-Dombois and Ellenberg, 1974: 220). The outcome, as seen in figure 3-2, is a larger number (nine) of community types or transect zones. Closer inspection shows that the increased number of zones resulted from a further breakdown of the mountain parkland zone (III on figure 3-1) into three subzones: an upper, middle, and lower mountain parkland zone (zones III, IV, and V on figure 3-2, upper diagram).

*
$$IS = \frac{\Sigma\,(Mw{:}Mg)}{a + b + c} \times \frac{Mc}{Ma + Mb + Mc} \times 100$$

where IS = index of similarity (Spatz), Mw = smaller quantitative value of a species common to relevés A and B, Mg = greater quantitative value of a species common to relevés A and B, a = number of species occurring in relevé A only, b = number of species occurring in relevé B only, c = number of species common to both relevés, A and B, Ma = sum of the quantitative values of the species unique to relevé A, Mb = sum of the quantitative values of the species unique to relevé B, Mc = sum of the quantitative values of the species common to relevés A and B. In this index a similarity of 25 percent is very high.

The fourth zonation diagram is derived from the most discriminatory rule for the two-way table analysis, the 66/10 rule. This rule requires that species to be recognized as a spatially associated group must occur together in at least 66 percent of the relevés to be grouped, and they must occur in not more than 10%

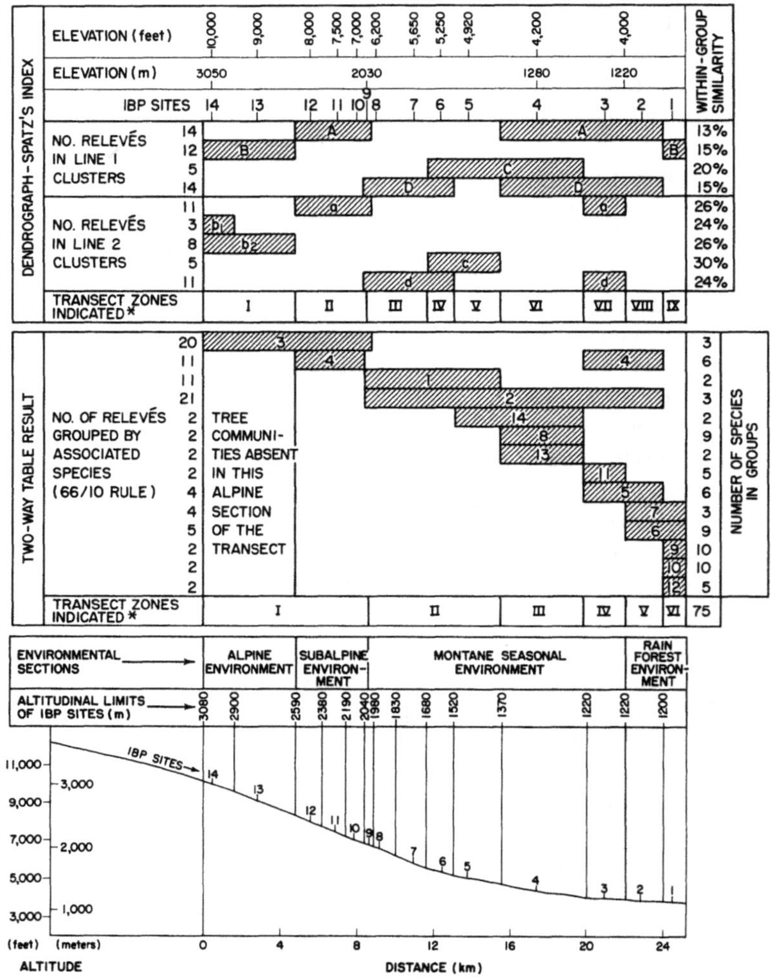

FIGURE 3-2. *Altitudinal transect zones derived from alternative options in dendrograph and two-way table techniques. The nine transect zones resulting from dendrograph analysis with Spatz's index are: I, sparse alpine heath scrub; II, subalpine* Metrosideros *scrub forest; III, upper mountain parkland; IV, middle mountain parkland (climatic station); V, lower mountain parkland; VI, savanna; VII, open* Metrosideros *dry forest; VIII, open* Metrosideros *rain forest; IX, closed* Metrosideros *rain forest.*

of the relevés outside the group. The 66/10 rule resulted in only six transect zones. The reason is the absence of a second cluster or group boundary between sites 12 and 13 on the lower diagram on figure 3-2. The other three zonation diagrams always showed two or more cluster limits within this section of the transect. Certainly there is an important physiognomic boundary between sites 12 and 13, since the upper tree line occurs at this level. Moreover, this physiognomic boundary is also an important species boundary because the tree species *Metrosideros collina* subsp. *polymorpha* does not occur upward of site 12. The reason why this is not strongly indicated by the 66/10 rule is that there are only a few high-altitude species, and these range through the alpine and subalpine environment. The breakdown into an alpine and subalpine zone, as indicated on the first three zonation diagrams, seems more appropriate. The remaining five zones on the lower diagram on figure 3-2 coincide closely with those on the first two diagrams (figure 3-1). Here also, the mountain parkland is not further subdivided.

In spite of the different techniques and criteria used in developing the four zonation diagrams, the classification results are rather similar. Except for one boundary variation, they differ only by subdivisions of broader zonal segments. Three of the four diagrams separate the alpine from the subalpine environment. One of the four subdivides the mountain parkland into three smaller segments. Thus the classification into seven transect zones or community types as indicated in figure 3-1 seems most appropriate. It also coincided closely with the subjectively defined ecosystem types in figure 2-2.

Species Distribution Trends in Tree Communities

Another problem is the interpretation of the individual clusters in figures 3-1 and 3-2. It is relatively easy to identify a common characteristic for some (but not all) major dendrograph clusters, if one is familiar with the data. For example, cluster E in figure 3-1 relates to the transect area occupied by *Acacia koa* tree communities, while cluster A includes certain *Metrosideros* tree communities. Other *Metrosideros* communities were sampled at sites 2 and 3, but these were not included in cluster A. Therefore a more detailed inspection of the data would become necessary to explain cluster A. However, since the dendrograph printout is based merely on a similarity matrix, there is no strictly analytical way to determine which species or species combinations and their quantities are specifically responsible for any given cluster. One would have to go back to the original relevé data and then superimpose on the clusters those species populations with their quantities that provide the closest fit, a time-consuming trial-and-error method. Moreover, comparison of the dendrograph clusters in figures 3-1 and 3-2 shows that they have almost no resemblance, further increasing the complexity of the dendrograph cluster interpretation.

TABLE 3-2 *Seventy-five Plant Species Found in Fourteen Spatially Associated Groups, Mauna Loa Transect.*

Species	Life form
Group 3	
Vaccinium peleanum[a]	High-altitude shrubs, native
Styphelia douglasii[a]	(High-altitude shrubs, native)
Pellaea ternifolia	Small, xerophytic fern, native
Group 4	
Coprosma ernodeoides[a]	Creeping shrub, native
Coprosma montana	Small tree, native
Raillardia ciliolata var. *ciliolata*[a]	Shrub, native
Machaerina gahniaeformis	Sedge, native
Luzula hawaiiensis	Rush, native
Pleopeltis thunbergiana	Fern, native
Group 1	
Styphelia tameiameiae[a]	Shrub, native
Panicum tenuifolium[a]	Grass, native
Group 2	
Pteridium aquilinum var. *decompositum*	Fern, endemic variety
Holcus lanatus[b]	Grass, exotic
Acacia koa var. *hawaiiensis*[b]	Tree, dominant endemic species
Group 14	
Oxalis corniculata	Small forbs, exotic
Anagallis arvensis	(Small forbs, exotic)
Group 8	
Sapindus saponaria[a]	Tall tree, native
Hibiscadelphus giffardianus	Small tree, native
Ipomoea congesta	Vine, native
Bromus catharticus	Grass, exotic
Geranium carolinianum	Forbs, exotic
Verbena littoralis	(Forbs, exotic)
Veronica plebeia	(Forbs, exotic)
Commelina diffusa[b]	(Forbs, exotic)
Sonchus oleraceus	(Forbs, exotic)
Group 13	
Myoporum sandwicense[a]	Tree, native
Pipturus albidus	Tall shrub, native
Group 11	
Dianella sandwicensis[a]	Native graminoid, Liliaceae
Erigeron canadensis	Forb, exotic
Bulbostylis capillaris	Annual sedge, exotic
Festuca octiflora	Grasses, exotic
Aira caryophylla	(Grasses, exotic)

TABLE 3-2 *(Continued)*

Species	Life form
Group 5	
Wikstroemia phillyraefolia var. *phillyraefolia*	Small tree, native
Rubus penetrans	Creeping thorn shrub, exotic
Andropogon virginicus	Grasses, exotic
Andropogon glomeratus	
Sporobolus africanus	
Sacciolepis indica	
Group 7	
Dicranopteris (Gleichenia) emarginata[a]	Stoloniferous, matted fern, native
Sadleria cyatheoides	Tall woody fern, native
Lycopodium cernuum	Club moss, native
Group 6	
Cibotium glaucum[a]	Tree fern, native
Coprosma ochracea var. *rockiana*	Trees, native
Ilex anomala	
Vaccinium calycinum	Shrub, native
Astelia menziesiana	Epiphytic forb, native
Machaerina angustifolia	Tall sedge, native
Isachne distichophylla	Grasses, native
Uncinia uncinata	
Eupatorium riparium[b]	Forb, exotic
Group 9	
Myrsine lessertiana	Tree, native
Clermontia parviflora	Shrubs, native
Cyrtandra platyphylla	
Hypericum degeneri	Small forb, exotic
Phajus tankervilliae	Forb, exotic
Grammitis tenella	Epiphytes, native
Mecodium recurvum	
Psilotum complanatum	
Microlaena stipoides	
Group 10	
Metrosideros collina subsp. *polymorpha* var. *macrophylla*[a]	Tree, widespread species, but restricted variety, native
Cibotium chamissoi	Tree ferns, native
Cibotium hawaiiense	
Athyrium sandwichianum	Tall, herbaceous fern, native
Stenogyne calaminthoides	Vine, native
Peperomia hypoleuca	Epiphytes, native
Adenophorus sarmentosum	
Hymenophyllum obtusum	
Hymenophyllum lanceolatum	
Grammitis hookeri	

TABLE 3-2 *(Continued)*

Species	Life form
Group 12	
Cheirodendron trigynum	} Trees, native
Coprosma rhynchocarpa	
Psidium cattleianum[b]	Tree, exotic
Elaphoglossum reticulatum	} Epiphytes, native
Adenophorus tripinnatifidus	

Note: Generated with two-way table 66/10 rule from 169 species in 48 relevés. Computer-derived number sequence from top to bottom as on zonation diagram, figure 3-2.
[a]Distribution diagrammed in figure 2-12 (native species).
[b]Distribution diagrammed in figure 3-3 (exotic species).

The two-way table technique offers a considerable advantage in that it specifies the species responsible for the clusters in the printout. Table 3-2 lists the seventy-five species that were clustered into the fourteen associated species groups in figure 3-2. The order of listing follows the group arrangement in the figure, from sparse alpine scrub to closed rain forest. Species from separate groups that define exactly the same transect section, such as groups 8 and 13, 7 and 6, and 9, 10 and 12, can be considered as spatially associated also. The reason why these spatially associated species were extracted into separate groups is related to the sample size. Larger relevés or more relevés probably would have combined these groups. Groups 9, 10, and 12 contain a number of rain forest epiphytes. Their presence on certain trees gave further weight to a mathematical splitting of groups within the same vegetation type.

From the various groups in table 3-2, the quantitatively more important species were selected for diagrammatic representation of their individual species distributions. Species identified by footnote "a" in the table are the quantitatively more important native species, which are shown in figure 2-12. Group membership of the species may refer to either the 66/10 or 50/10 rule or to both. This is not a critical point. In this case both options generated spatially associated groups of species, and their degree of spatial association can now be compared diagrammatically. For representing a more complete picture, the distributions of two important ungrouped tree species (*Metrosideros collina* subsp. *polymorpha* var. *incana* and *Sophora chrysophylla*) were included on the species distribution diagram in figure 2-12.

Zones I and II (alpine and subalpine transect zones) extend through areas of lava rock outcrop with only small amounts of fine soil. These transect zones are occupied primarily by shrub species. The tree species *Metrosideros collina* subsp. *polymorpha* appears in zone II together with several other shrub species (*Coprosma ernodeoides*, *Raillardia ciliolata*, and *Dodonaea sandwicensis*) (see figure 2-12).

The lava substrates of zones III and IV (mountain parkland and savanna) are covered downslope with an increasingly deeper and more continuous blanket of fine soil from volcanic ash. Here, herbaceous life forms, particularly grasses (*Deschampsia australis* and *Panicum tenuifolium,* among the native forms), are more abundant. This change coincides with an increase in tree species diversity (*Acacia koa, Sophora chrysophylla, Myoporum sandwicense,* and *Sapindus saponaria*) and a decrease in shrub species diversity (*Dodonaea sandwicensis* and *Styphelia tameiameiae* are the more important ones, while several others drop out). In zone IV (savanna), which has rather deep ash soil (more than 70 cm deep) and no rock outcrops, even *Dodonaea* and *Styphelia tameiameiae* disappear.

In zone V (*Metrosideros* dry forest), most of the shrubs (except the high-altitude species *Vaccinium peleanum* and *Styphelia douglasii*) reappear, while the other tree species disappear.

In transect zones VI and VII (rain forest), ferns become prevalent, and the rock outcrop-associated shrubs decrease with increasing soil depth in zone VII. Here, *Metrosideros* retains a dominant position on deeper soil only in the rain forest, where its variety *incana* decreases in abundance and coexists with the *Metrosideros* variety *macrophylla*.

Thus we can recognize on this diagram along the Mauna Loa transect interesting bimodal trends for certain woody plant species (*Metrosideros, Coprosma, Raillardia,* and *Styphelia tameiameiae*). Note also that there are two cogeneric bimodalities (in *Styphelia* and *Vaccinium*). These are all pioneer woody plants in primary succession, growing well on rocky lava substrates of intermediate ages (at least two hundred years old).

Species identified by note "b" in table 3-2 are the quantitatively more important grouped exotic species. Only four of these were spatially grouped by either the 66/10 or 50/10 rules. Therefore most of the more important exotics exhibited individualistic distributions and were selected from the ungrouped species in the two-way table and presented in the species distribution diagram (figure 3-3). The cosmopolitan bracken fern *Pteridium aquilinum* on Mauna Loa is represented by an endemic variety *decompositum*. It was included among the exotics because of the similarity of its distribution to that of an important exotic grass from Europe, *Holcus lanatus*.

Most of these exotics show their best development in zone IV (savanna) on deep, rich kīpuka soils. This site had a history of disturbance by fire and cattle (Mueller-Dombois and Lamoureux, 1967); also ground disturbance by pigs is a most prevalent mechanical factor in this ecosystem. In addition, the more rocky soil-adapted native woody species are absent in this ecosystem so that there has been less resistance to the invasion of exotic plant species. Thus the savanna can be seen as the center of exotic plant species establishment, from where the exotics seem to have spread into the neighboring zones up and down the transect, depending on their habitat adaptation (climatic and edaphic) and competitive capacity to cope with the resistance offered by the native plants

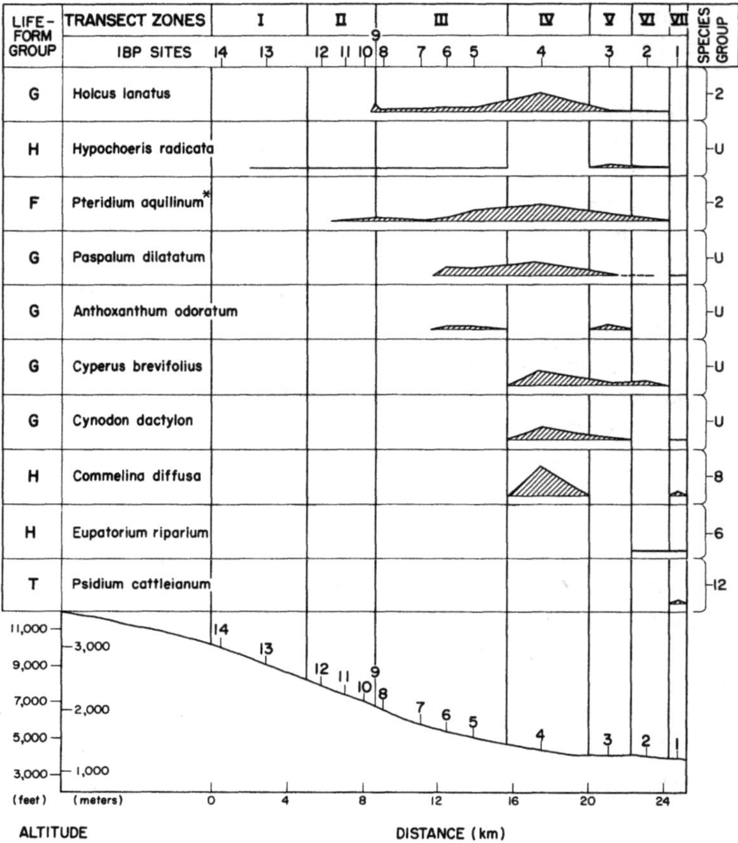

FIGURE 3-3. *Distribution of dominant exotic plant species along Mauna Loa transect.* Pteridium aquilinum *occurs as an endemic variety* decompositum. *Group numbers resulted from 66/10 rule. Curve heights relate to species quantities in percent cover plotted at equal intervals, from straight line* $= <1$–25 *percent cover maximum for* Commelina diffusa *at site 4. Life forms: T = tree, F = fern, G = grass or graminoid herb, H = herbaceous plant other than grass or graminoid (forb).*

along the transect. One important case of penetration of an exotic grass, *Holcus lanatus,* was experimentally demonstrated as caused by pig digging (Spatz and Mueller-Dombois, 1975). It was found that upon such disturbed ground, little resistance was offered by a native grass of similar life form, *Deschampsia australis.* Figure 3-3 indicates that even certain exotics are eliminated by competition. This applies to the low-growing composite weed *Hypochoeris radicata* and the low-growing grass *Anthoxanthum odoratum,* both of which apparently have been replaced from under the trees in the savanna by the taller-growing plant life forms. The herbaceous layer in the savanna (zone IV)

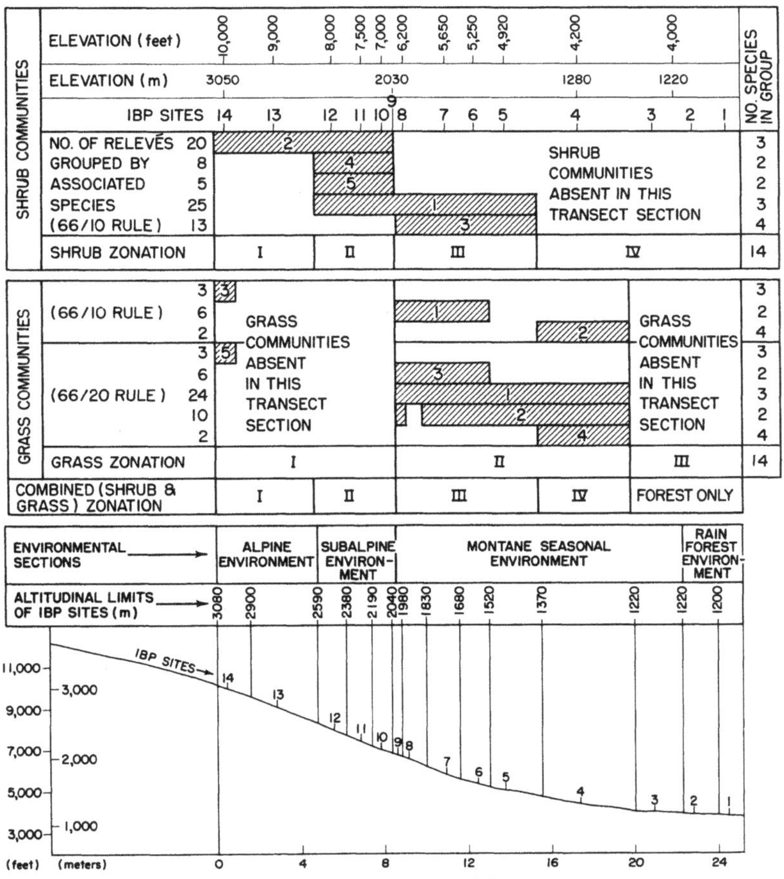

FIGURE 3-4. *Altitudinal transect zones derived from two-way table technique for thirty-eight shrub and thirty-two grass community samples (separately and combined). The zonal boundaries are based on at least two cluster limits. Combined zonation: I = sparse alpine scrub, II = open subalpine scrub forest, III = mountain parkland, IV = savanna.*

is made up primarily of tall-grass species and herbs (*Paspalum* and *Pteridium*) and mat formers (*Cynodon, Cyperus brevifolius,* and *Commelina*) that can maintain a dominant position under the prevailing conditions.

Shrub and Grass Community Zonation

Because of the similarity of zonation patterns that resulted from the dendrograph and two-way table methods, only the latter are represented in figure 3-4 for both the shrub and grass community data.

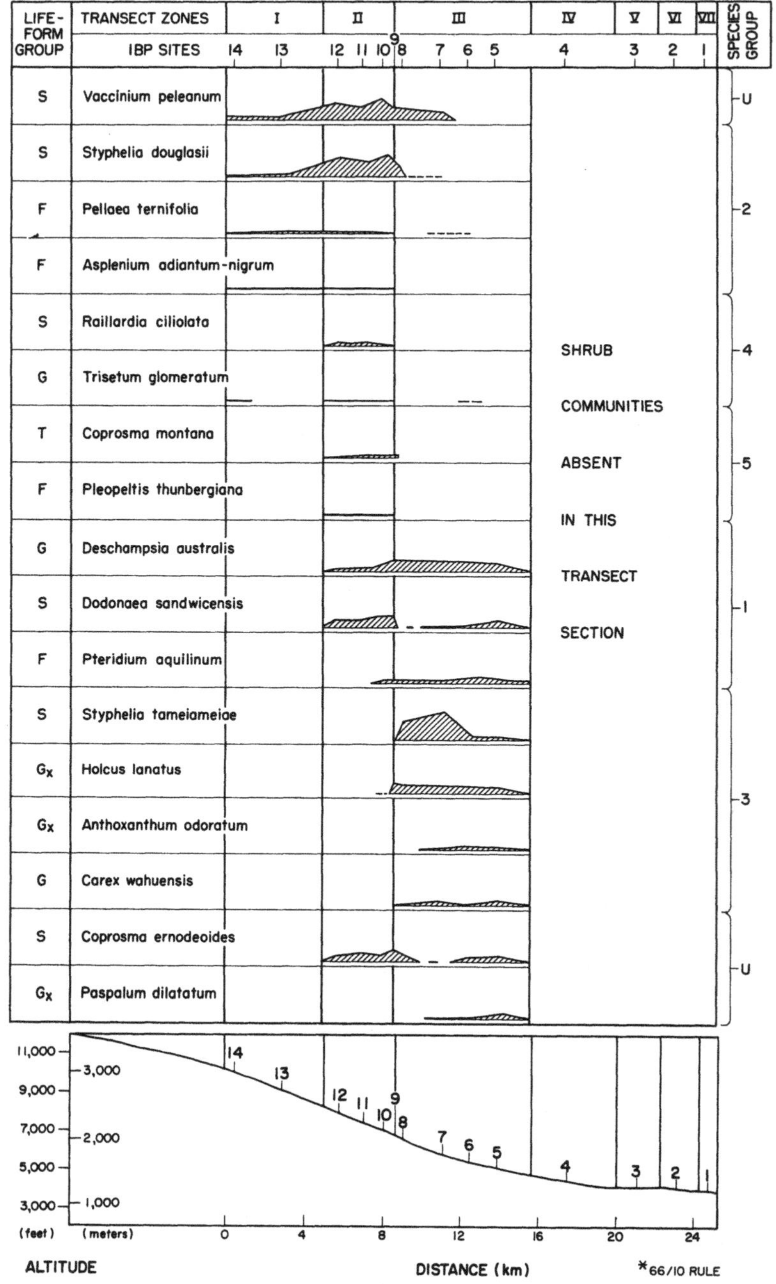

LIFE-FORM GROUP
TRANSECT ZONES
I
II
III
IV
V
VI
VII
IBP SITES
14 13 12 11 10 9 8 7 6 5 4 3 2 1
SPECIES GROUP
S Vaccinium peleanum
S Styphelia douglasii
F Pellaea ternifolia
F Asplenium adiantum-nigrum
S Raillardia ciliolata
G Trisetum glomeratum
T Coprosma montana
F Pleopeltis thunbergiana
G Deschampsia australis
S Dodonaea sandwicensis
F Pteridium aquilinum
S Styphelia tameiameiae
Gx Holcus lanatus
Gx Anthoxanthum odoratum
G Carex wahuensis
S Coprosma ernodeoides
Gx Paspalum dilatatum
U
2
4
5
1
3
U
SHRUB
COMMUNITIES
ABSENT
IN THIS
TRANSECT
SECTION
11,000
9,000
7,000
5,000
3,000
(feet)
3,000
2,000
1,000
(meters)
0 4 8 12 16 20 24
ALTITUDE
DISTANCE (km)
*66/10 RULE

The shrub community analysis, which is based on thirty-eight relevés with fifty species, resulted in four transect zones. Relevés describing the high-altitude sites 13 and 14 above the tree line were the same as those used for the tree community analysis. However, those defining zones II and III are independent shrub community samples (see table 3-1). The first three shrub zones show identical limits to those obtained from the tree community samples (figure 3-1), resulting in an alpine (I), subalpine (II), and mountain parkland (III) zone. Shrub communities were absent from the savanna (site 4) downslope to the closed rain forest (site 1). Since absence is also a diagnostic criterion, this transect section can be designated as zone IV.

The grass community analysis, which is based on thirty-two relevés with forty-nine species, gave no zonation pattern with the more discriminatory 66/10 rule according to the boundary criteria adopted. However, the 66/20 rule resulted in three general zones: an alpine-subalpine zone (I), a mountain parkland-savanna zone (II), and a zone in which grass communities (according to our definition) were absent (III). A relaxation in our boundary criterion from two to one cluster limit would have permitted a separation of the savanna from the mountain parkland. Then also the mountain parkland itself would have to be subdivided further into an upper, middle, and lower section. However, there is no need to relax the original boundary criterion. The species distribution trends remain the same and show quite clearly a good distinction of the savanna grass communities from those of the mountain parkland by species group 4.

When both the shrub and grass zonations are combined, the resulting zonation pattern is almost identical to that obtained by the tree community relevés (figure 3-1). The only difference occurs at the lower elevation end of the transect, where the tree community zonation results in three zones, one for each site (1, 2, and 3). Here the shrub-grass zonation could not give any further differentiation.

Species Distribution Trends in Shrub and Grass Communities

The number of species (fifty) sampled in the shrub relevés was less than one-third that encountered in the tree communities. Of these fifty species, fourteen were clustered into five species groups (figure 3-4, upper diagram).

FIGURE 3-5. *Species distribution diagram of all grouped and three ungrouped species of shrub community samples. Groups were generated by two-way table technique 66/10 rule and correspond to those shown on figure 3-1, upper zonation diagram. Life forms: T = tree, S = shrub, F = fern, G = grass or graminoid herb, Gx = exotic grass. Curve heights represent species quantities in percent cover plotted at equal intervals, from dashed line = <0.1% to solid line = <1% to 25% cover maximum for* Styphelia tameiameiae *at site 7.*

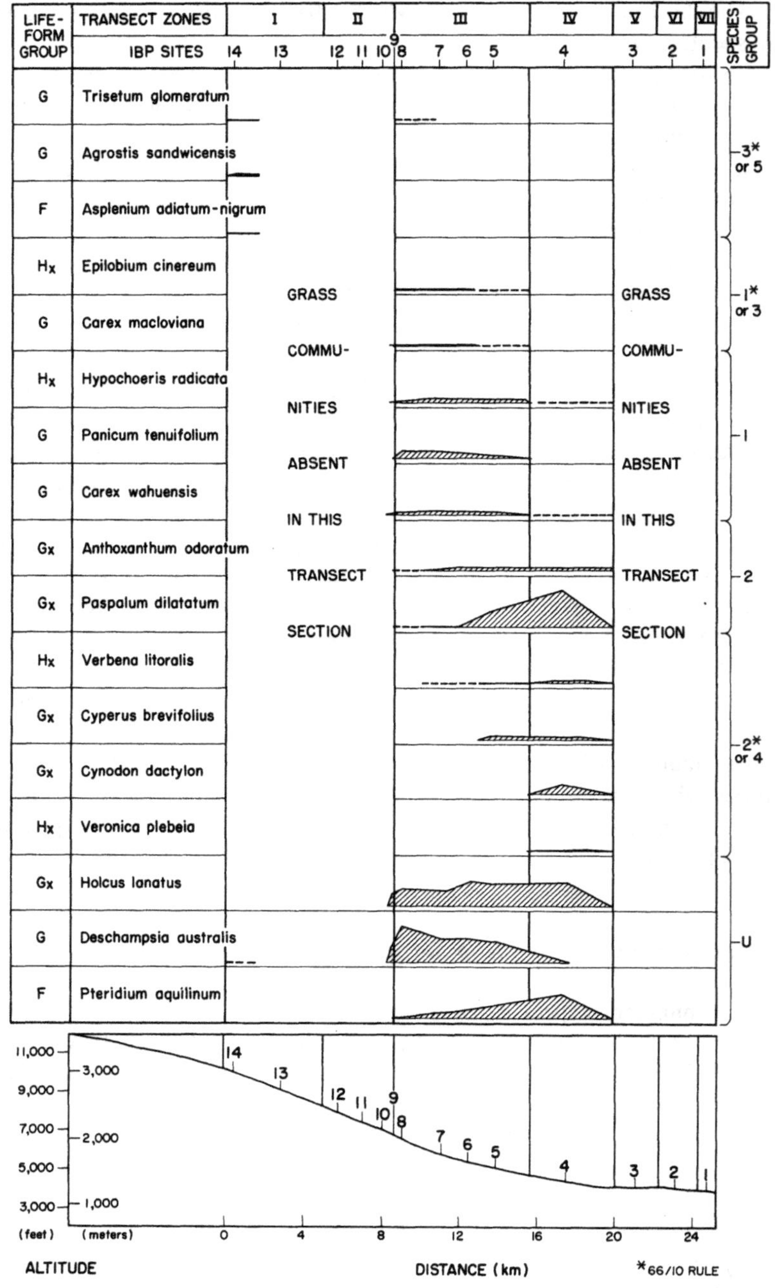

LIFE-FORM GROUP
TRANSECT ZONES
I
II
III
IV
V
VI
VII
IBP SITES
14 13 12 11 10 9 8 7 6 5 4 3 2 1
SPECIES GROUP
G Trisetum glomeratum
G Agrostis sandwicensis
F Asplenium adiatum-nigrum
Hx Epilobium cinereum
G Carex macloviana
Hx Hypochoeris radicata
G Panicum tenuifolium
G Carex wahuensis
Gx Anthoxanthum odoratum
Gx Paspalum dilatatum
Hx Verbena litoralis
Gx Cyperus brevifolius
Gx Cynodon dactylon
Hx Veronica plebeia
Gx Holcus lanatus
G Deschampsia australis
F Pteridium aquilinum
GRASS COMMU- NITIES ABSENT IN THIS TRANSECT SECTION
GRASS COMMU- NITIES ABSENT IN THIS TRANSECT SECTION
3* or 5
1* or 3
1
2
2* or 4
U
11,000
9,000
7,000
5,000
3,000
3,000
2,000
1,000
(feet)
(meters)
0 4 8 12 16 20 24
ALTITUDE
DISTANCE (km)
*66/10 RULE

These fourteen species are diagrammed with their individual quantitative distributions in figure 3-5, along with three of the quantitatively more important ungrouped species.

The high-altitude shrub *Vaccinium peleanum* extends downslope halfway into the mountain parkland (zone III) in the shrub communities, whereas it extends very little into zone III in the tree communities (figure 2-12). This difference is related to the substrate variation under the two life-form communities in this transect section. The shrub communities occur usually on lava rock outcrops with very shallow and discontinuous ash-pocket soil, while the *Acacia koa* tree communities are mostly associated with shallow but continuous ash soil in the mountain parkland (Mueller-Dombois, 1967). Similarly *Coprosma ernodeoides* shows a wider transect distribution in the shrub communities than in the tree communities. Other shrub species, such as *Styphelia douglasii, Raillardia ciliolata, Dodonaea sandwicensis,* and *Styphelia tameiameiae,* have similar distribution ranges in the tree communities (figure 2-12), but they are generally present with greater cover in the shrub communities. The grass species *Deschampsia australis, Holcus lanatus, Anthoxanthum odoratum,* and *Paspalum dilatatum* also show similar distribution ranges in the two life-form communities. A few new species are diagrammed in figure 3-5 that also occupy similar transect portions in the tree communities. These include the high-altitude fern *Pellaea ternifolia,* the subalpine tree *Coprosma montana,* and a fern, *Pleopeltis thunbergiana* (see table 3-2). The fern, which on the diagram is restricted to the open subalpine scrub forest, also occurs at lower elevations in similarly structured woody vegetation types.

Figure 3-6 shows fourteen grouped and three quantitatively important ungrouped species of the forty-nine species encountered in the grass communities. In the alpine zone near 3,000 m elevation, a few sparse grass patch communities occur on cinder with two native grasses, *Trisetum glomeratum* and *Agrostis sandwicensis*. *Trisetum glomeratum* is sparsely distributed downslope through the subalpine shrub communities (see figure 3-5) into the middle of the mountain parkland (zone III). The dominant native mountain parkland grass *Deschampsia australis* is also present in the high-altitude grass patch communities, but with insignificant quantity. *Deschampsia* is particularly dominant in the upper mountain parkland, from where it decreases gradually in quantity downslope. A similar but more restricted range is occupied by

---

FIGURE 3-6. *Species distribution diagram of all grouped and three ungrouped species of grass-community samples. Groups were generated by the two-way table 66/10 and 66/20 rules and correspond to those shown on figure 3-4, the two lower zonation diagrams. Life forms: G = grass or graminoid herb, F = fern,* H_X *= exotic herb or forb,* G_X *= exotic grass. Curve heights represent species quantities in percentage cover plotted at equal intervals, from dashed line = <0.1% to solid line = <1% to 30% cover maximum for* Paspalum dilatatum *at site 4.*

the species of group 1, which includes another native grass, *Panicum tenuifolium,* and the native sedge, *Carex wahuensis*. The exotic grass *Holcus lanatus* is quantitatively important in both the mountain parkland and savanna, occupying a range similar to that of the other European exotic grass, *Anthoxanthum odoratum*. The savanna is clearly separated from the mountain parkland by the species of group 4, which includes the grass *Cynodon dactylon*. However, as shown in figure 3-6, *Paspalum dilatatum, Holcus lanatus,* and *Pteridium aquilinum* var. *decompositum* are clearly the dominant herbaceous savanna plants. Their distribution ranges are similar in the two other life-form communities, but their quantities are generally greater in the grass communities.

Discussion and Conclusions

The four plant species distribution diagrams permit an evaluation of our data with the species distribution hypotheses stated in chapter 2, particularly with those of Whittaker (1975). His first hypothesis conveys the idea that species may combine into spatially associated groups that give way to other associated groups at relatively sharp boundaries along an environmental gradient. His second hypothesis conveys the idea that species do not combine into associated groups but that certain species exclude one another by competition along the gradient, resulting in sharp boundaries with little overlap between those competing species. His third hypothesis conveys the idea that species do not exclude each other by competition along the gradient but combine into associated groups that may overlap broadly with other associated groups. His fourth hypothesis conveys the idea that species do not combine into spatially associated groups and that the species turnover along the gradient is more or less random or scattered. Whittaker has concluded that his environmental gradient analyses on temperate continental mountain slopes (Smoky Mountains, Siskiyou Mountains, and Catalina Mountains) always supported the fourth hypothesis.

Our diagram of native species from tree communities (figure 2-12) shows that there are species with closely similar distributions along the transect. For example, the first two high-altitude species, *Vaccinium peleanum* and *Styphelia douglasii,* have closely parallel distributions. And the next two species, *Coprosma ernodeoides* and *Raillardia ciliolata,* show closely similar distributions. Group 3* is less similar, but group 7* (the two tree species *Sapindus* and *Myoporum*) occupies the same segment along this transect. The two tree species at the bottom of the diagram remained ungrouped since they occupied rather individualistic distribution ranges. The tree *Metrosideros collina* subsp. *polymorpha* var. *incana,* which is prominent in the subalpine forest and at the low end of the transect, almost complements the distribution range of the other major tree species *Acacia koa,* which is most prevalent in the mountain parkland and savanna. *Acacia koa* was planted at site 3 (in zone V, the open

Metrosideros dry forest) by the National Park Service. It did not occur there naturally.

This species distribution pattern supports in part the first and third hypotheses of Whittaker in that certain species occur in spatially associated groups and also in that certain associated groups give way to other associated groups at relatively sharp boundaries (for example, species group 3** is replaced by group 3* downslope at site 9, group 7* is replaced by group 8* at the boundary between sites 4 and 3), but also that groups overlap broadly (for example, groups 3** and 4** or groups 4* and 3*).

If we eliminated the spatially associated species from each group, the diagram could be used to support Whittaker's second and fourth hypotheses. The second hypothesis, which states that individual species exclude one another along sharp boundaries, may find support in comparing the boundaries of the cogeneric shrubs *Styphelia douglasii* and *S. tameiameiae* (at site 9) and *Vaccinium peleanum* and *V. reticulatum* (near site 7). However, the *Styphelia* separation is somewhat arbitrary because of vaguely defined species characteristics in the area of overlap at site 9. The fourth hypothesis would find support in the broad spatial overlap displayed in some species. Even the three species of group 3* (*Acacia koa, Panicum tenuifolium,* and *Styphelia tameiameiae*) may support the fourth hypothesis because their within-group distribution is almost at the limit of what may be called spatially associated. However, the three species coincide closely in their upper distribution limits at site 9.

Thus we can conclude that our tree community data with regard to native plant species give support to all four of Whittaker's hypotheses, but because of the occurrence of species with closely similar amplitudes, the data support hypotheses 1 and 3 more strongly than hypotheses 2 and 4.

The exotic species in the tree communities (figure 3-3) display a different pattern. The majority of these have their quantitative peak in the savanna (site 4). This deep-soil habitat, among all others, has been most severely subjected to ground disturbance by pigs. One important case of penetration of an exotic grass, *Holcus lanatus,* was experimentally demonstrated as caused by pig digging in the mountain parkland (Spatz and Mueller-Dombois, 1975). Upon such ground disturbance little resistance was offered by a native grass of similar life form, *Deschampsia australis.* In the savanna, where native grasses were very rare, pig-scarified places were immediately recovered by *Commelina diffusa* and *Cynodon dactylon.* Apparently *Deschampsia* has been displaced from the savanna under trees as indicated on figure 2-12, probably by shading. It is only present with small cover in the open grass savanna, indicating some displacement by competitive exclusion through introduced grasses. Likewise certain exotics seem to have been displaced by competitive exclusion as determined partly from experiments (Spatz and Mueller-Dombois, 1975). This applies to the low-growing composite weed *Hypochoeris radicata* and the tufted weed grass *Anthoxanthum odoratum,* although the latter is still present in the open grass communities of the savanna.

The exotic herbaceous species under trees—*Holcus lanatus, Paspalum dilatatum, Cyperus brevifolius, Cynodon dactylon,* and *Commelina diffusa*—conform somewhat to Whittaker's third hypothesis insofar as they form a group with coinciding population centers in the savanna, but their ranges extend from here in a highly individualistic pattern up or down the slope. A minor group of similar sort (*Psidium cattleianum* and *Eupatorium riparium*) seems to develop from the rain forest.

Such groups of species with coinciding quantitative peaks but noncorrelated ranges are not really recognized in Whittaker's hypotheses. Exactly coinciding quantitative peaks are rarely to be expected in spatially associated species. Quantitative peaks may even be inversely related (*Dodonaea* and *Acacia* in zone III). However, associated species usually show their best quantitative development over their common distribution range. That this is not a contradictory statement can be seen by comparing the population curves within the associated species groups in figures 2-12, 3-5, and 3-6.

Competition and evolution (*sensu* group adaptation) have been suggested in Whittaker's hypotheses as the general mechanisms causing the various spatial patterns. However, these general mechanisms or processes cannot be read directly from the distribution diagrams. The diagrams merely permit an interpretation of spatial occurrences, not spatial adaptations. For an interpretation of competition as a probable cause, it is possible to use one's knowledge of the plant life form and the functional performance of the species in question. For an absolute proof, one has to perform experiments. Knowledge of evolutionary adaptation can be obtained from intensive comparative vegetation studies of many localities in a much wider area than the transect itself and from succession studies. Fortunately such studies preceded or were done simultaneously with the Mauna Loa transect analysis (Mueller-Dombois, 1967; Mueller-Dombois and Lamoureux, 1967; Mueller-Dombois and Krajina, 1968; Spatz and Mueller-Dombois, 1973, 1975; Smathers and Mueller-Dombois, 1974). Therefore, the following explanatory statements of underlying causes of certain peculiar distribution trends can be made with confidence.

The spatial extensions of *Vaccinium peleanum* and *Coprosma ernodeoides* from the subalpine *Metrosideros* scrub forest (zone II) into the mountain parkland (zone III) in shrub communities but not in tree communities are related to the different substrate relations of the two life-form communities in the mountain parkland. The shrub communities occur on shallow rock outcrop soils (sometimes with very little or almost no soil). In contrast, the *Acacia koa* tree communities are usually associated, as are the grass communities, with the more continuous, deeper soils (10–70 cm) overlying lava rock.

The bimodal occurrences of certain shrub species in the tree communities along the transect (*Coprosma ernodeoides, Raillardia ciliolata, Dodonaea sandwicensis,* and *Styphelia tameiameiae*) are also related to the rock outcrop substrate in these different transect segments. Were it not for the presence of the ash-soil blanket in the mountain parkland and savanna, these shrubs would range throughout the whole transect, from the subalpine zone II to the open *Metrosideros* forest zone VI. Obviously these shrubs are pioneer species in primary succession throughout a wide range of climate. It is quite clear that the typical mountain parkland and savanna trees (*Acacia koa, Sophora chrysophylla, Myoporum sandwicense,* and *Sapindus saponaria*) as well as the grasses (*Deschampsia australis, Panicum tenuifolium, Holcus lanatus, Anthoxanthum odoratum, Paspalum dilatatum,* and *Cynodon dactylon*) and sedges (*Carex macloviana, Carex wahuensis,* and *Cyperus brevifolius*) are present in the mountain parkland and savanna because of the fine soil accumulation there. While the fine soil from ash may be geologically as recent or even younger than some of the lava rock outcrop substrate, the transect distribution of these species groups can be explained in terms of substrate adaptation. The shrubs clearly represent a more rockland-adapted vegetation while the parkland and savanna trees, grasses, and sedges represent a more fine-soil-adapted vegetation. Thus several of the spatial groups of species shown on the species diagrams have certainly become adapted to grow together in the different zones along the transect. The exotics are also largely concentrated in the fine-soil habitats because their preadaptations from their countries of origin are generally to root in fine soils rather than xeric rocky substrates. Thus the plant species distributions and adaptations along the Mauna Loa transect are primarily caused by differing soil-water relations associated with the degree of substrate coarseness. Climatic gradient adaptations can be expected in the species at the more extreme ends—for example, in the high-altitude shrubs (*Vaccinium peleanum* and *Styphelia douglasii*), grasses (*Trisetum glomeratum* and *Agrostis sandwicensis*), and ferns (*Pellaea ternifolia* and *Asplenium adiantumnigrum*), which are adapted to cope with frequent night frosts. The low-end transect ferns (*Dicranopteris emarginata* and *Cibotium glaucum*) are adapted to year-round high rainfall.

The species diversity along this recent island mountain slope would be very much simplified (probably the species number would be cut in half) were it not for the ash blanket in the mountain parkland and savanna zone that has produced some habitat complexity along the Mauna Loa transect.

BIRDS*

Sampling Information

Avian populations were sampled sixteen times at seven IBP focal sites (1, 3, 4, 5, 7, 9 and 12, table 2-2) from the closed rain forest (site 1) to the tree line ecosystem (site 12). Four sampling trips were made to the treeless alpine scrub sites 13 and 14. The sampling units were belt transects of 250 m width and

*By Sheila Conant

1,600 to 2,000 m length. Site 2 was not sampled because it was not large enough for a standard-sized transect except in an area where bird populations were undoubtedly influenced by the two heavily trafficked roads bounding either side of the study site. Sampling was done at monthly intervals from December 1972 through July 1973 and again from August 1974 through March 1975.

For bird censusing, Emlen's (1971) count × detectability method was used with a few modifications described in TR 74. The method consists of walking quietly along a belt transect and recording visual and auditory information in a highly systematic fashion for all birds that can be detected. Accurate distance records of bird signs are an important feature of the method. The number of detections and the distances allow for a rough density estimate of the bird populations. This quantity (abundance) and the number of times a bird species was encountered in the same belt transect at each repeated survey (frequency) was used to arrive at a quantitative estimate of each bird population along the Mauna Loa transect.

Bird Zonation Patterns

Twenty-two bird species were recorded along the Mauna Loa transect. Community analyses with these twenty-two species on the seven repeatedly surveyed focal sites resulted in zonation or community patterns similar to those derived from the vegetation data. Application of the Sørensen community coefficient based on presence-absence data gave the most satisfactory result, with six well-defined bird communities along the Mauna Loa transect. These are indicated as transect zones I through VI in the top line on the bird species distribution diagrams (figures 3-7 and 3-8). Further details of the bird community analysis are reported in TR 74. These bird zonation patterns could be named by those bird species that are important for differentiating these six bird communities, but because of their similarity to most of the vegetation zones, it is expedient to retain the vegetation names for reference purposes. The content or species diversity of each of the six bird zones becomes evident from the species distribution diagrams and following discussion.

Species Distribution Trends

Of the twenty-two bird species observed along the Mauna Loa transect during this study, seventeen are represented on the species distribution diagrams. Figure 3-7 shows the quantitative distribution of one indigenous *(Pluvialis dominica)* and eight endemic birds, and figure 3-8 that of eight exotic birds. These were the quantitatively more important bird species, whose frequency and abundance permitted the establishment of rather definite distribu-

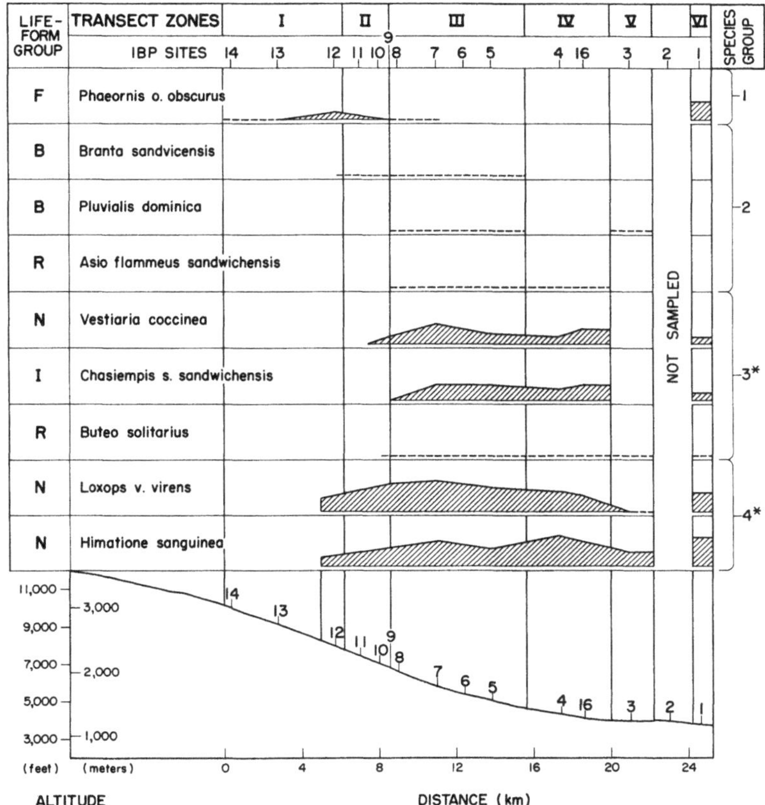

FIGURE 3-7. *Distribution of native bird species along Mauna Loa transect. Transect zones were established with the Sørensen index. Life-form groups: F = fruit-insect eater; B = browser; R = raptorial predator; N = nectar-insect eater; I = insect eater. Abundance scale (birds/40 ha) in units: maximum is 7 units (example,* Himatione *at site 4) = 160–300 birds, 6 units = 120–160, 5 = 70–120, 4 = 30–70, 3 = 15–30, 2 = 5–15, 1 = 2–5, dashed line = one bird per 40 ha. Numbers with asterisks are computer-generated species groups.*

tion trends. The five bird species of lesser importance were Erckel's francolin *(Francolinus erckelii),* the spotted dove *(Streptopelia chinensis),* the barred dove *(Geopelia striata),* the barn owl *(Tyota alba),* and the common Indian myna *(Acridotheres tristis).* These five are introduced species. No trends can be established as yet from their sporadic sightings, which were made primarily in the mountain parkland (zone III on the diagrams). The diagrams allow the recognition of spatially grouped and spatially unique bird species by their coincidence in distributions. Bird abundance (in terms of approximate numbers of individuals per 40 ha) is indicated by the height of the curves.

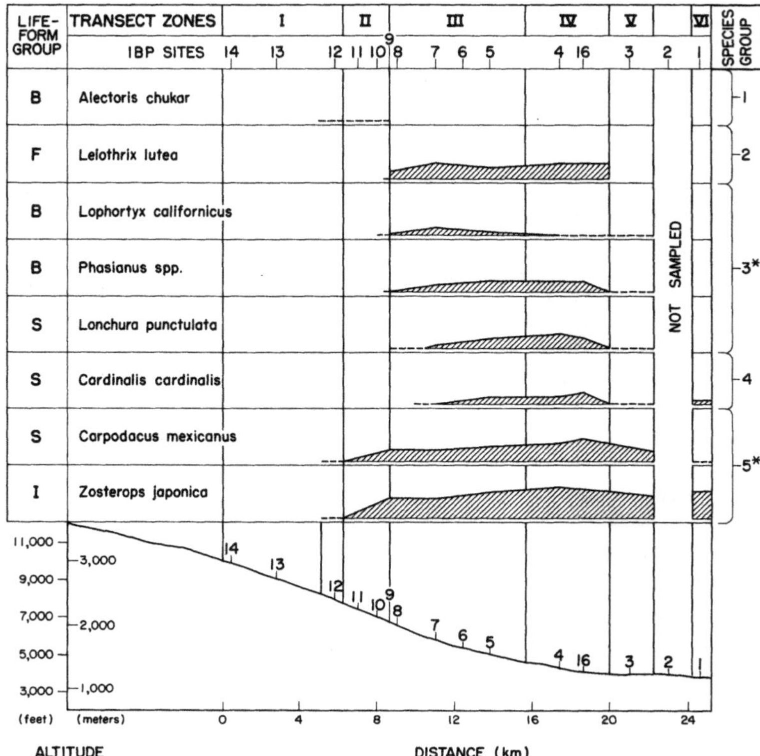

FIGURE 3-8. *Distribution of exotic bird species along Mauna Loa transect. Transect zones were established with the Sørensen index. Life-form groups: B = browser; F = fruit-insect eater; S = seed eater; I = insect eater. Abundance scale (birds/40 ha) in units: maximum is 7 units (example,* Zosterops *at site 4) = 160–300 birds, 6 units = 120–160, 5 = 70–120, 4 = 30–70, 3 = 15–30, 2 = 5–15, 1 = 2–5, dashed line = one bird per 40 ha. Numbers with asterisks are computer-generated species groups.*

On this basis, four species groups can be recognized among the native birds in figure 3-7. Their group numbers are shown on the right-hand side of the diagram. Species groups 3* and 4* were selected as spatially grouped by the computer in the two-way table analysis.

The 'ōma'o or Hawaiian thrush *(Phaeornis obscurus obscurus)* shows a unique distribution and therefore is in a group by itself (group 1). It was the only bird found with any population (5–15 birds per 40 ha) above the tree line (site 12) in the treeless sparse alpine scrub (site 13 and even sparsely at site 14). Moreover, the 'ōma'o was only rarely seen in the upper mountain parkland (zone III), and it was absent throughout the lower mountain parkland, the savanna (zone IV), and *Metrosideros* dry forest (zone V). But it reappeared with considerable abundance (30–70 birds per 40 ha) in the closed rain forest (zone VI).

Group 2, here shown as including three species, is a spatially rather heterogeneous group of open area birds. Nēnē, the Hawaiian goose *(Branta sandwicensis)*, occurred from the tree line (site 12) throughout the subalpine scrub forest (zone II) and the mountain parkland (zone III). The golden plover or kōlea *(Pluvialis dominica)*, a migratory and therefore indigenous bird, was seen particularly in the mountain parkland (zone III) but also in the rather open *Metrosideros* dry forest. The tall grass in the savanna (zone IV) may have prevented its presence there, since the golden plover is known to prefer short grassland. The Hawaiian owl or pueo *(Asio flammeus sandwichensis)* was seen throughout the mountain parkland and savanna (zones III and IV) but not in the *Metrosideros* forests above (zone II) or below (zone V).

Group 3 includes three endemic species of wider distribution that ranged from the lower part of the subalpine scrub from near the upper border of the mountain parkland to the rain forest. However, the ʻiʻiwi *(Vestiaria coccinea)* and the Hawaiʻi ʻelepaio *(Chasiempis sandwichensis sandwichensis)* showed a distribution interrupted by the *Metrosideros* dry forest (zone V), while the Hawaiian hawk or ʻio *(Buteo solitarius)* was found throughout this transect range. This indicates a difference in general behavior and feeding habits of these species. In the mountain parkland and savanna area, the ʻiʻiwi and ʻelepaio were found associated mostly with *Acacia koa* tree colonies or other forest groves with closed canopies. Such tree-stand structures were absent in the open *Metrosideros* dry forest (zone V), where trees are narrow crowned, have generally sparse foliage, and grow singly rather than in groves. The Hawaiian hawk is not limited by such patterns of fine structure of the tree vegetation since it occurred in open (zone V) as well as closed (zone VI) tree vegetation.

Group 4 includes the two most abundant native honeycreepers, the Hawaiʻi ʻamakihi *(Loxops virens virens)* and the ʻāpapane *(Himatione sanguinea)*. Their wide-ranging distribution from the tree line (site 12) to the closed rain forest (site 1) shows that they have considerable environmental tolerance. At least they occur over a rather wide range of temperature and rainfall regimes. Between the two species, the ʻamakihi may have a somewhat greater dependence on closed tree canopy structures than the ʻāpapane. This is indicated by the reduction in density of the ʻamakihi (25 birds per 40 ha) in the open *Metrosideros* dry forest (zone V) relative to the much more abundant (70–120 birds per 40 ha) ʻāpapane in this ecosystem. Certainly the most tolerant native Hawaiian bird in terms of temperature and rainfall regime variation along the Mauna Loa transect appears to be the ʻōmaʻo.

Five spatial groups are identified among the more common exotic birds on figure 3-8. The Chukar partridge *(Alectoris chukar)* was seen with low abundance (less than 1 bird per 40 ha) only in the subalpine scrub forest (zone II) above the mountain parkland. It was the bird species with the most restricted amplitude. Its distribution may indicate a beginning naturalization here or merely a transitory presence. The Chukar partridge probably is not yet offering any serious competition to the endemic browsing bird, the nēnē, in this habitat. The leiothrix *(Leiothrix lutea)* also showed a unique distribution. It was here restricted to the mountain parkland and savanna, a grassy terrain with bush and tree groups.

Group 3, a computer-generated spatially correlated species group, was formed by the California quail *(Lophortyx californicus)*, the pheasant (*Phasianus* sp.), and the spotted munia *(Lonchura punctulata)*. Like the leiothrix, these birds were found throughout the mountain parkland and savanna, mostly in the grassy terrain, but also in the open *Metrosideros* dry forest (zone V). In this forest they occurred only sparsely (approximately 1 bird per 40 ha).

The remaining three exotic birds extend into the closed rain forest (zone VI). The American cardinal *(Cardinalis cardinalis)* is shown as a unique (one species) group in figure 3-8. Actually it is spatially associated with the computer-derived group 3 in figure 3-7, the Hawai'i 'elepaio, the 'i'iwi, and the Hawaiian hawk. However, because the American cardinal is primarily a seedeater, it occupies a new niche, not filled by any of these three endemic birds.

Group 5 includes the house finch *(Carpodacus mexicanus)* and the Japanese white-eye *(Zosterops japonica)*. Along the Mauna Loa transect, these two introduced birds now occupy the same range as the two endemic honeycreepers, the Hawai'i 'amakihi and the 'āpapane. However, they also belong to different life-form groups. The two endemic birds are nectar and insect feeders, while the house finch is primarily a seedeater and the Japanese white-eye an insect feeder. Comparison of the diagrams shows that all four species are among the most common birds in the mid-range of the transect (the mountain parkland and savanna). At the extremes they segregate quantitatively. The two endemic birds are quite abundant throughout the subalpine scrub (with 15–30 birds each per 40 ha), up to the tree line (site 12), while the two exotic birds are seen only sporadically in this area. They stop, more or less, short of the tree line. Although all four species overlap also in the closed rain forest at the other transect end, the house finch is rather sparse in this habitat. Along the Mauna Loa transect, only the Japanese white-eye may infringe on the habitat resources shared by the endemic forest birds, the Hawai'i 'amakihi and the 'āpapane.

Discussion and Conclusions

The patterns of bird distribution raise several questions. The most important are these:

1. Is the Mauna Loa transect made up of sufficiently distinct areas that are large enough to allow spatial segregation of the bird species?
2. If so, what aspect of the bird distribution patterns can be explained as caused by environmental factors, including the vegetation?

3. What sort of relationship exists, if any, among the bird species that overlap or are grouped along the Mauna Loa transect?

The Mauna Loa transect, 25 km in length, ranges from 1,190 to 3,050 m and passes through vegetation types similar in kind and range to those encountered by Terborgh (1971) and Terborgh and Weske (1975) in their studies of bird distribution along altitudinal gradients in the Peruvian Andes. Terborgh's (1971) study area traversed 3,000 m elevation on a 40 km transect, and Terborgh and Weske (1975) conducted studies on an elevational gradient of 1,600 m over an unspecified length. Data collected in all three studies (including this one) indicate that distinct patterns of bird distribution emerge and can be shown to be related to the distribution of the major vegetation types that occupy large expanses along the transects.

Patterns of bird distribution on the Mauna Loa transect are obviously correlated to environmental factors, including the vegetation. This can be interpreted from both figures. For example, in figure 3-7, the upper limits of *Himatione* and *Loxops* end abruptly at the tree line. No birds of these species were seen above the tree line, while below they occurred with high densities. Similarly *Vestiaria* and *Chasiempis* exhibited abrupt density changes where the savanna vegetation merged with the open *Metrosideros* dry forest (boundary between zones IV and V). In contrast, in figure 3-8 the two exotic birds (*Carpodacus* and *Zosterops*) show more gradual density changes at the tree line, which indicates that the vegetation change at this point is not so discriminating a factor as in the case of *Himatione* and *Loxops*.

Species whose densities fall gradually as one or both distribution limits are approached appear to be responding more strongly to factors other than vegetation structure because their density changes do not correspond closely to the changes in vegetation structure. This applies to at least one distribution limit in the following ten species: *Phaeornis, Vestiaria, Loxops, Chasiempis, Lophortyx, Phasianus, Lonchura, Cardinalis, Carpodacus,* and *Zosterops*. Such other factors could be gradually decreasing tolerance to physical environmental changes, food limitations, or, in the case of some exotic birds, incomplete naturalization. The last may apply, for example, to the upper limit of *Cardinalis* in the middle of the mountain parkland and to the limited distribution of *Alectoris* in the alpine zone. Food limitations are probably responsible for the distribution patterns of the two native predatory birds, *Asio flammeus sandwichensis* and *Buteo solitarius*. Their main distribution center is in the mountain parkland and savanna, where the rodents were also most abundant. Lack of tolerance for cool alpine environments seems to be indicated as a cause for the upper distribution limits of *Carpodacus* and *Zosterops*.

In a study of the distribution of Peruvian forest birds along an altitudinal gradient, Terborgh and Weske (1975) found that competition and physical environmental factors were more important in limiting bird distribution than were abrupt vegetation boundaries. Although examples of species whose distribution limits are imposed by competitive exclusion cannot be found on the Mauna Loa transect, diffuse competition (defined by Terborgh and Weske, 1975, as competition among species exploiting a common resource) may influence species densities on the Mauna Loa transect where competitors occur together. For example, *Himatione* and *Loxops* compete for nectar in *Metrosideros* flowers. These species overlap in their ranges but show a significant density difference in zone V, the open *Metrosideros* dry forest. It seems that *Himatione* is the more successful competitor for *Metrosideros* flowers where they are of low abundance (as in the open forest of zone V). This interpretation is supported by Carpenter and MacMillen (TR 63), who found *Himatione* a successful defender of nectar feeding grounds against *Loxops*.

One other native bird, *Vestiaria*, and one exotic bird, *Zosterops*, also exploit *Metrosideros* nectar. Carpenter and MacMillen (1976a) reported that *Vestiaria* defends its feeding territory by excluding its competitors. This observation seemingly contradicts the data in figure 3-7. However, it is quite possible that the feeding territory of *Vestiaria* refers to a mosaic of small-area patterns (a few trees per territory), which would not be reflected in this relatively large-area analysis. The other three species reduce competition by shifts in foraging strategies (according to Carpenter and MacMillen, TR 63). For example, where they occur together, *Loxops* feeds primarily on insects and *Himatione* on *Metrosideros* nectar, whereas in the absence of *Himatione*, *Loxops* exploits nectar heavily.

Competition for food among the four nectar-insect feeders apparently does not affect their distribution limits. However, where nectar is limited, species densities may be affected in favor of the more aggressive competitor (for example, *Himatione* over *Loxops* and both over *Vestiaria* in zone V).

Perhaps the most puzzling distribution pattern is that of *Phaeornis*, which is characteristically a rain forest species (Berger, 1972:105). Only on Mauna Loa has it been found in subalpine scrub and at the tree line, as shown in figure 3-7. This bimodal distribution may be explained by competition for food from *Leiothrix*, the only other frugivore on the transect (figure 3-8). It appears that *Leiothrix* displaces *Phaeornis* along the mid-section of the transect. However, both species occur together in zone VI. Baldwin (1953) observed both species twenty years ago in the same rain forest area.

Although there are five browsers (the native birds *Branta* and *Pluvialis* and the exotic birds *Alectoris*, *Lophortyx*, and *Phasianus*) and three seed eaters (the exotics *Lonchura*, *Cardinalis*, and *Carpodacus*) which occur in the same transect zones, they probably do not compete for food. Food preferences among these browsers have been reported to differ (Woodside, 1958). The three seed eaters have different bill sizes, which suggests that they do not compete for food (Hespenheide, 1966; Willson, Karr, and Roth, 1975). However, since they are exotic species, they may not behave in the same way in this environment as in those in which they evolved.

Only three of the exotic bird species along the Mauna Loa transect have invaded the closed rain forest, and only one of these *(Zosterops)* with considerable abundance. Thus the exotic bird distribution pattern is similar to that of the exotic plants insofar as most of the latter also have their ecological optimum in the mid-section of the transect.

A major conclusion from this study is that changes in vegetation structure influence the bird distribution along the transect profoundly and that competition appears of little importance in controlling bird distribution limits. This finding stands in contrast to that of Terborgh and Weske (1975), who concluded that most of the bird species distribution limits along a similar altitudinal gradient in the Peruvian Andes were controlled by competition. This difference may relate to the continental setting of the Peruvian transect, where bird species diversity was very high. In contrast, bird species diversity is rather low along this island ecosystem gradient.

RODENTS*

Four species of commensal rodents (Muridae) are established in Hawaii. They were all introduced by humans. No rodents are known to have reached the islands by natural dispersal such as rafting. These rodents are variously adapted to urban and rural dwelling places, as well as to cultivated lands and wilderness habitats. Another characteristic of these mammals is their wide distribution in the world, which demonstrates a high degree of preadaptation for reaching and exploiting new territory.

The Polynesian rat *(Rattus exulans)* is an ancient arrival that traveled in vessels of Polynesian settlers and their forebears as they colonized Pacific islands by eastward migrations from homelands in Southeast Asia. The roof rat *(Rattus rattus),* Norway rat *(Rattus norvegicus),* and house mouse *(Mus musculus)* are derivatives of stocks that had invaded Eurasia and the Americas in generally westward dispersals, along the ancient terrestrial and marine routes of commerce, from centers of evolutionary development in Asia. The Hawaiian Islands were colonized only recently by unspecified sources of these stocks, after European discovery of the islands in 1778. Johnson (1962), Tomich (1969a), and Berry (1970) have reviewed these subjects.

On the island of Hawai'i, rodents generally are most abundant in lowland areas where they have become a public health concern and a factor in the economics of sugarcane production (Kartman and Lonergan, 1955; Tomich, 1970). For *R. rattus,* at least, there is evidence of adaptation to local habitats through natural selection, reflected across the island in clinal distributions of phenotypes for coat color (Tomich, 1968). *R. exulans* is almost exclusively a lowland species and is seldom a strict commensal. *R. norvegicus* thrives best in close association with humans wherever foodstuffs are in surplus. *M. musculus*

---

*By P. Q. Tomich

is the most widespread species and has lived as high as 3,780 m, near the summit of Mauna Kea.

Sampling Layout

Rodents were collected by systematic removal trapping covering all of the IBP sites. Nine traplines with thirty-nine traps each were established between 1,190 m (IBP site 1, rain forest) and 2,440 m (IBP site 12, tree line). There were twenty-six cage-style traps and thirteen rat-size snap traps in each line; every third trap was a snap trap. Spacing of traps was at 15 m along the contour of the land. Sampling at each site covered a subtransect 570 m in length. Squares of fresh coconut were used as bait. Each line was run for three consecutive nights per quarter, with the trapping effort (setting and checking traps) distributed evenly during four days each month for two annual cycles, beginning in October 1971. Some lines were run for one year only, others during both years of sampling. Thus each line represents one or two standard annual samples resulting from 468 trap nights (39 traps run for three nights each quarter).

Above the tree line, at 150 m altitudinal increments between 2,590 m and 2,900 m in the sparse alpine scrub (IBP site 13), an additional three lines were run during only the second year of the study, October 1972 through September 1973. These lines each consisted of nine traps: three rat-size snap traps, three mouse-size snap traps, and three Sherman metal box traps. Traps were spaced at 15 m and arranged in a repeated sequence by type. Bait was a mixture of peanut butter and rolled oats. These modifications in technique were required by the necessity of overnight packing to the sites above 2,440 m. Traps were set quarterly on the way up to the Red Hill cabin at 3,050 m (IBP site 14) and checked the next day on the return trip; then the traps were left open and untended until the next quarterly expedition to Red Hill. The bait remained effective for an undetermined period after the first-day check, and rodents caught thereafter were recovered in mummified condition three months later. Each of these three traplines covered a subtransect of 120 m. Because of the prolonged trap exposure, trapping effort was probably comparable to that on the lower sector of the mountain. The line and procedures at 3,050 m were the same except that traps were placed only about the cabin and closed after the first night of trapping.

All rodents caught were transported daily to the laboratory for identification and processing. Vegetation descriptions of the subtransects are available from TR 48. Transect zones are those defined by the vegetation analysis.

Species Distribution Trends

The sampling results are plotted in figure 3-9. Only three of the four species of rodents were captured along the Mauna Loa transect. *Rattus norvegi-*

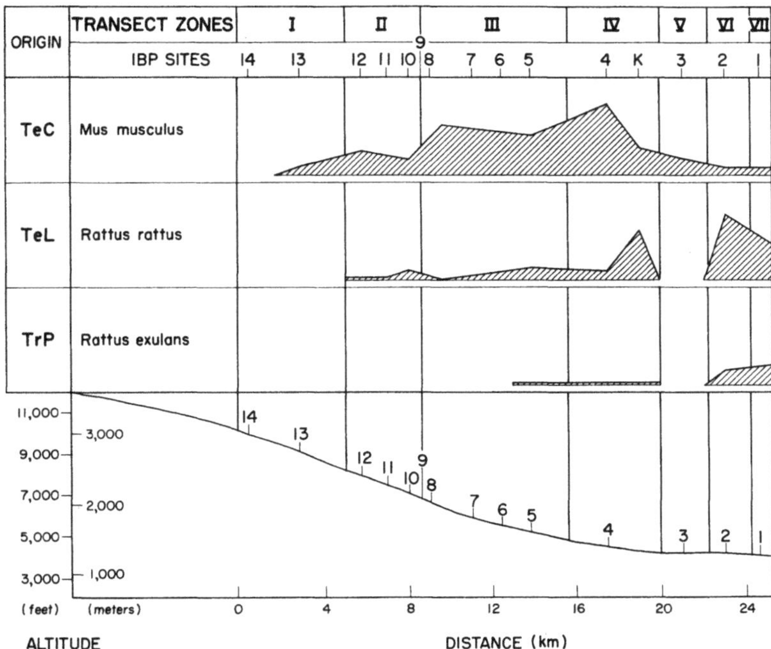

FIGURE 3-9. *Distribution of three rodent species sampled along Mauna Loa transect. Abundance scale in equal units of 5 animals trapped per year: Maximum is 9 units = 41–45 animals (*M. musculus *at site 4), 8 units = 36–40 animals, 7 units = 31–35, 6 = 26–30, etc., horizontal line = 1 animal. K = closed kīpuka forest in savanna zone IV. Origin: TeC = temperate-continental Asia, TeL = temperate-littoral Asia, TrP = tropical-Pacific Asia.*

cus has been recorded from near the lower end of the transect (site 2) in earlier work. At that time it was found only sparsely in the vicinity of the National Parks buildings in habitats lacking natural vegetation. *R. norvegicus* was not recorded during 1971 to 1973. Numbers of the three species captured varied considerably between the two trapping years. Both *R. rattus* and *M. musculus* were more abundant in the second year. The variation was well within the expected range of fluctuations for field rodent populations and does not interfere with the spatial assessments made here. For this reason the two-year rodent sampling was averaged and the spatial population trends plotted in the figure can be considered as the normalized rodent ranges and densities for the study years.

The diagram shows that the three introduced rodent species do not form a spatially associated group but overlap broadly in their ranges along the Mauna Loa transect. Also their population modes occur at different places. Clearly the most abundant species is the common house mouse *(M. musculus)*. This small rodent ranged from the alpine zone with its very sparse scrub vegetation at the upper limit of site 13 (from 2,900 m) continuously downslope into the closed rain forest (site 1 at 1,190 m). From here on it is found, as are the other two rodent species, downslope to sea level. Trapping near the higher alpine site 14, above 3,000 m at the Red Hill cabin, was quickly identified as a futile exercise upon discovery of a bin of stored barley there, easily accessible to rodents but untouched by them. In general, the distribution of mice resembles a bell-shaped curve indicating a broad peak in the middle section of the transect. This broad population mode coincides with an important structural vegetation change along the transect. From sites 4 to 8 the vegetation is characterized by closed herbaceous or grass cover with scattered trees (savanna, zone IV) or tree groups (mountain parkland, zone III). The tree groups (or *Acacia koa* tree colonies) in the mountain parkland usually form a mosaic together with small-area tall-shrub communities within the grassland matrix. The habitats at both ends of the bell-shaped mouse population curve are characterized by a predominance of woody vegetation. The rather abrupt decline within the savanna zone (IV), from site 4 to site K,* correlates with a vegetation change from a grass-covered site with scattered trees to a closed mesic kīpuka forest. The rather abrupt population change from near site 8 to site 10 correlates with the vegetation change from the mountain parkland with closed grass cover patches to the open subalpine scrub forest, which has almost no herbaceous undergrowth. Thus the mice clearly show their ecological optimum in vegetation with sun-exposed grassy habitats, and their population levels are lower in closed, shaded forests (site K and site 1 in particular) or in open, woody vegetation lacking any significant herbaceous undergrowth (zones I and II).

The roof rat *(R. rattus)* was trapped only up to the tree line (site 12). It occurred sparsely throughout the subalpine scrub forest (zone II), the mountain parkland (zone III) and the savanna proper at site 4. In the closed kīpuka forest within the savanna zone, it showed an abrupt increase in abundance (at site K). The roof rat was absent (or so sparse that it was not trapped) in the open *Metrosideros* dry forest (site 3) and showed a second peak in the rain forest along the transect (sites 1 and 2). Thus, the roof rat has its peak abundance along the transect in closed forest vegetation or in open forest with dense fern undergrowth (open rain forest site 2). Thus although *M. musculus* and *R. rattus* overlap broadly along the transect, they seem to displace each other in their respective ecological optima.

The Polynesian rat *(R. exulans)* clearly displayed its upper distribution limit along the transect. It was trapped with greater abundance below the transect in warmer rain forest habitats. Along the transect it was very rare above the rain forest and extended only about halfway into the mountain parkland to about 1,500 m altitude. Like the roof rat, the Polynesian rat was not trapped in

---

*Closed kīpuka forest

the open dry *Metrosideros* forest, which may lack sufficient food resources for both rat species.

Discussion and Conclusions

Each member of the complex of rodents found on the Mauna Loa transect requires fulfillment of two basic needs: an adequate supply of food and shelter from heat, cold, excessive moisture, and drought. When these two factors of food and shelter are favorable at any site, we can expect with some probability that a particular species will reside there. Predation was not considered in this study. The cooler temperatures at high altitudes and the more exposed substrates found there are not suspected of imposing directly limiting conditions on the versatile exotics that originated in the temperate zone.

M. musculus was the only species that occupied the sparse alpine scrub of transect zone I. Here it is able to build secure nests in the pāhoehoe lavas and apparently to survive on a lean supply of *Vaccinium* and *Styphelia* fruits and on insects associated with the vegetation. Low temperatures at this altitude are unlikely to be a detriment to this species, some populations of which thrive in cold storage lockers at −15°C (Laurie, 1946). Food is perhaps the single limiting factor for *M. musculus* in zone I.

The subalpine scrub forest of zone II appears to impose the same limitation on *M. musculus,* but to a lesser degree. The zone seems to permit the bare existence of *R. rattus* up to the tree line. Both species, at the 2,130 m level (IBP site 10), fed on fruits of *Coprosma ernodeoides* and a variety of insects. Low temperatures may impose some hardship on *R. rattus* because the species is adapted primarily to tropical and coastal temperate climates. The record high elevation for the rat in Hawai'i is 3,000 m (in a heated observatory building) at Haleakalā National Park on Maui. Scarcity of food for a mammal of this size may be the most restrictive factor in zone II.

R. rattus was similarly sparse in numbers in the mountain parkland of zone III, where the climate continues to be relatively dry and cool. *M. musculus* here attains a moderate density and feeds predominantly on fruits of *Styphelia,* insects, and seeds of grasses. Addition of abundant grasses to the vegetation is of suggested importance to *M. musculus,* probably primarily for shelter but also for food.

The single specimen of *R. exulans* in the mountain parkland zone IV, at 1,500 m, represents a record for altitude for this species in Hawai'i and reflects the generally restrictive warm-tropical distribution of this form. Conditions in zone IV may duplicate in part those at Stewart Island (New Zealand) at 47°S, the extreme latitudinal range of *R. exulans* (Watson, 1956). This rat may be unable to survive low ambient temperatures even where shelter for other murids is adequate. *M. musculus* was abundant in the savanna habitat of zone IV. A variety of fruits and insects and grass seeds figures prominently in its diet. *R.*

rattus was common in the closed kīpuka forest of the savanna zone IV, where a variety and abundance of fruits is available in a lush vegetation.

The site of trapping for zone V, in the open *Metrosideros* dry forest with sparse low shrubs, shows an unexpected absence of the *Rattus* species. The substrate of raw ash-covered pāhoehoe apparently does not produce sufficient food materials for rodents of the larger sizes. Ten *M. musculus* were caught in the 468 trap nights at this site. Insects and peels of *Vaccinium* fruits were identified in stomach contents.

Zones VI and VII (the rain forests) at about 1,200 m were similar as providers for rodents. Here *M. musculus* was reduced to a token population, *R. exulans* maintained a token population, and *R. rattus* responded well to the prevailing rain forest conditions. Fruits of *Ilex, Myrsine,* and *Perrottetia* and the pith of *Cibotium* are all prominent foods of *R. rattus*. *M. musculus* ate predominantly insects, and *R. exulans* ate a combination of fruits and insects.

The complete absence of *R. norvegicus* from the transect sites may be accidental. Four specimens were collected during two years of intensive trapping in the *Acacia-Metrosideros-Cibotium* rain forest site of Kīlauea Forest Reserve (where *R. rattus* is abundant and *M. musculus* is scarce, as in zones VI and VII). The Kīlauea Forest Reserve is at 1,600 m, only slightly lower than the high altitude record of 1,770 m for *R. norvegicus* in Hawai'i. *R. norvegicus* is clearly a temperate zone rat, able to withstand severe winters of coastal Alaska as far north as 64°30′, at the cost of frostbitten ears, feet, and tails (Schiller, 1956). Thus it is not the coldness or wetness of Hawaiian uplands that restricts *R. norvegicus* but more likely the absence of human waste foods.

ECTOPARASITES OF RODENTS*

Ectoparasites were taken from the three principal, introduced species sampled along the Mauna Loa transect: *Mus musculus, Rattus rattus,* and *Rattus exulans*. The three host species are widely distributed outside the Hawaiian Islands, as are most of their parasites recorded from Hawai'i. Taxonomically the parasites that were analyzed in this study are included in the class Insecta and the subclass Acari (mites) of the class Arachnida. The insects represent two orders: two species of Anoplura (lice) and two species of Siphonaptera (fleas). The Acari also represent two orders: three species of nest mites in the suborder Mesostigmata of the order Parasitiformes and three species of fur mites in each of the suborders Prostigmata and Astigmata of the order Acariformes.

---

*By F. J. Radovsky and J. M. Tenorio

There have been many other studies of the distributional association of the parasites of these rodents, including several in Hawai‘i (Eskey, 1934; Mitchell, 1964a, 1964b; Haas, 1969). This study differs from previous ones both in using an altitudinal transect and in choosing sites that are not influenced by humans. The second difference is particularly significant since the host species that were most frequently sampled, *M. musculus* and *R. rattus*, are most often observed as commensals of humans. Sampling along an altitudinal gradient that does not have human-created habitats has permitted an analysis of the other major ecological factors that influence the ectoparasite distributions.

Sampling Procedure

The sampling for the ectoparasites and the collection of the rodents were combined, with most of the rodents trapped in the rodent study being examined for ectoparasites. A representative sample was examined in some cases where many rodents were taken at a single site in one night, and some rodents in very poor condition were not examined for ectoparasites. Of those examined, the host condition at the time of parasite recovery was evaluated. Since the hosts in poor condition (for example, those overrun by ants or partially eaten by predators) were found to have a generally lower mean parasite incidence and intensity, hosts in this condition category were not included in the analysis. This amounted to less than 10 percent of the total number of hosts examined.

The locations and frequency of sampling on the Mauna Loa transect have been given. In addition, samples were taken from rodents trapped in the Kīlauea Forest and on five sites below the transect, in the elevation range of 840 to 900 m; they included sites exhibiting marked contrasts in the amount of annual rainfall (wet sites and dry sites). The Kīlauea Forest and one of the sites below the transect were sampled for two years. The remaining four sites below the transect were sampled only over a one-year period.

The number of rodents that were sampled for inclusion in this analysis along the transect were as follows: *Mus musculus*, 188; *Rattus rattus*, 123; and *Rattus exulans*, 8. The additional sampling sites provided the following samples: *M. musculus*, 182; *R. rattus*, 299; and *R. exulans*, 29.

The field processing of the rodents involved removing each animal from the trap and placing it individually in a plastic bag with a chloroform-soaked cotton. In the laboratory, the rodents were washed in a detergent solution. The washing procedure was standardized to achieve consistency in sampling efficiency. This provided a high rate of recovery for most parasite species. For example, respective numbers of two species of mites in the genus *Laelaps* were comparable to recovery with a skin-digestion method used in Hawai‘i by Mitchell (1964b). Recovery of some fur mites and of lice was less reliable by the washing method but sufficient to show distributional trends and relative incidence of infestation.

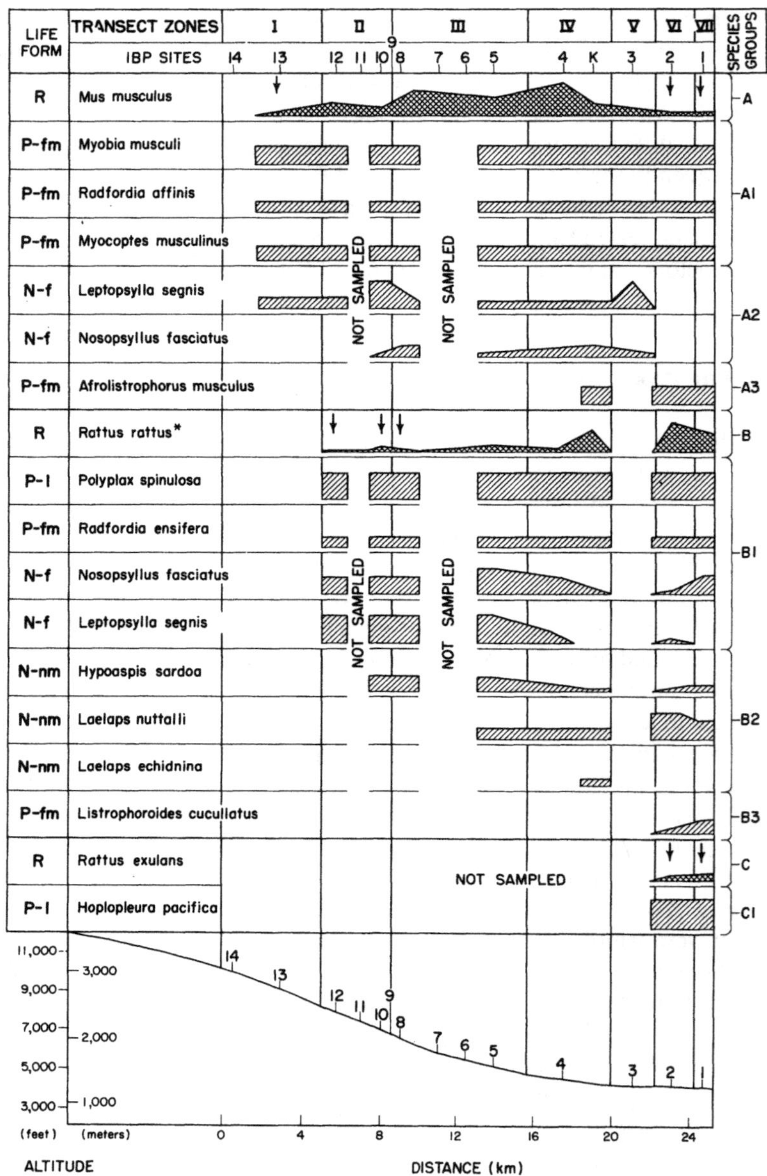

↓ LESS THAN FIVE RODENTS EXAMINED FOR PARASITES

* ALL PARASITES SHOWN FOR *Rattus rattus* WERE COLLECTED AT COMPARABLE FREQUENCIES FROM *Rattus exulans* AT THE LOWER END OF THE TRANSECT AND BELOW THE TRANSECT WHERE THAT SPECIES OCCURRED.

Species Distribution Trends

A total of twenty-six species of arthropod parasites and commensals was recovered from rodents in this study. Only thirteen selected species are treated here. For information on other species and additional data on all species, readers are referred to TR 58, Radovsky et al. (1979) and Tenorio and Goff (in press).

The thirteen parasite species selected for this treatment were those that were recovered with sufficient frequency or numbers of individuals from the rodent samples along the transect, so that their distribution trends could be portrayed with some confidence as shown in figure 3–10. The smallest sample along the transect was one rat at site 12. Here the numbers of individuals of the four parasite species found were considered great enough so that the frequency estimate of the lower sites, where more than four rats were sampled, was extrapolated to site 12. Transect sites with fewer than five rodents examined are indicated by an arrow above the respective rodent-host curve in figure 3–10. The primary basis of the parasite curves in the figure is parasite frequency, determined by the number of times the parasite was recovered from the total number of examined rodents of the host species. However, where fewer than five rodents were examined at a site, the height of the parasite curve relates to a frequency estimate based on results at proximal sites. Furthermore, curve height for fixed or permanent (P) parasites (the two species of lice, P-1, and the six fur mites, P-fm included) is shown as a constant level (within the respective distributional range) equivalent to the mean frequency for that species throughout the transect. These parasites, adapted for clinging firmly to hairs, were recovered irregularly by the washing technique that was employed, so that use of the frequency counts for individual sites to construct a curve would be misleading.

On *Mus musculus,* ectoparasite species group A1 is clearly associated with the entire range of its host on the transect—from the closed rain forest (site 1) to the cool alpine scrub desert (site 13). These three fur mites (*Myobia*

FIGURE 3-10. *Distribution of thirteen ectoparasite species on three rodent hosts along Mauna Loa transect. Curve heights for R = rodent species (A, B, C) were halved (from figure 3-9). Ectoparasite quantities are shown in percentage frequency (number of times species was recovered on total number of host specimens examined). One curve height unit indicates 10 percent frequency. Maximum frequency was 80 percent exemplified by* Hoplopleura pacifica. *Life form symbols for ectoparasite species: P = permanent parasite, N = nest parasite, fm = fur mite, nm = nest mite, f = flea, l = louse.* Rattus exulans *was sampled only at sites 1 and 2. Sites with fewer than five rodents examined are indicated by an arrow above each rodent curve. K = kīpuka forest, an additional sampling site in the savanna zone IV.*

musculi, Radfordia affinis, and *Myocoptes musculinus*) are permanent ectoparasites, and it was expected that they would show such an even distributional correlation with the host. In contrast, the species in groups A2 and A3 are not associated with this host over its entire range. Group A2 is comprised of two fleas, of which *Leptopsylla segnis* is considered to be primarily a parasite of mice and *Nosopsyllus fasciatus* primarily of rats. The absence of both fleas from the rain forests (at sites 1 and 2) may be attributable to the small number of samples there (fewer than five mice examined). At the higher elevation end of the transect, the two flea species indicate a distributional difference. *Leptopsylla segnis* was found to be associated with mice to their upper distribution limit in the sparse alpine scrub (site 13), while *Nosopsyllus fasciatus* occurred on mice only up to site 10 in the subalpine scrub forest. This spatial difference seems to be more apparent than real, because the same two flea species were recovered from *Rattus rattus* at the tree line site 12, which formed the upper distributional limit for this second host species. *Nosopsyllus fasciatus* (the rat flea) occurred with much lower frequency on mice at site 10 than *Leptopsylla segnis* (the mouse flea). Therefore the absence of *Nosopsyllus fasciatus* on mice at sites 12 and 13 may be a sampling error caused by the probable low frequency of this flea on mice at these sites. The two fleas are nest parasites, and a total distributional association with their hosts was not expected.

A very different distribution is shown by *Afrolistrophorus musculus,** which was found associated with mice only in the rain forests (sites 1 and 2) and in the closed kīpuka forest (site K). This fur mite, like those of group A1, is also classified as a permanent ectoparasite. One would have expected a distribution similarly paralleling that of the host. The disappearance of this fur mite from its host at higher elevations, therefore, was a surprise. The transect distribution strongly indicates that this fur mite is associated with mice only in warmer climates.

Eight quantitatively important ectoparasites were recovered from the roof rat *(Rattus rattus)*. All eight species were recovered with comparable frequency from the Polynesian rat *(Rattus exulans)* at the lower end of the transect or at other sites below the transect. Among these rat-associated ectoparasites, four species extended over the entire host range or nearly so. These are combined here in group B1. Two of these are classified by their life form as permanent parasites; one is a louse, *Polyplax spinulosa* (the spined rat louse), and the other a fur mite, *Radfordia ensifera*. The two other species of spatial group B1 are again the flea species *Nosopsyllus fasciatus* and *Leptopsylla segnis,* which were recovered also on mice (group A2). At the lower end transect sites their distributions differ. As figure 3-10 (group B1) shows, *Leptopsylla segnis* was absent on rats in the closed kīpuka forest (site K) and in the closed rain forest (site 1), whereas *Nosopsyllus fasciatus* was well represented at both sites. This indicates a differentiation by habitat. Further evidence was obtained by the

---

*Recently transferred from *Listrophorus* by Fain (1970)

sample sites below the transect, from which it became clear that the mouse flea *(Leptopsylla segnis)* shows its ecological optimum in drier habitats, while the rat flea *(Nosopsyllus fasciatus)* shows its ecological optimum in the wetter (rain forest) habitats. This coincides with the ecological optima of the two rodent species.

Group B2 includes three nest mites (*Hypoaspis sardoa, Laelaps nuttalli,* and *Laelaps echidnina*) with rather individualistic distributions on rats along the transect. They have in common that they extend over fewer sites than does the host range. *Hypoaspis sardoa* ranges from the rain forest to the subalpine scrub forest on *Rattus rattus*. Its absence in the open dry *Metrosideros* forest at site 3 is caused merely by the absence of rats at this site. Its apparent absence at the tree line site 12 may reflect that the sample here consisted of only one rat. Thus further sampling would probably have placed *H. sardoa* in spatial group B1. The situation is quite different with regard to the two other nest mites of the genus *Laelaps*. Their distributions are definitely more host independent and restricted to the lower and warmer part of the transect. Yet their specific distributions are very different. *Laelaps nuttalli* (the domestic rat mite) shows a greater frequency in the rain forest along the transect (sites 1 and 2) than in the mountain parkland and savanna sites (5, 4, K). Sampling at other sites clarified that its ecological optimum is below the transect, where the climate becomes still warmer. In contrast, *Laelaps echidnina* (the spiny rat mite) was restricted to the kīpuka forest along the transect, where it overlapped with *Laelaps nuttalli*.

Group B3 shows only one species, *Listrophoroides cucullatus (= Listrophorus expansus),* that was restricted to the rain forest sites (1 and 2) along the transect. This species, surprisingly, is a permanent fur mite that departed from the strong host-associated pattern that was expected for this life form of ectoparasite in the same manner as the fur mite *Afrolistrophorus musculus* on mice (group A3). Thus these two species belong to the same ecological group in spite of the fact that they occurred on different hosts. Both species are present and generally abundant on their hosts up to about 1,220 m, but they are absent above this altitude, although their hosts extend from a warmer into a cooler climate upslope along the altitudinal transect.

The Polynesian rat *(Rattus exulans)* did not play an important role along the Mauna Loa transect. This is shown by its restricted distribution in figure 3-10, which merely represents an extension from the warmer tropical lowland to its upper limit in the lower mountain parkland (site 5). Ectoparasite samples were taken only from the transect at the rain forest sites, and these were from fewer than five animals per site. The only ectoparasite species found on this rat (but not on *Rattus rattus*) along the transect was *Hoplopleura pacifica,* a louse and permanent parasite, which is common in the lowlands on *R. exulans*. *Hoplopleura pacifica* has been recorded from *Rattus rattus* as well as other *Rattus* species in earlier studies. The high level of specificity found for this louse species in this study indicates the need for further taxonomic study of what may be a species complex.

Discussion and Conclusions

A recognized ecological classification of arthropod ectoparasites separates them into three categories based on the permanency of association with the host and the nest of the host. Permanent parasites, sometimes called host restricted, remain on the host throughout their lives, passing between hosts when they are in close contact. Examples in this study are lice and fur mites. Nest parasites spend part of their time on the host, for feeding and dispersal, but also are found in the host's nest (or den, roost, bedding ground, and so forth) and have at least one developmental stage off the host. Examples in this study are fleas and mesostigmatic mites. Field parasites are more widely dispersed in the host's environment, acquiring a host by waiting for its passing; they feed, usually for a sustained period, and then drop off, more or less at random. Examples are chigger mites and some ticks, but none were involved in this study. It was anticipated that the permanent parasites would be uniformly distributed throughout the range of the host or hosts with which each was associated, while the nest parasites would tend to have distributions more restricted than those of the hosts because of climatic influences on the nest environment.

The two species of lice (*Polyplax spinulosa* and *Hoplopleura pacifica*) displayed precise correlations with host distribution by extending to the upper distributional limit on their respective hosts. Also four of the fur mites (*Myobia musculi, Myocoptes musculinus, Radfordia affinis,* and *R. ensifera*) showed exact distributional correlations. This conforms to the predicted relationship and supports the concept that small, permanent parasites dwelling in the host's pelage are protected from the ambient environment by the homeothermic mechanisms of the host.

In striking contrast and unexpectedly, the two other fur mites, *Listrophoroides cucullatus* and *Afrolistrophorus musculus,* were altitudinally restricted and unable to survive above 1,220 m on the Mauna Loa transect. *L. cucullatus* is a common parasite of *Rattus* species in tropical and subtropical regions. *A. musculus* was described from mice in Hawai'i and has been recorded elsewhere only on the same host in Puerto Rico. Temperature appears to be the factor limiting altitudinal distribution, to our knowledge the first demonstration of such a local effect for a permanent parasite. The mean annual temperature at 1,220 m in the study area was 16°C. Conceivably the effect of temperature could operate through some influence on host physiology. However, we believe that temperature acts directly on these parasites. Both species have a very elongate body form in contrast to the more compact form of the other fur mites in the study. Elongation increases body surface area and hence reduces the ability to conserve heat. It may also bring part of the mite further from the host skin. One possible advantage of elongation is to facilitate movement up and down hairs, and this behavioral tendency would also remove the mites from proximity to the host skin. Thus morphological and behavioral modifications with adaptive value in a mild climate could limit the extension of the parasite ranges into more rigorous climates.

As was expected, the nest mites showed some degree of independence of their hosts. This was borne out by the distributional results shown for the nest mites in group B2 on the diagram. Of these, *Hypoaspis sardoa* became increasingly abundant at higher elevations. The species is known primarily from a variety of small mammals and their nests in cold-temperate regions. Its absence at the tree line site probably was a sampling error, since only one rat was examined there. More significant is its quantitative decrease at the lower end of the transect. At sites below the transect (between 840 to 900 m) it was nearly absent. This suggests that *H. sardoa* cannot maintain itself in the warmer climate at low elevations in Hawai'i and this in turn may explain the absence of any previous record for this species in the state.

In contrast, the nest mite *Laelaps nuttalli* appears to be restricted by the cooler temperatures prevailing at higher elevations. At several sites below the transect, *L. nuttalli* was common in both wet and dry habitats. This species is widely distributed, primarily as a parasite of *Rattus* species, but it has most frequently been recorded in tropical and warm temperate areas. Our results suggest that the presence of *L. nuttalli* in cool-temperate areas is dependent on its living on rodents that use man-made buildings for their habitation. The third nest mite, *Laelaps echidnina,* which along the transect was recovered from rats only in the kīpuka forests, is elsewhere a widely distributed parasite on a variety of rodents. It has been recorded from both the tropics and the cold-temperate zone. In this study, *L. echidnina* was most abundant in the cool and moist Kīlauea Forest (1,645 m) where *L. nuttalli* was rare. It was found at four of the five sites below the transect, but invariably with lower frequency than *L. nuttalli. L. echidnina* appears to do better at cooler temperatures. Further sampling and analysis is necessary to clarify factors influencing its distribution on the transect.

The two flea species *Nosopsyllus fasciatus* and *Leptopsylla segnis,* which were found abundantly on both mice and rats in this study, are more abundant in temperate than in tropical regions. Haas et al. (1972) reported that only occasional specimens of *N. fasciatus* were found as low as 366 m on Hawai'i Island in winter, and the lowest record from a rodent nest was 473 m. The same authors note that *L. segnis* is not often found at sea level. Within the range of the present study, moisture appeared to be a more significant determinant than temperature. However, it was probably the combination of moderately high temperature and high moisture that excluded *L. segnis* from all rodent hosts at the wet sites below the Mauna Loa transect. Clearly *N. fasciatus* does better in rain forest habitats and *L. segnis* does better in drier habitats. The results show that *Rattus rattus* can be a maintaining host for the mouse flea *(L. segnis)* and *M. musculus* for the rat flea *(N. fasciatus)* under favorable conditions.

CANOPY-ASSOCIATED ARTHROPODS*

An important habitat of many native arthropod species is the tree canopy. The objective of this study was to investigate the arthropod communities associated with the two major native tree species (*Metrosideros collina* subsp. *polymorpha* and *Acacia koa* var. *hawaiiensis*) as they occur along the Mauna Loa transect. The particular focus in this analysis was to establish their quantitative species composition, to evaluate the canopy arthropods in terms of their functional group membership, to determine the degree of penetration of exotics, and to study the variation of these attributes throughout this environmental gradient.

Sampling Information

Eight IBP sites were sampled, from the beginning of the Mauna Loa transect in the closed rain forest (site 1) to the tree line in the subalpine area (site 12) (see table 2-2; Gagné, canopy insects). Both the lower-elevation sites (1, 2, 3) and the upper-elevation sites (10, 12) were occupied by 'ōhi'a-lehua *(Metrosideros)* tree communities, and the mid-elevation sites (4, 6, 9) were occupied by koa *(Acacia koa)* tree communities.

Sampling at these eight sites was done by fogging representative portions of the tree canopies with an insecticide (pyrethrum synergized with piperonyl butoxide) and by collecting the arthropods underneath on cloth sheets of 2.4 m by 2.4 m size. From here they were transferred into mason jars. These were then filled with alcohol for storage until further processing. Samples were replicated at each site. Larger cloth sheets (4.8 m by 4.8 m) were used for taller tree groups. All results were converted to the same unit area.

Sampling was repeated every two months from March 1971 through December 1973. Further details of the sampling technique are given in TR 77.

Canopy Arthropod Zonation Patterns

Cluster analysis by the dendrograph technique with Motyka's similarity index resulted in three general canopy arthropod community zones. These coincided with the major tree species distributions in the form of a subalpine 'ōhi'a zone (I), a koa-tree-dominated mountain parkland and a savanna zone (II), and a montane 'ōhi'a zone (III). The latter zone was further subdivided into an open and a closed 'ōhi'a forest according to the two-way table technique. Further details are given in TR 77. Thus, while the canopy arthropod data

*By W. C. Gagné

resulted in distinct community zonations, their community patterns were more generalized than those obtained from the floristic and avian analyses.

Species Distribution Trends

The Mauna Loa transect study yielded 159 canopy-associated arthropod taxa. A checklist is given in TR 77. Many were species collected only sporadically, so they could not be used for indicating definite distribution trends. Therefore, for this spatial distribution analysis, a selection was made of those species that by their constancy of presence from repeated samplings indicated that they were established canopy residents at the sites.

Knowledge gained from the literature, other sources, and field observations permitted a sorting of the canopy arthropods into different groups with similar ecological roles in the ecosystem (life form groups). An assessment of their origin as either endemic or exotic members of the communities was also possible. On this basis of grouping, the distribution trends of thirty-four species were diagrammed. Figures 3-11, 3-12, and 3-13 show the transect zones as identified by the vegetation analysis. The canopy sampling sites are indicated beneath in relation to the quantitative distribution of the two host-tree species. The canopy arthropods are spatially ordinated on each diagram, but their distribution trends can also be compared between the diagrams. The quantity of each species is based on a count of individuals (density), which was standardized for the sample size and converted to a log base 2 index (for example, one to two specimens is one unit; three to four specimens is two units; five to eight specimens, three units, and so forth).

Endemic Herbivores

Thirteen arthropod species are diagrammed in figure 3-11 and form six spatial groups. Group 1 appears to be host specific on ‘ōhi‘a trees over a wide range of environmental conditions. The first two species (*Sarona adonias* and *Oceanides pteridicola*) extend from the tree line site 12 (where they occur in great abundance) into the closed ‘ōhi‘a rain forest, while the third species (*Oceanides vulcan*) seems to be cool-temperature limited by not extending into the tree line ecosystem. However, as evidenced by sampling a few isolated ‘ōhi‘a trees (TR 77) in the mid-section of the transect (zones III and IV), the three species of group 1 occur throughout this altitudinal range, but they are absent on koa. All three species are Heteroptera, but they belong to two functional groups. The mirid, *Sarona adonias,* feeds on the cell sap of ‘ōhi‘a leaves, while the two lygaeids, *Oceanides* species, are seed predators of ‘ōhi‘a. It can be concluded from their distribution that these are the quantitatively most important native insects in this area on *Metrosideros* in their respective ecological niches.

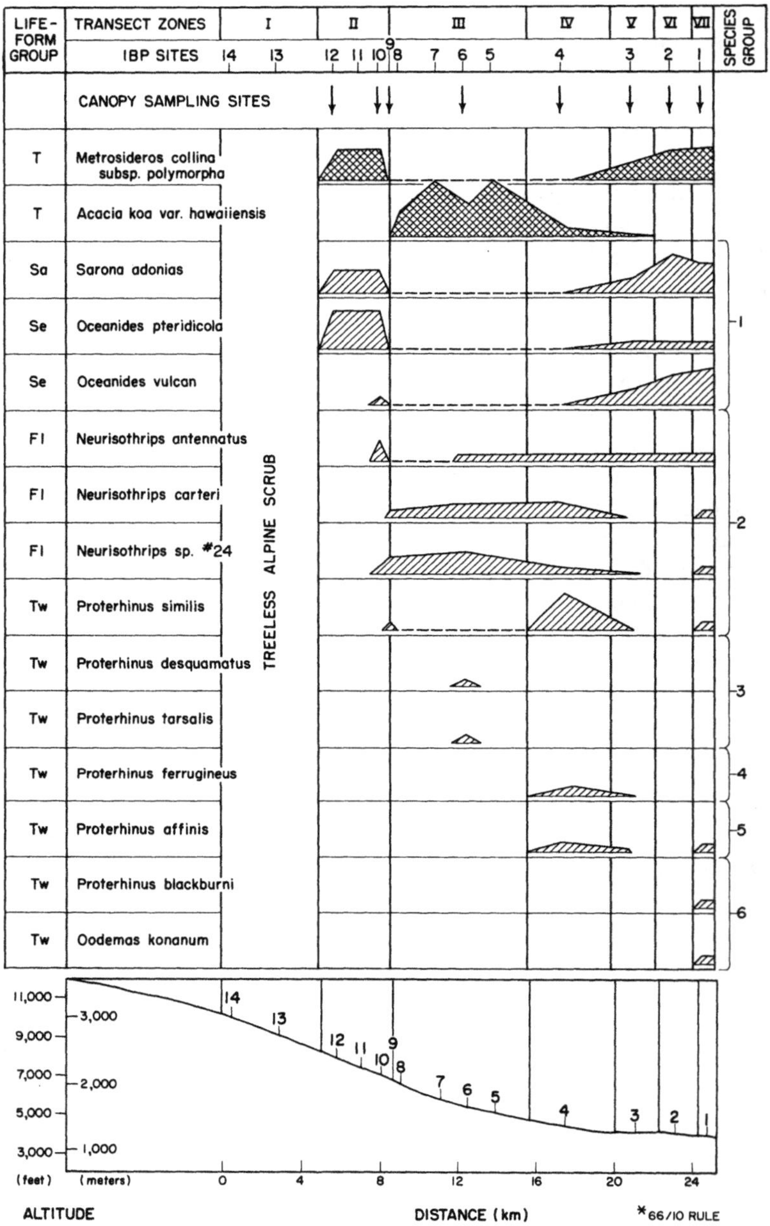

LIFE-FORM GROUP
TRANSECT ZONES I II III IV V VI VII
IBP SITES 14 13 12 11 10 9 8 7 6 5 4 3 2 1
CANOPY SAMPLING SITES
SPECIES GROUP

T Metrosideros collina subsp. polymorpha
T Acacia koa var. hawaiiensis
Sa Sarona adonias
Se Oceanides pteridicola
Se Oceanides vulcan
Fl Neurisothrips antennatus
Fl Neurisothrips carteri
Fl Neurisothrips sp. *24
Tw Proterhinus similis
Tw Proterhinus desquamatus
Tw Proterhinus tarsalis
Tw Proterhinus ferrugineus
Tw Proterhinus affinis
Tw Proterhinus blackburni
Tw Oodemas konanum

1 2 3 4 5 6

TREELESS ALPINE SCRUB

11,000 9,000 7,000 5,000 3,000 (feet)
3,000 2,000 1,000 (meters)
14 13 12 11 10 9 8 7 6 5 4 3 2 1
0 4 8 12 16 20 24

ALTITUDE DISTANCE (km) *66/10 RULE

Group 2 includes four herbivorous insects that are more prevalent on koa than on ʻōhiʻa trees. They were absent at the tree line (site 12), but all were present in the closed ʻōhiʻa rain forest (at site 1). Three of them (*Neurisothrips antennatus, N. carteri, N.* sp. #24) are flower-feeding thrips of the order Thysanoptera, while the fourth one is a twig-boring proterhinid weevil (*Proterhinus similis,* a Coleoptera). The absence of *Proterhinus* from site 6 (in the mid-part of the mountain parkland, zone III) is hard to explain. It may have been caused locally by the competition of the two other proterhinid twig borers of group 3 (*Proterhinus desquamatus* and *P. tarsalis*) that were present at site 6 but absent on the koa trees at sites 4 and 9. This hypothesis of competitive displacement of congeneric species in the same functional group would be interesting to test. The single species group 4 (*Proterhinus ferrugineus*) coexists with *P. similis* and *P. affinis* as the three quantitatively most important twig borers of koa trees in the savanna zone IV, where twig borers are more abundant than elsewhere along the transect. *P. affinis,* however, occurs also on ʻōhiʻa trees in the closed rain forest (zone VII) and thus has a unique distribution. Here it coexists with the two twig borer weevils of group 6 (*Proterhinus blackburni* and *Oodemas konanum*) that appear to be more restricted to ʻōhiʻa trees.

The quantitative distributions of the thirteen endemic herbivorous canopy arthropods as diagrammed in figure 3-11 support the zonation into five arthropod communities derived from the two-way table analysis. This can be visually judged from the diagram; *Sarona* and *Oceanides pteridicola* define by their upper-elevation distributions the subalpine zone. The mountain parkland (zone III) is defined largely by the two *Neurisothrips* species, *N. carteri* and *N.* sp. #24, and by the simultaneous absence of the twig-boring weevils, *Proterhinus ferrugineus* and *P. affinis,* which define the savanna (zone IV), at least as an arthropod subcommunity of the koa-dominated grassland area (zones III and IV). The open ʻōhiʻa forests (zones V and VI) on the diagram are defined largely by the absence of twig borer species, while the closed ʻōhiʻa forest (zone VII) is defined by the presence of several twig borer species.

Exotic Tree Residents

Eleven quantitatively important introduced canopy arthropods are diagrammed in figure 3-12. These fall into six spatial groups.

FIGURE 3-11. *Endemic herbivorous arthropods residing in tree canopies. These are the thirteen quantitatively more important species collected in this study. Life form symbols: T = tree, Sa = sapsucker, Se = seed predator, Fl = flower feeder, Tw = twig borer. Tree quantities are in percentage crown cover (transferred from figure 2-12). Arthropod curve heights are given in the* log_2 *scale: 1 unit curve height = 1–2 specimens, 2 units = 3–4 specimens, 3 units = 5–8, 4 = 9–16, 5 = 17–32, etc. Maximum units here are 10 = 513–1,028 specimens, exemplified by* Proterhinus similis *at site 4.*

Iridomyrmex humilis is an ant (Formicidae) with a uniquely ubiquitous distribution. It occurs from the closed rain forest to the openly structured tree line and is found in considerable abundance on both host trees. It therefore

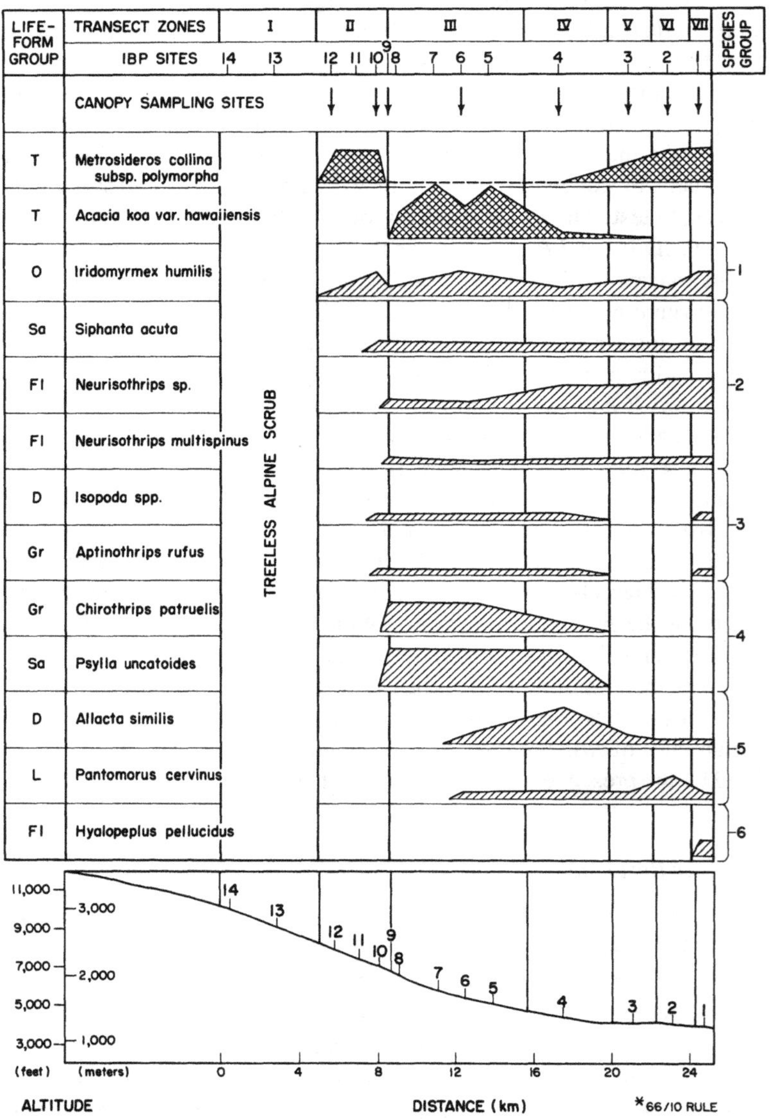

FIGURE 3-12. *Exotic tree-residing arthropods. The eleven more dominant taxa in this group. Life form symbols: T = tree, O = omnivore, Sa = sapsucker, Fl = flower feeder, D = detritivore, Gr = grass feeder, L = leaf chewer. Curve heights based on same criteria as given in figure 3-11.*

represents the most widely penetrating exotic tree-residing arthropod species in this area. This is not surprising since *Iridomyrmex,* as an omnivore, is in a life form group by itself, which is not represented by any canopy-associated native arthropod. This ant, therefore, has occupied an empty niche in the ecosystems along the Mauna Loa transect.

Group 2 includes three species of wide distribution that are similarly unspecific in their host-tree relationships in that they occurred with comparable abundance on both *Metrosideros* and *Acacia koa*. However, their distribution upslope extended only a little beyond the upper boundary of the mountain parkland (zone III). One of these, *Siphanta acuta,* is an introduced sapsucking arthropod that overlaps in its transect distribution widely with the endemic sapsucker *Sarona adonias*. But it is unlikely that the exotic *Siphanta* has become a strong competitor of the endemic *Sarona*. The latter is strongly host specific on *Metrosideros,* while *Siphanta* does not exhibit any host specificity. The situation may be different with the two other species of spatial group 2, both of which are exotic flower feeders of the genus *Neurisothrips* (*N. multispinus* and *N.* sp.). These two introduced flower feeders overlap on both host trees with three native species of *Neurisothrips* (figure 3-11).

Group 3 includes two taxa that were absent in the openly structured *Metrosideros* forests (zones V and VI). One of these, Isopoda (which here includes more than one species), is a detritus-feeding taxon of the pillbug group (Crustacea), whose area of activity is more beneath the trees than in the canopy. The same applies to the grass-feeding thrips *(Aptinothrips rufus),* which merely resides on tree branches. This species also fills a niche that apparently was not occupied by any native species of thrips, which may explain its widespread occurrence.

The group 4 species, *Chirothrips patruelis* and *Psylla uncatoides* (Homoptera), obviously prefer *Acacia koa* as host species. They were the most abundant exotic canopy arthropods in the parkland and savanna zones (III and IV).*

The group 5 species, *Allacta similis* (a cockroach, Blattaria) and *Pantomorus cervinus* (a Coleoptera weevil), have no particular preference for either of the two host species, but the cockroach was distinctly more abundant in the koa savanna (zone IV) than in the ‘ōhi‘a rain forest (zone VI and VII), while the quantitative distribution of the Coleoptera weevil was opposite; it was more abundant in the ‘ōhi‘a rain forest and less so in the savanna.

Group 6, here represented by only one species, *Hyalopeplus pellucidus* (a mirid bug, Heteroptera), is the only introduced flower feeder that has become host specific on ‘ōhi‘a trees (according to this analysis). Its restriction to the closed rain forest along the Mauna Loa transect may indicate either a beginning naturalization on ‘ōhi‘a from lower to higher elevations or a sensitivity to cooler

---

**Psylla uncatoides* was studied in more detail by Leeper and Beardsley (TR 23).

climates at higher elevations, since it was not found in the subalpine 'ōhi'a forest (zone II).

The introduced tree-residing arthropods exhibit less host specificity and less segregation into community zonations than the endemic herbivorous tree residents. As an overview of figure 3-12 shows, only three or four community zones are indicated by the distribution trends of these exotic arthropods. The subalpine zone II is characterized by a near absence of exotics, except for the omnivorous ant *(Iridomyrmex humilis)*. In contrast, the mid-elevation grassland and koa tree area (zones III and IV) is characterized by an abundance of exotic tree-residing arthropods, and there is no clear separation indicated of the mountain parkland (zone III) from the savanna (zone IV). Only four species (those of groups 3 and 4) do not extend their range through the two open 'ōhi'a forest zones (V and VI). Thus, the exotic arthropod species distribution supports the more generalized zonation pattern identified by the dendrograph analysis.

Endemic and Exotic Insect Predators

Ten arthropod species are diagrammed in figure 3-13, which represent only one general life form group. They are all predators, but five of them are endemic and five exotic. The endemic species are by far more abundant than the exotic ones. Both can be divided into two spatial groups.

Group 1 includes only one species, *Koanoa hawaiiensis* (an endemic mirid bug, Heteroptera), that shows the same ubiquitous tendency by its distribution and abundance as *Iridomyrmex* in figure 3-12. Both are in the same spatial group.

Group 2 includes the remaining four endemic predators (all are Heteroptera) that are distributed from the closed rain forest to the upper limit of the mountain parkland or slightly into the subalpine scrub forest (at site 10). Three of these are also mirid bugs: the two *Psallus* species (*P. luteus* and *P. sharpianus*) and *Orthotylus azalais,* and one is a nabid bug *(Nabis oscillans)*. The *Psallus* species were not collected in the open 'ōhi'a rain forest (zone VI) and were also much less abundant in the rain forest. Their closely matched quantitative distribution patterns suggest almost direct competition for prey, an interesting point for testing. *Nabis oscillans* is the most abundant insect predator in the 'ōhi'a rain forest.

None of the five exotic predators shows any correlation in distribution with one another. Their division into two spatial groups only distinguishes the higher-elevation species of group 3, the two ladybird beetles, *Cryptolaemus montrouzieri* and *Scymnus varipes,* from the lower-elevation species of group 4, *Karnyothrips flavipes, Rhizobius ventralis,* and *Scymnus loewii*. The last two are also ladybird beetles. Thus, the endemic predators are all Heteroptera, while the exotic predators are comprised mostly of Coleoptera (the four ladybird beetles) and one Thysanoptera (the thrips species).

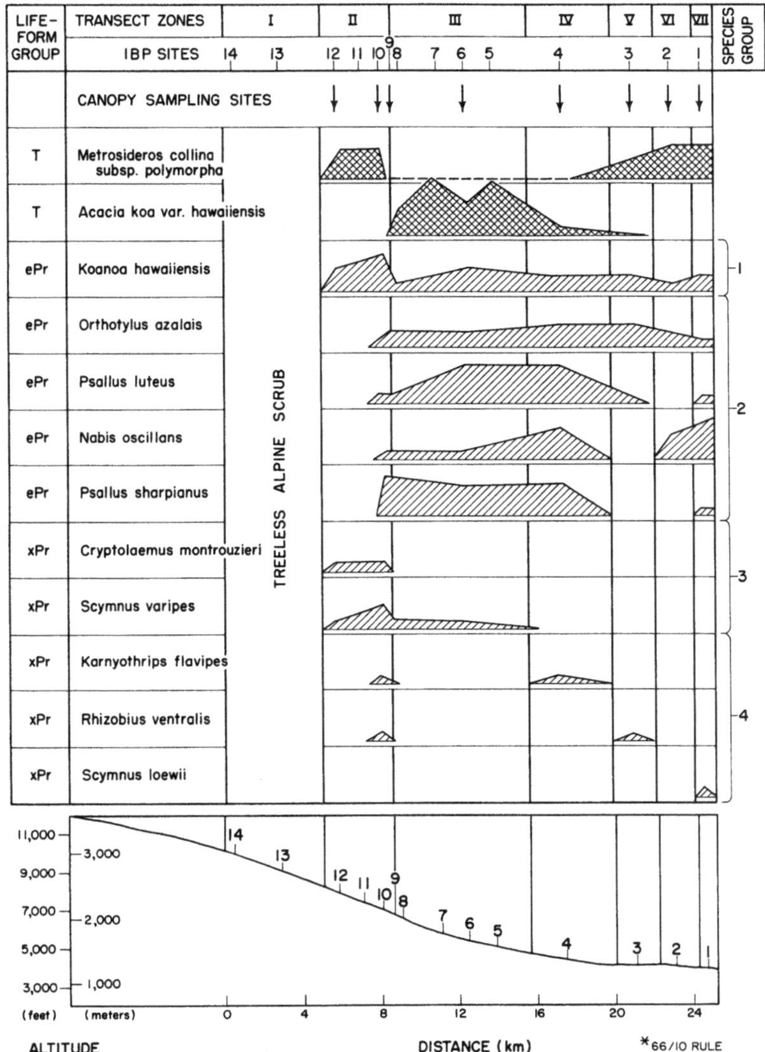

FIGURE 3-13. *Endemic and exotic insect predators residing in tree canopies. The ten more dominant species. Life form symbols: T = tree, ePr = endemic predator, xPr = exotic predator. Curve heights based on same criteria as given in figure 3-11.*

The predators (endemic or exotic) show little or no host plant specificity as indicated by their distribution patterns. The exotic species do not indicate any community zonation, while the endemic predators support the general three-way arthropod zonation obtained from the dendrograph analysis.

Discussion and Conclusions

Several canopy-associated arthropods show wide distributions along the transects; therefore they could not be used to test Whittaker's four distribution hypotheses. However, very wide ecological distributions as here displayed by three endemics (*Sarona adonias, Oceanides pteridicola,* and *Koanoa hawaiiensis*) are characteristic for one group of successful island species, while very narrow ecological amplitudes are characteristic for another (usually more specialized) group of island species. One can interpret the wide distribution as expected of evolutionarily more recent native species because the trend resembles that displayed by some successfully naturalized exotics, such as the ant *Iridomyrmex humilis*. However, the evolutionary time scale is not the only criterion. The distribution range of a species depends also on its ecological function and the presence of competitors (species with closely similar ecological roles).

The three widely distributed native arthropods all have different feeding niches; *Sarona* feeds on leaves, *Oceanides pteridicola,* on seeds and *Koanoa* on other arthropods. Competition by similarly adapted species is limited. Specialization to ʻōhiʻa has occurred in the first two species. Therefore the two phytophagous exotic sapsuckers (*Siphanta acuta* and *Psylla uncatoides*) may offer only little if any competition to the native sapsucker *Sarona*. These food adaptations combined with limited competition and apparently wide physiological tolerances may explain such wide distributions.

Spatially closely associated species with more limited ranges that conform to Whittaker's distribution hypotheses 1 and 3 are shown on each diagram. Examples are the two flower-feeding *Neurisothrips* species in figure 3-11 and the two *Proterhinus* species of group 3 (on that diagram). Further, the exotic arthropods in groups 3 and 4 in figure 3-12 and the two endemic predators of *Psallus* (*P. luteus* and *P. sharpianus*) in figure 3-13 form species groups each with closely matched distribution ranges and similar areas of optimum development. The last two species show greater boundary overlap and thus conform in their distribution pattern more to Whittaker's third hypothesis.

Nonassociated distributions are displayed particularly by the exotic predators (groups 3 and 4 in figure 3-13). These conform to Whittaker's fourth hypothesis of a more random spatial arrangement. Congeneric competitive exclusion along the gradient as postulated in Whittaker's second hypothesis may be indicated by the distribution pattern of the *Proterhinus* weevils in figure 3-11. Thus the canopy arthropods along the Mauna Loa transect display all four pattern trends recognized by Whittaker, with associated species groups forming the more prevalent trends.

The degree of penetration of exotics is related to some extent to their origin and feeding attributes. Those species of temperate origin—for example, some ladybird beetles (*Cryptolaemus, Scymnus,* and *Rhizobius*)—appear preadapted to higher and cooler portions of the transect. Omnivores—for example, the Argentine ant *(Iridomyrmex humilis)*—and the sap feeders—for example, the torpedo bug *(Siphanta acuta)*—also demonstrate penetration to higher elevations, as does a species of koa psyllid *(Psylla uncatoides)* originating from phyllode-bearing *Acacia* species in Australia. Species of tropical or subtropical origin such as the roach *(Allacta similis)*, show less altitudinal penetration, indicating a cool temperature limitation. Ecologically more specialized arthropods, such as the flower-feeding plant bug *Hyalopeplus pellucidus,* appear to be limited by host species distribution and humidity.

Thus the Mauna Loa transect analysis of canopy-residing arthropods has revealed a distribution pattern of endemics and exotics that can serve as a predictive tool for other similar environmental gradients where the life form groups of such arthropods have been identified and quantified.

ARTHROPODS ON REPRODUCTIVE ORGANS OF 'ŌHI'A-LEHUA TREES*

A number of arthropods occur on the reproductive organs of the 'ōhi'a trees (*Metrosideros collina* subsp. *polymorpha*) (Zimmerman, 1948a–d, 1958a–b; Swezey, 1954). Moreover, certain insects have been suspected to function as seed predators of this important native tree species. The integrated transect approach provided an opportunity to identify the arthropod fauna on 'ōhi'a reproductive structures (buds, flowers, and seedpods) throughout a variety of environments. The specific objective of this spatial analysis was to determine whether these reproductive-organ-associated arthropods were distributed concomitantly with 'ōhi'a trees or whether their presence was related to other environmental variables.

Sampling Information

Five IBP sites (3, 7, 9, 10, and 12) were sampled twenty-four times at monthly intervals from January 1973 through December 1974. This high sampling frequency was considered necessary for determining which arthropods were most likely to be closely associated with reproductive structures. Another purpose was to find out what sort of changes (if any) occur in the reproductive-organ-associated arthropod fauna with the phenological changes associated with seed formation.

At each of the five IBP sites fifteen 'ōhi'a trees were tagged. These fifteen trees were treated as replicate samples of the site. Thus the study was done on seventy-five sample trees, and these were each sampled twenty-four times. The repeated sampling gave an estimate of the frequency with which each arthropod species was found. On each sample tree, reproductive organs were plucked off,

*By B. M. Brennan, W. C. Mitchell, and D. Springer

put immediately into plastic bags, labeled, and sealed for further processing in the laboratory. When possible, at least twenty flower buds or seedpods and four inflorescences were collected from each tree at each sampling period. Buds, pods, and flowers were placed into separate bags for determining phenological trends.

In the laboratory, samples were kept between 4.5°C and 13°C no longer than seventy-two hours before they were processed through a modified Tullgren funnel. After forty-eight hours the collected taxa were separated and the number of their individuals counted. Spiders were identified to at least the family level. Immature Lepidoptera were not identified, although they may play an important role in seed predation. Currently a study is being undertaken to follow up on this aspect.

Species Distribution Trends

The sampling yielded fifty-four identifiable arthropod taxa. Only those that were found repeatedly in association with reproductive organs of ʻōhiʻa are discussed here. The minimum acceptance criterion for species useful as indicators of distribution trends was established as being a species obtained in at least two collections at the same IBP site. This reduced the number of taxa to be analyzed to sixteen, which are diagrammed in figure 3-14. Six distribution groups can be recognized on this diagram.

Group 1 includes three insect species: a sapsucking mealybug, *Pseudococcus* sp., a flower-feeding (anthophagous) thrips, *Neurisothrips antennatus*, and a predatory mirid bug, *Psallus luteus*. These insects are almost equally frequent on ʻōhiʻa trees in both the cool subalpine (zone II) and the warmer montane seasonal (zone V) ʻōhiʻa forests, and they occur consistently along the mid-section of the transect wherever scattered ʻōhiʻa trees occur.

Group 2 includes three insect taxa: a seed-feeding lygaeid bug (*Oceanides pteridicola*), a parasitic wasp (*Telenomus* sp.), and the native bee (*Nesoproso-*

FIGURE 3-14. *Distribution of arthropods collected frequently from ʻōhiʻa reproductive structures along Mauna Loa transect. Percentage frequency is indicated by the height of each curve. Darkened areas represent distribution trends corresponding to that of* Metrosideros *tree communities. Blank areas under curves indicate frequency of arthropod species on a few scattered* Metrosideros *trees in the transect area. Dashed lines refer to species present in less than 10 percent of the total monthly samples; 100 percent frequency is indicated for* Pseudococcus *sp. over its observed range. Life-form symbols: Fl = flower feeder, O = omnivore, Pa = parasite, Pr = predator, D = detritivore, Sa = sapsucker, Se = seed predator, T = tree, ? = unknown function. *Exotic species.*

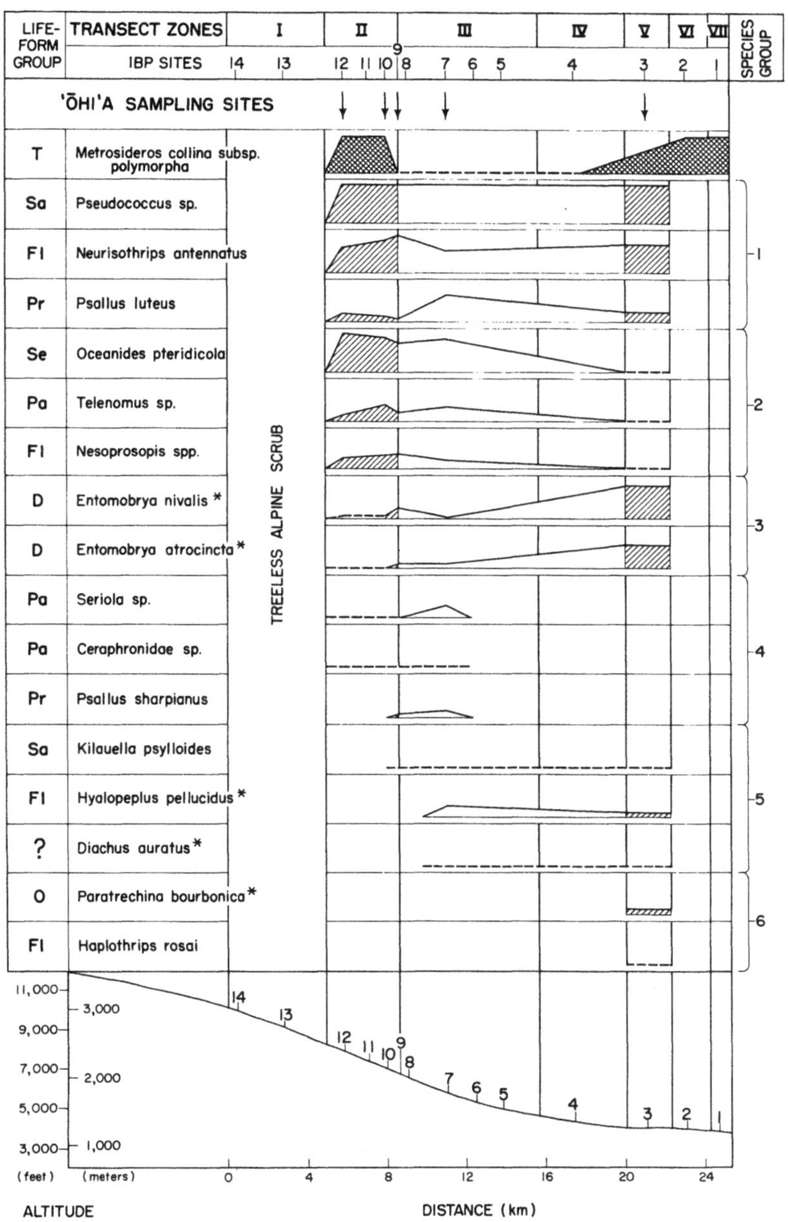
LIFE-FORM GROUP
TRANSECT ZONES
I
II
III
IV
V
VI
VII
IBP SITES
14 13 12 11 10 9 8 7 6 5 4 3 2 1
SPECIES GROUP
'ŌHI'A SAMPLING SITES
T Metrosideros collina subsp. polymorpha
Sa Pseudococcus sp.
Fl Neurisothrips antennatus
Pr Psallus luteus
Se Oceanides pteridicola
Pa Telenomus sp.
Fl Nesoprosopis spp.
D Entomobrya nivalis *
D Entomobrya atrocincta *
Pa Seriola sp.
Pa Ceraphronidae sp.
Pr Psallus sharpianus
Sa Kilauella psylloides
Fl Hyalopeplus pellucidus *
? Diachus auratus *
O Paratrechina bourbonica *
Fl Haplothrips rosai
TREELESS ALPINE SCRUB
1
2
3
4
5
6
11,000
9,000
7,000
5,000
3,000
(feet)
3,000
2,000
1,000
(meters)
0 4 8 12 16 20 24
ALTITUDE
DISTANCE (km)

pis spp.), which also feeds on nectar and pollen of the ‘ōhi‘a blossoms. These three taxa have a similar distribution to group 1, but their abundance pattern is decidedly different.

Group 3 includes two detritus-feeding insect species, both exotic collembolans (springtails): *Entomobrya nivalis* and *E. atrocincta*. These insects were more abundant in the montane seasonal ‘ōhi‘a forest (zone V).

Whereas the first three groups were found at all five sites, groups 4, 5, and 6 include species with restricted distributions. Group 4 contains two parasitic Hymenoptera (*Seriola* sp. and a cerephronid) and a mirid predator, *Psallus sharpianus*. These were restricted to higher elevations on the transect.

In contrast, group 5 contains three insect species, a psyllid *(Kilauella psylloides),* a flower-feeding mirid bug *(Hyalopeplus pellucidus),* and a chrysomelid beetle *(Diachus auratus),* which were not found on subalpine ‘ōhi‘a trees (zone II) but were present on scattered ‘ōhi‘a trees throughout the mountain parkland (zone III), savanna (zone IV), and montane seasonal ‘ōhi‘a forest (zone V).

Group 6 shows two species of insects: an ant *(Paratrechina bourbonica)* and a flower-feeding thrips *(Haplothrips rosai)* that were restricted to the ‘ōhi‘a forest at lower elevations (zone V).

Discussion and Conclusions

Five of the sixteen frequently occurring arthropods collected on ‘ōhi‘a reproductive structures were also collected in the canopy samples by Gagné. Three of the five (*Neurisothrips antennatus, Oceanides pteridicola,* and *Hyalopeplus pellucidus*) specifically feed on reproductive structures, primarily flowers. The other two species are mirid bugs (*Psallus luteus* and *P. sharpianus*) and are known to be insect predators. None of the remaining twenty-nine taxa collected by Gagné were collected in our samples, indicating that they were not found on ‘ōhi‘a, or that they were primarily associated with some other ‘ōhi‘a plant structure, or that there were significant differences between sampling methods. The differences in the collections are especially apparent in two of the species in our group 1 (*N. antennatus* and *P. luteus*), which were collected by Gagné in both *Metrosideros* and *Acacia koa* canopy samples (figures 3-11 and 3-13). We found *N. antennatus* particularly abundant in the subalpine (zone II) on ‘ōhi‘a reproductive organs, whereas Gagné did not collect this thrips at the upper tree line (site 12). In this case the apparent difference in distribution may be due to differences in sampling methods. Possibly small insects, such as thrips, on upper branches or reproductive organs, incapacitated by the pyrethrum-piperonyl butoxide fogging, were blown away from the canvas catchment sheet used by Gagné. It is also possible that such small insects were caught in crevices on the branches of the host plant and did not reach the canvas sheet. Another insect that occurred with extremely high frequency at every site, which apparently was not collected by Gagné, was the sapsucking *Pseudococcus* sp. This relatively sessile mealybug would not likely drop from its feeding site when killed by insecticide fogging, and therefore it would be found only with our sampling method. These differences call attention to the need for a careful examination of the efficiency of sampling for particular problems.

The eleven remaining arthropods that were absent in Gagné's 'ōhi'a canopy samples but consistently present on 'ōhi'a reproductive structures are considered to be the taxa most strongly associated with 'ōhi'a buds, flowers, or pods. Two of these, *Nesoprosopis* spp. and *Paratrechina bourbonica,* were found predominately associated with the flowers. Perkins (1899) stated that larval food of *Nesoprosopis* contains comparatively little pollen, as these bees have no special pollinigerous apparatus. Perkins confirmed this by examining hive cells where he found larvae floating on liquid food. *P. bourbonica* may also have been attracted to the blossoms for the abundant nectar or to honey-dew-producing mealybugs on the flowers. The remaining nine species may be conveniently divided into phytophagous and entomophagous insects, including parasites and predators. The more detailed ecological role of each species is given in figure 3-14.

Although *Nesoprosopis* spp. and *Paratrechina bourbonica* were collected predominately at the nectar-rich blossoms, the nine other insects were most abundant and frequently found in the pod samples. Four species, including two hymenopterous parasites (*Seriola* sp. and a cerophronid) and an insect predator *(Haplothrips rosai),* were absent from at least one of the reproductive structures examined. Arthropod distributions therefore may be directly or indirectly influenced by the 'ōhi'a reproductive phenology.

The distributions of some of these insects may also be influenced by the distribution of other host plants. For example, at least eight of the sixteen taxa diagrammed have also been found on *Acacia koa,* and three have been collected on at least three other hosts commonly found along portions of the transect (Zimmerman, 1948a–d; Swezey, 1954; Gagné, TR 77). Of the fifty-four taxa identified, twenty have been recorded on *Acacia koa, Sophora chrysophylla, Myoporum sandwicensis,* or *Dodonaea sandwicensis.* Thus distributions and frequencies may also have been influenced by the specific vegetational composition within or close to the sampling sites.

Climatic factors, including temperature, humidity, and wind, affect different groups of insects to varying degrees. Insects that were collected as immatures (thrips and *Pseudococcus*) or as wingless adults (*Pseudococcus* and *Entomobrya*) made up approximately two-thirds of the total number of those identified. Most of these insects were very small (2–3 mm). Some of these occurred on each of the phenophases examined, indicating that they may have been intimately associated with the microhabitat provided by the reproductive structures. These insects may be more influenced by microclimatic and physiological changes that occurred at the transect site. The distribution and activity of insects that exhibit high degrees of mobility—for example, parasitic and predatory hymenopterans, chrysopids, coccinellids, and spiders—would be more strongly influenced by macroclimatic factors.

In a general sense, temperature does not appear to be an important factor in the distribution of the insects in the 'ōhi'a trees. Specific temperature relationships are difficult to determine, especially where the insects live in protected microhabitats and where there is not a strongly pronounced seasonal change in the temperature. Temperature, however, may have an important indirect effect on the distribution of these insects as it influences the phenological pattern of the host plants and thus controls the availability of habitat.

Other environmental factors may also have some influence on the distribution of the insects, although in these cases, too, there is little direct evidence of controlling relationships. Heavy rains may cause flooding of reproductive organs of plants and cause some elevated mortality rates. High humidities may act indirectly by providing the necessary conditions for microbial diseases, which will reduce insect populations. This was found by Conant (TR 64) with the nuclear polyhedrosis virus-regulating mamane moth *(Uresephita polygonalis)* populations at IBP site 4. The virus apparently is confined to areas that are frequently covered by clouds. Moisture also acts indirectly by influencing the growth and phenology of the host plants. This relationship has been shown by an analysis of the tagging of the blossoms of the host plants. These data show that there were both more 'ōhi'a blossoms at the drier site, as well as a greater abundance of individuals. There was not, however, a significant difference in the number of taxa found in these samples. Rainfall records made during 1973 and 1974 (Bridges and Carey, TR 38, TR 59) indicate that site 7 (1,720 m) was the driest site sampled. Monthly records made at the time of each sample collection show that this site had over five times as many blossoms per tagged tree ($\bar{x} = 24.37$) as the next most reproductively active site (site 3, $\bar{x} = 4.43$). Although there were more insects collected at site 7, the number of taxa found at each site varied little. Therefore rainfall may be an important factor in determining the abundance of insects along the transect.

Another limiting factor of insect populations may be associated with the presence of sulfur fumes. Although no measurements were made of the number of sulfur particles in the air, the strong irritating odor of sulfur was noted during several collecting periods at IBP site 3 (1,220 m). Both sulfur dioxide and carbon bisulfide have insecticidal qualities (Metcalf et al., 1962).

With respect to endemism, the following pattern seems of interest. The two most abundant species, *Pseudococcus* sp. and *Neurisothrips antennatus,* are endemic and have high populations along the entire transect. The endemic lygaeid seed feeder *Oceanides pteridicola* and the endemic bee *Nesoprosopis* spp. were more abundant at the higher elevations. In contrast, the exotic collembolans, *Entomobrya nivalis* and *E. atrocincta,* were most abundant and frequent at IBP site 3 (1,220 m). Three other exotic insects (*Hyalopeplus pellucidus, Diachus auratus,* and *Paratrechina bourbonica*) also have distributions limited to sites in the montane seasonal environment. Although *Haplothrips rosai,* an endemic thrips, was found only at IBP site 3, Zimmerman (1948b) states that this insect is common on several other host plants, including *Vaccinium reticulatum,* which is abundant at site 3 but not at the upper elevations. Thus, our analysis shows the exotic insects to be more prevalent at the lower end of the transect, while the endemic insects are more abundant at the upper end. The influence of exotic insects on the distribution of endemic insects is unknown. The transect pattern indicates that more detailed studies on competitive displacement among these arthropods versus their habitat adaptation should be pursued.

Along the Mauna Loa transect it appears that the degree of endemicity, the presence or absence of other host plants, and moisture, may directly influence the distributions of insects associated with the reproductive structures of ʻōhiʻa. Occasional sulfur fuming near sites 2 and 3 may also depress insect populations there. Temperature may be more important in the seasonal distribution of ʻōhiʻa associated insects. All of those factors that influence the reproductive phenology of ʻōhiʻa indirectly affect the distribution of ʻōhiʻa associated arthropods.

WOOD-BORING CERAMBYCID BEETLES*

The native Hawaiian fauna in the beetle family Cerambycidae is comprised of nearly 140 species. These belong to three subfamilies (Cerambycinae, Parandrinae, and Prioninae), with all but two species belonging to the plagithmysine complex. The Parandrinae and Prioninae are primitive subfamilies, each represented in Hawaiʻi by a single large endemic species. They are *Parandra puncticeps* and *Megopis reflexa* and bore primarily in quite dead or rotting wood in rain forests at lower and middle altitudes. These two species were not sampled. By contrast, the plagithmysine complex belongs to the tribe Clytini in the more advanced subfamily Cerambycinae and contains some 136 species and subspecies, all placed now in the genus *Plagithmysus.* All members of this complex evolved from one immigrant ancestor. Currently each species or subspecies is found only on a single island. Even though plagithmysines are associated with most of the larger common native Hawaiian woody plants, they are seldom seen, and many species are considered rare. They usually bore in dying branches or stem tissue of living plants that are damaged by wind, vulcanism, disease, humans, and so forth. In many cases it is difficult to determine whether the beetles are primary or secondary in relation to the death of branches or trees. In either case they usually appear to be the first insects to bore in branches or trunks of trees that are physiologically weakened. Associations of plagithmysines with their host plants are largely at the generic level and sometimes at the species level, but some plagithmysine species attack two or

*By G.A. Samuelson and J.L. Gressitt

more plant genera. In some cases different species of plagithmysines occupy one host genus or species in the same location of the same island. Of some forty-six species and subspecies of plagithmysines found on the Island of Hawai'i, at least fourteen are now known to occur along the Mauna Loa transect.

The objective of this study was to determine the plagithmysine borer-host associations and the incidence of borers throughout a major environmental gradient. Earlier collections and rearings of borers from sites within the present transect area disclosed most of the species of borers to be expected, with information on the abundance levels of some, but different sites were never compared systematically along a gradient.

Sampling Information

Host trees and woody shrubs were examined along the Mauna Loa transect at and between designated IBP sites. Samples consisted of host branch and stem segments, each 1 m long. These segments were examined for signs of boring. The sampling consisted of recording the presence of larval galleries and emergence holes of adults. Live specimens of larvae and adults were used for confirmation of the sampling data. Because of the general rareness and sporadic distribution of these borers, only two frequency classes were established. A borer species was considered frequent where 30 percent or more of the host samples showed positive signs of boring; a borer species was considered infrequent when its signs were found in less than 30 percent of the host samples. Moreover, in species for which the gallery and emergence hole frequency was very low (less than 5 percent), the borer was marked with an asterisk in figure 3-15 to indicate possible sampling error.

Species Distribution Trends

Altitudinal ranges for plagithmysine borer species and their host plants along Mauna Loa transect are shown in figure 3-15. In most cases, the distribution of the borers is closely associated with the distribution of their host species. The borers can be divided into four spatially associated groups, which correspond to the major vegetation zones.

FIGURE 3-15. *Altitudinal ranges of plagithmysine borers and their host plants along Mauna Loa transect. B = borer, S = shrub, T = tree, ST = shrubby tree, V = vine. Cross-hatch, frequent >30% presence in wood samples; diagonal lines, infrequent <30%. Solid line = regular (±continous) distribution of host species. Interrupted line = sporadic distribution of host species. K = kīpuka forest. *Frequency not confidently established.*

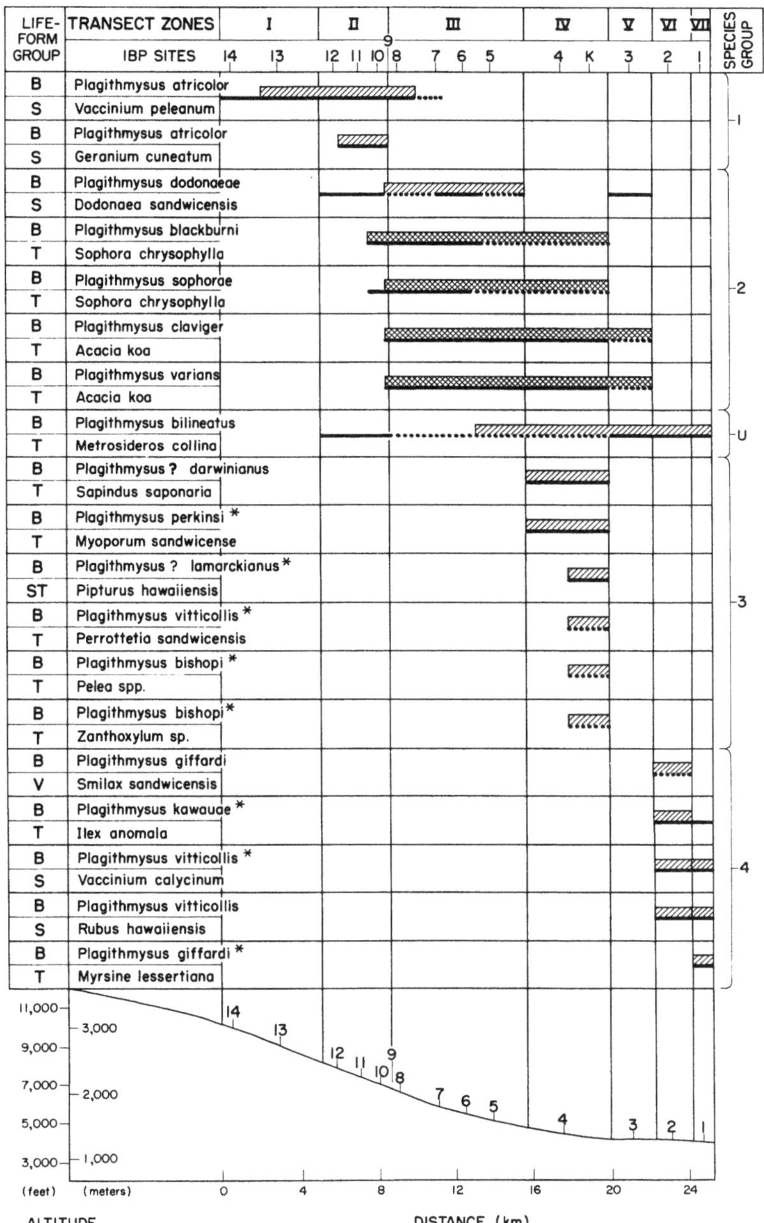
LIFE-FORM GROUP
TRANSECT ZONES
I
II
III
IV
V
VI
VII
IBP SITES
14 13 12 11 10 9 8 7 6 5 4 K 3 2 1
SPECIES GROUP
B Plagithmysus atricolor
S Vaccinium peleanum
B Plagithmysus atricolor
S Geranium cuneatum
B Plagithmysus dodonaeae
S Dodonaea sandwicensis
B Plagithmysus blackburni
T Sophora chrysophylla
B Plagithmysus sophorae
T Sophora chrysophylla
B Plagithmysus claviger
T Acacia koa
B Plagithmysus varians
T Acacia koa
B Plagithmysus bilineatus
T Metrosideros collina
B Plagithmysus ? darwinianus
T Sapindus saponaria
B Plagithmysus perkinsi *
T Myoporum sandwicense
B Plagithmysus ? lamarckianus *
ST Pipturus hawaiiensis
B Plagithmysus vitticollis *
T Perrottetia sandwicensis
B Plagithmysus bishopi *
T Pelea spp.
B Plagithmysus bishopi *
T Zanthoxylum sp.
B Plagithmysus giffardi
V Smilax sandwicensis
B Plagithmysus kawauae *
T Ilex anomala
B Plagithmysus vitticollis *
S Vaccinium calycinum
B Plagithmysus vitticollis
S Rubus hawaiiensis
B Plagithmysus giffardi *
T Myrsine lessertiana
1
2
U
3
4
11,000
9,000
7,000
5,000
3,000
3,000
2,000
1,000
(feet)
(meters)
0 4 8 12 16 20 24
ALTITUDE
DISTANCE (km)

Group 1: High-altitude species representative of alpine and subalpine vegetation zones (I and II). This group includes only one borer species *(Plagithmysus atricolor),* which occurred in two shrubs (*Vaccinium peleanum* and *Geranium cuneatum*) in this cool segment of the altitudinal gradient.

Group 2: Mountain parkland-savanna species. This group includes five wide-ranging borers (*P. dodonaeae, P. blackburni, P. sophorae, P. claviger,* and *P. varians*) on their respective hosts, which extended through both the mountain parkland (zone III) and savanna (zone IV). The range of the two borers on *Acacia koa* (*P. claviger* and *P. varians*) extended even farther into the dry *Metrosideros* scrub forest of zone V, in which a few *Acacia koa* trees had been planted by the Park Service.

Group 3: Savanna and kīpuka species. These include five borers of more restricted distribution in the montane seasonal environment of the transect. Two of these borers (*P. ?darwinianus* on *Sapindus saponaria, P. perkinsi* on *Myoporum sandwicense*) occurred on trees growing in the grassland of this savanna segment, and three borers (*P. ?lamarckianus* on *Pipturus hawaiiensis, P. vitticollis* on *Perrottetia sandwicensis,* and *P. bishopi* on *Pelea* spp. and *Zanthoxylum* sp.) were associated with hosts in the closed kīpuka forest (Kīpuka Puaulu). Hosts of the latter subgroup were components of a forest island within the savanna (zone IV).

Group 4: Rain forest species. This group includes three borer species (*P. giffardi, P. kawauae,* and *P. vitticollis*) that were found on forest plants in the open rain forest (zone VI, *Smilax sandwicensis,* and *Ilex anomala*), in both open and closed rain forest (*Vaccinium calycinum* and *Rubus hawaiiensis*), and in the closed rain forest (zone VII, on *Myrsine lessertiana*).

Two borer species that are particularly important in the rain forest extended upslope beyond zone VI into the seasonal environment. One of these was the ungrouped species *P. bilineatus* on *Metrosideros collina* subsp. *polymorpha,* which extended from the closed rain forest zone VII upslope into the middle of the mountain parkland (zone III). Its distribution was unique, and thus *P. bilineatus* could not be grouped spatially with any of the other borer species. The second borer, which extended through environmentally quite different zones, was *P. vitticollis*. It occurred on two rain forest shrub hosts (*Vaccinium calycinum* and *Rubus hawaiiensis*) from both the open and closed rain forests and on a tree species *(Perrottetia sandwicensis)* that along the transect was found only in the mesic kīpuka forest (zone IV).

Discussion and Conclusions

Certain borers were noticeably more restricted in distribution than their plant hosts. This was seen in both single-host- and multiple-host-associated borers.

Plagithmysus atricolor occurred entirely within the upper transect zones on its two shrub hosts *Geranium cuneatum* and *Vaccinium peleanum*. The borer covered the whole altitudinal range of *Geranium*, which parallels a portion of the broader range of *Vaccinium*. However, the borer did not extend to the extreme distribution limits of *Vaccinium peleanum*. We can assume that factors associated with host physiology are involved in determining the limits of *P. atricolor*. The shrub species probably loses vigor near both ends of its range.

Two single-host-associated borers stand out because of limited distribution on a broadly distributed host. One is *P. dodonaeae*, which was found on *Dodonaea sandwicensis* only in the mountain parkland (zone III) but not on *D. sandwicensis* in the subalpine scrub (zone II) or in the open ʻōhiʻa *(Metrosideros)* dry forest (zone V). Its pattern of association with the host is similar to that of *P. atricolor* on *Vaccinium peleanum*, which suggests that the borer in each case is found where the hosts are rather well developed. The second single-host-associated species of more limited distribution than its host was *P. bilineatus* on *Metrosideros collina*. Here the borer extended only halfway through the host range on the transect. We can assume that a combination of more limited environmental tolerance of the borer (as compared to its host) and a change in the physiology of the host itself may limit the borer.

In two cases, a single-host plant species was found to harbor two different species of host-specific borers. These included two of the larger tree species along the transect, *Acacia koa* and *Sophora chrysophylla;* only the former was sampled in more detail. The two borers *P. claviger* and *P. varians* were present on koa over its broad altitudinal range. Both borers commonly occurred in the same host individual. Cross-section and emergence-hole analyses of host stems, branches, and twigs showed resource partitioning of these borers by branch size. *P. claviger* was found in twigs and smaller branches and *P. varians* in small to large branches and stems. Comparative Malaise trap collections in the Kīlauea rain forest and along the transect showed an increase of both koa borers under progressively more lighted situations that prevailed along the transect. The borers also increased in number when the trap was elevated well above ground level over dense understory. Placement of the trap at ground level in an open koa colony along the transect (zone III) most closely approximated conditions within the koa crowns of the taller Kīlauea rain forest because of the low-spreading crown habit of the transect colony. The even proportions in the catch of *P. claviger* and *P. varians* reflected this, since both small and large branches occurred in the vicinity of the trap. Placement of the trap at ground level under mature koa in the dense Kīlauea rain forest, where there is a greater vertical distance between the ground and tree crown, resulted in a smaller proportion of *P. claviger* caught relative to *P. varians*, as well as in a smaller catch in general.

Two of the rarer borers, *P. giffardi* and *P. vitticollis*, indicated a more diverse pattern of distribution. They were found to be associated with different hosts that occurred in different plant communities, sometimes in different zones. *P. giffardi* occurred on two taxonomically very different host species, one a liliaceous vine (*Smilax sandwicensis,* zone VI) and the other a myrsinaceous tree (*Myrsine lessertiana,* zone VII). *P. vitticollis* was associated with two kinds of shrubs in the rain forest (*Vaccinium calycinum* and *Rubus hawaiiensis*) and with a tree *(Perrottetia sandwicensis)* in the mesic kīpuka forest. These borers should be studied further to determine discreteness of populations and degrees of preferences or tolerances to the different hosts. In a contrasting pattern, *P. atricolor* occurred on both of its hosts in the same general vegetation.

Since the plagithmysine borers evolved toward a close association with their plant hosts, their spatial groupings are closely reflected in the spatial groupings of their hosts. Along the Mauna Loa transect, these borers were classifiable into a high-altitude species, mountain parkland-savanna species, savanna and kīpuka species, and rain forest species. No borer was found on more than three host species on the transect. Most borers occurred in association with a single-host species. Also in two cases where two borer species occurred on the same host tree species, the borers were found to be single-host-associated species. The two borers occurring together on *Acacia koa* showed resource partitioning by branch size, with one borer being restricted to twigs and small branches and the other to larger branches and stems. Most borers were observed to have distributions exactly paralleling that of their hosts along the transect, but other borers departed significantly in their distributions from that of their hosts. The latter noncorrelated distributions suggest that environmental factors exert a control either directly or indirectly through changes in host physiology. Further studies are indicated that aim at searching for the more detailed underlying causes of these differing distribution patterns among the plagithmysines.

SELECTED DETRITOPHAGES (DIPTERA: SCIARIDAE)*

Adult Sciaridae are small, usually black, flies commonly found around decomposing plant and animal materials. They are popularly known as fungus gnats. Immatures (larvae) generally feed on fungi, decomposing plant and animal materials, or animal excrement. Some species are known to feed on living plant tissues. Most of the sixteen species found on the Mauna Loa transect are probably mycetophagous, phytosaprophagous, or both.

The life histories of fourteen Hawaiian species (including eight of the species found on the Mauna Loa transect) have been studied in the laboratory, where they were reared on bacto-agar/cornmeal-agar slant cultures in constant temperature cabinets of 20°C. Mean development time from oviposition to adult emergence ranged from 16.2 to 34.0 days (Steffan, 1974; TR 25).

---

*By W.A. Steffan

Quantitative studies of distribution of Sciaridae at the species level along altitudinal gradients have not been carried out previously even though Sciaridae are ubiquitous and important components of many ecosystems. They are inconspicuous and difficult to identify to the species level.

Sampling Procedure

Several sampling techniques with different specific objectives were combined for the transect analysis to recover all Sciaridae species over the entire gradient from IBP sites 1 through 14. For a more detailed seasonal analysis, modified Malaise traps (Gressitt and Gressitt, 1962) were set up on the three IBP sites at which climatic data were recorded—sites 4, 6, and 9. The Malaise trap is an interception trap designed to collect flying insects more or less at random. It is excellent both for obtaining an index of population abundance and for studying seasonal population fluctuations. Several factors, including cost of traps and time required to process and identify the vast amount of material collected, prohibited the use of Malaise traps on the other IBP sites. As an indication of the magnitude of the processing required, 85 percent of all Sciaridae (11,300 specimens) collected on the transect by various investigators came from the Malaise traps.

Starting on January 20, 1971, the Malaise trap at site 6 was in place continuously, with the catch collected weekly; after July 5, 1971, site 6 was sampled on alternate weeks only, until August 1973, for a total of eighty-one sample weeks. Site 4 was sampled from April 11, 1972, to August 4, 1973, on alternate weeks for a total of thirty-three sample weeks. Site 9 was sampled from March 27, 1972, to August 1, 1973, on alternate weeks for a total of thirty-seven sample weeks.

Sampling methods employed by other investigators yielding Sciaridae used in this analysis included pitfall traps and pyrethrum spray sampling of foliage. An initial analysis of all these data indicated several gaps in the distribution of the Sciaridae. To check this, IBP sites 2, 3, 5, and 7 were sampled by sweeping with an insect net during a continuous five-day period. As a result, the gaps were found to be artificial. IBP sites 13 and 14, at 2,740 and 3,050 m respectively, were sampled by Delfinado, and no Sciaridae were found there.

Species Distribution Trends

Sixteen species of Sciaridae were found on the transect. All of these species are shown in figure 3-16 which gives a diagrammatic representation of the species distribution trends along the transect. Even though samples were collected with different techniques, all sample data were treated together in this figure. Population abundance levels were coded subjectively with values from 1 through 9 for display; these levels were based on the quantitative enumeration of specimens of each species obtained from the different sampling techniques.

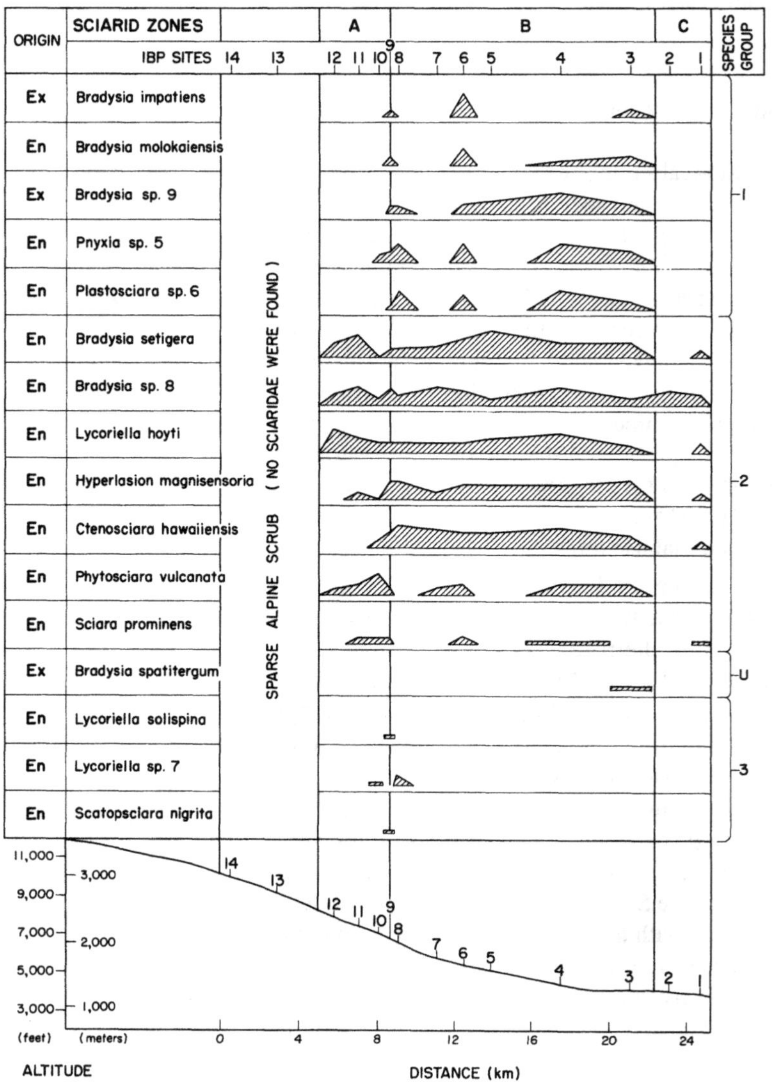

FIGURE 3-16. *Distribution of Sciaridae populations along Mauna Loa transect as caught in Malaise traps and pitfall traps, by sweep and pyrethrum sampling. Curve heights represent subjective ratings of population densities from 1 to 9 based on the combined counts from all samples per site. Maximum abundance (rating of nine units) is indicated, for example, for* Bradysia setigera *at site 5.*

TABLE 3-3. *Specimen Numbers of Adult Sciaridae*

Code	Malaise trap	Pitfall trap	Pyrethrum fogging	Sweep sampling
1	1		1	
2	2–20	1	2–9	1
3	21–40	2–3	10–19	2–3
4	41–100	4–6	20–29	4–6
5	101–300	7–9	30–39	7–9
6	301–600	10–14	40–49	10–14
7	601–1,000	15–24	50–59	15–24
8	1,001–2,000	25–50	60–70	25–50
9	2,001–4,000			

Based on the sampling yields obtained by the four different methods, the specimen numbers of adult Sciaridae presented in table 3-3 represent approximately equivalent population levels. The distribution analysis indicated that the sixteen species of Sciaridae collected on the Mauna Loa transect form three groups. Group 1 is a spatially associated group of five species (*Bradysia impatiens, Bradysia molokaiensis, Bradysia* sp. 9, *Pnyxia* sp. 5, and *Plastosciara* sp. 6) that extends from the dry ‘ōhi‘a forest (1,220 m) to the upper limits of the mountain parkland (2,080 m). The absence of these species from IBP site 7 is probably artificial. More intensive or additional sampling from other areas within this zone undoubtedly would reveal these species. Two of these species, *Bradysia impatiens* and *Bradysia* sp. 9, are exotic. *Pnyxia* sp. 5 is a wingless species and therefore is a good indicator species for site-residence relationships, since movement by winged adults between sites would not have to be considered. Both *Bradysia impatiens* and *Bradysia molokaiensis* are predominantly lowland species.

Group 2 contains seven abundant species. Four of these (*Bradysia setigera, Bradysia* sp. 8, *Lycoriella hoyti,* and *Phytosciara vulcanata*) extend to the tree line through the subalpine forest, and all except *P. vulcanata* extend into the rain forest. The three other species of this group (*Hyperlasion magnisensoria, Ctenosciara hawaiiensis,* and *Sciara prominens*) do not extend as far up the transect as the tree line ecosystem, stopping in the subalpine scrub. All of the species of this group are endemic.

Group 3 is composed of three species: *Lycoriella solispina, Lycoriella* sp. 7, and *Scatopsciara nigrita*. These were collected only in the upper part of the mountain parkland (sites 8–10). These three species are also endemic. One species, *Bradysia spatitergum,* could not be grouped. It was collected only in the seasonally dry ‘ōhi‘a forest (site 3) below the savanna. This species is an exotic.

Computer analysis of the species distribution trends yielded three transect zones (A, B, and C in figure 3-16). Zone A consists of the subalpine environment (sites 10–12); zone B extends through the entire central portion of the transect from the upper boundary of the mountain parkland (site 9) through the open *Metrosideros* dry forest (site 3); and zone C has two sites (1 and 2), which are both located in the rain forest ecosystem. These three zones correspond to separate environmental sections (figure 2-1).

Discussion and Conclusions

Sciarid distribution supports Whittaker's (1970, 1975) third hypothesis, here applied to one group of small arthropods rather than plants. According to our data, sciarid species do not exclude each other by competition along the altitudinal gradient that was sampled. One group of five species (group 1) does not extend into the more extreme environments along the transect—the rain forest and subalpine environments. At these limits they drop out as a group. As a group, they also broadly overlap with the distribution range of the other main distribution group (number 2), which consists of seven abundant species and shows an even wider distribution. In a general way, the limits to the distribution of the species of Sciaridae along the transect correspond to the distribution changes of major environmental and substrate factors.

Species of group 1 have a narrower range with more gaps in their distribution than do the species of group 2. Group 1 seems to be more sensitive to environmental variations along the transect, which would include both physical and biotic factors. One of the difficulties in assessing determinants of sciarid distribution is the mobility of most Sciaridae versus the relatively short distances involved between sites. The entire transect spans a distance of only 25 km. Except for the wingless species, *Pnyxia* sp. 5, the samples could consist of individuals with different mobility characteristics: those breeding at the sample site, those moving into the site, or those passing through.

Pnyxia sp. 5 represents the distribution of most species of group 1. The range of this species coincides with those fungi defining the mesic montane soil-fungus zone B, where overall fungal species were found to be both quantitatively and qualitatively most abundant. *Pnyxia* sp. 5 is probably a litter inhabitant whose larvae feed on fungi living in decomposing organic matter. The ecological role of Sciaridae has never been clearly defined except that larvae occur in various organic substrates. Kennedy (1974) has shown that *Bradysia impatiens* has a higher mean survival rate and develops faster on a fungal diet as opposed to a nonfungal diet. He also summarized food habits of sciarid larvae and indicated that developing sciarid larvae may utilize a fungus as a primary food source instead of the substrate on which they occur. The close correlation of the sciarid zones to the soil-fungus zones would be expected.

Therefore, as in soil-fungus distributions, factors that govern the distribution of vital organic substances would be expected to influence sciarid distribution.

The upper limit of group 1 is probably determined by a combination of factors, including low temperature, low relative humidity, and a decrease in the availability of decomposing organic substrates that characterize the subalpine environment. The lower limit of this group probably is determined by high relative humidity, which characterizes the rain forest environment. Many of the fungi are also absent from this zone because of excessive moisture.

The gap in distribution of IBP site 7 is probably due to sampling design. The absence of *Pnyxia* sp. 5 can be easily explained since a pitfall trap was not set up at this site, and this species was collected only in the pitfall traps. The sweep samples were taken only during a continuous five-day period, and, due to extreme fluctuations of sciarid populations during the year, absence of the group 1 species at site 7 would not necessarily indicate that they were not present at other times. The gap at IBP site 5 possibly could be similarly explained, although a pitfall trap was in operation there for a short period of time.

With the exception of *Bradysia* sp. 8, five of the seven species of group 2 are relatively similar in distribution to the species in group 1 in that they either do not extend into the rain forest environment or are relatively rare in this zone. Their extension into the subalpine environment could be due to migration rather than indicating actual breeding in this zone. This is one of the difficulties in explaining spatial distribution of flying insects within a relatively short distance.

Bradysia sp. 8 was the most widely distributed and one of the most abundant species on the Mauna Loa transect. It is the most common species in the rain forest (sciarid zone C). This species is unusual in that no males have as yet been found, although hundreds of females have been collected by various sample methods. Attempts to rear this species have not been successful. This species is either parthenogenetic or the males are wingless. In any case, apparently it can thrive under a broader range of environmental conditions than any of the other species found on this transect.

Generally the sciarid populations are both quantitatively and qualitatively most important in zone B, which is equivalent to the mesic montane soil-fungus zone B. This zone is characterized by seasonal, mesic conditions, moderately deep to deep soils, and a variety of plant communities, including scrub, grassland, savanna, and forest. The soil-fungus study indicates a positive general correlation between high overall fungal populations and moderate levels of moisture as well as organic matter content. McColl (1975), in a recent study of the invertebrate fauna of the litter surface of a *Nothofagus truncata* forest floor in New Zealand, reported a significant positive correlation between number of adult Sciaridae and the frequency of 70 to 60 percent relative humidity. Our laboratory studies have also indicated that moderate moisture levels were most favorable for the development of Sciaridae. Such moderate moisture levels are characteristic for the mid-section of the Mauna Loa transect, from sites 3 to 9.

BAIT-TRAPPED DROSOPHILIDS*

Adults of many *Drosophila* species are attracted to a wide variety of fermenting substances, although some are more specialized in their feeding and breeding sites (Carson, 1971). It was possible to use such an attractant to survey the composition of the ecologically more generalized *Drosophila* fauna along the Mauna Loa transect. The study was undertaken primarily to collect *Drosophila immigrans* and other exotic species of the genus *Drosophila* for use in the studies of chromosomal polymorphisms.

Sampling Information

The *Drosophila* were collected by exposing plastic buckets, 33 cm in diameter and 38 cm deep, each of which contained about 1 kg of yeasted bananas as bait. These traps were placed on shaded ground, and the flies attracted to the bait were collected with the aid of a collecting net. Occasionally it was necessary to trap the flies by placing an empty vial directly over the flies as they fed on the bait. In almost all collections, five to seven traps were exposed for four days, and the collections were made from the second through fourth days, visiting the traps once between the hours of 9 and 11 A.M. and again between 4 and 6 P.M. The captured flies were transferred to food vials (Spieth, 1966) and returned to the laboratory for identification and counting.

Collections were obtained from IBP sites 2 through 12, except site 5. This general area affords a wide range of ecological conditions. At three sites (2, 6, and 9) nine collections were made (April and December 1971; October 1972; April, July, August, September, October, and December 1973). At the other sites, repeated collections were made during 1973.

Species Distribution Trends

Twenty-seven species of fruit flies were collected, eighteen species of the genus *Drosophila* and nine species of the genus *Scaptomyza*. Seven of the *Drosophila* species were exotic. All others were endemic.

There were great differences in population densities. By far the most abundant species was the exotic *Drosophila simulans* with over 55,000 individuals trapped. Then came another exotic, *Drosophila immigrans,* with over 6,000 specimens. Third most abundant was the endemic *Scaptomyza paralobae* with 451 individuals trapped during the entire sampling period. All other species were represented with fewer than 200 individuals in the total collection. For the analysis of distribution trends, only the quantitatively more important

*By Y. K. Paik and K. C. Sung

species are diagrammatically shown in figure 3-17. Included are the thirteen species for which at least seven individuals were caught during the sampling along the transect.

The thirteen dominant species fall into five overlapping distribution

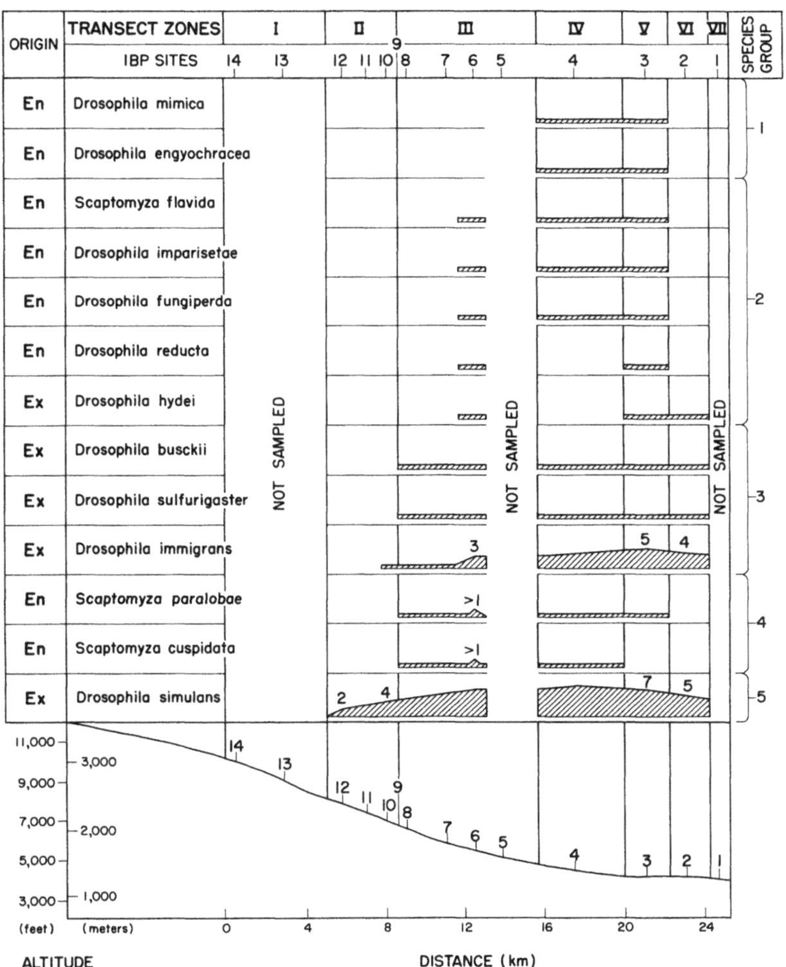

FIGURE 3-17. *Distribution of thirteen species of drosophilid flies caught by the banana-baiting technique along Mauna Loa transect. These were the quantitatively more important species. Their densities were adjusted for trap nights and are shown by the $\log_2$ scale. Only* Drosophila immigrans *and* D. simulans *were caught with greater abundance: 2 units = 3–4 specimens per trap night, 3 = 5–8, 4 = 9–16, 5 = 17–32, 6 = 33–64, 7 = 65–128, 8 = 129–256 (= maximum at site 4 collected of* D. simulans). *En = endemic, Ex = exotic species.*

groups. Except for *Scaptomyza cuspidata,* they all overlap in the seasonally dry forest site 3, but they extend upslope to different elevations.

Group 1, which includes two endemic species, *Drosophila mimica* and *D. engyochracea,* extends upslope into the savanna. Group 2 with the five species—*Scaptomyza flavida, Drosophila imparisetae, D. fungiperda, D. reducta,* and *D. hydei*—extends farther upslope into the middle of the mountain parkland (site 6). The only species in this group that is an exotic is *D. hydei*. It differs from the others by showing a range extension into the open rain forest at site 2. The distribution gaps at site 4 for *D. hydei* and *D. reducta* can be attributed to sampling error because of the small number of specimens collected at adjacent sites.

Group 3 includes three exotic species: *D. busckii, D. sulfurigaster,* and *D. immigrans*. This group extends upslope from the rain forest through the mountain parkland to its upper limit (at site 9) or even a little beyond that into the subalpine scrub forest. The range extension was shown by the most abundant species of group 3, *D. immigrans*.

Group 4, which includes the two endemic *Scaptomyza* species, *S. paralobae* and *S. cuspidata,* shares its upper distribution limit with group 3. However, both of these *Scaptomyza* species were not collected in the rain forest (site 2), and they displayed a common abundance peak in the mountain parkland (site 6).

Group 5 is represented by only one species, *Drosophila simulans,* which was found to extend up to the tree line ecosystem (at 2,500 m elevation).

The fourteen rarer species include all those for which fewer than seven individuals were trapped. Of these, the following ten probably belong to group 1 or 2: *D. ananassae, D. basisetosa, D. digressa, D. mercatorum, D. murphyi, D. hawaiiensis, S. parva, S. exigua,* and *S. articulata*.* These species only extend into the middle of the mountain parkland.

The remaining four rare species—*D. pectinotarsus, S. bryanti, S. hackmani,* and *S. adunca*—extended to the upper boundary of the mountain parkland. As a result of these collections, these species appear to belong to group 4. With all of these rare species, however, more sampling is necessary before definite conclusions can be reached regarding their spatial distributions.

Discussion and Conclusions

The species collected in this study are a case example of broadly overlapping groups whose distributions approximate the model of Whittaker's third hypothesis. Most of the *Drosophila* and *Scaptomyza* species caught by this artificial baiting technique can be considered as generalists. Exceptions are the two endemic *Drosophilas* of group 1, *D. mimica* and *D. engyochracea,* which

**D. ananassae* and *D. mercatorum* are exotic species.

do not seem to stray far from their host tree *(Sapindus saponaria)* in the mesic kīpuka forest.

The endemic species clearly show the narrower distribution ranges in this group of organisms. Among the five species of spatial group 2, only the exotic *Drosophila hydei* falls somewhat out of line by showing a wider range, which extends into a different climatic zone, the rain forest (at site 2). The four endemic species (*Scaptomyza flavida, D. imparisetae, D. fungiperda,* and *D. reducta*) exhibit a narrower range and a better distributional association with one another. Their ecological optimum, like that of spatial group 1, appears to be in the seasonally dry and mesic environments rather than in the wetter areas along the transect. The exotic *D. hydei* may be ecologically more similar to spatial group 3 in that it can tolerate a greater climatic variation, particularly with regard to rainfall. Group 3 includes only exotic *Drosophila* species (*D. busckii, D. sulfurigaster,* and *D. immigrans*), and this group clearly displays an environmentally generalistic distribution. In contrast, group 4 consists again of endemic species, the two more abundant *Scaptomyza* (*S. paralobae* and *S. cuspidata*) that were caught along the transect. Their range is environmentally more restricted than that of group 3. Moreover, the two *Scaptomyza* species are spatially integrated with the endemic *Drosophila* species in the sense that their ecological optima are in different habitats. These *Scaptomyza* species clearly extend farther up the transect to the lower limit of the subalpine forest. They seem to find their ecological optima in the mountain parkland (zone III), while the optima of the endemic *Drosophila* species are farther down the transect in the more mesic savanna (zone IV). The spatial integration by these endemic genera is less obvious only because of the presence of the five exotic *Drosophila* species (*D. hydei, D. busckii, D. sulfurigaster, D. immigrans,* and *D. simulans*). Unlike the endemic species caught in this study, they all extend into the transect rain forest (at site 2), and they show greater tolerance to environmental variations along the transect.

Thus, among the drosophilid flies sampled in this study, the endemics and exotics, each considered as a group, are ecologically quite well differentiated.

LITTER-INHABITING DIPTERA*

From the background work that has been done on the leaf breeding Drosophilidae (Heed, 1968; Carson et al; 1970; Montgomery, 1975), it is obvious that large numbers of Diptera, as well as other arthropods, breed in plant litter. Sampling these litter-inhabiting organisms with the generally used Berlese or Tullgren funnel technique, however, is not satisfactory since these funnels recover only a small percentage of the Diptera, or other flying or delicate insects. The adults and immature stages of most holometabolous insects are

*By D. E. Hardy, M. D. Delfinado, and D. Fujii

highly sensitive to the heat and desiccation generated in the funnels. Techniques that have been developed based on the emergence of adults have proved to be successful for recovering flies breeding in the litter and were used in this study.

This study was undertaken to determine the kinds and abundances of Diptera that breed in litter along the Mauna Loa transect, to establish the altitudinal range of the species along this elevational gradient, and to make comparisons with the species found at nearby sites.

Sampling Information

Litter collection sites were chosen along the entire Mauna Loa transect, with the exception of IBP site 7 (1,720 m). IBP site 4 was sampled in the nearby kīpukas. In addition, samples were taken in three other rain forest areas: the Kīlauea Forest Reserve, the 'Ōla'a Forest, and two sites off the Saddle Road at 1,580 m (kīpukas 9 and 14) between Mauna Loa and Mauna Kea. The two Saddle Road sites are of interest because the elevation, climate, and flora are all similar to the Kīlauea Forest Reserve site.

Five samples of litter, preferably moist, were collected above the soil once a month (occasionally once in two weeks) at each of eleven IBP sites. The litter lying on an area 30.5 by 30.5 cm was collected to a depth of about 4 cm and placed in a plastic bag, which was then stored in a cold box for transport back to the laboratory. The use of an insulated box with a coolant was essential to the sampling since the Diptera are so sensitive to heat and desiccation.

Temperatures in the laboratory were maintained at 16 to 18°C with near absolute humidity. Here the samples were transferred to gallon jars or cans for rearing. Each jar or can contained 8 to 10 cm of damp, washed, and sterilized sand upon which the sample was placed. The container was covered with a sheet of plastic containing several holes plugged with cotton to allow for air exchange. When the larvae living in the decaying vegetation mature, they burrow into the sand and pupate. Collections of the emerging adults were made twice weekly for a period of three months. This technique has proved highly effective for recovering flies breeding in the litter. The period of time in getting the samples from the field to the laboratory is very critical. Some of the blanks or gaps occurring in the sample record may be due to the loss of specimens caused by temperature changes. Ideally such rearings should be done near the sample areas. The sampling period for this study was from July 12, 1971, to August 22, 1972.

Species Distribution Trends

Forty-five species of litter-inhabiting Diptera were reared from samples collected along the Mauna Loa transect. These species represent thirteen families and twenty-one genera; all but six are endemic species. Another six of these are new species, which still need to be described. Many were present with rather small populations. We considered a minimum of five specimens reared from the same site over the one-year sampling period as sufficient evidence that a species was residing at that site. This criterion yielded twenty quantitatively more important species, whose distributions are diagrammed in figure 3-18. No specimens were reared from litter of alpine shrubs at sites 13 and 14. According to their distribution trends along the transect, the litter-inhabiting Diptera can be divided into five spatially associated species groups.

Group 1 includes four species residing at higher altitudes along the transect in the subalpine scrub forest. One of these (a new genus and species of Cecidomyiidae) has a wider distribution, from about 1,800 m to 2,300 m, while the three others (*Orthocladius mauiensis, Lycoriella hoyti,* and *Scaptomyza inequalis*) occur over very narrow ranges at 2,130 m (site 10) and 2,030 m (site 9). However, *Orthocladius mauiensis* and *Lycoriella hoyti* were also collected in one of the three other rain forest sites sampled away from the transect: the ʻŌlaʻa Forest which is north of IBP site 1 at 1,200 m elevation. These two species therefore are not restricted to high altitudes. In each case, their population sizes at the ʻŌlaʻa site were very low (two and one specimens, respectively) so that the higher altitude site on the transect appears more favorable to them. The two other species of group 1 (the new species of Cecidomyiidae and *Scaptomyza inequalis*) were found exclusively along the transect and thus can be designated as high-altitude Diptera.

Group 2 includes two species that were found only at mid-elevation (1,600 m) along the transect in the litter of an *Acacia koa* tree colony (IBP site 6). One of these, *Forcipomyia* n. sp., was found also at all three rain forest sites away from the transect, where it was considerably more abundant (29 specimens at the ʻŌlaʻa site, 109 at the Kīlauea Forest site, and 23 at the Saddle Road kīpukas). Thus the residence of *Forcipomyia* in the mountain parkland (site 6) represents only an extension of this species into a drier climate. It may still be moving into other suitable habitats. In contrast, *Scaptomyza* near *cuspidata* was found only along the transect at this site and thus can be considered a mid-elevation, mesic forest species.

Group 3 includes two species present at two places along the transect, at site 6 and in the closed mesic kīpuka forest (K) in the savanna (zone IV). Both species, *Scaptomyza cuspidata* and *Drosophila simulans,* occurred also at one of the three rain forest sites outside the transect, in the Kīlauea Forest (at 1,600 m). Thus the two species form a truly associated group with similar optima, their best representation being in the kīpuka forest.

Group 4 includes nine species that occurred together in the closed mesic kīpuka forest in transect zone IV. Six of these (*Drosophila mimica, D. carnosa, D. chaetopeza, Homoneura unguiculata,* and *Scaptomyza articulata*) were found only here, while three of the species occurred also at other rain forest sites. Of these, *Drosophila imparisetae* definitely finds its ecological

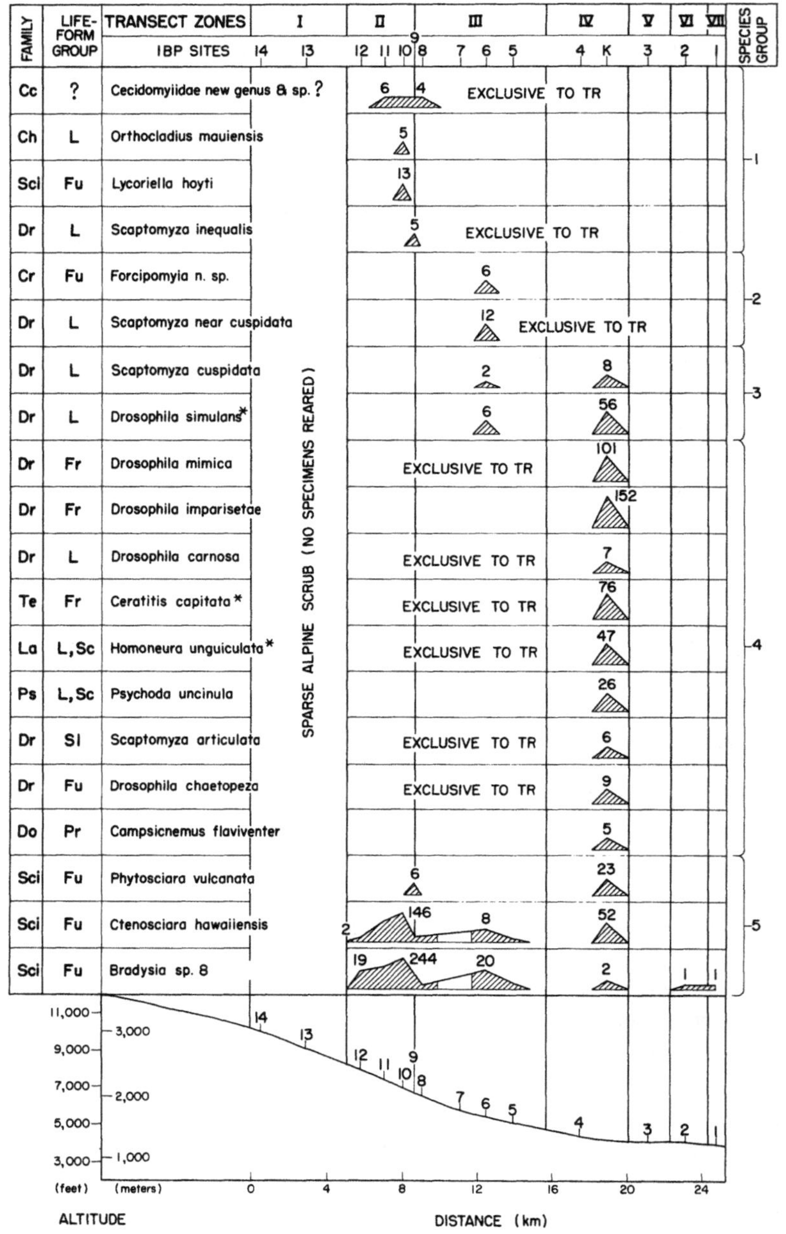
FAMILY
LIFE-FORM GROUP
TRANSECT ZONES
I II III IV V VI VII
IBP SITES 14 13 12 11 10 9 8 7 6 5 4 K 3 2 1
SPECIES GROUP
Cc ? Cecidomyiidae new genus & sp. ?
Ch L Orthocladius mauiensis
Scl Fu Lycoriella hoyti
Dr L Scaptomyza inequalis
Cr Fu Forcipomyia n. sp.
Dr L Scaptomyza near cuspidata
Dr L Scaptomyza cuspidata
Dr L Drosophila simulans*
Dr Fr Drosophila mimica
Dr Fr Drosophila imparisetae
Dr L Drosophila carnosa
Te Fr Ceratitis capitata*
La L, Sc Homoneura unguiculata*
Ps L, Sc Psychoda uncinula
Dr Sl Scaptomyza articulata
Dr Fu Drosophila chaetopeza
Do Pr Campsicnemus flaviventer
Sci Fu Phytosciara vulcanata
Sci Fu Ctenosciara hawaiiensis
Sci Fu Bradysia sp. 8
EXCLUSIVE TO TR
SPARSE ALPINE SCRUB (NO SPECIMENS REARED)
ALTITUDE
(feet) (meters)
11,000 9,000 7,000 5,000 3,000
3,000 2,000 1,000
DISTANCE (km)
0 4 8 12 16 20 24

optimum in the mesic kīpuka forest along the transect because it occurred in considerably smaller numbers in two of the three rain forest sites (with two specimens at the 'Ōla'a site and seven at the Saddle Road). In contrast, *Psychoda uncinula* occurred at all rain forest sites outside the transect with greater numbers than in the mesic kīpuka forest (241 specimens at the Saddle Road, 61 in the 'Ōla'a Forest, and 33 in the Kīlauea Forest). Its ecological optimum is therefore in the rain forest. For this reason it is surprising that *Psychoda uncinula* was not found at IBP site 1. Possibly this species is rather sensitive to the sulfur fumes that occasionally may reach this site. The third species, *Campsicnemus flaviventer,* was present in two other rain forest sites, in the 'Ōla'a Forest with eight specimens and at the Saddle Road with fifteen. Thus this *Campsicnemus* species shows only a slightly better development in these rain forests than in the mesic kīpuka forest on the transect.

Group 5 includes three Sciaridae species, which all have wider distributions along the transect than any of the species discussed above. *Bradysia* sp. 8 ranges from the tree line forest (site 12) to the closed rain forest (site 1). Thus it shows truly ubiquitous tendencies. This was the only quantitatively important Diptera found in the transect rain forest sites 1 and 2, which apparently were avoided by all other litter-inhabiting Diptera. The influence of occasional sulfur fumes may be the discriminating factor for these species. Like the *Bradysia* sp. 8, *Ctenosciara hawaiiensis* shows its optimum in the subalpine scrub forest (at site 10), but it has a second optimum also in the mesic kīpuka forest together with the third species of the wider-ranging ones, *Phytosciara vulcanata*. As expected, these three wide-ranging Sciaridae were also present at the rain forest sites outside the transect. However, here their quantities were much smaller than at most transect sites. An exception was the *Bradysia* sp. 8, which was represented by 50 specimens at the 'Ōla'a site.

The Mauna Loa transect sampling also yielded twenty-five quantitatively less important Diptera: Two species at transect site 1 in the closed rain forest: *Eurynogaster maculata* (Dolichopodidae), one individual; *Drosophila dasyc-*

FIGURE 3-18. *Litter-inhabiting Diptera along Mauna Loa transect. Included are only those species of which at least five specimens were reared from a site sample of litter. Quantities are given in numbers of specimens reared and are shown diagrammatically by the* log_2 *index. All sites except 7 were sampled. Family symbols: Cc = Cecidomyiidae, Ch = Chironomidae, Cr = Ceratopogonidae, Sci = Sciaridae, Dr = Drosophilidae, Te = Tephritidae, La = Lauxaniidae, Ps = Psychodidae, Do = Dolichopodidae. Life form symbols: L = litter breeders usually saprophagous, Fu = fungus feeders and breeders (usually), Fr = fruit feeders and breeders, Sc = scavengers, Sl = slime flux breeder on trees (*Myrsine *sp.), Pr = predator of small animals, ? = life form unknown. K = kīpuka forest. TR here refers to Mauna Loa transect. *Exotic species.*

nemia (Drosophilidae), one individual. One species at transect site 2 in the open rain forest: *Giardomyia furvescens* (Cecidomyiidae), one individual. Nineteen more species at transect site 4 in the mesic kīpuka forest: *Campsicnemus* sp. 1 (Dolichopodidae), three individuals; *C.* sp. 2, one individual; *Drosophila fungiperda,* two individuals; *D. immigrans,* three individuals; *D. pectinitarsis,* two individuals; *D.* sp. ♀♀, three individuals (this species recurred at site 9 with two individuals); *D.* n. sp. aff. *molokaiensis,* two individuals; *Scaptomyza flavida* (Drosophilidae), one individual; *S. hackmani,* two individuals; *S. intricata,* one individual; *S. lobifera,* one individual; *S. scoloplichas,* one individual; *Lispocephala dexioides* (Muscidae), two individuals; *Chonocephalus simiolus* (Phoridae), one individual (this species recurred at the tree line site 12 with another individual); *Megaselia heterodactyla* (Phoridae), two individuals; *Neoexaireta spinigera* (Stratiomyidae), one individual; *Dacus dorsalis* (Tephritidae), two individuals; *Limonia grimshawi* (Tipulidae), three individuals (over 50 individuals were found at each of the three rain forest sites outside the transect); *L. stygipennis,* one individual.

Three more Sciaridae were found at higher elevation transect sites: *Scatopsciara nigrita,* one individual at site 10; *Bradysia triticus,* one individual at the tree line site 12; and *Sciara prominens,* three individuals at site 12.

Discussion and Conclusions

The litter-inhabiting Diptera portrayed in figure 3-18 include four species of Sciaridae that were reported in this chapter by Steffan. These are *Lycoriella hoyti, Phytosciara vulcanata, Ctenosciara hawaiiensis,* and *Bradysia* sp. 8. Except for the last two species, which in both studies showed wide-ranging and almost continuous distributions along the transect, the distributions differ considerably in detail. This is not surprising in view of the different sampling methods employed. Steffan used four methods to sample Sciaridae, all related to above-ground areas where the flies roam freely in their adult stage. The litter-rearing method employed in this study was a sample of their breeding localities. According to this method *Lycoriella hoyti* was found only at site 10 in the subalpine scrub forest, whereas Steffan's samples located the same species from this vegetation type downslope into the rain forest. This makes it very probable that some breeding localities downslope were not revealed by the litter-rearing method, although they can be expected there. However, the larvae samples obtained in the subalpine scrub forest, the low quantity of *Lycoriella* flies sampled by Steffan in the rain forest, and his population mode shown for the higher elevation sites indicate that *L. hoyti* has its ecological optimum in the drier climate above the rain forest that prevails along the transect. The same applies also to *Phytosciara vulcanata*. There is a close similarity in the distribution range of this species recorded in both studies. The litter-rearing method shows two locations, site 9 (at the upper limit of the mountain parkland) and site

K (the mesic kīpuka forest) in the savanna (zone IV). Steffan's sample gives a wider range about these two sites and a location in between (site 6, in the middle of the mountain parkland) but no specimens from the rain forests (sites 1 and 2). Thus *P. vulcanata,* as well as *Ctenosciara hawaiiensis,* show their ecological optimum outside the rain forest as revealed by both study methods (aboveground samples and litter rearing).

Two other Sciaridae species—*Scatopsciara nigrita* and *Sciara prominens*—were found in both studies. These two species yielded so few individuals by the litter-rearing method that their distributions were not portrayed on the diagram. *Scatopsciara nigrita* was reared from litter in the subalpine scrub forest (site 10) with only one individual, while *Sciara prominens* yielded three specimens from the tree line site (number 12). Inspection of Steffan's figure 3-15 shows a much wider distribution for *Sciara prominens,* extending with a few interruptions from the tree line site into the rain forest, while *Scatopsciara nigrita* was sampled above the ground at nearly the same subalpine site as it was reared. Thus, while the litter-rearing method indicated more localized distributions in general, the approximate transect range was revealed for the more abundant species.

The diagram of the litter-inhabiting Diptera (figure 3-18) also includes four species caught by the banana-baiting technique. These are the two species in spatial group 3, *Scaptomyza cuspidata* and *Drosophila simulans,* and the two most abundant species in spatial group 4, *Drosophila mimica* and *D. imparisetae*. The two species of spatial group 3 were both reared from litter samples in the middle of the mountain parkland (site 6). The banana-baiting technique supports this upslope distribution in both cases. It shows that *Scaptomyza cuspidata* ranges throughout the mountain parkland and savanna zones, while *Drosophila simulans* was found even up to the tree line site. Thus the baiting technique indicates a considerable range extension upslope for this exotic *Drosophila* species. However, the number of flies caught above the mountain parkland in the subalpine scrub forest was very much fewer than those caught in the mountain parkland from where it was reared also. This numerical difference, and the probability that the successful exotic *Drosophila simulans* is less specialized and thus may roam over a wider elevational segment than the other *Drosophila* species, may explain the variation in sampling results obtained from the two techniques. The other two *Drosophila* species recovered as quantitatively important in both studies, *D. mimica* and *D. imparisetae,* have similar distribution ranges within each sampling technique. However, as one may expect, both species indicate a wider range by the baiting technique, which shows them distributed throughout the savanna (zone IV) and reaching into the neighboring vegetation zones. The litter-rearing technique indicates the mesic kīpuka forest within the savanna zone as the primary breeding habitat of these two abundant endemics. Both are known to lay eggs and complete larval development in the fruits of the indigenous tree *Sapindus saponaria,* which occurs in the kīpuka and is also scattered throughout the savanna.

Four other species were recovered by both the baiting and rearing techniques: *Drosophila immigrans, D. fungiperda, Scaptomyza flavida,* and *S. hackmani*. *Drosophila immigrans* is an exotic that was found as the second most abundant species in the banana-baiting technique, according to which it extended from the *Metrosideros* dry forest (site 3) into the subalpine forest (site 10). Only three individuals were obtained by the litter-rearing method from the mesic kīpuka forest, too few even to indicate their distribution trend in figure 3-18. *D. immigrans* and *D. simulans* are abundant in temperate and tropical areas throughout the world. They are associated with rubbish heaps but are also known to breed on a wide variety of fallen and fermenting fruits and flowers in sites far from human habitation (Carson, 1965). Such locally enriched substrates for yeast growth appear to be scarce along the Mauna Loa transect, although Carson (1965) reports that *D. immigrans* was reared from a slime flux of *Acacia koa* in that area. Montgomery (1975) reared both species from natural areas in Hawai'i and has shown that they may breed on endemic as well as exotic substrates.

Only one individual of *Drosophila fungiperda* was reared from litter in the kīpuka forest while with the banana-baiting technique it was found roaming beyond the borders of this mesic forest in the savanna and above, into the center of the mountain parkland (site 6) and below, into the *Metrosideros* dry forest (site 3). A similar range extension was found by the baiting technique for the adult flies of the remaining two *Scaptomyza* species (*S. flavida* and *S. hackmani*), which were reared only from the mesic forest. Perhaps such a range extension indicates the more generalized feeders among the endemic *Drosophila* and *Scaptomyza* flies. *Scaptomyza flavida* and *S. hackmani* have been known to be strongly associated with three native woody plant genera: *Myrsine, Clermontia,* and *Cheirodendron*. Except for *Myrsine,* which also occurred at site 1, these trees are restricted to the mesic kīpuka forests, where they are quite sparsely represented. The range differences of *S. flavida* and *S. hackmani* as found with the two sampling techniques were, however, less dramatic than those found for the two important exotic *Drosophila* species, *D. simulans* and *D. immigrans*.

On the basis of this comparison and the distribution pattern shown in figure 3-18, the litter-inhabiting Diptera can be divided into three ecological distribution groups. First are the Diptera best represented in moderately dry forest communities. These include the wide-ranging Sciaridae in spatial group 5 on the diagram but also those Diptera of spatial group 1, which were sampled from the subalpine scrub forest. The two group 1 species designated as exclusive to the transect, the new Cecidomyiidae species and *Scaptomyza inequalis,* were not recorded in any rain forest litter sample. The other two species, *Orthocladius mauiensis* and *Lycoriella hoyti,* were recorded at the 'Ōla'a Rain Forest site, but with fewer specimens than on the transect. Thus, these are similarly wide ranging as spatial group 1 but also seem to find their ecological optimum in moderately dry forest communities. Of spatial group 2, only one species belongs in this ecological group: *Scaptomyza* cf. *cuspidata,* which was found only on the transect (in the mountain parkland, site 6). Of the other species of group 2, the new species of *Forcipomyia* was much more abundant in the rain forest and thus belongs in the third ecological distribution group.

The second distribution group consists of the Diptera that find their ecological optimum in the mesic kīpuka forests. These include the two species of spatial group 3 (*Scaptomyza cuspidata* and *Drosophila simulans*) and the six species reared only from the kīpuka forests along the transect (*Drosophila mimica, D. carnosa, Ceratitis capitata, Homoneura unguiculata, Scaptomyza articulata,* and *D. chaetopeza*). Of these, *Ceratitis capitata* is known to prefer feeding on the exotic shrub *Solanum pseudocapsicum,* which occurs abundantly in Kīpuka Kī.

The third group consists of the Diptera best represented in rain forest habitats. In addition to the new species of *Forcipomyia* (of spatial group 2), this group includes the remaining three species of spatial group 4: *Drosophila imparisetae, Psychoda uncinula,* and *Campsicnemus flaviventer.* All four species were much more abundant at the rain forest sites outside the transect so there is little doubt about their ecological optimum being in rain forest habitats. Their absence from the two rain forest sites along the transect is attributable possibly to the occasional volcanic sulfur steaming there. The only taxonomic group of the Diptera less affected by such fuming appears to be the Sciaridae, since members of this group were recorded from these transect sites by both the free-roaming adult and litter-reared samples.

Little definite information is available on the ecological role of the litter-inhabiting Diptera other than the life-form designations noted in figure 3-18. From this, one can recognize certain functionally interesting tendencies. For example, among the Diptera that have their ecological optimum in the moderately dry forest communities are several fungus feeders (the Sciaridae) but also a few saprophagous flies, such as *Orthocladius mauiensis, Scaptomyza inequalis,* and *S.* cf. *cuspidata.* A greater variation in life forms is exhibited by the mesic forest-associated Diptera, which in addition to fungus feeders and saprophages, also include fruit feeders, slime flux breeders, scavengers, and predators. A similar complement of at least six general life-form types can also be found among the rain-forest-associated Diptera, however, they are not well represented along the Mauna Loa transect. Although a high species packing was observed in the mesic kīpuka forest along the transect and in the rain forests outside the transect, there appears to be little direct competition among the litter-inhabiting Diptera because of their resource partitioning, which usually turned out to be highly specialized when examined in more detail. For example, *Sapindus saponaria* and *Sophora chrysophylla* were dominant trees in the kīpuka forests, and a few individuals of *Cheirodendron trigynum* occurred. Each of these native trees supports a more or less specialized fauna of Diptera on their slime flux, leaves, rotting bark, stems, and fruits. Several species commonly breed together in the same decaying leaf, piece of stem, bark, fruit, or other parts of various endemic plants. Apparently many can coexist in the same substrate. We have seen eggs of as many as six species in a single *Cheirodendron* leaf. Heed (1971: 118) reared a record total of nine species, out of seventeen specimens, from one small sample of *Ilex* leaves collected in a rain forest site.

Although the Diptera species examined here do not seem to compete with one another significantly, the presence of Diptera predators has been considered important. Spieth (in Carson et al., 1970: 491, and Spieth, 1974: 101) has suggested that persistent predation was a major driving force in the evolution of the great diversity of Hawaiian Drosophilidae. He further speculated that the development of lek behavior in *Drosophila* arose as a means of escaping predation.

SOIL ARTHROPODS*

The fauna of most soils is extremely rich in species and life forms. These range from microfauna, such as Protozoa and Rotifera to macrofauna, such as earthworms and, in continental areas, gophers and moles. Most of the biomass is represented by microscopic to small (macroscopic) organisms, among which the arthropods are dominant in many situations. Soil arthropods have received considerable attention in temperate and subtropical Europe. Little work has been done in Hawai'i thus far and has been limited primarily to systematic surveys at low elevations.

Our sampling was done with pitfall traps and with simple Berlese (or Tullgren) funnels, resulting in recovery of the arthropod mesofauna, ranging from mites less than 1 mm in length to beetles and spiders about 1 cm in length.

Species identifications have been obtained for many of the higher categories in this study and are in progress for others. In this section, we have restricted our distributional analysis for the Mauna Loa transect to the springtails (Collembola) and the soil mites (Acari) of the suborder Mesostigmata. Each of these two groups is represented by many species in the Hawaiian Islands, and they are dominant and ecologically important among the soil mesofauna. The primitively apterous hexapods of the order Collembola typically are found in soil and litter. These are first-order consumers, feeding primarily on fungi, algae, and decomposing plant or animal matter. The soil mites include a wide range of consumer types, but those analyzed here are from groups that are largely predatory. Other soil mesofauna, such as oribatid and prostigmatid mites, spiders, and many orders of insects, were also found in abundance and will be treated elsewhere.

---

*By F. J. Radovsky and J. M. Tenorio. We would like to acknowledge the assistance of Dr. Peter Bellinger in providing preliminary identifications of Collembola and relevant information as to their distribution and habitats.

Sampling Procedure

Except for the open rain forest (site 2) and the alpine zone (sites 13 and 14), most of the IBP sites along the Mauna Loa transect were sampled, often with more than one sampling location. Site 7 was left out because of the habitat uniformity in this middle section of the transect. For comparison, two additional sites were sampled outside the transect in the Kīlauea rain forest at 1,600 m elevation.

Two quite different sampling methods were applied at the same sites and generally at the same times: pitfall trapping and Berlese extraction. Pitfall trapping for soil organisms is considered a relative sampling method. Trap results are influenced to some extent by the mobility and other behavioral traits of the organism. Moreover, the results are cumulative depending on the duration of a trapping period. However, pitfall traps are frequently used because they yield greater numbers of some species, they can sample a wider area for the more vagile species (thus being less influenced by the exact site where a sample is taken), and they may assist in interpreting cyclical patterns. Pitfall traps similar in design to that described by Fichter (1954) were used. The effective perimeter of the trap—at ground surface—was 1 m.

Berlese extraction is considered an absolute sampling method because it is capable of yielding an almost unbiased sample of soil arthropods (vagile and nonvagile). Soil samples are taken in the field, transported in plastic bags to the laboratory, and placed on wire screening in a funnel. The soil is dried from above by the heat of a lamp. Desiccation of the substrate, heat, and light force the arthropods down to the metal screening, through which they fall into a container of fluid preservative below the funnel. Theoretically all arthropods present in a sample will be collected in this way. However, only highly sophisticated extractors, which are not in general use, will recover all organisms. In addition, the efficiency of the method is influenced by time and conditions at which samples are held between collection and extraction, the method of collection, and the method of handling the samples when they are placed in the extractor.

Pitfall traps were operated for two years (October 1971 through September 1973). During the first year, traps and Berlese sampling sites were located at twelve sites on the transect and two sites in the Kīlauea Forest. During the second year, four of the sampling sites on the transect were relocated, three to other points on the transect and one below the transect.

Berlese samples were taken at least monthly throughout the two-year study period. The method of collecting samples was changed between the first and second year of sampling. During the first year, a circle covering a 0.05 m^2 area was set down, and all litter within the circle was taken as one sample; the underlying soil was dug up with a trowel and an amount equal to 0.5 liter in volume was taken as another sample.

During the second year, soil sampling involved a coring device. A 1 m square was marked. Litter from approximately one-fourth of the square consti-

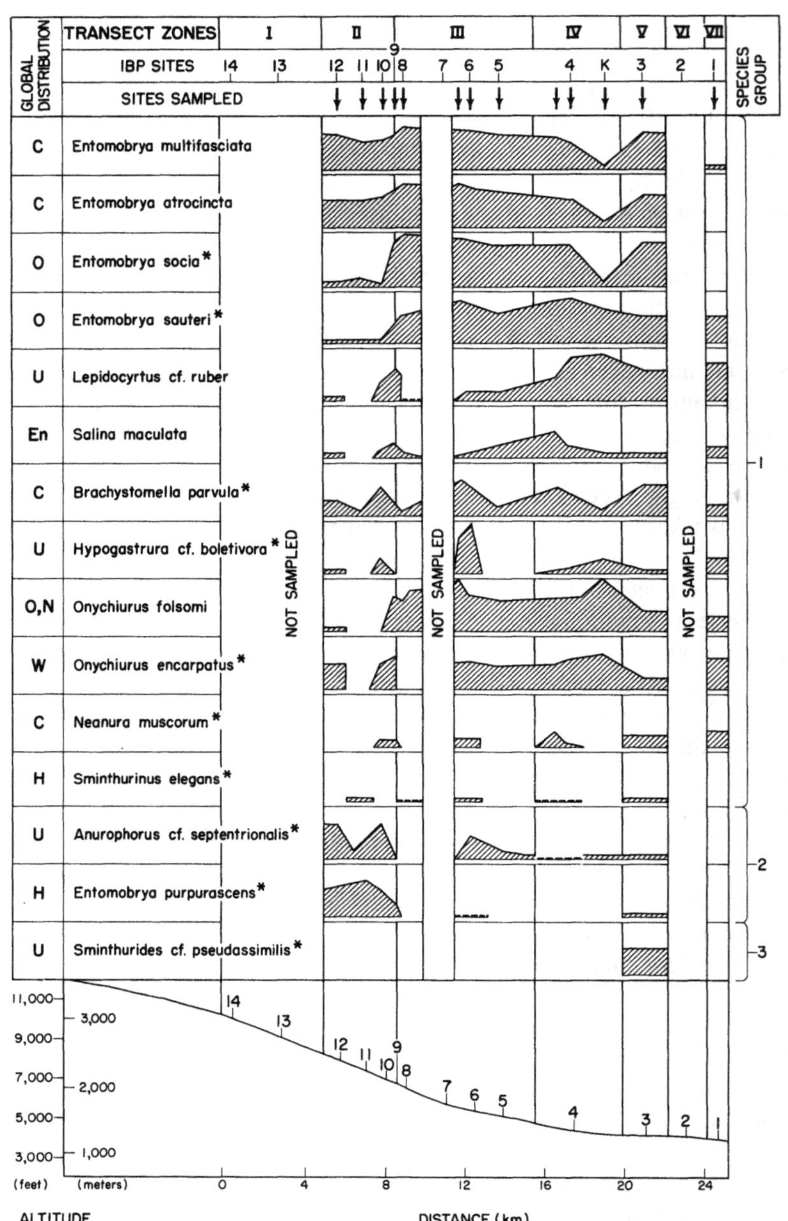
GLOBAL DISTRIBUTION
TRANSECT ZONES
I II III IV V VI VII
IBP SITES
14 13 12 11 10 9 8 7 6 5 4 K 3 2 1
SITES SAMPLED
SPECIES GROUP
C Entomobrya multifasciata
C Entomobrya atrocincta
O Entomobrya socia*
O Entomobrya sauteri*
U Lepidocyrtus cf. ruber
En Salina maculata
C Brachystomella parvula*
U Hypogastrura cf. boletivora*
O,N Onychiurus folsomi
W Onychiurus encarpatus*
C Neanura muscorum*
H Sminthurinus elegans*
U Anurophorus cf. septentrionalis*
H Entomobrya purpurascens*
U Sminthurides cf. pseudassimilis*
NOT SAMPLED
NOT SAMPLED
NOT SAMPLED
1
2
3
11,000
9,000
7,000
5,000
3,000
(feet)
3,000
2,000
1,000
(meters)
0 4 8 12 16 20 24
ALTITUDE
DISTANCE (km)
* NEW RECORD FOR HAWAIIAN ISLANDS

tuted one sample. Soil cores were taken at each of the four corners of the square. The corer allowed separation of each coring into three depths: 0–3 cm, 3–6 cm, and 6–9 cm. The same depth ranges from the four corners were combined into one sample, giving four samples for extraction from each site: litter and three soil depths. Each soil sample (made up of parts of four corings) amounted to a volume of 275 cm^3.

The change in sampling method was necessitated by the obvious inadequacy of the technique used during the first year. Apparently because of the disturbance of the normal soil interstices in the first-year method, many smaller arthropods were trapped in the soil, and recovery of arthropods was grossly qualitative. In the second-year method, the coring technique left the soil fauna relatively undisturbed, and sampling results were improved.

Soil Arthropod Distribution Trends

For the transect analysis, we found it best to sum the individual counts made per site for each arthropod species collected from both pitfall traps and Berlese extractions. This gave a truer representation of the total transect range of each species than the use of either method alone. The collections from the first-year Berlese extraction were excluded from the quantitative summation, but they were used to round out distribution gaps. Thus a species recovered from a site only by Berlese extraction during the first year is indicated as present at that site in figures 3-19 or 3-20. In testing the data we found that summation of the two different collection yields (pitfall years 1 and 2, plus Berlese year 2) did not bias the quantitative trends along the transects for the species selected.

Collembola

Figure 3-19 shows the distribution of fifteen Collembola species (springtails) collected along the Mauna Loa transect. These fifteen were selected from a total of twenty-four species taken on the transect, as both sufficiently abundant to show meaningful distribution trends and identified with a high level

---

FIGURE 3-19. *Distribution of fifteen common Collembola species (springtails) along Mauna Loa transect. Curve heights represent numbers of individuals summed from pitfall traps (two years) and Berlese funnel extractions (second year only) and converted to log_2 scale. Maximum quantity is thirteen units = 4097–8912 individuals. Collembola found at a site only in year 1 Berlese extractions are indicated by a broken line. Global distribution: C = cosmopolitan, H = holarctic, N = nearctic, W = Western Hemisphere, O = Oriental, U = uncertain, En = endemic species.*

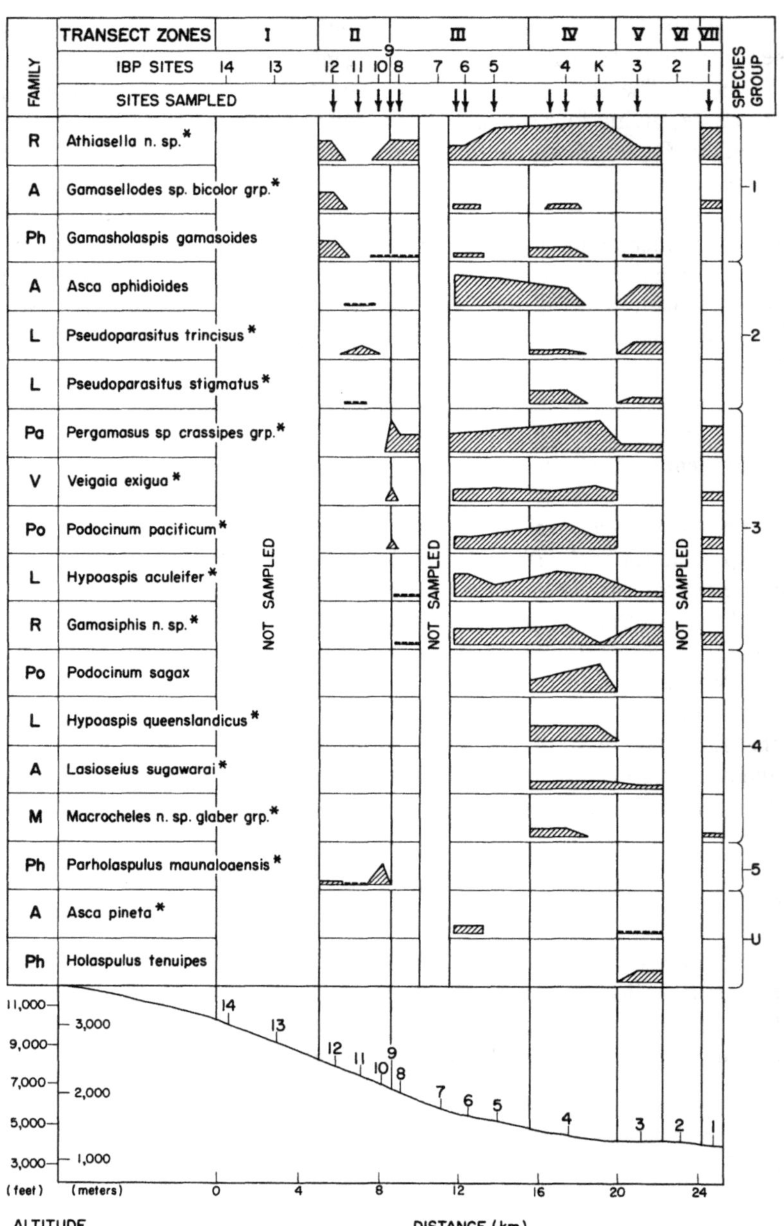

* NEW RECORD FOR HAWAIIAN ISLANDS

of confidence. Ten of the species reported here are new records for the Hawaiian Islands; these are identified on the figure with an asterisk. The fifteen springtails can be divided into three spatial distribution groups. Group 1 combines twelve species that are distributed throughout the sampled transect or nearly so. However, differences in the abundance peaks of these species suggest some ecological differentiation along the transect. The two *Entomobrya* species (*E. multifasciata* and *E. atrocincta*) were particularly abundant in the subalpine scrub forest (zone II) upward to the tree line site 12; they showed a common low population level in the closed kīpuka forest (site K), and they were either absent or present in low numbers in the closed rain forest (site 1). Thus both species were abundantly represented in open-structured woody vegetation (regardless of altitude), but they were very much reduced in population levels in closed forests, regardless of whether these were mesic or rain forests. A third species, *Entomobrya (Homidia) socia,* shows the same pattern of low numbers or absence in the closed forests but was much reduced in the open subalpine scrub forest. Its higher elevation distribution parallels closely that of *Entomobrya (Homidia) sauteri,* which also was sparsely present in the open subalpine scrub forest. However, unlike *E. socia, E. sauteri* was abundant in both the closed mesic forest (site K) and the closed rain forest (site 1). Interestingly both of these species were relatively abundant also in the closed, wet Kīlauea Forest (where *E. multifasciata* and *E. atrocincta* were rare). *Lepidocyrtus* cf. *ruber* showed extremely high population levels in both closed forest sites (K and 1); however, it was barely detected in the Kīlauea Forest. These first five species, all in the family Entomobryidae, show distribution modes that trend downslope in the order shown on figure 3-19.

The next three species—*Salina maculata* (Entomobryidae), *Brachystomella parvula* (Hypogastruridae), and *Hypogastrura (Ceratophysella)* cf. *boletivora* (Hypogastruridae)—indicate rather erratic distribution peaks along with their wide spatial range. These erratic peaks may be correlated with more subtle substrate or vegetation patterns.

The two Onychiuridae, *Onychiurus folsomi* and *O. encarpatus,* were found abundantly at most sites. The *Onychiurus* species, both of which are

---

FIGURE 3-20. *Distribution of eighteen Mesostigmata species in soil and leaf litter along Mauna Loa transect. Curve heights represent numbers of individuals summed from pitfall traps (two years) and Berlese funnel extractions (second year only) and converted to* log_2 *scale. Maximum quantity is 10 units = 505–1008 individuals for* Athiasella *n. sp. at site K = kīpuka forest; minimum quantity is 1 unit = 1–2 individuals. Species found at a site only in year 1 Berlese extractions are indicated by a broken line. Family symbols: R = Rhodacaridae, A = Ascidae, Ph = Parholaspidae, L = Laelapidae, Pa = Parasitidae, V = Veigaiidae, Po = Podocinidae, M = Macrochelidae.*

deep-soil arthropods, range throughout the transect and have very broad abundance peaks throughout the middle portion of the transect from the closed kīpuka forest (site K) to the upper limit of the mountain parkland (site 9). *O. folsomi* shows some reduction in population levels in the ʻōhiʻa forest zones at the transect extremes, zone VII (rain forest) and zone II (open subalpine scrub-forest), while *O. encarpatus* shows rather high population levels in these zones.*

The last two species of group 1, *Neanura muscorum* (Hypogastruridae) and *Sminthurinus elegans* (Sminthuridae), show an interrupted distribution along the transect. Their population levels were relatively low and uniform. Their interrupted occurrences probably are a reflection of characteristically low population numbers.

Group 2 includes two species, *Anurophorus* cf. *septentrionalis* (Isotomidae) and *Entomobrya (Entomobryoides) purpurascens* (Entomobryidae), which are also widely distributed along the transect. Both species have distinct population peaks at the higher elevations (2,100–2,300 m) in the subalpine scrub forest (zone II). Their absence in the rain forest (zone VII at 1,200 m) is probably real in view of their decreasing population levels downslope from the subalpine environment to the open *Metrosideros* dry forest (zone V).

Group 3 is represented by only one species, *Sminthurides* cf. *pseudassimilis* (Sminthuridae). It was restricted to the open *Metrosideros* dry forest (1,200 m), where it was found in moderate numbers. However, this species was also collected in the Kīlauea Forest. *S.* cf. *pseudassimilis* departs in its narrowly restricted range from the distributional behavior of the other Collembola considered here.

Mesostigmata

The second group of soil arthropods treated here are the mesostigmatic soil mites shown in figure 3-20. These mites are frequently predators on springtails as well as other small arthropods. We collected at least forty-two species, of which eighteen were sufficiently numerous and reliably identified to warrant portrayal of their distribution trends along the transect. Fourteen of these species are new records for the Hawaiian Islands. These are identified by an asterisk in figure 3-20. Probably all eighteen species are exotics. They were divided into five spatially associated species groups.

Group 1 includes three species of nearly ubiquitous distribution along the transect: *Athiasella* n. sp., *Gamasellodes* sp. (*bicolor* group), and *Gamasholaspis gamasoides*. *Athiasella* n. sp. was by far the most abundant of these soil mites, particularly along the lower half of the transect.

**O. folsomi* was further investigated in its consumer relationship to algae (TR 55).

Group 2 includes three species: *Asca aphidioides, Pseudoparasitus trincisus,* and *P. stigmatus*. These were restricted at the two ends of the transect by being absent in the closed rain forest (zone VII) and at the tree line (site 12). They were only sparingly represented in the subalpine scrub forest (zone II). They were well represented in the low-elevation *Metrosideros* dry forest (zone V) and in the savanna (zone IV). *A. aphidioides* was particularly abundant in the mountain parkland and savanna but significantly absent, together with *P. trincisus* and *P. stigmatus,* in the closed mesic kīpuka forest (site K). Thus the three mites of group 2 show a similarly negative response to areas of closed woody vegetation.

Group 3 includes five species: *Pergamasus* sp. (*crassipes* group), *Veigaia exigua, Podocinum pacificum, Hypoaspis aculeifer,* and *Gamasiphis* n. sp. These species were not collected in the subalpine scrub forest (zone II). They were all relatively abundant in the mountain parkland and savanna (zones III and IV) and well represented in the closed rain forest (zone VII). Thus these species occurred abundantly in grassy and woody vegetation types, and in the latter they seemed to be more restricted to those woody vegetations with abundant herbaceous undergrowth. Both the *Metrosideros* dry forest (zone V) and the subalpine scrub forest (zone II) lack a significant herbaceous undergrowth layer.

Group 4 is represented by four soil mites that show a relatively narrow distribution in the lower part of the transect: *Podocinum sagax, Hypoaspis queenslandicus, Lasioseius sugawarai,* and *Macrocheles* n. sp. (*glaber* group). They were all represented in the savanna (zone IV), but they were not collected in the mountain parkland. They were also absent or only sparsely present (*Macrocheles* n. sp.) in the rain forest (zone VII).

The three remaining mesostigmatic soil mites—*Parholaspulus maunaloaensis, Asca pineta,* and *Holaspulus tenuipes*—have individual distributions. Because of its definite restriction to the subalpine zone II, *Parholaspulus maunaloaensis* represented a group by itself, in this case group 5. The other two species were designated as ungrouped (u). *Asca pineta* was taken in small numbers at two sites in the mountain parkland (zone III) and only once (in year one, Berlese sample) in the *Metrosideros* dry forest (zone V). Its plotted distribution appears to be a reflection of its sparseness along the transect. Its actual distribution may be wider. The same can be said of *Holaspulus tenuipes,* which has been taken in other collections on Hawai'i and O'ahu Islands at high and low elevations in both native and exotic forest habitats (Garrett and Haramoto, 1967; unpublished specimens in Bishop Museum collection). It is probably widely distributed in the Hawaiian Islands.

Discussion and Conclusions.

The two major groups of soil arthropods investigated here show different distribution patterns. Except for *Sminthurides* cf. *pseudassimilis,* the springtails (Collembola) tended to range along the whole length of the sampled transect, and ecological differentiation was indicated by abundance peaks in different vegetation zones. In contrast, the mesostigmatic soil mites generally were more restricted spatially. Only group 1 (*Athiasella* n. sp., *Gamasellodes* sp., *bicolor* group, and *Gamasholaspis gamasoides*) was as widely distributed as most of the springtails.

These observations might be explained by considering a combination of various factors (see, for example, Wallwork, 1970; Butcher et al., 1971): emergent vegetation, edaphic factors such as pore volume, microflora and soil moisture, the climatic gradient, and mechanisms of dispersal.

Very few of the springtails showed correlations with the vegetation zone boundaries. However, three of the *Entomobrya* species (*E. multifasciata, E. atrocincta,* and *E. socia*) exhibited a clear correlation with the structure of the vegetation. They showed distinctly low levels of density or even absence in the two closed forest vegetation types on the transect (sites 1 and K) and in the Kīlauea Forest, and they were very abundant in the open woody vegetation types. It would be of some interest to determine whether shade or some other factor associated with a closed canopy is the cause for depressing population numbers of these species. *Entomobrya atrocincta* was also collected in considerable numbers on 'ōhi'a reproductive structures. This collembolan therefore extends its niche into the tree stratum, where it apparently congregates in positions exposed to the sun. The species was not recovered in the general canopy samples.

While only one species of springtail, *Sminthurides* cf. *pseudassimilis,* was restricted to the lower elevations (1,200 m) along the transect, restriction to lower elevations was common among the mesostigmatic mites. The upper distribution limit for the five species in spatial group 3 coincided with an important vegetation boundary—the upper limit of the mountain parkland, which is characterized here by a transition from closed grassy undergrowth to more open shrubby undergrowth in the subalpine scrub forest. Possibly these mites are restricted by this structural vegetation change. However, their upper limit may also be controlled by the decrease in temperature upslope. The temperature gradient seems more clearly responsible for the upper limit of the four species in spatial group 4 because the closed tall-grass cover continues from the savanna (their upper boundary) into the mountain parkland without any abrupt change. Soil moisture appears of lesser importance because the sites were all on well-drained substrates.

A striking difference is evident between the soil arthropods analyzed in this section and the foliar arthropods, as well as the arthropods on 'ōhi'a reproductive structures. On the transect, the latter two groups have a very high level of endemicity. This difference may be related to two factors: the different nature of their substrates and the problems of dispersal.

In the broader geographic context, the soil is certainly a much more homogeneous medium as an arthropod substrate than are the endemic tree species. Thus it is not too surprising to find a higher endemicity among arthropods utilizing primarily endemic trees as hosts than among those living in the soil. Only one soil arthropod species is clearly an endemic, the springtail *Salina maculata*. Several others, those designated "cf." in figure 3-19 and "sp." or "n. sp." in figure 3-20, may prove to be endemics.

It seems extremely improbable that long-distance dispersal has contributed significantly to the distribution of soil arthropods. Most species were probably introduced through human influence, such as in soil material and plants. Distribution of the species here reported upslope on Mauna Loa also can be attributed to dispersal primarily in soil fragments or on the feet of introduced animals (pigs, goats, rodents) and humans and their pack animals.

Whether endemic soil arthropods have been displaced as a consequence of human activity, direct or indirect, is an interesting question that may never be fully answered. Certainly Hammer's investigations (for example, 1972, 1973) of oribatid mites in southern Polynesia throw doubt on previous suppositions of many oribatid endemics on different Pacific islands and support the existence of an oribatid fauna that is generally widely distributed in Oceania. However, we agree with Hammer's emphasis on the importance of humans in the present distribution of soil arthropods. This is suggestive, at least, of a reduction in endemicity through displacement and dispersal.

Considerable new information has been obtained through this investigation, as is evident from the many species of soil arthropods not recorded in Hawai'i previous to this study: ten of the fifteen Collembola and fourteen of the eighteen Mesostigmata reported here.

TERRESTRIAL ALGAE*

The study of the ecology of terrestrial algae is in its infancy. In spite of an early start, ecological information, particularly of tropical terrestrial algae, is only fragmentary and suggestive. This is due both to severe taxonomic problems and to problems in sampling. This paucity of knowledge has caused some difficulties in the quantitative distribution analysis attempted here.

Terrestrial algae are very mobile organisms. They are generally distributed as wind-blown particles (spores or vegetative cells) that fall on all surfaces (Brown et al., 1964) from where they are washed or swim into the soil. Not all living algal particles that arrive at the soil surface grow and colonize or form communities. Many are eaten by insects (France, 1913, fide Petersen, 1935; McGurk, 1974, TR 55) or are removed by wind (Brown, 1971), rain (Petersen, 1935), or other means.

According to Petersen (1935), algae have been found in virgin soils to depths of 2 m, where they may occur at densities of 20,000 to 56,000 particles

*By M. S. Doty and Linda-Lee (McGurk) Watson

per surface square centimeter below the 5 cm level. Still more were found where the soil was fertile. Those from subterranean soils were the same algae as those in the surface layers. France (1913) reported (fide Petersen, 1935: 25) a change in dominance from Chlorophyceae through Bacillariophyceae to Cyanophyceae at successively greater depths to 50 cm. However, the algae at these depths may not multiply significantly in nature (Petersen, 1935). Yet algae were found to remain alive in dry soil samples for at least seventy-three years (Bristol, 1919).

The roles of terrestrial algae have been discussed widely (Petersen, 1935) for a long time, but Treub (1888) was possibly the first to note that they play a pioneering role on new tropical volcanic materials. Their pioneer functions have been described also for new Hawaiian volcanic substrata (Doty, 1967; Smathers and Mueller-Dombois, 1974) and those on Surtsey in the temperate zone (Behre and Schwabe, 1970). The oxygen, carbon dioxide and food-producing roles, as well as their suppression and stimulation of other soil organisms, have received attention, but little or no measurement seems to have been done (Parker et al., 1961). For temperate areas, their role as pioneer humus formers on new soil has long been recognized (Graebner, 1895). Terrestrial algae may also play this role in tropical calcareous atoll soils (Doty, 1954).

The generic criteria for tropical terrestrial algae are being developed. Species criteria are rudimentary. Even the foremost specialists working with tropical terrestrial algae use only generic names in their ecological publications. There are only a few publications on tropical terrestrial algae giving other than taxonomic information. Most of these publications are on the algae common to highly calcareous soils. The isolation techniques used appear to favor the blue-green algae.

In Hawai‘i, Brown (1971) studied wind-borne and soil-surface algae on a transect across the island of O‘ahu parallel to the prevailing wind direction. He found such algae as *Chlorococcum, Chlorella, Phormidium, Nostoc, Lyngbya, Scytonema, Chroococcus, Calothrix,* and *Plectonema* at all stations. In paired soil and aerial collections, the presence of *Stichococcus, Nannochloris, Botrydiopsis,* and *Brachyococcus* were highly correlated. On a presence-absence basis of algal genera, Brown recognized five different spatial distribution groups across the island. The present study incorporates algal population quantities in the analysis of their spatial distribution along the Mauna Loa transect.

Sampling Procedure

All IBP transect sites were used for repeated sampling except the open rain forest site (2) and the two sparse alpine scrub sites (13 and 14), each of which was sampled only once. These three sites have not been included in the analysis of the data. IBP sites 3, 6, and 10 were particularly heavily sampled, with monthly records from August 1972 through October 1973. The other IBP sites were each sampled at least twice.

Two techniques were used to obtain samples. Large surfaces, such as soil and rocks, were sampled by pressing the polyethylene glycol coated end of a wooden or plexiglass rod against the surface and then storing the sampling rod in a sealed screw-capped vial. In the laboratory, these samples were prepared for incubation by dissolving the polyethylene glycol and quantitatively spraying a portion of the suspended sample onto 10N BBM agar plates (Lee and Bold, 1974). Small soil samples and litter fragments were sampled by sonication in liquid BBM with aliquots of the liquid sprayed on the same kind of agar plates. A vigorous effort was made to ensure that the time between sample collection in the field and plating in the laboratory was less than twelve hours.

Incubation was carried on for a few weeks at 25°C and 200 foot-candles under cool-white fluorescent lights. Counting and sorting algal colonies was done on each plate when new colonies of the species being counted no longer appeared on the plates. The number of colonies was tabulated for each species, with these counts being put into generic groups for the analysis. Recognition of species was attempted at first but found impossible because of the lack of taxonomic information.

Algal Distribution Trends

Figure 3-21 portrays the distribution trends of fifteen of the thirty-eight algal genera recovered from the Mauna Loa transect. These fifteen accounted for 98.4 percent of the total number of algal colonies obtained in this study. The number of algal colonies was summed for each generic entity per sampling site. This sum was then divided by the number of samples taken from the site and converted to a square centimeter unit basis. The taxa accepted for this analysis were those that were represented with at least half a percent relative colony density in the sum total of all algal colonies recovered.

Repeated sampling of the same site yielded additional algal genera. Upon plotting the cumulative occurrence of genera over the number of repeated samples for the same site, it was found that the curve leveled off after about ten samples. This means that ten repeat samples were necessary for obtaining a near-complete representation of the algal genera present at a site. This has been recognized in the algal distribution diagram in figure 3-20, where the sites that were sampled more than ten times are indicated by an arrow. However, the diagram does show only the dominant fraction of the algal community. Less intensely sampled sites thus are not expected to distort the between-site comparison significantly. With these limitations in mind, we can recognize three general spatial distribution groups along the Mauna Loa transect.

Group 1 includes seven algae that showed generic population modes in the upper part of the transect. The first of these, *Polycystis*, exhibited a distinct peak at the tree line (site 12). The next three genera—*Nostoc*, *Navicula*, and *Scytonema*—can be described as bimodal. They showed one population mode in the open subalpine *Metrosideros* scrub forest (site 10) and another at lower

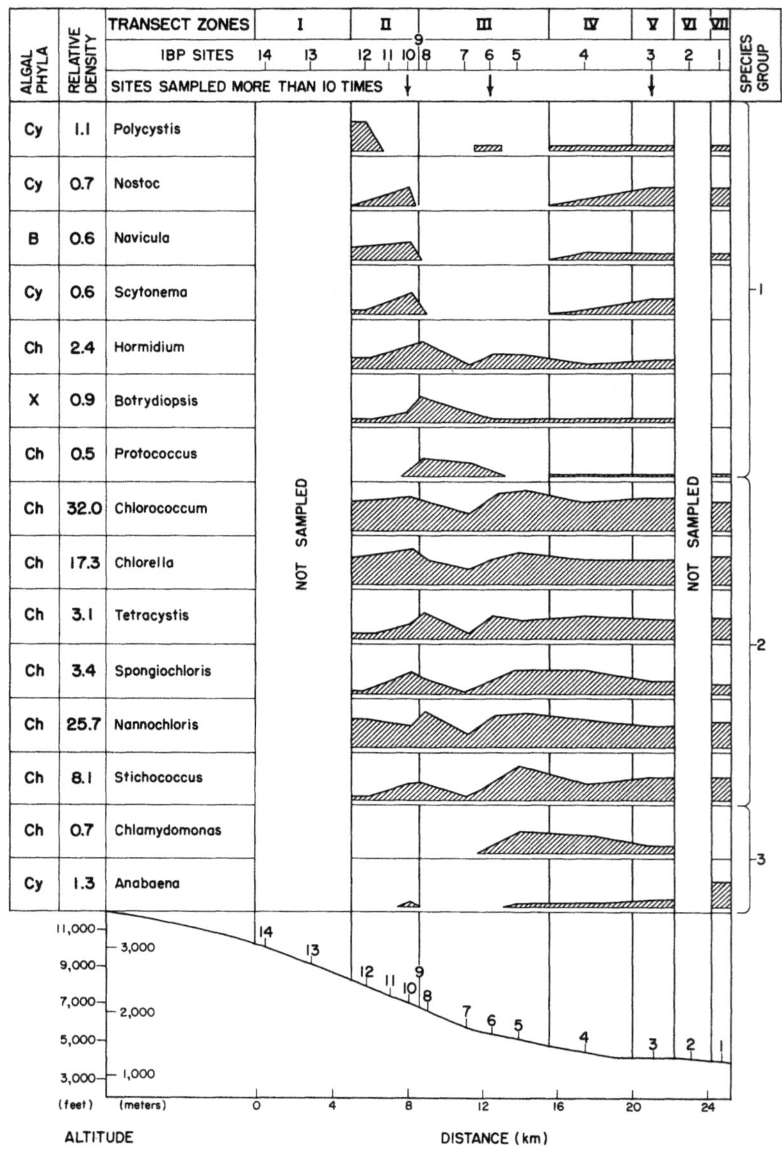

FIGURE 3-21 *Distribution of fifteen algal genera along Mauna Loa transect. These are the quantitatively more important taxa, which together accounted for 98.4 percent of the total number of algal colonies (relative density) recovered from the samples. A taxon was considered quantitatively important if it was recovered with at least half a percent relative density. The curve heights are based on the number of colonies recovered per cm² for the taxon at a particular site divided by the number of samples taken from the*

elevation, centered in the open *Metrosideros* scrub forest at site 3. The absence of these three genera from the middle of the transect, the mountain parkland, appears to be real since the mountain parkland (site 6) was one of the very intensely sampled sites. The next three genera of group 1—*Hormidium, Botrydiopsis,* and *Protococcus*—were best represented in the upper part of the mountain parkland.

Group 2 includes the six quantitatively most important algal genera, which were dominant throughout the sampled transect from the subalpine scrub forest (zone II) to the closed rain forest (zone VII). Their dominance, which was established in percentage of the total algal colony count and which ranged from 32 to 3 percent, was in this order: *Chlorococcum, Nannochloris, Chlorella, Stichococcus, Spongiochloris,* and *Tetracystis*.

Group 3 includes two species with population modes in the lower part of the transect. *Chlamydomonas* was important in the lower part of the mountain parkland and savanna and *Anabaena* in the closed rain forest (site 1). *Chlamydomonas* was never recovered at the intensely sampled high elevation site 10 while *Anabaena* was recovered there, but with very low quantity.

Discussion and Conclusions

All dominant terrestrial algae recovered in this study were spread widely along the Mauna Loa transect. This is not surprising for two important reasons: the samples relate to spores and vegetative cells and the distributional trends are displayed for genera rather than species. The absence of a clear algal community zonation pattern may largely be attributable to these reasons. It may also be possible that these algae form only one broadly defined community.

While recognizing these limitations, the generic distribution trends displayed here offer the basis for at least two testable hypotheses: that the generic population modes are a reflection of the fallout pattern of their wind-borne particles, and that the population modes are indicative of favorable growth conditions occurring at the sites where the respective genera were found to be quantitatively important.

Brown (1971) determined a distinct fallout pattern of airborne terrestrial algae along the 22 km long transect that he sampled on O‘ahu. At sea level on

---

site. These values have been converted to their log_2 values for plotting: 1 unit = 1–2 mean number of algal colonies per cm^2, 2 units = 3–4 colonies, 3 units = 5–8, etc. The maximum number of units shown is 11 (= mean no. of colonies per cm^2 from 1025–2048) for Chlorococcum *at site 5. Sites 8 and 11, in the upper transect range, were not sampled. For these, the curve extrapolations represent the best estimate. Algal Phyla: Cy = Cyanophyta, Ch = Chlorophyta, X = Xanthophyta, B = Bacillariophyta.*

the windward side of the island, he found no airborne algae. This led to his conclusion that the terrestrial algae are derived primarily from the island on which they are sampled. A low mode occurred on the windward midslope, a depression on the very windy summit, and a pronounced peak on the upper leeward slope. Thus the fallout pattern was by no means random. In some cases, he found a good correlation of airborne algal genera with the soil-borne ones. However, a clear sorting of algal genera by the wind pattern was not indicated, implying that the distribution of algal genera may still be random within the modes or depressions of their summation curve.

The generic population curves derived from the present study indicate an algal diversity mode at site 3, where all fifteen genera were recovered. Possibly site 3 was a high fallout area since it is at a point on the transect where the trade winds lose their momentum. However, the combined algal colony counts at this point along the Mauna Loa transect did not differ significantly from those at several other places.

This renders the second hypothesis as a more likely explanation. In this case, the generic population modes indicate a certain spatial differentiation that may be interpreted as their ecological optima along the transect. It is of some interest here that the truly ubiquitous spatial group 2 is comprised totally of green algae (the Chlorophyta); only two green algae (*Hormidium* and *Chlamydomonas*) were placed into other spatial groups. In contrast, the blue-green algae (the Cyanophyta) developed population modes at the transect extremes—*Polycystis* in the tree line ecosystem and *Anabaena* in the closed rain forest—or they showed otherwise restricted distributions—*Nostoc* and *Scytonema*. *Scytonema* is known as an important pioneer alga on new volcanic surfaces in Hawai'i (Doty, 1967). It is quite conceivable that the bimodal population trend of *Scytonema* relates to its prevalence on rock surfaces that dominate the ground in both the open *Metrosideros* scrub forests in the subalpine (zone II) as well as in the lower elevation seasonal environment (zone V). The other two bimodally distributed algae, the blue-green *Nostoc* and the diatom *Navicula*, may find optimal conditions in the same sites for other reasons. *Nostoc* may thrive best there because of minimal competition by grasses and other vigorous herbs. Its good representation on the shaded rain forest floor may also be for the same reason. Vegetative undergrowth is dense only in the shrub and tree fern layers. The same or several other explanations may apply for the parallel optima of *Navicula*. It is not clear how permanent the ecological optima are that are indicated from this study.

One pattern clearly has emerged from this study: the Chlorophyta are the broadest generalists among the terrestrial algae along the Mauna Loa transect, and the Cyanophyta are the most specialized, probably because of their ecological tolerance to environmental extremes and their poor competitive ability in communities occurring in more favorable site conditions. The two other algal phyla, the Bacillariophyta (diatoms) and Xanthophyta (yellow-green algae), were each represented only by one dominant genus on the transect. Their distribution trends suggest that they survive in a less broad range than the Chlorophyta as a group, but no general conclusion like that for the Cyanophyta can be drawn for these other two phyla.

SOIL AND LEAF FUNGI*

The altitudinal distributions of both soil- and leaf-surface-borne fungi along the Mauna Loa transect were studied by separate research teams. Research on leaf-surface-borne or phylloplane fungi was initiated by Baker, Dunn, and Sakai (TR 42). Later research on soil-borne fungi was conducted by Stoner, Stoner, and Baker (TR 75). Although these two studies were conducted within separate time frames and according to different designs, their results are examined together here to outline our current ecological understanding of fungal distribution.

The data on soil-borne fungi indicate clearly a differential distribution of fungal communities along the altitudinal gradient, and the determined distributional patterns support strongly the delineation of distinct fungal zones along the gradient.

Research on phylloplane fungi produced no clear evidence of altitudinal zonation among those organisms. However, this was primarily attributable to the limited sampling of the Mauna Loa transect (1,280–2,440 m for phylloplane versus 1,190–3,050 m for soil fungi) and the restricted distribution of the tree species sampled in the phylloplane research. Both the soil and phylloplane studies elucidated factors affecting the distribution of fungi.

Sampling Procedure for Soil Fungi

The altitudinal distribution and ecological significance of soil fungi and fungal communities along the Mauna Loa transect were determined by studying the individual and group occurrence of a limited number of selected reference genera and species (Stoner 1976, TR 75). Reference fungi are defined as those genera or species (1) that are sufficiently represented and distributed along the transect to support statistical and subjective analyses; (2) that belong to major taxa generally associated with established ecological roles (for example, cellulose versus simple sugar decomposition) or habitats (for example, root inhabiting versus humus decomposing) in the soil-plant-root environment; (3) that support intersite comparisons; (4) that can be detected accurately by methods that support a realistic, feasible approach to the extensive sampling area; and (5) that support comparisons of island and continental ecosystems based on available literature. This nonfloristic approach, which focuses attention on

*By M.F. Stoner and Gladys E. Baker

predetermined reference fungi and employs selective methods that ensure their isolation, is a substantial, successful departure from previous methodologies in soil-fungus ecology.

In the first phase of the research, a general survey was performed to determine what fungi existed in A_1-root zone soils at intermediate sites (elevations) along the transect. It was assumed that the survey would provide a reasonable sampling of representative fungi for the purpose of selecting reference genera for a more complete study of the transect. According to the results of the survey and the requirements of the basic research plan, the reference genera selected were *Absidia, Cylindrocarpon, Fusarium, Gliocladium, Gliomastix, Mortierella, Mucor, Paecilomyces, Penicillium, Pythium, Rhizopus, Trichoderma,* and *Verticillium*.

Soils were collected within 10 m^2 areas at sixteen sites along the transect from 1,190 m to 3,050 m elevation. The composite soil sample for each site was prepared from five subsamples, each of which was taken from the A_1 or equivalent mineral surface horizon in the root zone of dominant vascular plants of an area. Sampling was accomplished during one week in July 1973.

The fungal content of each sample was determined by a soil-dilution spread-plate technique whereby soil diluted in a carrier of 0.1 percent water agar was distributed over the surface of solidified, selective agar media in petri dishes. Fungal culture media contained antibacterial antibiotics. Among the selective media used, diet-food medium (Stoner, 1967) was very effective in the isolation of *Gliocladium, Penicillium,* and allied genera; peptone-pentachloronitrobenzene agar for fusaria (Nash and Snyder, 1962); and V-8 juice agar with benomyl for selected Oomycetes and Zygomycetes (TR 75).

Populations of fungi were determined by counting colonies on spread plates and were expressed quantitatively as propagules per gram oven-dry soil.

Sampling Procedure for Leaf Surface Fungi

Phylloplane fungal communities were determined for the two endemic trees, *Metrosideros collina* subsp. *polymorpha* and *Acacia koa,* using a floristic approach. Leaves were collected at five sites along the transect at 1,280 m (IBP site 4), 1,600 m (site 6), 2,030 m (site 9), 2,130 m (site 10), and 2,440 m (site 12). The two tree species did not occur together at all sites. As a result, the data are more reflective of tree-related rather than altitude-related distributional influences.

The phylloplane mycoflora was determined by direct observation of leaves incubated in moist chambers and by the culture on selective media of fungal samples obtained from leaves by washing, maceration, and swabbing. Selective media such as sodium caseinate agar and rose bengal agar contained antibacterial antibiotics. Comparisons of isolates from washed and unwashed leaves were interpreted to indicate differences between the transient and more tenacious resident populations of fungi.

The anatomy and morphology of leaves was studied by light and scanning electron microscopy to reveal features that could influence the distribution of phylloplane fungi.

Soil Fungal Zonation Patterns

Altitudinal zonation of the Mauna Loa transect according to soil-borne fungal communities was determined largely on the basis of the differential

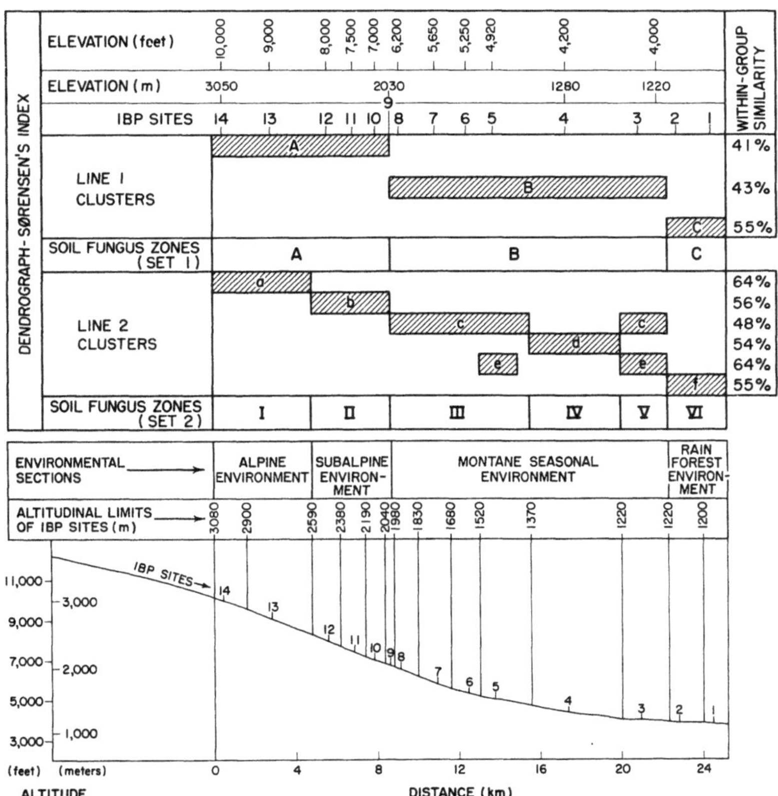

FIGURE 3-22 *Zonation diagram showing fungal zones along Mauna Loa transect derived from dendrograph technique with Sørensen's qualitative similarity index. The analysis was based on seventeen soil collection sites (fungal community samples) each with five subsamples. Zone set 1: A = dry, cool, high-altitude scrub, B = mesic montane parkland and savanna,* C = Metrosideros *rain forest. Zone set 2: I = alpine scrub, II = subalpine scrub forest, III = mountain parkland, IV = montane kīpuka and savanna,* V = Metrosideros *dry forest, VI =* Metrosideros *rain forest.*

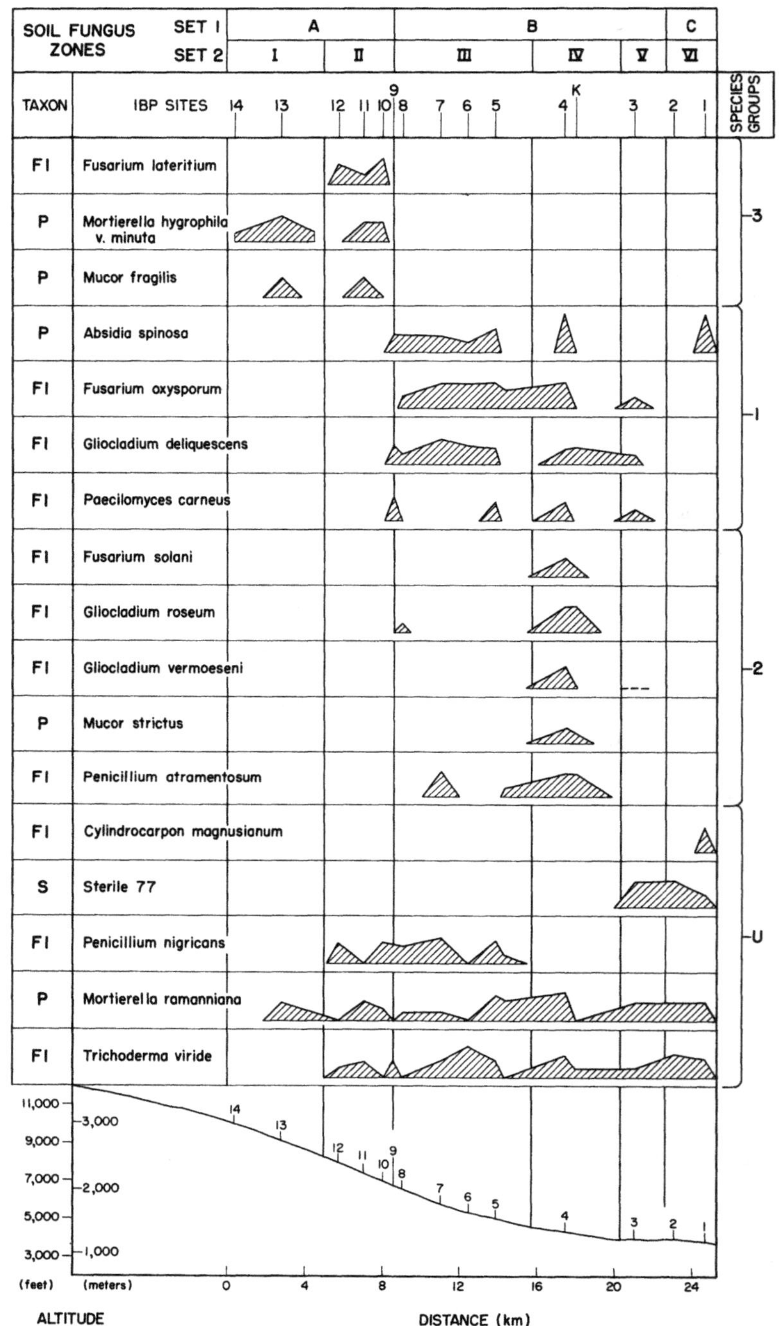
SOIL FUNGUS ZONES
SET 1
SET 2
A
B
C
I
II
III
IV
V
VI
TAXON
IBP SITES
14 13 12 11 10 9 8 7 6 5 4 K 3 2 1
SPECIES GROUPS
Fl Fusarium lateritium
P Mortierella hygrophila v. minuta
P Mucor fragilis
P Absidia spinosa
Fl Fusarium oxysporum
Fl Gliocladium deliquescens
Fl Paecilomyces carneus
Fl Fusarium solani
Fl Gliocladium roseum
Fl Gliocladium vermoeseni
P Mucor strictus
Fl Penicillium atramentosum
Fl Cylindrocarpon magnusianum
S Sterile 77
Fl Penicillium nigricans
P Mortierella ramanniana
Fl Trichoderma viride
3
1
2
U
11,000
9,000
7,000
5,000
3,000
3,000
2,000
1,000
(feet)
(meters)
0 4 8 12 16 20 24
ALTITUDE
DISTANCE (km)

group distribution of reference species. Only reference species were employed in the objective analyses carried out by the dendrograph and two-way table techniques. Reference fungi as well as other species isolated from soil samples were considered in subjective mycological-ecological evaluations of the results.

Two related sets of soil fungus zones (figure 3-22) were determined according to different levels of interpretation of the dendrograph (qualitative Sørensen index). The two-way table analysis (50/10 rule) gave closely similar results (TR 75). Both of these zone sets or patterns are useful in understanding the environmental factors affecting fungal distribution. Each zone has a characteristic fungal community. Soil-fungus zone set 1 (zones A, B, and C) includes relatively broad areas of fungal distribution that reflect more general environmental parameters related to soil, vegetation, and climate. These broader soil-fungus zones are designated as the dry, cool, high altitude scrub (A), mesic montane parkland and savanna (B), and *Metrosideros* rain forest (C). Set 2 (zones I–VI) includes relatively narrow zones of distribution that reflect more localized environmental influences, which tend to associate fungal groups with specific soil-plant-climate complexes. These soil-fungus zones are designated as the (I) alpine scrub, (II) subalpine scrub forest, (III) mountain parkland, (IV) montane kīpuka and savanna, (V) open *Metrosideros* dry forest, and (VI) *Metrosideros* rain forest.

With one exception, the broad zones of set 1 encompass two or three of the more narrow zones in set 2. Zone C of set 1 is identical with zone VI of set 2, the *Metrosideros* rain forest soil fungus zone. Fungal distribution patterns do not indicate clear zonal distinction of open and closed rain forest soil areas within zone C.

Distribution Trends of Soil Fungi

The ecological significance of the soil-borne fungi that exist along the Mauna Loa transect is not in their individual identities or taxonomic affinities per se but in the composition and spatial distribution of the communities they form and in their association with other elements of the ecosystems. The species distribution diagram (figure 3-23) relates transect zones to the distribu-

---

FIGURE 3-23 *Distribution diagram of populations of selected fungal species from soils along Mauna Loa transect. Curve heights are based on relative population levels rated from very low to high. Dashed lines imply presence at very low levels. Species group numbers along the right-hand column refer to spatially associated groups in the two-way table (50/10 rule). Ungrouped species are at the bottom of the diagram. FI = Fungi Imperfecti, P = Phycomycete, S = Sterile Isolate. The closed kīpuka forest is indicated by K near IBP site 4.*

tion of representative populations of fungi. Included on the diagram are reference fungi grouped in the two-way table (50/10 rule) and some ungrouped species that were considered important in the confirmation of zones. The ordering of species in the diagram supports a visual association of representative fungal groups and their associated zones along the transect. Since the isolation methods employed in this research were selective for certain reference groups, the species in the distribution diagram were not chosen to carry any particular, individual significance in a floristic sense. Instead these fungi were selected to represent zonal communities.

The genera *Fusarium, Gliocladium,* and *Penicillium,* which occurred widely in communities along the transect, are important indicators of fungal zones. The genus *Fusarium* displayed very clear zonation according to the differential distribution of three species: *F. lateritium* for the subalpine scrub forest, *F. oxysporum* for the montane seasonal environment as a whole (largest populations correlated with abundant grasses), and *F. solani* for the kīpuka and savanna within the montane seasonal environment. Species of *Absidia, Mortierrella,* and *Mucor* also were useful indicators that, in groups with other reference genera, clarified the zonal boundaries.

Certain fungi such as *Mortierella ramanniana, Trichoderma viride,* and *Cylindrocarpon magnusianum,* which had very broad or narrow ranges of distribution, were useful in subjective evaluations of zonation patterns indicated by objective data.

Species diversity and populations generally were greatest in fungal communities of the mesic, montane regions along the transect. Conversely the least diversity and the smallest populations were noted at the extreme ends of the transect in the alpine scrub (I) and rain forest (VI) soil-fungus zones. The rain forest zone is of particular interest because of its apparent lack of important genera such as *Gliocladium, Fusarium, Mucor,* and *Penicillium* that have representatives in most other zones of the transect.

The validity of the established soil-fungus zones is supported by the varied composition of the fungal groups employed in zonation. The zonation was determined according to the distribution of fungi with different ecological roles, growth habits, and positions within the soil ecosystems. Therefore the fungal groups representing zones, are important functional as well as taxonomic elements.

Distribution Trends of Leaf Surface Fungi

Qualitative and quantitative differences were noted in the phylloplane communities of *Metrosideros collina* subsp. *polymorpha* and *Acacia koa*. Of 150 fungi isolated from leaves, 69 were obtained from *Metrosideros* only and 21 from *Acacia koa* only. About 14 percent of the fungal species isolated were shared by *Metrosideros* and *Acacia*. According to isolations obtained by the maceration of washed leaves, 40 species were designated as actual residents of *Metrosideros,* and 30 as residents of *Acacia. Metrosideros* had more diverse phylloplane communities than *Acacia;* however, the major taxa, such as Fungi Imperfecti, had about the same proportional representation on *Metrosideros* and *Acacia*.

Ninety percent of the fungi isolated from leaves were Fungi Imperfecti. About 50 percent of the fungi were isolated only once; most taxa were not isolated more than seven times throughout the study. A few fungi were isolated fifty times or more; *Aureobasidium pullulans* and *Cladosporium cladosporioides* are examples. *Aureobasidium pullulans* was isolated the greatest number of times at all elevations and on both trees. The largest numbers of isolates recorded for *Metrosideros* leaves occurred at an elevation of 1,600 m (5,250 ft, near IBP site 6) and for *Acacia,* at 2,030 m (6,650 ft, IBP site 9).

Fifty-two of the 150 species recovered from leaves were isolated also from air at field sites, indicating that phylloplane fungi are effectively carried along the gradient by air currents. Presumably both *Metrosideros* and *Acacia* are exposed to similar inocula; yet because of contrasting foliar characteristics, the two trees differ considerably in the number of fungi they trap and retain. Significant quantitative differences in the phylloplane communities for *Metrosideros* and *Acacia* correlated with the type of leaf surface. The leaves of *Metrosideros* and *Acacia,* as shown clearly by light and scanning electron microscopy (figure 3-24), provide different spore-trapping surfaces. Collected leaves of *Metrosideros* had well-developed, intertwined abaxial hairs. The phyllodes of *Acacia koa* lack hairs. The total number of fungal species trapped by the hairy leaves of *Metrosideros* was three times greater than the number recovered from the hairless phyllodes of *Acacia*. If the transient species (those easily removed from leaves by washing) are omitted from the counts, the ratio of resident fungi on *Metrosideros* and *Acacia* is the same. The pubescence of *Metrosideros* leaves is variable; the greater the degree of pubescence, the better the spore-trapping potential of the foliage.

Discussion and Conclusions

Zonation of the Mauna Loa transect is clearly indicated by the distribution patterns of soil-borne fungal groups. Furthermore the genera involved and their patterns of distribution provide insights into the involvement of various biotic and abiotic factors as determinants of fungal distribution. The interrelationships of fungal distribution, vascular plant communities, edaphic factors, climate, and other characteristics indicate that the zonation patterns are valid representations of ecosystem differences along the Mauna Loa transect.

Research on leaf surface fungi did not provide clear evidence for altitudinal distribution; however, it did indicate biotic factors that influence the distribution of phylloplane fungi among vascular plants along the transect.

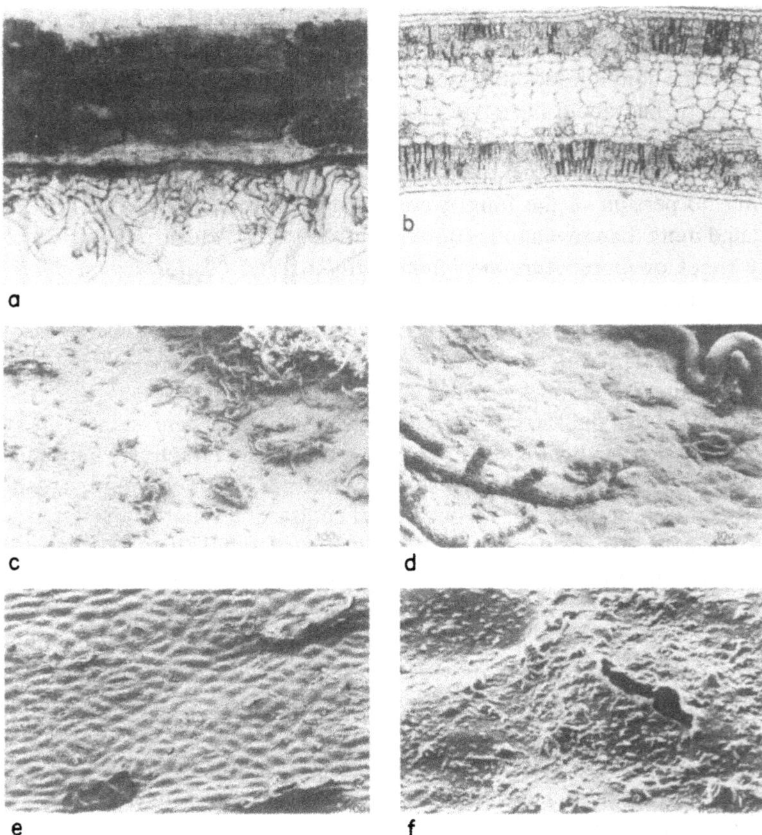

FIGURE 3-24 *Contrasting leaf surfaces affecting the trapping of fungi by tree foliage. (a) Free-hand section of fresh* Metrosideros *leaf with well-developed abaxial hairs. (b) free-hand section of* Acacia koa *phyllode showing lack of hairs. (c) SEM photo of* Metrosideros *abaxial surface showing sooty molds and hairs. (d) Detail of c. (e) SEM photo of* Acacia koa *phyllode showing stomata. (f) Detail of e.*
Photographs by W. A. Sakai.

The distribution patterns and species composition of soil-fungal communities along the Mauna Loa transect are more representative of temperate and subtropical woodlands and grasslands than of warmer, tropical regions. The predominance of *Penicillium* and the paucity of *Aspergillus* species is just one indication. Many of the species identified from the transect have been reported frequently from temperate or subtropical areas. The subtropical-temperate character of the fungal communities along the transect, in spite of the latitude of Hawai'i, is attributed to the selective influence of this particular insular environment. Foremost among the determinants of this selective environment are the east flank position of the area on Mauna Loa, and the 1,190 m to 3,050 m elevation range, contributing to the relatively mild, cool climate along the transect. This location, together with the young volcanic soils, is considered the underlying basis for the nature and distribution of soil-borne fungi. The ecosystem types along the Mauna Loa transect provide the additional biotic and abiotic environmental dimensions that determine the specific distribution of fungi.

Analyses indicate that the distribution of fungi along the transect is governed by interacting environmental factors and that their combined influences result generally in an altitudinal zonation of fungal communities. Individual factors are secondary in importance to the combined forces of soils, vascular plant associates, and climate. The former two factors seem especially influential. The basic heterotrophic nature of fungi naturally necessitates their direct or indirect association with other organisms through parasitism, commensalism, decomposition of organic substrata, and other relationships. Factors that govern the distribution of vital organic substrata therefore can be expected to influence fungal distribution. This overall consideration points to the most basic determinants of fungal distribution.

A good example of the involvement of interacting factors in distribution was revealed in regard to the genus *Fusarium* (Stoner, 1974). The species of *Fusarium* showed a definite altitudinal zonation along the transect (figure 3-23). However, the distribution was not correlated with any specific, individual factor such as soil pH or organic matter. Instead there was a broader correlation with soils, plants, and climatic conditions. There appeared to be three centers of *Fusarium* distribution: in the montane kīpuka and savanna zone, in the montane seasonal environment as a whole, and in the subalpine zone. *Fusarium* was not detected in the rain forest soils of fungal zone C.

The influence of general climatic factors on distribution is suggested by the fact that while certain vascular plant associates (for example, *Metrosideros* and *Holcus*) have relatively wide ranges or widely spaced occurrences on the transect, some fungi (such as certain *Fusarium* and *Penicillium* species) are more limited in range. The restriction of some fungi to certain regions of the transect is considered more an effect of general climatic factors than of vegetation types. For example, *Fusarium solani, Mucor strictus,* and *Gliocladium vermoeseni* were found only in the lower, warmer third of the transect; *Fusarium oxysporum, Gliocladium roseum,* and *Paecilomyces carneus* in the intermediate climatic range (fungal zones (IV–III); and *Mortierella hygrophila* v. *minuta* and *Mucor fragilis* in the still drier and cooler upper third of the transect (zones II–I).

There is a strong similarity between the transect zones determined from the distribution of vascular plants and those determined independently from the soil fungi. This similarity is a strong indication of the interrelated distribution of the two groups of organisms. Frequent associations of certain fungi with specific vegetations, such as *Fusarium* species with grass-containing communities (Thornton, 1960; Mueller-Dombois and Perera, 1971; Domsch and Gams, 1972), have been cited elsewhere and were noted also on the Mauna Loa transect.

Distribution could be governed in some cases by special relationships such as parasitic specialization between fungi and vascular plants; in other instances, distribution could involve broader influences, such as the dependence of a fungus on the overall biotic and abiotic conditions prevailing in the root zone of a particular higher plant community.

It is not known to what extent fungal distribution along the transect is determined by interactions of fungi themselves, such as competition for nutrients or the production of antibiotics. Such interactions are possible, although they probably would be secondary to the more basic distributional determinants. Although it is difficult to assess the complete ecological roles of individual fungi, possibly those species represented in zone-delimiting groups along the transect are organisms with reasonably distinct roles or niches and, therefore, minimal within-group competition.

Soils of the rain forest zone C (figure 3-22), with their quantitatively and qualitatively restricted fungal populations, may have decomposition regimes quite distinct from those of other soils along the transect (TR 75). This and other differences could contribute significantly to the unique environmental character of these soils and, thus, affect species distribution.

Temperature, moisture, and organic matter appear to be the most influential soil factors. The largest populations of *Fusarium solani,* for example, were found in the warmest kīpuka soils studied. *Aspergillus,* a genus frequently associated with warmer, tropical soils, was detected infrequently in the relatively cool soils of the transect. *Penicillium,* a characteristic genus of cooler latitudes, was well represented along the transect.

Surely the movement of fungi across zonal boundaries along the Mauna Loa transect has occurred by various agents, such as feral animals, cattle, humans, and wind. The grazing of cattle in the area continued to 1948. Foraging and other actions of feral pigs and other wild animals, including arthropods, continues today. The activities of pigs alone should support the widespread movement of fungi. Research on the phylloplane communities (TR 42) demonstrated that the propagules of many fungi, including some species isolated also from soil, were carried by air currents along the altitudinal gradient. Still in spite of these possibilities for free fungal movement, the data on soil-borne fungi demonstrate the existence of distinct fungal zones along the transect, indicating that the nature of fungal communities at this time in the evolution of the island is determined more strongly by environmental parameters than by the distributional range of propagules and that the established systems are not generally susceptible to major alteration by isolated invasions or disturbances by extrazonal elements.

4

Spatial Integration of the Organisms Studied along the Transect

D. Mueller-Dombois

Here spatial integration is recognized as a pattern indicating some degree of order in the species distributions within and among the various organism groups studied, including their relationship to the transect environments. We began the altitudinal gradient analysis with a number of questions and hypotheses relating to the distributional characteristics of island species in each organism group, their spatial and functional interrelationships, their taxonomic composition in different segments along the transect, the degree of penetration of introduced biota and their possible interaction with the native biota, and the fragility or stability of the species and community patterns observed. We were also interested in finding out whether there is anything peculiar about the spatial distribution of island biota. To answer these questions and evaluate our hypotheses, we collected basic data for thirteen organism groups, including both a species composition record at a number of predetermined sites along a mountain transect and an attempt to quantify the populations of each species at each of the sampling sites.

In terms of field techniques, this proved to be a relatively easy task in the vascular plant records, but it was more complicated to differing degrees in the other organism groups. In these groups, several repeated records at the same sites were necessary to establish a reasonably reliable distribution trend. In some organism groups—for example, the terrestrial algae—repeated sampling turned up new taxa each time, and it was found that approximately ten repeat samples were required to obtain a satisfactory record of taxa at a sampling site. Moreover major taxonomic problems became evident in a number of organism groups because they are as yet so little known. Terrestrial algae could not be identified to the species level, and among the soil arthropods a very large number of species (over 70 percent) turned out to be new records for the Hawaiian Islands. Several new endemic species were found among the Diptera.

Quantification problems were encountered in most organism groups. In general, the more ideal density count proved futile for this analysis in most organism groups. Instead the less time-consuming frequency enumeration could be applied to most groups as a measure of population quantity. This presents no real problem in the interpretation of their distribution trends.

For data processing we used two techniques developed in vegetation science and numerical taxonomy. One relates to a site comparison of species contents (Q-type analysis), the other to a species-to-species comparison (R-type analysis). The two analyses types are complementary in the search for community zones and boundaries. The computerized site comparison was applied only to five organism groups (plants, birds, foliar arthropods, Sciarids, and soil fungi) in which deviations in community boundaries appeared to be indicated. The species-to-species comparison was applied to all organism groups.

For the R-type analysis, we devised the species distribution diagram, which was usable for all organism groups as a uniform format of display. In four cases (plants, birds, Sciarids, and soil fungi) the associated species groups are those identified by the computer, but the ordering of the groups on the diagrams was done by hand. In all other organism groups, the total species order was hand arranged. In this way, the species distribution ranges and quantitative modes could be ordinated more clearly than was possible with the modified Ceska-Roemer program, which was used merely as a search tool for the important species. We also deviated from the program by giving certain individual species a group number when these displayed an important community-differentiating trend.

Another point must also be reemphasized, which was brought out with Stoner's idea of using reference fungi. Usually only a certain fraction of the species content at a site can be used to indicate reliable distribution trends. Most organism groups, with the exception of three (in our case)—the introduced rodents, the Sciarid flies, and the plagithmysine borers—contained a number of accidental species, which are those that are only represented once or a few times sporadically with low population quantities at one or a few sites in the set being studied. Among birds, these accidentals were only five of the twenty-one species recorded along the transects. In most other organism groups the proportion of accidentals was greater. We resolved this problem by concentrating our species diagram analysis on the quantitatively more important or repeatedly encountered species along the transect—the reference species.

ZONATION TRENDS IN ORGANISM GROUPS

In the context of this transect study, the concept of zonation trends is synonymous with that of community trends. According to our analysis a transect zone or transect community can be defined as a species assemblage that shows some degree of homogeneity along a certain segment of the transect. However, the critical point is the community boundary rather than the degree of homogeneity. Therefore we may ask what sort of community boundaries emerged from our species distribution analysis.

The Mauna Loa transect was laid out through four of the six previously defined environmental sections in Hawai'i Volcanoes National Park. The four environmental sections traversed by the transect were the alpine, subalpine, montane seasonal, and rain forest sections. In the plant zonation analysis, different computerized cluster techniques resulted in similar community boundaries with only minor variations. The sectional boundaries were confirmed, and additional boundaries were indicated, resulting in seven well-defined transect communities: an alpine and subalpine community, a parkland, a savanna, a seasonally dry 'ōhi'a forest, and two kinds of 'ōhi'a rain forests, a structurally open one and a closed one.

The bird zonation analysis resulted in a differentiation of transect bird communities that coincided closely with those of the vegetation. However, in the bird zonation analysis, the alpine zone boundary was shifted downslope to include the upper tree line because an important species turnover was recorded just below the line. In the zonation analysis of the canopy arthropods, the alpine zone was omitted because of the absence of trees there. The subalpine zone was well differentiated. The parkland and savanna zones were grouped, and the three 'ōhi'a zones at the lower end of the transect were again well separated. Grouping the parkland and savanna zones in this analysis gave evidence of a fairly homogeneous canopy arthropod community on *Acacia koa,* which exists over an altitudinal range of 1,220 m to 2,040 m. The Sciarid zonation analysis gave results exactly coinciding with the four major environmental sections. The alpine section was characterized by a distinct absence of Sciarids; the other three sections became evident because of major species turnovers at the section boundaries. This broad zonation is a reflection of the broad distribution ranges of many Sciarid species. The soil-fungal zonation analysis resulted in a transect community breakdown almost exactly duplicating the plant zonation analysis. The only difference was that the two rain forest communities, the open and closed forest structures, were not reflected in a significant turnover of soil-fungal species.

Evidence from the independent computer-generated zonation analyses of the five different organism groups (plants, birds, canopy arthropods, Sciarids, and soil fungi) strongly indicates the reality of several community boundaries along the Mauna Loa transect. It is true, however, that a decision was required to determine the level of clustering that was ecologically meaningful. Yet our independent analyses should clearly discredit the often-expressed notion that communities are merely fictions. They exist in nature, but they may be conceived at different levels of generalization depending on the criteria used for their discrimination.

SPECIES PACKING IN THE TRANSECT ZONES

A further indication of zonation trends within and among the organism groups is obtained by making a zone-by-zone comparison of species numbers.

TABLE 4-1 *Number of Reference Species in Transect Zones*

Organism groups	Alpine I	Sub alpine II	Montane seasonal III	IV	V	Rain forest VI	VII	Total reference species
Terrestrial algae[a]	NS	14	12	15	15	NS	11	15
Rodents	1	2	3	3	1	3	3	3
Collembola	NS	14	14	13	15	NS	9	15
Ectoparasites								
On *Rattus rattus*	Host absent	4	6	7	Host absent	7	6	8
On *Mus musculus*	4	5	5	6	5	4	4	6
Canopy arthropods								
Natives	No trees	7	9	9	9	4	10	13
Exotics	No trees	8	10	10	6	6	9	11
Predators	No trees	9	6	6	4	3	6	10
Arthropods on ʻōhiʻa reproductive organs	No trees	12	14	11	13	NS	NS	16
Birds								
Natives	1	5	9	6	4	NS	6	9
Exotics	0	3	7	7	6	NS	3	8

Sciarids	0	9	15	11	12	1	6	16
Plants[b]								
Natives	2	8	13	7	10	11	2	18
Exotics	1	2	4	5	6	5	5	9
Mesostigmata	NS	10	10	15	11	NS	8	18
Soil fungi	3	6	9	11	7	3	5	17
Drosophilids	NS	2	11	11	12	5	NS	13
Litter Diptera	0	7	9	14	0	1	1	20
Plagithmysines	1	2	6	10	3	4	3	14
Total reference species	13	127	159	170	129	46	95	239
Total sample	10	19	19	19	18	13	17	
Adjusted species richness[c]	18.2	127	159	170	129	67.2	106.1	

[a]Numbers for algae refer to genera, not species.

[b]The plant record relates to the tree communities only; shrub and grass community records were excluded because their ranges were more limited.

[c]Number of reference species expected if the same number of organism groups had been sampled at each site (for example, for alpine site I $(13 \times 14)/10 = 18.2$; for rain forest site VI $(46 \times 19)/13 = 67.2$). Organism groups that do not occur on the site due to lack of host are not included in this total.

NS = not sampled.

Table 4-1 lists the number of reference species of each organism group by transect zones. This provides for an index of species packing or the richness of biota in the different transect zones. The list was obtained from the species distribution diagrams in the preceding chapters, which show only the quantitatively more important or reference species of each organism group rather than the total community. The list gives fourteen organism groups instead of thirteen. This enlarged number results from treating the two groups of soil arthropods, the Collembola (springtails) and Mesostigmata (soil mites), separately because of their different distributional characteristics.

By scanning along the row of species numbers in each organism group, zonation boundaries can be considered most firmly established wherever there is a proportionately important change in reference species numbers. For example, algal community boundaries are only vaguely indicated between zones II and III, III and IV and V and VII. Four algal zones may be distinguished along the transect: a subalpine (algal rich), a mountain parkland (algal poor), a savanna and 'ōhi'a dry forest zone (algal rich), and a rain forest zone (algal poor). However, the proportionate changes are rather small (less than 30 percent), which in this particular case refer to the number of genera. On the basis of taxonomic richness, therefore, one may recognize only one algal community along the transect. Among the rodents a significant drop in species is seen in zone V, the 'ōhi'a dry forest, which lacks both rat species (*Rattus rattus* and *R. exulans*). The rodent species diagram in figure 3-9 clearly supports the distinction of a separate, rodent-poor zone in this 'ōhi'a dry forest. The same is true for the alpine environment (zone I), where only *Mus musculus* was found. However, at the upper transect end, the rodent zonation is less well indicated. Here it is more continuous. The Collembola, the springtails among the soil arthropods, indicate only one clear zonation boundary, between the rain forest and the rest of the transect. Four of the Collembola species drop out in the rain forest, which can be considered relatively species poor in this group. Thus two Collembola zones are indicated by species numbers: a rain forest (species poor) and an almost transect-wide montane seasonal and subalpine (species rich) zone. Similarly broad zonations (only two to three) are indicated by the ectoparasites on rodents, the native canopy arthropods, and the arthropods on 'ōhi'a reproductive organs. The other groups show greater variation in species numbers along the transect and thus more transect boundaries. However, change in species number is not the only indication for community boundaries, qualitative changes are equally important. For example, two adjacent gradient segments may have the same or similar numbers of species, but they may include different kinds, thereby resulting in a certain species turnover.

The lowest number of reference species was recorded in the alpine zone. All organism groups sampled in this zone showed reductions in species numbers from those found in the subalpine zone. Two groups were represented with relatively high numbers in the subalpine zone, the terrestrial algae and Collembola. Their taxonomic richness throughout the transect suggests that these two organism groups probably would decrease minimally in taxonomic richness in the alpine environment, in contrast to most other groups. The terrestrial algae encounter less competition by higher plants there, and the springtails (Collembola) would still be able to find sufficient food material in the form of algae and soil fungi. In contrast, the other group of soil arthropods, the soil mites *(Mesostigmata),* show a definite taxonomic mode in the savanna zone (IV). By drawing a curve from this mode and extrapolating the curve to the alpine zone, a comparatively much smaller number of soil mites can be expected there than springtail species. The soil mites are known as predators on the springtails. The reduction of species numbers upslope in the soil mites may be functionally related to the stable trend of springtail species numbers upslope. The environmental harshness increases, while the predator pressure decreases, perhaps resulting in a balance of stresses for the Collembola upslope.

In the subalpine zone, the total species number is increased greatly over that in the alpine zone because of the presence of 'ōhi'a trees in the subalpine environment. This is obvious for the thirty-eight species of tree-residing arthropods recorded in this zone, but it is also the major cause for the sudden increase in birds, plants, and Diptera flies. The nectar-seeking endemic honeycreepers, the Hawai'i 'amakihi and the 'apapane, are certainly there because of 'ōhi'a. The other bird species, several of the plant species, and the Diptera flies appear in the subalpine zone primarily because of the ameliorated microclimate associated with the 'ōhi'a trees. Here dense shrub aggregations occur as fog-drip communities under the isolated and scattered 'ōhi'a trees (see figure 2-2).

A further significant increase in species richness was recorded in the mountain parkland (zone III). The closed grass cover with shrub and tree islands provide a much lusher vegetation as a whole. Here the fungus gnats (the Sciarids), most of them endemic species, showed their taxonomic peak. A further, less significant increase in overall species richness was recorded in the savanna zone (IV). Here the soil mites and the litter-breeding Diptera displayed their taxonomic peaks in the mesic kīpuka forest. There may be an evolutionarily significant predator-prey relationship between these two organism groups. The high diversity among the litter-breeding Diptera has been attributed to predator pressure, and the soil mites along the transect, thought originally to be mostly exotic species, are probably largely indigenous. Another important peak in species richness is displayed by the plagithmysine borers in this zone because a variety of specific host trees is found in the mesic kīpuka.

A significant drop in overall species richness is associated with the change from savanna to the 'ōhi'a dry forest (zone V). This open forest contained several of the same plant species also found in the open subalpine forest, including, of course, 'ōhi'a. Sparser undergrowth in this open forest, which may reduce the number of breeding sites, can be seen as a contributing factor for species loss among the litter-inhabiting Diptera. The sparse undergrowth may also contribute to the absence of the two rat species (*Rattus rattus* and *R. exulans*). The number of plagithmysine borer species is much reduced because only two host tree species occur here. One of these, *Acacia koa,* was planted by the National Park Service, and the ranges of the two borer species associated with koa (*P. claviger* and *P. varians*) are thus also extended beyond their natural ranges into this zone.

A further overall decrease in species richness is indicated for the rain forest along the transect, which may be somewhat surprising. Part of the reason is an artifact, because most of the plant species characteristic of this transect segment were not portrayed on the diagram simply for lack of space. A rough count from the species list provided in table 3-2 indicates that at least twenty species (fifteen natives and five exotics) should be added to the open rain forest zone VI and at least thirty-two (thirty natives and two exotics) to the closed rain forest zone VII. This would result in a species richness of the open 'ōhi'a rain forest approximating that of the open 'ōhi'a dry forest and in a species richness of the closed 'ōhi'a rain forest approximating that of the savanna zone with the kīpuka forest. The two rain forest zones occur on geologically young substrates. The low-stature tree, undergrowth, and epiphytic floras were depauperated relative to the older rain forest, the Kīlauea forest, studied by our group and reported on in the next chapters. In addition, the tree arthropod fauna was apparently reduced because some volcanic sulfur steaming affected this rain forest segment occasionally. This may also have influenced the litter-breeding Diptera, which were represented with only one species each in the two rain forests. The soil fungi were also significantly poorer in the two rain forests as compared to the mountain parkland and savanna zones. Correlated with the fungus species reduction in the rain forests was a species loss in the two fungus-related insect groups, the fungus gnats (Sciaridae) and the springtails (Collembola).

Thus along our transect, the overall greatest number of species occurred in the mid-section. This result coincides with the altitudinal distribution of arthropod species richness as found by Janzen et al. (1976) in the Venezuelan Andes.

GRADIENT SENSITIVITIES AMONG ORGANISM GROUPS

Differences in sensitivity to the mountain gradient are clearly displayed by the various organism groups. The term *gradient sensitivity* applies here to the differing proportions of wide-ranging and narrow-ranging species among the organism groups. For the purposes of this analysis, wide-ranging species were defined as all those that range from the montane rain forest to the subalpine environment; narrow-ranging species were defined as those ranging through one or two transect zones only. These two range criteria, which are interpreted here as gradient-sensitivity indexes, are relatively independent of one another. They do not add to 100 percent within an organism group because there is a remainder of species that can be considered of intermediate range. Table 4-2 shows an approximate ranking of the organism groups from low to high gradient sensitivity as recognized by their proportion of wide-ranging species.

TABLE 4-2 *Gradient Sensitivity among Organism Groups along Mauna Loa Transect*

Organism group	Total reference species on diagrams	Proportion of wide-ranging reference species	Proportion of narrow-ranging reference species	Number of Reference Species: Truly ubiquitous	Number of Reference Species: With distinct bimodality or multi-modality	Gradient sensitivity	
Terrestrial algae[a]	15	73%	0%	9	6	Low	
Rodents	3	67	0	1(*Mus*)	1(*R. rattus*)	Low	
Collembola	15	60	7	8	9	Low	A
Ectoparasites							
On *Rattus rattus*	8	63	33	3	0	Low	
On *Mus musculus*	6	50	0	3	1	Low	
Canopy arthropods							
Natives	13	54	31	2	8	Moderate	
Exotics	11	55	9	1(*Iridomyrmex*)	2	Moderate	
Predators	10	50	50	1(*Koanoa*)	5	Moderate	
Arthropods on ʻōhiʻa reproductive organs	16	50[b]	31	3	0	Moderate	B
Birds							
Natives	9	55	11	3	3	Moderate	
Exotics	8	25	13	1(*Zosterops*)	0	Moderate	
Sciarids	16	38	25	1(*Brad.* #8)	11	Moderate	
Plants[c]							
Natives	18	33	33	1(*Metrosideros*)	5	High	
Exotics	9	22	22	1(*Hypochoeris*)	3	High	
Mesostigmata	18	28	28	1(*Athiasella*)	9	High	C
Soil fungi	17	18	47	1(*Mortierella*)	6	High	
Drosophilids	13	15[b]	23	1(*Drosoph. sim.*)	2	High	
Litter Diptera	20	5	85	1(*Brad.* #8)	5	Very high	D
Plagithmysines	14	0	71	0	0	Very high	

[a]The algal sample relates to genera, not species.
[b]Ranging from zones II to V, all others from II to VI or VII.
[c]The species diagrams for the shrub- and grass-community analyses were excluded because of the limited transect ranges of these communities.

The terrestrial algae are listed first as the group showing the least sensitivity to the gradient. However, this organism group is an exception insofar as the inividual taxa could be identified only to the genus level. It is likely that their gradient sensitivity would increase if the algal genera were identified to the species level. However, an even more important factor is the sampling procedure. The algal sample probably relates to both algal disseminules (airborne and then soil-deposited vegetative cells and spores) and algal colonies actively growing at the site. If only the actively growing algal colonies were sampled, one might expect a much greater gradient sensitivity in this group, perhaps similar to that of the soil fungi or higher plants.

Listed next on table 4-2 are the rodents, their ectoparasites, and the Collembola (the springtails of the soil arthropods). The rodents were represented with only three species along the transect, and two of them, *Rattus rattus* and *Mus musculus,* ranged from the rain forest to the subalpine scrub. They are exotic species of relatively recent introduction, which have met with no competition and with very little interference from predators. Three predators potentially could be of some significance: the mongoose, the Hawaiian hawk, and the Hawaiian owl. However, the near absence of competition among the rodents combined with their great physiological tolerances are probably the major factors responsible for their wide-ranging distributions along this island mountain. As expected, many of the ectoparasites associated with the rodents stay on them over their transect range. Yet half of the ectoparasite community on *Mus musculus* displayed distributions more restricted than that of their host, thereby indicating a somewhat greater gradient sensitivity.

The Collembola show a majority of their species (60 percent) relatively insensitive to the transect gradient. Their general distribution pattern is quite similar to that of the terrestrial algae. However, since the algae were not identified to the species level, we cannot say how ubiquitous they are. So within the limits of our data, the Collembola are, next to the rodents, the most ubiquitous organism group encountered in our analysis. Only one Collembola species, *Sminthurides* cf. *pseudassimilis,* was restricted to one transect community—the open, seasonally dry ‘ōhi‘a forest at the lower end of the transect. Nine species showed disrupted (or multimodal) distributions, which may reflect some seasonality in the population densities in these species. The dominantly wide-ranging characteristic of the Collembola indicates their independence of the vegetation pattern. There was only one endemic species, *Salina maculata,* among the fifteen common Collembola encountered along the transect. This endemic species, however, also belongs to the wide-ranging Collembola. It is therefore probable that the ubiquitous character of the Collembola as found along our transect is a general characteristic of this group rather than a reflection of their being mostly introduced species.

Four organism groups—the algae, rodents, their ectoparasites, and the Collembola—can be combined into one category or spatial sensitivity type, which may be characterized as showing a low degree of gradient sensitivity.

The ranking order of the remaining organism groups in table 4-1 shows that they can be combined into three further gradient-sensitivity types according to their proportion of wide- and narrow-ranging species. Organism groups with a moderate degree of gradient sensitivity include the canopy arthropods, the arthropods found on the ʻōhiʻa reproductive structures, the birds, and the Sciarids (the fungus gnats). These four groups contained generally more wide-ranging than narrow-ranging species in our analysis. The organism groups with moderate gradient sensitivity contained a lower proportion of wide-ranging (from 25 to 55 percent) and a higher proportion of narrow-ranging (from 9 to 50 percent) species than found in the low sensitivity groups. The reasons for their moderate gradient sensitivity differ among these organism groups. The two groups of tree arthropods contain a high proportion of strongly host-associated species. Since their major host tree, the ʻōhiʻa, ranges from the montane through the subalpine environment, many of the tree arthropods display a similarly wide range of distribution. In contrast, the birds and Sciarids are relatively more independent of the plant distribution. Both are very mobile, airborne organism groups, which can roam freely through larger segments of the altitudinal transect. However, several of the endemic honeycreepers among the birds are known as nectar feeders on the ʻōhiʻa trees. This explains, in part, the relatively high proportion (55 percent) of wide-ranging species among the native birds.

Organism groups with high gradient sensitivity include the plants, the soil mites (Mesostigmata), the soil fungi, and the bait-trapped drosophilids because the proportion of narrow-ranging species among them is generally equal to or greater than their proportion of wide-ranging species. Moreover, the proportion of wide-ranging species is moderately low in these four organism groups, from 15 to 33 percent. However, among this low percentage of wide-ranging species are significant island species—for example, the ʻōhiʻa tree. This gradient-sensitivity ranking is interesting in that the soil fungi and soil mites (Mesostigmata) respond similarly to the environmental gradient as do the higher plants and drosophilids. Both the soil fungi and soil mites differ from the drosophilids and higher plants in lacking endemic species. The soil microfungi and soil mites are characterized by wide dispersability, which may even be of intercontinental dimensions. Yet they display the same gradient sensitivity as do the more locally unique and island-specific higher plants and drosophilids. This suggests that the soil microfungi and soil mites may be used as biological indicators of environmentally equivalent habitats in different islands and continental areas, where both the higher plants and drosophilids would display entirely different species assemblages.

A very high degree of gradient sensitivity was displayed by the litter-breeding Diptera and plagithmysine wood borers. These two organism groups contained a very high proportion (over 70 percent) of narrow-ranging species and almost no wide-ranging species. Both organism groups have a very high proportion of endemic species, which evolved in adaptation to certain native woody plant species. Their very high degree of gradient sensitivity therefore is largely a reflection of the gradient sensitivity of their host plants. However, the 'ōhi'a borer, *Plagithmysus bilineatus,* did not, as one could expect, accompany its host tree over the entire range from the rain forest to the tree line. Instead it occupied the host only half way up the mountain slope, thereby displaying a greater gradient sensitivity than its host.

In summary, the proportions of wide-ranging and narrow-ranging species encountered in this transect analysis were found to be primarily a function of the ecological properties of a given organism group. However, endemism is important on a secondary level. The organism groups described as displaying a low degree of spatial sensitivity (type A on table 4-2) contain mostly introduced species or species of wide geographic distribution on the world scale, while the organism groups described as displaying a very high degree of spatial sensitivity (type D) contain a very high degree of endemism.

The two additional columns in table 4-2 provide further parameters of gradient sensitivity extracted from the diagrams. The numbers of species truly ubiquitous refer to a subselection of the wide-ranging species: those species whose ranges were cut off or whose population numbers were high at both ends of the transect. The numbers of these species support the grouping order into the four sensitivity categories. It is, however, noteworthy that the organism groups classified as displaying high gradient sensitivity each contain at least one truly ubiquitous species along the transect.

The final column, giving species with distinct bimodality or multimodality, bears no direct relationship to the gradient sensitivity order. All species with distinctly interrupted distributions along the transect were included in this total. The two groups of soil arthropods, the Collembola and Mesostigmata, and the Sciarids showed the highest number of these species. This indicates patchiness in distribution as a characteristic of the soil arthropods and fungus gnats. Patchiness was also common among the terrestrial algae, the soil fungi, the canopy arthropods, and the litter-inhabiting Diptera. It is not entirely clear whether this characteristic is of ecological significance. In some cases it may merely reflect an undersampling of a species that is not very dominant. In other cases it may truly reflect patchy distributions as influenced by local site factors. The second explanation may be more generally applicable since all of the organism groups can be expected to react more sensitively to local site variations and momentary environmental fluctuations (such as soil moisture or weather factors) than the perennial organisms, including the plants, birds, and rodents. However, the plants also showed several cases of interrupted distributions. The most obvious is the bimodality of the 'ōhi'a tree along the transect. This relates to a similarity of the physical substrate in the subalpine and lower montane seasonal environments, an effect that is transmitted to some extent by the plants to the other organism groups, notably the canopy arthropods.

SPATIAL ASSOCIATION TRENDS

Absolute similarity or matching of the distribution ranges of two or more species along an environmental gradient can hardly be expected within an organism group. Between certain organism groups, however, excellent distributional correlation has been shown along the transect. The two most host-dependent organism groups in our study, the ectoparasites on introduced rodents and the plagithmysine borers on native woody plants, contained several such species. Among the ectoparasites on *Mus musculus* and *Rattus rattus,* six of fourteen (43 percent) species were perfectly correlated; among the plagithmysine borers nine of fourteen (64 percent) were perfectly correlated. Several species with near-perfect spatial correlations to host plants were also found among the canopy arthropods, including the endemic phytophages *Sarona adonias* and *Oceanides pteridicola* on the ʻōhiʻa tree and the exotic phytophages, *Chirothrips patruelis* and *Psylla uncatoides,* on the koa tree *(Acacia koa).* Furthermore, eight of sixteen (50 percent) of the arthropod species found associated with ʻōhiʻa reproductive structures had spatially coinciding distribution ranges with ʻōhiʻa along our transect. Another closely plant-dependent group in our analysis was the litter-breeding Diptera. However, in this group, the site environment, as conditioned by the tree-canopy structure and the type of litter, leads to a more complex spatial interaction in this group.

Even the most host-dependent organism group of our study, the plagithmysine borers, displayed a certain degree of independence or individuality in their spatial distribution along the transect gradient. It has also been demonstrated within this organism group that two borer species adapted to the same host (in this case, *Plagithmysus claviger* and *P. variens* to *Acacia koa*) separate on a finer spatial level than the altitudinal gradient. The two borer species have exactly coinciding spatial distributions along the transect, but they occupy different spatial segments, with little overlap, on their host tree. Such a finer-level spatial segregation probably could be demonstrated for many of the organism groups and species studied along the transect. Such an analysis was done for the Kīlauea rain forest.

An important objective in our Mauna Loa transect analysis was to determine the spatial distribution patterns of the species along the elevational gradient and in relation to one another. The results are summarized in table 4-3. Two columns of species group numbers have been extracted from the species diagrams of each organism group. The first column refers to the total number of differentiating species groups recognized on each diagram. In some cases community-differentiating trends were recognized by single species rather than by groups of species. An example, among the avifauna, is the ʻōmaʻo, or Hawaiian thrush *(Phaeornis o. obscurus),* which by its presence in the alpine environment contributed a community-differentiating pattern to the native birds. However, since the ʻōmaʻo has a unique spatial distribution, which is not

TABLE 4-3. *Spatial Association Trends among Species of Each Organism Group*

Organism group	Number of differentiating species groups on diagrams	Number of spatially associated species groups	Proportion of spatially associated to all species on diagrams	Adjusted for truly ubiquitous groups
Terrestrial algae[a]	3	2	8/15 (53%)	(0%)
Rodents	0	0	0/3 (0%)	
Collembola	3	2	14/15 (93%)	(40%)
Ectoparasites				
On *Rattus rattus*	3	1	3/6 (50%)	(0%)
On *Mus musculus*	3	1	4/8 (50%)	(13%)
Canopy arthropods				
Natives	6	5	8/13 (62%)	(46%)
Exotics	6	4	9/11 (82%)	
Predators	4	1	4/10 (40%)	
Arthropods on ʻōhiʻa reproductive organs	6	4	14/16 (88%)	(67%)
Birds				
Natives	4	2[b]	5/9 (56%)	
Exotics	5	2[b]	5/8 (63%)	
Sciarids	3	3[b]	15/16 (94%)	
Plants				
Natives	7	7[b]	15/18 (83%)	
Exotics	0	0	0/8 (0%)	
Mesostigmata	5	4	15/18 (83%)	
Soil fungi	3	3[b]	12/17 (71%)	
Drosophilids	5	4	12/13 (92%)	
Litter-Diptera	5	4	17/20 (85%)	
Plagithmysines	4	3	11/14 (79%)	

[a]The algal sample relates to genera, not species.
[b]Computer-generated species groups

related to any other bird species encountered along the transect, it is excluded from the count of associated native bird species in the third column in table 4-3. Other differentiating group numbers served merely for convenience in discussing the individual species diagrams in chapter 3. Therefore not much significance can be attached to the first column of numbers in table 4-3 other than that they are usually greater than the number of spatially associated species groups. This second column refers to the number of species groups, each containing two or more species that displayed closely coinciding distribution ranges along the transect. Except for the computer-derived species groups, indicated by note "b" on the table, all other associated groups were determined visually on the diagrams. In the visual method, more discriminatory criteria for coinciding species ranges were applied than by the computer method. Therefore the designation of numbers of associated species groups for each organism group is more conservative than the normal spatial association criteria of requiring at least 50 percent range correlation and not more than 10 percent noncorrelation.

The results in table 4-3 show that spatially associated species groups were evident in all but two of the organism groups studied along the Mauna Loa transect. The two exceptions were the introduced rodents and exotic plants. The three species of rodents each had unique and broadly overlapping ranges along the transect, and the same applied to the eight important exotic plant species. However, among the latter, four species (*Holcus lanatus, Commelina diffusa, Eupatorium riparium,* and *Psidium cattleianum*) were grouped by the computer (using the 66/10 rule) with other native plant species. The names of these spatially associated native species are listed on table 3-2 within their respective group number. Therefore even among the spatially noncorrelated exotic plant species, half of them become spatially correlated if they are pooled with all other plant species in this test.

All other organism groups displayed at least one spatially associated group of species. However, those with only one spatially associated species group (the rodent ectoparasites and the predators among the canopy arthropods) contain in this group the species of transect-wide distribution. This implies that in these organism groups, there is only one strong community of species indicated, which is of at least transect-wide distribution. It is not known how these species segregate beyond the limits of the transect. Therefore the three organism groups with only one associated species group can also be considered as untested for their spatial association.

This leaves nine organism groups that displayed two or more spatially associated species groups in our transect analysis. Those with only two groups each were the terrestrial algae, the Collembola, and the native and exotic birds. In each of these, one of the two spatially associated groups is also of transect-wide distribution. The second group, however, displays restrictions along the gradient that are very similar for the species in that group. For example, among the algae, this second group contains three genera, *Nostoc, Navicula,* and *Scytonema*. All three show distinct bimodality by displaying a population peak each in the subalpine ‘ōhi‘a scrub zone II and in the lower-elevation ‘ōhi‘a scrub zone V and by being distinctly absent in the mountain parkland zone III. At least along this transect, this indicates strong commonality in distribution. Among the Collembola, the second spatially associated and transect-restricted group includes two species, *Anurophorus* cf. *septentrionales* and *Entomobrya purpurascens*. Both displayed their ecological optimum in the subalpine environment (zone II), and their population densities decreased from here on downslope to an interrupted distribution, ending in complete absence in the rain forest. This also indicates strong commonality in distribution.

The spatial association trends in table 4-3 show only a very general correlation to the gradient sensitivity trends in table 4-2. This very general correlation refers to the four organism groups (algae, rodents, rodent-ectoparasites and springtails) that were characterized as having low gradient sensitivity. As one

TABLE 4-4. *Proportion of Spatially Associated Species in Organism Groups*

Organism group	Proportion of spatially associated species with ranges confined within transect	Tendency for spatial group formation
Algae	0%	Low
Rodents	0	Low
Ectoparasites	0	Low
Collembola	0	Low
Birds, natives	0	Low
Plants, exotics	0	Low
Canopy arthropods		
Natives	15	Moderate
Exotics	18	Moderate
Mesostigmata	28	Moderate
Arthropods on ʻōhiʻa reproductive organs	38	Moderate
Birds, exotics	38	Moderate
Sciarids	50	Moderate
Drosophilids	69	Frequent
Soil fungi	71	Frequent
Plants, natives	72	Frequent
Plagithmysines	79	Frequent
Litter Diptera	85	Frequent

would expect, these biota displayed only few spatially associated species groups (up to two per organism group). The other biota, which showed greater gradient sensitivity, displayed more associated species groups, but they also varied greatly (from none to seven). This implies that community trends (in the sense of several species with closely coinciding distributions occurring together) along a gradient can be quite different among biota of similar gradient sensitivity.

The third column in table 4-3 gives the proportion of spatially associated species of each organism group. These values are biased in favor of spatial association because some species form associated groups simply because their ranges are cut off by the transect. For this reason a column for adjusted values has been added by removing all of those spatially associated species whose ranges have been cut off at both ends of the transect. This adjustment results in a closer correlation of spatial association with gradient sensitivity among the organism groups. If one wishes to go a step further and remove all spatially associated species whose ranges have been cut off at either both or one end of the transect, the proportion of spatially associated species is decreased much further. According to this more drastic adjustment, none of the four organism groups of low sensitivity has spatially associated species. To these, three others have to be added: the predators among the canopy arthropods, the native birds, and the exotic plants. The rest of the biota still show spatially associated species (see table 4-4).

This second, more drastic adjustment increases the differences among the organism groups, but changes little in the general ranking of gradient sensitivity. One can now recognize three categories according to spatial group formation:

1. Organism groups that did not show spatial groups within the confines of the transect (designated as showing low tendency for spatial group formation).
2. Organism groups with moderate spatial group formation (from 15 to 50 percent) within the transect confines.
3. Organism groups with frequent spatial group formation (from 69 to 85 percent) within the transect confines.

One can conclude that species-grouping tendencies in the spatial gradient sense differ among organism groups. Of course, the measure of spatial association among species varies with the criteria applied. Hence our data show that these differentiations are real and measurable.

Spatial associations of species along a gradient, however, do not provide all of the criteria that result in the structural segregation of biotic communities. For example, the soil fungi with only three spatially associated species groups resulted in nearly the same segregation of transect communities as did the native higher plants with seven associated species groups. Species group overlap and the absence of species groups from certain transect segments are equally important, additional criteria. Furthermore, in species-poor biota, individual, quantitatively well-represented species may very well indicate trends of community segregation along a gradient.

DISTRIBUTIONAL RELATIONSHIPS OF NATIVE AND EXOTIC SPECIES

The organism groups studied along the Mauna Loa transect can be divided into four types according to their origin in the Hawaiian Islands. Only two organism groups, the rodents and their ectoparasites, are purely exotic or human introduced. The second type is the soil arthropods, the Collembola and the Mesostigmata, which are believed to be largely exotic in origin. Only one of the Collembola was described as an endemic species, *Salina maculata*. The Collembola and the Mesostigmata each contained a high proportion of new species records for the Hawaiian Islands. Radovsky and Tenorio believe that some of these new records may turn out to be endemic species upon further taxonomic clarification. They contend that the soil arthropods are largely human-introduced. However, according to the investigations of Hammer, there is a good possibility that some of the soil arthropods are of wider oceanic island distribution, which implies an indigenous origin. Thus, according to the best current information available, the soil arthropods can be described as containing both exotic and indigenous species.

The third origin type, the terrestrial algae and soil fungi, contains dominantly or purely indigenous forms. It is doubtful that introduction by humans played an important role in the island establishment of these two organism groups. Both are characterized by having species with very wide, almost global, dispersability. Doty believes, however, that this group might contain a number of tropical species, once they are properly identified. This possibility notwithstanding, both the terrestrial algae and soil fungi can be considered as having arrived largely without the aid of humans; they are thus indigenous to the Hawaiian Islands.

This leaves a fourth type of organism group: those that displayed various degrees of endemism along the transect. Table 4-5 lists these organism groups in a ranking order from low to high endemism. This order is based on the ratio of endemic to all species shown on the diagrams.

The ranking order is not entirely different from the gradient sensitivity ranking shown in table 4-3. The Collembola were among the organism groups displaying low gradient sensitivity, and the litter-inhabiting Diptera and plagithmysine borers were those with the highest gradient sensitivity. The remaining six organism groups in table 4-5 showed either moderate or high gradient sensitivity. Thus at a general level, gradient sensitivity and endemism go together.

For a better understanding of the distributional relationships of native and exotic species, it is necessary to examine their ecological roles or their life

TABLE 4-5. *Organism Groups with Endemic Species Ranked from Low to High Endemism as Displayed along Mauna Loa Transect*

Organism group	Proportion of endemic species	Endemism
Collembola	7%	Low
Birds[a]	47	Moderate
Canopy arthropods	53	Moderate
Drosophilids (bait-trapped)	62	High
Vascular plants[a]	66	High
Arthropods on ‘ōhi‘a reproductive organs	69	High
Sciarids	81	Very high
Litter-inhabiting Diptera	85	Very high
Plagithmysines	100	Very high

[a]These groups each contain an indigenous non-endemic species, which were excluded in the count of endemics.

forms in more detail. Among the nine organism groups ranked for endemicity (in table 4-5), five were further characterized by species life forms. This information is summarized in table 4-6.

Among the vascular plants, exotic species were prevalent only among herbaceous life forms, especially the forbs. These appeared particularly in the rain forest *(Eupatorium riparium)* and savanna zones *(Commelina diffusa, Verbena littoralis, Veronica plebeia)* of the transect, but one exotic forb extended its distribution into the alpine environment *(Hypochoeris radicata)*. More significant in biomass, however, were the exotic grasses *(Holcus lanatus, Paspalum dilatatum, Cynodon dactylon),* which clearly displayed their current ecological optima in the savanna zone. This zone among all others along the transect showed the most thoroughly scarified topsoil from feral pig activity. From this zone many exotic herbaceous plants (including a sedge, *Cyperus brevifolius*) seem to spread in both or one direction up and down the transect. The transect ranges of the exotic plant species showed little spatial commonality other than the two distribution centers. A second, less pronounced center was the rain forest, where the exotic tree *Psidium cattleianum* and the exotic forb *Eupatorium riparium* showed a low quantitative importance. Their importance, particularly that of the tree species, may increase in the future. Also in the rain forest, feral pig activity can be seen as the major factor promoting the spread of exotic plants. In addition to being more dominant along the transect, particularly among the woody life forms, the native plant life forms display a more complete life-form spectrum. There were no exotic shrubs or ferns in our samples.

The native birds, like the native plants, displayed a richer life-form spectrum than did their exotic counterparts. Specialized nectar seekers and raptorial predators were missing among the exotic birds. However, a new general bird niche has become established by the presence of three species of seed feeders. All three were quantitatively well represented along the mid-section of the transect, and one, the house finch *(Carpodacus mexicanus),* extended its range from the rain forest into the alpine environment, almost coinciding with the range of the most widespread exotic plant, the composite *Hypochoeris radicata*. There is no recognized functional relationship between these widespread exotics of these two different organism groups, but the two species react somewhat similarly to the underlying environmental gradient. Both are not characteristic rain forest species. They are instead indicators of disturbances in rain forests, such as those caused by roads and other openings. An equally widespread exotic bird is the Japanese white-eye *(Zosterops japonica),* which is designated as a primary insect feeder. Thus the two widespread exotic birds do not seem to interact in an important way. However, a certain interaction of the Japanese white-eye with the three endemic nectar-insect feeders and honeycreepers, the ‘i‘iwi *(Vestiaria coccinea),* the Hawai‘i ‘amakihi *(Loxops v. virens),* and the ‘apapane *(Himatione sanguinea),* was noted by Conant. The Japanese white-eye may also exploit nectar of the ‘ōhi‘a tree, which appears to

TABLE 4-6. *Life Forms of Native and Exotic Species in Five Organism Groups*

Life forms on diagrams	Number of species	
	Endemic	Exotic
Plants		
Woody		
Trees	6	1
Shrubs	7	0
Tree ferns	1	0
Herbaceous		
Ferns	5	0
Grasses	4	4
Sedges	2	1
Forbs	1	6
Total	26	12
Birds		
Fruit-insect eaters	1	1
Browsers	2	3
Nectar-insect eaters	3	0
Insect feeders	1	1
Raptorial predators	2	0
Seedfeeders	0	3[a]
Total	9	8
Canopy arthropods		
Sapsuckers	1	2
Seedeaters	2	0
Flower and nectar feeders	3	3
Twig borers	7	0
Omnivore	0	1[a]
Leaf chewer	0	1[a]
Detritivores	0	2[a]
Grass feeders	0	2[a]
Predators	5	5
Total	18	16
Arthropods on ʻōhiʻa reproductive organs		
Sapsuckers	2	0
Seedeaters	1	0
Flower and nectar feeders	3	1
Omnivore	0	1[a]
Detritivores	0	2[a]
Predators	2	0
Parasites	3	0
Unknown	0	1
Total	11	5
Litter-inhabiting Diptera		
Saprophagous litter breeders	5	1
Scavenging litter breeders	1	1
Fungus feeders and breeders	6	0
Fruit feeders and breeders	2	1
Slime flux breeder	1	0
Predator	1	0
Unknown	1	0
Total	17	3

[a]Occupants of new general niches.

be an important food source of these endemic honeycreepers along the transect. In this case, similar distribution along the transect is associated with a distinguishable interaction. In many other cases, similarly distributed species are functionally unrelated or independent.

Both the canopy arthropods and the arthropods collected from 'ōhi'a reproductive organs (flowers, fruits, capsules, and buds) display a number of the same life-form types. They contain species in six common life forms: sapsuckers, seedeaters, flower and nectar feeders, predators, omnivores, and detritivores. In several life-form categories, even the species collected in the two organism groups were the same. This is not surprising since the arthropods found associated with 'ōhi'a reproductive organs were merely collected from a more localized place on the host tree. Even among these, the life-form spectrum shows that only two forms are functionally specialized on the reproductive organs of 'ōhi'a: the seedeaters and the flower and nectar feeders. The two collecting methods—total canopy fogging and plucking from reproductive structures—showed that only endemic seedeaters of one genus, *Oceanides,* are active on 'ōhi'a. However, the nectar-feeder niche was invaded by exotic arthropods. Both collecting methods resulted in finding in this niche the exotic *Hyalopeplus pellucidus,* and the canopy-fogging method also yielded two exotic *Neurisothrips* species. Endemic species of the same genus *(Neurisothrips)* appear to be the main original occupants of this general niche.

The two collection methods for tree-residing arthropods also yielded similar new life forms—exotic omnivores and detritivores—that were not represented among the endemics. In this case, however, the two methods yielded complementary information; the omnivore collected by fogging the canopies was the transect-ubiquitous ant, *Iridomyrmex humilis,* and the omnivore plucked from 'ōhi'a reproductive structures was the spatially very restricted ant, *Paratrechina bourbonica.* Among the exotic detritivores the two methods yielded different species. Those obtained through fogging the canopy were the cockroach *Allacata similis* and pillbugs (probably more than one species) of the taxon Isopoda. These two taxa were collected mostly from *Acacia koa* but also from 'ōhi'a trees. The exotic detritivores obtained from plucking 'ōhi'a reproductive structures were two springtail (Collembola) species of the genus *Entomobrya* (*E. nivalis* and *E. atrocincta*). Two additional exotic life-form types not represented among the endemics were obtained from canopy fogging. These were a leaf-chewer species, the Coleoptera weevil *Pantomorus cervinus,* and two grass-feeding thrips *(Aptinothrips rufus* and *Chirothrips patruelis).* The leaf chewer was most prevalent on 'ōhi'a at the lower end of the transect in the rain forest, and the two grass feeders were most prevalent on *Acacia koa* in the savanna and mountain parkland zones.

A number of insect predator species were obtained from both collecting methods. On 'ōhi'a reproductive structures, only two endemic predators, the mirid bugs *Psallus luteus* and *P. sharpianus,* were found. The canopy-fogging method resulted in each five endemic and five exotic species. Among the five endemics were the same two *Psallus* species plus two other mirid bugs (*Koanoa hawaiiensis* and *Orthotylus azalais*) and a nabid bug *(Nabis oscillans)*. The five exotic predators belonged to very different taxonomic groups. Four of them were ladybird beetles (*Scymnus varipes, S. loewii, Rhizobius ventralis,* and *Cryptolaemus montrouzieri*) and one was a thrips *(Karnyothrips flavipes)*.

The distribution pattern of these canopy-residing insect predators is almost opposite to expectation insofar as the endemics are all wide ranging and the exotics are all restricted along the transect. In this case the endemics are all successful, environmentally rather insensitive, dominants, while the exotics may have become more recently established and are still moving. However, it is also possible that the exotic ladybird beetles, being ecologically such a different group from the endemic mirid bugs, have already naturalized. In that case the restricted and rather segregated distribution pattern of the ladybird beetles may indicate some degree of competitive exclusion among their species—for example, *Scymnus varipes* in the subalpine and mountain parkland zone and *Scymnus loewii* in the rain forest. On the other hand, as Gagné noted, the exotic predator pattern may also be related to their temperate origin and preadaptation to cool environments. Three of the ladybird beetles species indicated ecological optima at the higher elevations.

The general distribution pattern of endemic and exotic arthropods collected from 'ōhi'a reproductive organs is somewhat similar insofar as it shows more endemics among the widespread gradient-insensitive species and more exotics among the gradient-restricted species. In many cases, one can assume that the exotics have not yet become fully naturalized.

Among the three groups of Diptera flies studied along the transect—the Sciaridae, the bait-trapped drosophilids, and the litter-inhabiting Diptera—only the last were differentiated by life forms (as summarized in table 4-6). The Sciaridae were all characterized as fungus feeders. Among the litter-inhabiting Diptera, only six of the twenty species were characterized as fungus feeders. Four of these were also collected among the Sciaridae samples. The bait-trapped drosophilids, whose life forms were not specified, also showed four species among the litter-inhabiting Diptera. Two of these (*Drosophila mimica* and *D. imparisetae*) were designated as fruit feeders and breeders by Hardy et al. and the other two (*Scaptomyza cuspidata* and *Drosophila simulans*) as saprophagous litter breeders. Thus we can assume that the life-form spectrum is very simple for the Sciaridae (just one life form, fungus gnats), while it includes at least two life forms for the bait-trapped drosophilids. If investigated further, the spectrum may be as rich as that shown for the litter-inhabiting Diptera, which include several specialists breeding in the slime flux, fruits, and leaves of particular native tree species.

The life-form spectrum of the litter-inhabiting Diptera (table 4-6) shows only three general niches invaded by exotics. Three others, and possibly a fourth, are occupied only by endemic flies. This indicates a low competitive pressure from exotics within this organism group, which also showed very high (85 percent) endemism. However, the exotic *Drosophila simulans,* which invaded the saprophagous litter-breeding niche, is the most successful exotic *Drosophila* species captured along the transect by the banana-baiting technique. This species overlaps in its distribution range with the five endemic species of this life-form group. However, the feeding substrate is rather generally defined for this group so that there may be very little interference in feeding and breeding sites among the endemic and exotic flies. Hardy et al. note, however, that predation on Diptera is considered to be an important evolutionary stress factor that may have been responsible for the great richness of endemic species in this group of flies.

The distribution pattern of the litter-inhabiting Diptera shows a distinctly high species packing in the mesic kīpuka forest (zone IV) and a good segregation of Diptera communities along the transect. A difference in the distribution pattern between exotic and endemic drosophilids is shown among those collected by banana bait. Here the exotic *Drosophila* species are distinctly more widely spread along the transect than are the endemic *Drosophila* species.

This distribution pattern between exotic and endemic Diptera flies is opposite to that found for the tree-residing arthropods. Moreover, the invasion of exotics did not result in the appearance of new life forms among the Diptera flies, whereas the invasion of exotics among the tree-residing arthropods was associated with the appearance of several new life-form types.

VERIFICATION OF DISTRIBUTION HYPOTHESES

What are our results in terms of the basic questions and hypotheses presented in chapter 2? We addressed our analysis to a fundamental question of MacArthur (1972) on plant distribution patterns in the tropics, to the four theoretical distribution patterns suggested by Whittaker (1970), to three similar models presented by Terborgh (1971), and to two specific island-biota-oriented hypotheses of our own.

MacArthur wondered whether there was something more to plant distribution patterns than the random appearance and disappearance of species, which Whittaker claims to have found along altitudinal gradients in temperate mountain regions. MacArthur suggested that the situation could be different in the tropics. He hypothesized that in the tropics, perhaps plant species might displace each other synchronously—that is, in spatially associated groups—along environmental gradients. MacArthur considered this question of fundamental importance and worthy of testing in different areas. The judgment of whether life zones are a reality in nature or a mere convenience of the investigator could then be rendered.

Whittaker (1970, 1975), who claims to have found only one realistic distribution model for plant distributions, that of random assortment, however, makes allowances for three other possible explanations. These, he says, appear as alternative distribution models in the literature. In extracting these ideas from the literature with such clarity, Whittaker gave an important stimulus to our island mountain analysis. Terborgh's models of bird distribution patterns are quite similar to those of Whittaker's plant distribution models except for some of the underlying reasons, which are of considerable importance.

The types of species distributions found in our analysis can be reduced to eight. Seven of them are sketched in figure 4-1. An eighth distribution type could be called nonoverlapping, associated species groups. This pattern was the least common. It was well expressed only among the litter-breeding Diptera and the plagithmysine borers. Even in these two organism groups, not all species were associated in nonoverlapping groups.

A nonoverlapping associated species group pattern is what Whittaker presented as hypothesis 1 in his theoretical review. If this is also what MacArthur had in mind, then his question can be answered negatively with our data from a tropical island mountain at least. Nonoverlapping spatially associated groups of species that displace each other along a gradient form the exception rather than the rule. In contrast, overlap of species ranges is far more prevalent. It applies to the other seven types of species distributions isolated from our results and sketched on figure 4-1.

Type 1, unrestricted wide-ranging groups of species, was common in our analysis. The proportion of these species in each organism group provided the key for ranking them in order of gradient sensitivity. These wide-ranging species are not necessarily spatially associated. They only appear to be associated because their ranges are cut off at the limits of the gradient. All but one organism group, the plagithmysine borers, contained at least one such wide-ranging species.

Type 2, interpenetrating or overlapping but spatially associated groups of species with restricted distributions, was also common in our analysis. It occurred in all organism groups except those with low gradient sensitivity. This pattern of spatially associated, interpenetrating groups of species is what Whittaker presented as hypothesis 3 in his theoretical review. However, the underlying causes, which Whittaker implies, cannot be read so easily from this pattern. Evolutionary adaptation of species to one another is certainly the less common reason for spatial association in our analysis. Close species interdependencies reflected in spatial association can be claimed only for the strongly host-dependent organism groups, the ectoparasites on rodents, and the plagithmysine borers on certain native woody plants. Even in these groups, some species did not display perfect spatial association with their hosts along the gradient.

In most cases, the phenomenon of spatial association cannot be reduced to a single cause. We can only say that the species that are spatially associated into a group along a gradient react similarly to a complex of environmental and biotic factors. This complex changes along the gradient at the starting or drop-out points of the associated species group. However, in many cases it is quite possible to untangle this complex for certain species combinations.

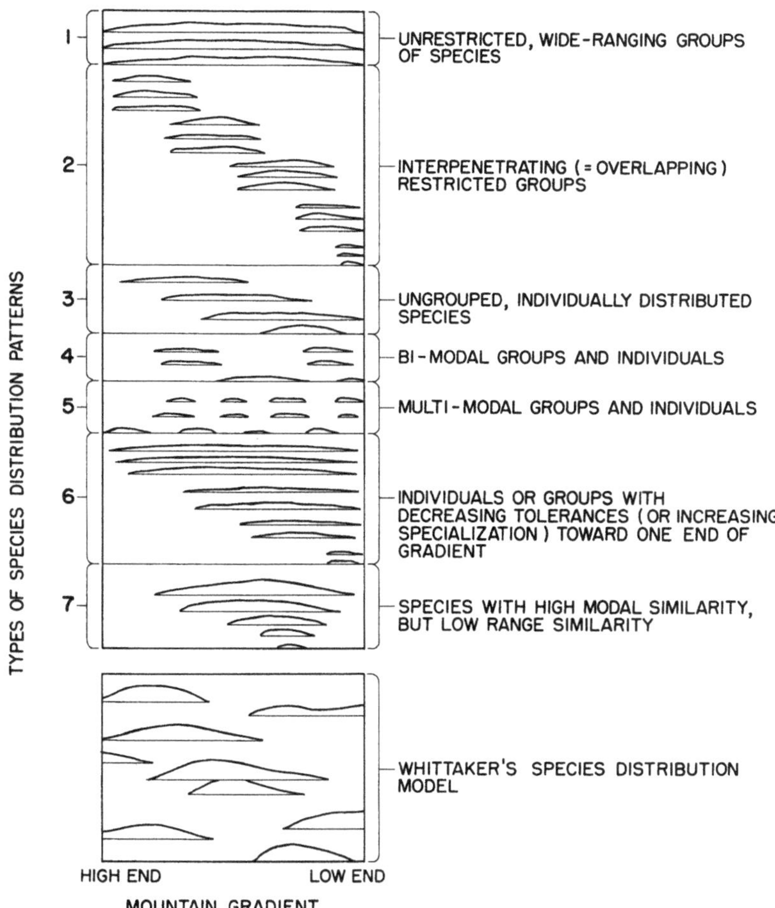

FIGURE 4-1. *Generalized trends of species distributions along Mauna Loa transect and comparison to Whittaker's distribution model.*

Type 3, ungrouped (nonassociated) and individually distributed species, represents another common pattern that occurred in our analysis. Among the native plants, there were two nonassociated dominants (*Metrosideros collina* and *Sophora chrysophylla*), which were unnumbered on the diagram. The type 3 pattern is almost the exact equivalent to Whittaker's own distribution model shown at the bottom of figure 4-1. It also corresponds to Terborgh's first bird distribution model. To make Whittaker's plant distribution model more easily comparable, his generalized species curves were lifted from the x-axis for displaying them individually. Whittaker's emphasis on species population quantity was maintained; it is indicated by the bell-shaped curves. One could easily rearrange his nine species curves with their ranges and modes so that a more structured order emerges. If, moreover, the life forms of the woody plant species were indicated, it would not seem that difficult to argue for certain, reasonably logical, community boundaries along Whittaker's generalized mountain gradient.

Type 4, a pattern of bimodal, spatially associated groups or individuals, occurred in several organism groups along our transect. We found it among species of native shrubs, exotic forbs and grasses, rodents, native birds, canopy arthropods, and others. In each case these bimodal patterns were quite explainable. For example, the two native shrubs, *Coprosma ernodeoides* and *Railliardia ciliolata,* which formed a spatially associated group, each showed two population modes, one in the subalpine environment and another in the open ʻōhiʻa forests at the lower end of the transect. Both sites had shallow ash-pocket soils with prevalent lava rock outcrop. Certain canopy arthropods had bimodal distributions because of the bimodal distribution of their major host tree, the ʻōhiʻa.

Type 5, a multimodal pattern, was exhibited by a number of species among the soil arthropods (the Collembola and Mesostigmata), the Diptera flies (the Sciarids, litter-breeding Diptera), and the soil fungi and algae. In most cases, these multimodal species are probably wide-ranging species whose abundances are so low that they were inadequately sampled in some parts of their range. Nevertheless the multimodal pattern is significant among those that showed up in our analysis. It also represents an interesting challenge for a better explanation in some cases, since the multimodal species show a low gradient sensitivity, but at the same time they may be more responsive to temporal variations in any local site environment along the transect.

Type 6, a pattern of individuals or spatially associated groups with decreasing tolerances toward the high end of the gradient, was characteristic for the rodents and their ectoparasites, the endemic birds, and the soil mites (Mesostigmata). This pattern shows a clear truncation at the low end of our transect. It indicates that for these organism groups, at least, the transect represented only a fragment of their life zone along this mountain. The general ecological optimum for each of these organism groups may be in the montane rain forest or even below.

Type 7 shows species with high modal similarity along the transect but with low range similarity. This pattern was characteristic of the exotic plants, exotic birds, and exotic Drosophilids. It indicates a center of distribution in the lower mid-section of the transect (the savanna zone), from where the species of these organism groups seem to have spread up and down the transect. It is doubtful that these exotics have as yet become fully naturalized; several may still be moving to fill the habitat area suitable for them. From the information available thus far, none of these exotics seems to be a serious competitor to the native species in their respective organism groups.

Our own two initial hypotheses relate to the propositions that the endemic island biota (birds and insects) are closely adapted to the native vegetation and thus form spatially associated species groups along major gradients and that climatic factors have significant effects on island ecosystem stability in terms of their resistance to the invasion of exotics.

The first proposition generally has been verified for the endemic birds and tree-residing arthropods but not for the endemic Diptera. Spatial adaptation to the vegetation pattern in the form of the seven recognized transect zones or communities was clearly shown by the native birds. Most of their upper-range boundaries and a few of their lower-range boundaries coincided with the vegetation boundaries, which were generated through computer analysis. Several of the lower bird-range boundaries were cut off artificially by the transect limit. Had the gradient analysis been continued downslope, more coinciding bird-range and vegetation boundaries may have emerged there also. The exotic birds displayed ranges that coincided similarly with major vegetation boundaries. However, their distribution pattern showed less spatial integration insofar as their use of the transect gradient was more or less restricted to the broader mid-section. In contrast, the endemic birds made more complete use of the transect gradient by being better represented in the alpine and rain forest environments and by showing less quantitative population overlap in the mid-section of the gradient. The tree-residing phytophagous arthropods also showed good correlations with the vegetation boundaries in general. As a group, the endemic species also showed better spatial integration along the transect than did the exotic species as a group. The same pattern trends as given for the birds apply to them; the endemic tree-residing arthropods are well represented at the high end of the transect (in this case, the subalpine scrub forest) and at the low end (in the montane ʻōhiʻa forests). They show a better species population balance than do the exotic tree-residing arthropods that were more prevalent or clustered in the broader mid-section of the transect (on *Acacia koa*) and less at the ends (on ʻōhiʻa).

The endemic tree-residing predator arthropods displayed less correlation with the vegetation boundaries. Several ranged only halfway upslope into the subalpine vegetation. Among the endemic Sciarids only a few showed ranges coinciding with the vegetation boundaries. Several ranged only halfway upslope into the subalpine vegetation, and many seemed to be particularly well represented across the subalpine-mountain parkland boundary. It appears that they may thrive in this ecotonal or vegetation-boundary environment. Even more so, the litter-breeding Diptera showed population distributions that appeared in most cases to be independent of the vegetation boundaries. In several cases their distributions ran across the subalpine-mountain parkland boundary. In most instances, their species population modes and ranges were more restricted than the vegetation zones along the transect, indicating that their relationship to the vegetation appears at a finer level of variation than was encompassed in the transect zones.

Our second hypothetical proposition—that climatic factors have significant effects on island ecosystem stability—cannot be answered from this analysis (but see part IV to this book); however, the second part of this proposition, which relates to the resistance of the endemic community to the invasion of exotics, can be answered to some extent. Few exotic species (except, perhaps, among the Collembola) were found in the alpine zone. In proceeding downward on the transect, exotics became more prevalent. Their species numbers and population sizes were greatest in the mid-section (birds, canopy arthropods, and exotic Drosophilas) or lower mid-section (plants) of the transect and decreased again somewhat into the rain forest. The transect climate, which can be described as mesic in the mid-section of the transect, is only one reason for this exotic species pattern. Many of the exotic species (plants, birds, and canopy arthropods) are of temperate-zone origin and thus most likely find their requirements best fulfilled in a climate that is not as wet as the rain forest or as continually cool as the alpine and subalpine zones along the transect. Thus the exotic species invasion in the mid-section of the transect, while related to the climatic gradient, is not necessarily a function of a lesser resistance to invasion by the native biotic community. In fact, very little competitive interaction has been noted among native and exotic organisms along the transect. An exception is the proven spread of the European grass *Holcus lanatus* and the corresponding displacement of the native *Deschampsia australis* in the mountain parkland (Spatz and Mueller-Dombois, 1975). However, this slow displacement of an endemic grass is not simply a case of superior competitive capacity on part of the exotic grass; the spread clearly is related to ground disturbance caused by the feral pig. The same applies probably to the other exotic forbs that have spread from the savanna. The widespread but quantitatively not prevalent occurrence of the composite weed *Hypochoeris radicata* may be an exception. This weed, which extends from the rain forest up into the alpine zone along the transect, can be found in habitats that are not scarified by pigs.

It thus can be predicted that the native species pattern as revealed by this analysis will remain relatively stable at the upper and lower ends of the transect, if the area is protected by the National Park Service from feral animals to the degree that it is now. A stable pattern has also been predicted for the fungi, which are native but not endemic. Changes can be expected among the exotic component of the organism groups through further spreading into available habitats of the now-existing species (for example, further range extensions can be expected among the various ladybird beetles) and through the appearance of new exotic species.

CONCLUSIONS

The analysis has shown that two of Whittaker's distribution models were well supported with our data. These relate to the distribution patterns 2 and 3 on figure 4-1, which correspond to Whittaker's third and fourth distribution hypotheses. His first hypothesis, which he did not believe himself, found very limited support. It relates to the occurrence of spatially associated species groups, which displace each other along a gradient with essentially no overlap. Whittaker's second hypothesis also found little support in our data. It relates to a pattern, in which the species are not spatially associated into groups but in which competing species (ones that are ecologically similar or congeneric) are displacing each other more or less abruptly along the gradient. It is the same hypothesis that Terborgh presented as his second model for bird distributions. We found something that may approach this pattern in the gradient order of four native shrubs, *Styphelia douglasii* being replaced by *S. tameiameiae* downslope, and *Vaccinium peleanum* being replaced downslope by *V. reticulatum*. A similar downslope replacement pattern can be claimed for the three species of *Fusarium* found in the soil fungal analysis and perhaps for the ladybird beetles of the genus *Scymnus* among the tree-residing predator arthropods. Similar gradient arrangements could be mentioned for some Diptera flies. However, particularly for those organisms, displacement through competition along this gradient is unlikely, since there are many spatial sites of interaction that are not associated with the gradient.

The most interesting pattern is that of spatial association of interpenetrating species groups (our type 2 pattern), which Whittaker's temperate mountain analyses did not reveal. This pattern was primarily responsible for the recognition of the seven transect zones or transect communities. It also gives an answer to MacArthur's question of whether species may change synchronously along gradients. In nearly all organism groups tested in our analysis, they do. Terborgh also may have implied this pattern with his third model of bird distribution rather than that of associated species groups with little overlap, which corresponds to Whittaker's first. Terborgh speaks of "ecotones" (transitions or diffuse boundaries) when he refers to those gradient segments where spatially associated species groups overlap. This is an interesting idea, which is also in some agreement with the usual concept of community boundary. However, this interpretation of ecotone would narrow the concept of community considerably. In many instances such ecotones would be much broader than the communities themselves, or the ecotones would even eliminate such narrowly defined "communities" altogether. In other words, a continuum would be recognized.

This brings up the point of definition. No fundamental issue is won by insisting on naming a phenomenon one way or the other. However, it is useful in science to use for the same phenomenon a name that most people have used or are familiar with. A community is usually recognized along a gradient where a certain number of species appear or disappear. A community, to be different from an adjacent community, need not be comprised of a totally different set of species. Only a certain proportion, about one-third, needs to be different, and even this may vary since there are no strictly objective rules. The width of the boundary is indicated by the degree of synchrony or spatial association with which a set of species appears or disappears along a gradient. If the degree of synchrony or spatial association is low, then one can recognize an ecotone in the sense of a diffuse boundary, and if there is no spatial association at all, community recognition becomes truly arbitrary. The latter is what Whittaker concluded from his gradient studies in temperate mountains. Is his conclusion valid also for our tropical island mountain analysis?

The answer, clearly, is no. Except for the few organism groups that showed low gradient sensitivity, we documented that spatially associated species groups exist along our mountain gradient. This is something other than documenting the existence of communities. The latter cannot be documented unless there is a universally acceptable definition. However, what we have documented is the underlying phenomenon for community recognition. Based on this phenomenon of spatially associated species groups, one can recognize communities at different levels of generalization or geographic scale.

One critical question remains: Was our analysis really a fair test of Whittaker's gradient analyses? The answer is yes and no. Whittaker arranged his altitudinal samples in the geologically older temperate mountain systems in such a way that he achieved a gradient, which seemed to be free of environmental discontinuities. We attempted to do the same but did not totally succeed. Although the geological substrate was from chemically uniform basaltic lava with a thin overlay of volcanic ash, the latter varied in depth perhaps more than would be desirable under ideal conditions. The ash blanket thinned down more or less abruptly at the mountain parkland to subalpine zone boundary and at the savanna to ʻōhiʻa dry-forest boundary. Even at the boundary between our alpine and subalpine zone where there is no recognizable substrate discontinuity, we may recognize a climatic boundary, the nocturnal ground-frost line, which appears to persist even during August at this elevation (2,600 m).

We conclude that our species association patterns were caused largely by marked environmental changes encountered along our mountain gradient. However, such environmental changes are probably as frequent in nature as are environmental continua. Therefore spatial association of species in response to pronounced environmental variations does not reduce the fundamental significance of this response phenomenon for the recognition of biological communities.

III
Community Structure and Niche Differentiation

5

Introduction

A. J. Berger, Sheila Conant, F. J. Radovsky, D. Mueller-Dombois and F. G. Howarth

CONCEPTUALIZATION AND OBJECTIVES

Insular ecosystems differ from those of continents in having relatively depauperate faunas and in lacking a number of higher taxa; this results both in functional gaps and in greater or lesser shifts in roles (or general niches) of many organisms or groups of organisms. Consequently there are differences between insular and continental ecosystems (representing the same biome types) not only in species assemblages but also in their structural and functional relationships.

Generalizations can be made about the ecology of islands, but the special geographic and historical characteristics of each island group must be considered. Moreover, although islands in general form a group distinct from continents, the Hawaiian archipelago is a particularly special case among islands because of its extreme geographic isolation. The integrated ecological studies reported here are the first of their kind in Hawai'i. By examining selected communities (ecosystems) as case examples, we hope to elucidate some general principles in community ecology through comparison with continental ecosystems and with other insular ecosystems.

The structure of a community may be described partially in terms of the distribution of its component organisms in space and in time. Spatial distribution extends in both horizontal and vertical dimensions. Structural relationships between the organisms are intraspecific (size, age, reproductive conditions, and so forth of species populations) as well as interspecific. This interspecific aspect includes trophic levels, food webs, pollination, and predation. Thus in our considerations here we are examining both the role of the species in the community as well as the division of the community into a set of potential (and realized) roles. These two aspects are, of course, interrelated.

To a large extent, productive studies in community ecology are now focusing on the general niche, a functional unit or species-group concept, apparently introduced by Elton in 1927 for subunits in animal communities. He observed that "the ground plan of every animal community is much the same" (Elton, 1966:63). In other words, he suggested a similar spectrum of functional roles that can be performed by different animal species in geographically separate but structurally similar biomes. Elton's concept allows for the recognition of "vacant niches" (Miller, 1967). About these functional roles, Odum (1971:234) states that "the ecological niche of an organism depends not only on where it lives but also on what it does (how it transforms energy, behaves, responds to and modifies its physical and biotic environment), and how it is constrained by other species."

Thus the niche concept is applicable to all organisms, including plants, and it can be defined as the total space occupied by an organism in an ecosystem and the functional role that it assumes in that space. In this context it is useful to distinguish between specific and general niche. The term *specific niche* refers to a species population within an ecosystem. *General niche* refers to a group of species of closely similar ecological characteristics occurring together in an ecosystem (Mueller-Dombois, 1975).

Milstead (1972) proposed that there is a "finite number of broad ecological niches" (that is, general niches) in any given type of biome and that these niches can be defined by comparing the biology of convergent animal forms on different continents. He noted, however, that "past natural history studies do not yield the detailed type of information needed to make such comparisons." Thus, we find, on the one hand, that the study of either organisms or niches actually involves consideration of every conceivable relationship between an organism and its niche (as well as to every other organism in the niche), and, on the other hand, that adequate information is not available for comparing niches on different continents or even on the same continent or large island group. Hubbell (1968), for example, wrote that "no complete inventory of a large island's plants and animals has ever been made. . . . But even in an archipelago supposedly so well known as Hawai'i, every intensive modern study of a group of native animals or plants not only turns up previously unknown species, sometimes in large numbers, but also often reveals unsuspected and sometimes surprising evolutionary and ecological phenomena." This prediction was corroborated several times during the course of this IBP project, especially with regard to the discovery of a previously unknown endemic fauna in Hawaiian caves.

The general niche or broad ecological niche concept relates to both the resource fragment and a functionally similar group of species that make use of this fragment in an ecosystem context (*sensu* Mueller-Dombois and Ellenberg, 1974:166). The functional group as defined by the IBP working group on ecosystem theory (proposal 6, Denver, January 1973) approximately parallels the botanical concept of synusia; that is, species of similar life form, broadly similar in structure and function, occurring together in the same habitat (for example, upper deciduous tree canopy synusia, evergreen shrub synusia, vascular epiphyte synusia, and so forth; Gams, 1918; Lippmaa, 1939). More closely comparable to the botanical term *synusia* is the term *guild* (Root, 1967; Reese, 1969) in animal ecology, which refers to a group of animal species occurring together in the same ecosystem that have similar gross morphologies and functions (for example, wood-boring beetles, rodents, and so forth). Traditionally animal ecologists have stressed functional similarity, while plant ecologists have placed a greater emphasis on structural similarity as an indicator of functional similarity. This difference is related to the mobility of animals versus the stationary life mode of plants. However, both organismic character traits, structure and function, are addressed in the terms *synusia* and *guild*.

Thus, a number of terms relate to the concept that broad classes of niches are both definable and useful in the development of ecosystem theory. This concept is an important basis for our approach to community structure and niche differentiation, in which we use both life-form groups (sensu synusiae) and functional groups (sensu guilds) as indicators of general niches.

SELECTION OF CASE EXAMPLES

We chose two types of communities or ecosystems for detailed analyses. The Kīlauea forest is less disturbed than the montane rain forest on the Mauna Loa transect and comprises a much larger and more homogeneous tract of forest than any on the transect. This choice was intended to complement the transect approach by undertaking an intensive study of a relatively undisturbed forest. Because this site is on private land and is not accessible to the public, the principal disturbing influences in the forest are introduced animals. The most significant vertebrate species in terms of general habitat alteration is the feral pig, but one species of rat is common, mongooses are spread throughout the forest, and other rodents and feral cats are present. The second example is an ecosystem type rather than a specific locality. Caves, more particularly lava tubes, became an important part of the IBP studies after July 1971 when cave-adapted animals were first discovered in Hawai‘i. This ecosystem has been added to those used for intensive study and community analyses because of its obvious advantages. Both caves and islands are favored for studies in evolutionary biology because of their relative isolation.

HYPOTHESES

The following hypotheses seemed worthy of consideration.

1. That there might be distinct differences in the numbers and kinds of niches in continental as compared with insular ecosystems.
2. That general niches are filled in many instances by forms in higher taxa not associated with these niches on continents (that is, adaptive shifts).
3. That certain general niches are not fully occupied by native species and that these are more easily invaded by exotic species.

FIGURE 5-1. *Kīlauea rain forest. The upper canopy trees (25–30 m tall) are of* Acacia koa *var.* hawaiiensis. *A* Metrosideros collina *subsp.* polymorpha *is leaning on left side, where it forms the subcanopy with several other individuals. Tree ferns,* Cibotium glaucum, *below. Their upper fronds reach to 5 m height.*

KĪLAUEA FOREST STUDY SITE

The Kīlauea Forest Reserve was chosen for study for several reasons. It was close to our field station, it is an important ecosystem type on one of the IBP transects, and it is an excellent example of a relatively undisturbed *Acacia koa-Metrosideros collina-Cibotium* spp. montane rain forest. This is an important ecosystem type that currently is found in only a few places on the island of Hawai'i. The other sites are on the southern and western slopes of Mauna Loa and Hualālai (Nelson and Wheeler, 1963), the east flank of Mauna Loa between 1,520 and 1,830 m (Mueller-Dombois, 1966), and the east flank of Mauna Kea (Mueller-Dombois and Krajina, 1968).

The Kīlauea Forest Reserve was made available for study by the Trustees of the Bernice P. Bishop Estate, which owns the reserve. The estate offered the reserve for study primarily because of its interest in *Acacia koa* as a commercial timber tree. The original intent was to study the structure and dynamics of the virgin forest ecosystem prior to its regeneration after harvesting koa, then scheduled to begin in 1972.

The forest was soon recognized by the IBP participants as the best intact example of this forest type remaining in the state. In addition, the forest provided habitat for a number of rare and endangered birds and plants. Therefore it was suggested to the trustees that logging be postponed indefinitely so that full advantage could be taken for scientific study, nature education, and aesthetic enjoyment of a rare forest type containing rare and endangered organisms. The estate has shown some interest in proposals to set aside a portion of the reserve as a natural area and to investigate the feasibility of developing *Acacia koa* as a commercial timber species on adjacent lands that already have been harvested and pastured.

Site Description

The Kīlauea forest is a mixed *Acacia koa-Metrosideros collina* forest with arborescent shrubs and *Cibotium* spp. (tree ferns) occurring on shallow to moderately deep, humus-enriched soil from volcanic ash (Mueller-Dombois, 1966:418), (see figure 5-1). The forest occurs in a year-round humid climate between 1,200 and 1,800 m on the southeast slope of Mauna Loa. It is a montane tropical rain forest. Although the Kīlauea forest has never been logged or grazed, it has been exposed to disturbance because of the long-term presence of exotic mammals, especially the feral pig *(Sus scrofa)*. Whitesell (1964) pointed out that the *Acacia koa-Metrosideros collina-Cibotium* spp. montane rain forest is now limited on the island of Hawai'i because extensive areas formerly covered by this forest type have been cleared and converted to pasture.

Location and Climate

The IBP study site in the Kīlauea Forest Reserve is located 8 km north of Hawai'i Volcanoes National Park. The reserve extends as a narrow strip from 1,160 to 1,900 m on the east flank of Mauna Loa. The site itself is a large plot, 80 ha in size and occurring at 1,590 to 1,650 m elevation.

The forest is in the cloud-fog zone and receives precipitation from both rainfall and fog drip. Based on extrapolation from Taliaferro's (1959) rainfall map, the median annual rainfall at the study area is about 1,900 mm. According to the fog-drip analysis (Juvik and Perreira, TR 32), we can expect that another 900 to 1,000 mm is added to the annual rainfall from fog interception. Thus the forest floor is almost continually moist to wet. The mean annual temperature at the study area, computed from adjacent station records, is 13°C.

Monthly rainfall and mean monthly temperatures at the IBP weather station from 1972 to 1974 are plotted in the form of a climatogram (figure 5-2). Total rainfalls for 1972, 1973, and 1974 were 1,701, 1,498, and 2,173 mm, respectively. The diagrams show several short dry periods when the monthly rainfall was less than 100 mm. These periods can occur at almost any time of the year, although they usually come during the summer. There are occasionally short drought periods—one occurred in July 1973—as defined by the climatogram relationships. To a certain extent, these dry periods may be counterbalanced by cloud interception. During all months, the relative humidity values are relatively high, with a monthly average generally above 90 percent (Bridges and Carey, TR 22, 38, 70).

No freezing temperatures were recorded in the meteorological shelter (1.5 m height) during the study period (1971–1974). The January minimum recorded daily temperature reached 2°C in 1973. The mean January temperature for the four-year period was 12.5°C. The mean July temperature for this same period was 14.3°C with the highest recorded daily July temperature reaching 27°C (Bridges and Carey, TR 22, 38, 70).

Strong winds, with speeds up to 40 meters per second can be expected with the *kona* storms each year during the winter, but hurricanes with winds above

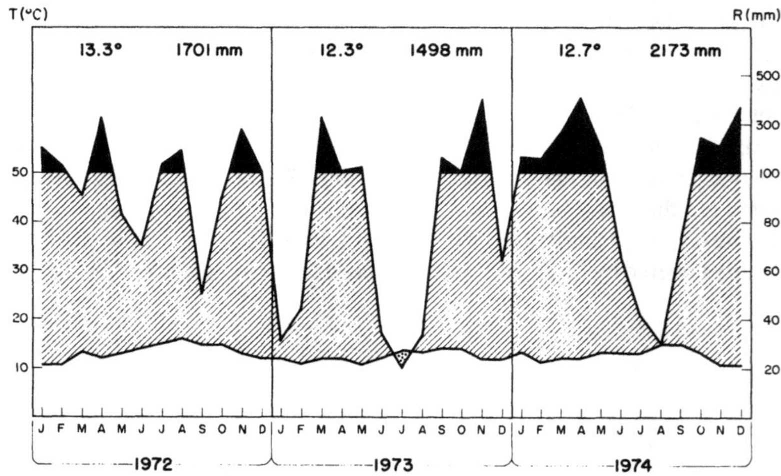

FIGURE 5-2. *Climatogram of Kīlauea forest at IBP study site showing monthly mean rainfall and temperature curves for 1972 through 1974 (method after Walter, 1971).*

75 mps occur only rarely at erratic intervals of perhaps several decades. Tropical storms with slightly lower wind speeds occur on the average every three or four years (Blumenstock, 1961).

Soil and Substrate

According to Stearns and Macdonald (1946), the geological substrate of the Kīlauea Forest Reserve is a Mauna Loa lava flow belonging to the prehistoric Ka'ū volcanic series, which includes lavas from the mid-Pleistocene to the present (Macdonald and Abbott, 1970). Although the lavas have not been dated, Macdonald (pers. comm.) estimates their age at less than 5,000 years.

The geological material is primarily 'a'ā lava (basaltic) with some ash overlay and microtopographical relief. This land form is similar to that found in the *Metrosideros-Cibotium* rain forest of 'Ōla'a, southeast of the Kīlauea Forest Reserve. Although well-defined drainage patterns are lacking, puddles and small pools are formed throughout much of the year, and small creeks run downslope during rain storms.

Cline (1955) mapped the study-area soils as rockland pāhoehoe lava with Pu Oo, Maile, or Olinda soil material. In the study site itself, the A_1 horizon, which is composed of ash with heavily admixed organic matter, lies directly on 'a'ā lava. For this reason, the soil is better described as rockland 'a'ā lava rather than rockland pāhoehoe lava. Some of the Olinda soils were interpreted by Mueller-Dombois (1966a) as shallow to moderately deep, discontinuous ash soil on 'a'ā, and this is characteristic of the study-site soils, which also have numerous semiweathered 'a'ā rock outcroppings. Average soil depths (from 800 samples) ranged from 20 cm to 65 cm (Maka, TR 31).

The dark brown soil is rich in organic material and has the texture of a clay loam. The organic matter content was found to be 13 percent (by ignition method; Stoner, Stoner, and Baker, TR 75). The pH was very acid and ranged from 4.65 (summer measurements; Stoner, unpublished data) to 4.1 (Becker, 1976, winter measurements). Nitrate nitrogen (NO_3-N) was found to be moderately high to high (44 to 160 ppm) and sulfate sulfur (SO_4-S) moderate (20 to 60 ppm) (by Stoner), while the exchangeable cations (calcium, magnesium, and potassium) occur only in very low quantities. Although there is no continuous, well-defined litter layer, rotting logs, fallen branches, and patches of leaf litter are common on the forest floor (Cooray, TR 44). Sato et al. (1973) have described the study-area soil as Pi'ihonua silty clay loam, which is said to dehydrate irreversibly into fine gravel-size aggregates.

Vegetation and Fauna

This forest type represents a transition between typical *Acacia koa* forest found from about 1,520 to 1,830 m and the *Metrosideros-Cibotium* rain forest found below 1,370 m. The upper tree canopy is dominated by tall (20–30 m) *Acacia koa* trees with large (80–150 cm) trunk diameters and full crowns. The fact that Krajina (1963) did not mention *Acacia koa* as a rain forest component underscores the point that this species is not typically found in all rain forests. The height of *Metrosideros* is slightly less than that of *Acacia koa,* and its crown and trunk diameters are somewhat smaller. The second tree layer is composed of low-stature tree species, such as *Myrsine lessertiana, Myoporum sandwicense, Cheirodendron trigynum,* and *Coprosma rhynchocarpa.* A third layer is composed primarily of tree ferns (*Cibotium* spp.). Occasional openings in the tree canopy are occupied by sedges and other herbaceous plants.

Of the tree species found at the study site, *Metrosideros collina* and *Cheirodendron trigynum* are the most common, followed by *Acacia koa* and *Ilex anomala.* Trees scattered throughout the forest are *Myrsine lessertiana* and *Myoporum sandwicense.* Species rarely reaching tree size (more than 5 m) include *Pelea* spp. and *Coprosma rhynchocarpa* (Maka, TR 31).

The montane rain forest is the habitat of a number of native bird and insect species and of several introduced mammal species. Berger (TR 11) reported nine species of endemic birds, including four endangered species, and two introduced species from the Kīlauea Forest Reserve. Conant (TR 74) observed four additional introduced bird species, as well as those native and exotic species recorded by Berger. Tomich (TR 2:168) found *Rattus rattus* (roof rat) to be abundant, *Herpestes auropunctatus* (small Indian mongoose) to be common, and *Mus musculus* (house mouse), *Rattus norvegicus* (Norway rat), and *Felis catus* (feral cat) to be rare in the forest. Cooray (TR 44) saw a feral dog *(Canis familiaris)* deep in the forest during September 1973. With its abundant food supply and cover, the montane rain forest is the favored habitat of the feral pig (Giffin, 1972). Cattle from the adjoining ranch enter the Kīlauea forest occasionally through broken fences, but they usually stay on or near the logging road. Molluscs, soil-dwelling arthropods, and earthworms are numerous (Cooray TR 44).

Sampling Design

An 80 ha study site was established more or less centrally within a large (800 ha) homogeneous tract of the montane rain forest. The layout of this area is shown in figure 5-3. Its location relative to the other IBP sites is shown on figure 1-6, and its location on the island of Hawai‘i can be seen on figure 1-5.

A climatic station was established at the northwest corner of the study site, which was maintained and serviced at weekly intervals from August 1, 1970, through August 1, 1975. The corner of the site was near a logging right-of-way, although it had not been used for operational logging. The 800 m baseline was run more or less parallel to this logging road at a direction of 55°NE. From this base line, four 1,000 m transects were established in a southeast direction of 145°. These transects were run parallel to each other, 200 m apart. Five plots were established along each transect at 200 m intervals, resulting in twenty plots. These transects and reference points served for all sampling programs, and transect widths, plot sizes, and intervals were adjusted to each investigator's needs.

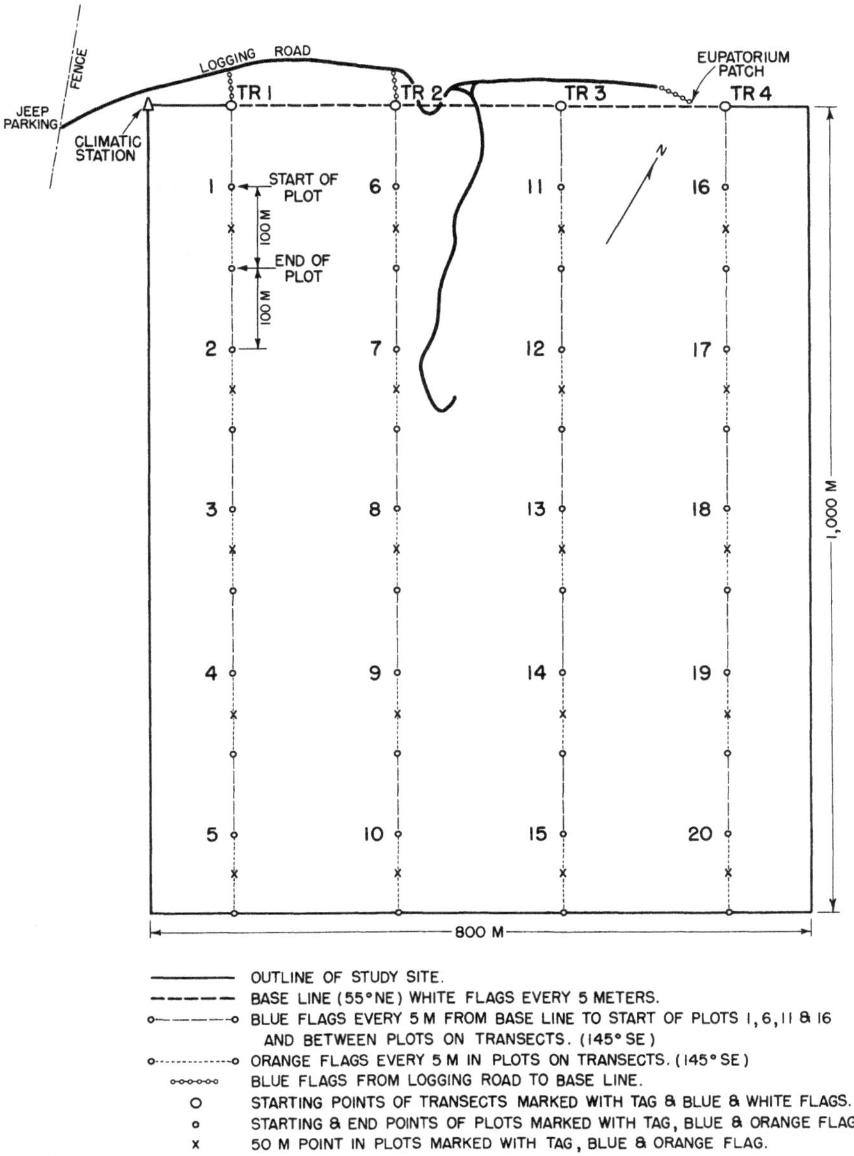

FIGURE 5-3. *Sampling layout of the 80 ha IBP study site in Kīlauea rain forest. Four major transects (TR 1, TR 2, TR 3, TR 4) each with four plot-starting points (1–20) were marked in the field at distances and directions shown.*

The basic sampling unit of the plant ecological survey was a belt transect or long rectangular plot, 6 m by 100 m (figure 5-4). Each plot was divided into 3 by 5 m subplots for the quantitative assessment of the forest undergrowth and into 5 by 5 m subplots for the enumeration of trees by species and sizes.

The forest animals were censused in the same sample units. Tree borers were studied repeatedly along all four transects by inspecting suggestive scars on the trunks and fallen trees and by searching for free-moving specimens in flight and on foliage. Small mammals were sampled on two transects monthly for two years, alternating among the four so that each transect was sampled bimonthly; snap and live traps were at 5 m intervals with every third one a snap trap. Birds were surveyed repeatedly by slowly walking each transect with frequent stops using Emlen's method, as described in chapter 2. The activity of feral pigs was surveyed periodically by recording activity signs (rooting, wallows, freshly felled tree fern trunks, feces, and so forth) along the transects. Foliar arthropods were sampled in the central area of each transect at plots 3, 8, 13, and 18 on dominant trees by bimonthly fogging of their crowns with synergized pyrethrum.

Transects and plots were flagged conspicuously not only for orientation in censusing on the transects but also for reducing the chances of investigators becoming disoriented in the forest. The dense, nearly homogeneous vegetation offered few clues for orientation, particularly when there was a complete, low cloud cover, sometimes at or below canopy level. Experienced hunters and hikers not infrequently become lost in Hawaiian rain forests, especially when a heavy fog rolls in.

LAVA TUBE ECOSYSTEM AS A STUDY SITE*

The cave ecosystem shares some attributes with an island ecosystem as a whole in that both are discretely defined, isolated geologic entities. The cave ecosystem is rather simple, with limited energy inputs and a relatively uncomplicated food web with few species. This simplicity is related to its youthfulness and the restrictions imposed by the rigorous environment. It represents an "island" ecosystem within an island ecosystem.

The discovery of cave-adapted arthropods in Hawaiian lava tubes was unexpected, and therefore it opened up a new area of investigation under the auspices of the IBP project (Howarth, 1972). Four long-held assumptions were thought to preclude the existence of a cave-adapted fauna in Hawai'i. First, obligatory cavernicoles were known mostly from limestone caves in temperate regions south of maximum glaciation and were considered glacial relicts, and most of the few known tropical forms are aquatic and were assumed to be relicts of changes in seawater level (Barr, 1968; Vandel, 1965). Second, lava tubes

*By F. G. Howarth

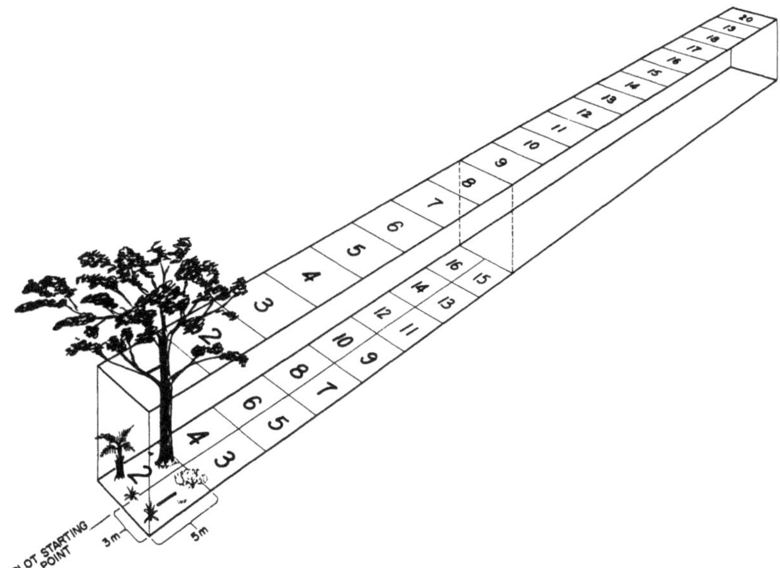

FIGURE 5-4. *Basic sampling unit of plant ecological survey, a long 6 by 100 m rectangular plot with subplots. See figure 5-3 for distribution.*

were considered too ephemeral and too often polluted with volcanic gases to allow evolution of cave forms (Torii, 1960; Barr, 1968). Third, islands were considered unfavorable sites for the evolution of cave faunas because the pre-adapted forms (the taxa represented in continental caves) are poorly represented on oceanic islands as a result of their lack of dispersability (Peck, 1974). And finally, Hawaiian lava tubes were known to lack the two main energy sources of continental caves: sinking streams and trogloxenes, those animals that roost in caves and return to the surface to feed.

Site Description

Lava Tube Formation

A lava tube is a discrete ecosystem, rigidly defined by the geological setting. Limestone caves can be considered an integration of the erosion and degradation processes coupled with the dissolution of limestone to form new passages in a dynamic, continuing process over a long geological period. Lava tubes, on the other hand, have an initial period of formation, after which the processes of erosion and siltation degrade the caves in a brief geologic time. There is no chance for the later enlargement of the passage in a lava tube.

Recently Peterson and Swanson (1974) described the formation of lava tubes in active flows. They discovered that the process of lava-tube formation (by insulating the flowing lava and transporting it great distances from the vent) is one of the most important factors in building shield volcanoes. This discovery increased geological interest in lava tubes.

Lava tubes form almost exclusively in pāhoehoe basalt. Pāhoehoe lava flows in rivers and nearly always crusts over and forms lava tubes. However, not all tubes drain, forming caves, and not all caves are open to the surface. The caves range in size from tiny, finger-sized tunnels to caverns more than 10,000 m long and 17 m in diameter.

Near the vent a highly fluid flow may move like a river. As it moves, the flow loses heat and gas and tends to crust over. The crust floats downstream. It may jam and then form a roof, or it may sink or break up and may form another crust. At the same time, a lava shelf grows inward from the edges by crystallization of the lava. If the flow level remains relatively constant, these two shelves may join and form an arched roof. The same crystallization process builds a roof downstream from a crustal jam or arched roof. Thus the surface of the lava river usually soon solidifies for most of its length.

During the eruption, the level of the flowing lava in the tube can vary considerably from small, sluggish flows to those completely filling the tube. The lava can even overflow the levee through "skylights." Such oscillations can leave linings 1 to 20 cm thick on the walls and ceiling of the tube.

Away from the vent, the front of a pāhoehoe flow advances much like an amoeba. A toe of molten lava breaks its surface crust, and the draining lava spreads like a thin sheet. This new surface solidifies and then breaks again, draining the interior and forming a new toe. Toes also form on older toes, layer on layer, as the front advances, building a levee in which the main river flows. These toes are fed by small distributary tubes from the main river. Fluid pressure and escaping gas often swell the still-flexible crust of pāhoehoe toes, forming lava blisters, some of which may be several cubic meters in volume. The pāhoehoe flow becomes deeper as the thin layers of lava, 5 to 40 cm in thickness, flow out as toes. Many of these layers are separated by spaces of 0.5 to 10 cm for part of their areas. The total thickness of the flow may be 15 m or more.

When the flow stops, the lava may drain from the main tube by gravity and leave a void or cave. The structure of the cave depends on the particular method of formation, on the amount of molten lava that drained from the cave, and on the deformation by spalling (breakdown) and slumping of the walls, both while the lava is flowing and during cooling. It is believed, but not proven, that the flowing lava river can erode the substrate and deepen and alter its channel.

Distribution

Lava tubes are an important land form in Hawai'i. They are common among the recent lava flows on four of the volcanic mountains of Hawai'i—

Kīlauea, Mauna Loa, Hualālai, and Mauna Kea—and on the Haleakalā volcano of Maui. They occur from sea level to 3,965 m, so they are distributed from the warm tropical lowland to the cool alpine environments. In age, they vary from those just formed to half a million years old. Most caves on the older volcanoes have been destroyed by erosion. A few caves do exist in the most recent lava flows of the older islands of O'ahu and Kaua'i. Remnants of buried lava tubes occasionally are exposed on the sides of river canyons on the older mountains, but these are rarely more than shelter caves.

A subterranean dispersal system exists for the cave fauna. During the tube or cave formation process, gas bubbles escaping from the liquid lava are often trapped by the solidifying lava, forming vesicles. Often these vesicles interconnect, forming channels 1 to 10 mm in diameter through the rock. As the lava flow cools and contracts, cooling cracks radiate throughout the flow, offering a network of dispersal routes for cavernicoles. Thus young basalt has numerous gas vesicles, fissures, small lava tubes, and other voids that allow subterranean intercave dispersal of cave animals. Therefore the existence of a significant cave fauna can be predicted in any area where lava tubes are numerous, vulcanism continual, and where the environment allows colonization of the caves. This has been verified in other places of the world (Peck, 1973; Briggs, 1973).

Cave Environment

A typical lava tube can be divided into four environmental zones: the entrance zone, which is very often richer in species than either the epigean (outside) or the hypogean (inside) zones; the twilight zone, where light is reduced and green plants progressively drop out; the transition zone, in constant darkness but where certain outside environmental factors are still effective; and

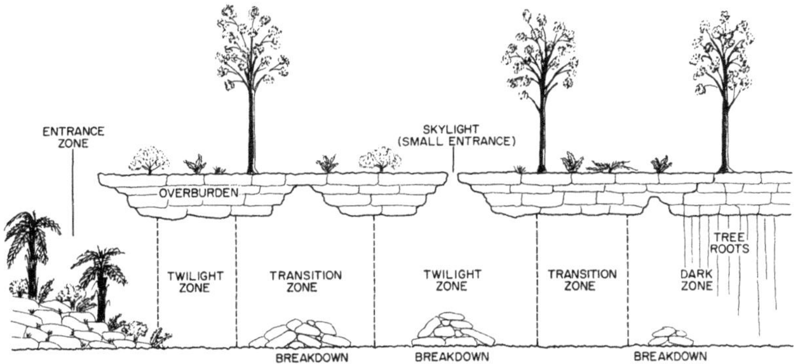

FIGURE 5-5. *Generalized profile of a Hawaiian lava tube with the four habitat zones that occur repeatedly in different caves. Height of cave to length of zones is exaggerated.*

the true dark zone, where constant darkness and constant climatic conditions usually prevail. A diagram of a lava tube with such zones is shown in figure 5-5.

The extent of these zones depends largely on the size and exposure of the entrance and on the size of the cave itself. Caves with a very small entrance may have no real entrance zone and only a small twilight zone. In caves with a large entrance, the twilight and transition zones may extend far into the cave. In lava tubes with two entrances, the passage between may be entirely a transition zone because a chimney effect between the entrances allows wind to pass through, which increases evaporation and introduces some other effects of the epigean environment.

Moisture, even a ceiling drip or a pool on the floor, can mitigate the environment and may extend the active area of colonization by troglobites (obligatory cave animals) well into the twilight zone, especially under and behind rocks.

The true cave ecosystem, in which the specialized cave animals occur, is restricted to the dark zone, as is true in continental caves. It is a rigorous environment conditioned by the absence of light, a relatively stable physical surrounding, a constant humidity above the physiological limits of most terrestrial animals, and an absence of the usual environmental cues.

There are few measurable environmental cues in the cave environment. The cave animals retreat into the cave as the transition zone enlarges somewhat in the cooler winter months due to colder air sinking into the cave entrance and, after warming to cave temperature, drying the cave. This winter effect is more pronounced and well documented for temperate continental caves, where the seasonal climatic change is much more extreme than in Hawai'i. Flushing of roots and an increase of ceiling drip during rainy periods may act as cues for some root feeders.

The cave temperature is approximately the average annual air temperature above ground for the area. Thus on Hawai'i cave temperatures range from 25°C in lowland caves to approximately 10°C at timberline and to below 5°C at 3,965 m.

Energy Sources

Three main energy sources have been identified in Hawaiian lava tubes: living plant roots, two kinds of slimes, and other organic substrates.

Tree roots provide the most important energy source in Hawaiian lava tubes. Lava tubes generally have little overburden; that is, they are close to the surface. Also some of the native Hawaiian trees, notably *Metrosideros collina* subsp. *polymorpha,* colonize relatively young lava flows (Smathers and Mueller-Dombois, 1974). The roots dangle into the caverns and supply food, either directly as living or decaying roots or by forming pathways for the percolation of organically enriched water.

The number of roots can be surprisingly great. In the Emesine lava tube, which occurs in the 1881 lava flow at 1,760 m on Mauna Loa, scattered groups of roots extend from the surface to about 5 m depth in the cave. Yet the surface of the lava flow is practically barren, with native shrubs 1 to 2 m tall, which are spaced 1 to 2 m apart in the lava fissures only. The roots are almost all of *Metrosideros,* although a few unidentified roots also penetrate the roof of the cave.

Metrosideros roots can be very large. The largest noted to date is in a lava tube at 400 m on the east slope of Kīlauea. It was 10 cm in diameter where it penetrated the roof; it could be followed for more than 30 m along the floor as it meandered downslope. Its branches entered cracks and climbed up the cave walls toward water sources. Rhizomes of the mat-forming fern *Dicranopteris linearis* occurred sparsely in the same cave.

In lowland caves in disturbed areas, roots of certain introduced trees are common. Those found in dry-zone caves in Kaua'i were of the legume *Pithecellobium dulce, Ficus* sp., and *Eugenia cumini.* Grass roots also may penetrate into lava tubes. In caves in pastures on the western slope of Mauna Loa at approximately 1,200 m, roots of the introduced kikuyu grass *(Pennisetum clandestinum)* were common in the shallow (approximately 1 m overburden) caves and supported the cave moth (*Schrankia* sp.) and the tree cricket (*Thaumatogryllus* sp.). Grass roots have been noted in other caves as well, such as at Kīpuka Puaulu (1,220 m, Hawaii Volcanoes National Park), where the overburden is 1 to 3 m.

Two kinds of slime—one white and one brown—have been observed in many Hawaiian caves. The white slime, the more common material, usually is found along moistened walls in lava tubes. It has an actinomycete soil odor and a moldy taste but contains very little organic material other than a few fungal hyphae and spores. The brown slime usually occurs in more localized places and is associated with dead roots. It smells like moist humus and contains a large amount of organic material, mostly from root tissues. Fungal hyphae and spores are also abundant. Stoner (unpubl. data) identified in a sample of the brown slime macroconidia of *Fusarium,* oospores of *Pythium,* and other spores of an unidentified fungus.

Other organic substrates or energy sources are provided in caves by percolating water, which carry with them some organic matter and by animals that blunder into caves but cannot survive there. The quantity of these other organic substrates is difficult to specify, but the high populations of cave-inhabiting predators and scavengers indicate that in some cases they may form an important additional energy source.

The kinds of energy sources and their magnitudes in Hawaiian lava tubes are quite different from those described for continental limestone caves. In the latter, two energy sources have been identified as most significant: sinking streams, which carry organic debris in greater quantities, and certain animals, such as bats, birds, and crickets that roost in the caves. This difference in energy sources undoubtedly also contributes to the difference in the fauna found in the two types of caves.

Cave Fauna

The animals found in caves can be classified into four groups. The troglobites, obligatory cave animals, are unable to survive outside of caves. The trogloxenes regularly roost in caves but normally feed outside. The accidentals are surface- or soil-inhabiting animals that wander into caves but cannot survive there. These are very general ecological categories. However, detailed knowledge of an animal's behavior and distribution is necessary to assign it to the correct category. As an aid for this categorization, some authors (Vandel, 1965; Barr, 1968) have used morphological characteristics for interpretation. Species that display anophthalmy, depigmentation, or other morphological cave adaptations are usually considered to be troglobites. Species not displaying such morphological characteristics are usually classified as troglophiles until more data become available, at which time they may be reclassified as troglobites.

More than 150 species of animals have been recorded in Hawaiian caves to date. About one-third of these are probably accidentals. The approximately 100 cavernicolous species are all arthropods. Twenty-seven of these are now considered troglobites and the rest troglophiles. There are no known trogloxenes in Hawaiian caves.

A complete list of the accidentals would include virtually all of the surface fauna of the area, but the most common species are soil and entrance animals. Most are arthropods; a few are vertebrates, earthworms, or exotic snails; and a planarian (flat worm) has also been found. Two exotic vertebrates, the roof rat *(Rattus rattus)* and the barn owl *(Tyto alba),* nest in suitable entrances and occasionally wander into the dark zone, but they are considered accidental there.

A majority of the troglophiles are exotic species and most are known from outside of caves; the American cockroach is an example. Among the native troglophiles are a crane fly, a biting midge, several fungus gnats, a *Drosophila,* and a few moths of the genus *Schrankia*.

Most interesting are the troglobitic species, for these have evolved from representatives of the native Hawaiian fauna and depart markedly from the norm of their groups. The following are the most notable: two species of blind, "big-eyed" hunting spiders, *Lycosa howarthi* and *Adelocosa anops;* an eyeless terrestrial amphipod, *Spelaeorchestia koloana;* two species of the native cambalid millipede genus *Dimerogonus;* four as yet undescribed crickets in two subfamilies, including representatives of the true tree crickets, *Thaumatogryllus* sp. (these crickets are not related to continental cave crickets); a remarkable undescribed eyeless earwig; two blind true bugs, a terrestrial water treader,

Speovelia aaa, and a thread-legged bug, *Nesidiolestes ana;* two species of cixiid planthoppers, *Oliarus polyphemus* and *O. priola;* and one or more undescribed species of cave moths, family Noctuidae, *Schrankia* sp.

Sampling Design

More than 100 lava tubes have been investigated, and records on their biotic communities and environmental conditions were made. Caves representing as many types and ages from as many climatic zones and islands as possible were surveyed in an effort to learn the parameters responsible for the distribution of cave organisms. Most caves were eroded remnant sections, too small or disturbed and lacking a specialized fauna. About thirty caves, which had an interesting fauna, were visited periodically and comparative analyses were made. Several of these occur along the Mauna Loa transect, near IBP sites 1 and 4, and between the latter and the Kīlauea rain forest site. An important low-elevation cave was studied in the rain forest below the Mauna Loa transect at 400 m elevation on the road to Hilo. This is the Kazumura cave. These more intensively studied caves were mapped to compare the spatial relationships of their habitat zones.

The caves were entered and searched by investigators using standard safety equipment, which includes hard hats, ropes, and independent backup lights. Arthropods and other biota were sought throughout each cave, but intensified observations were made in the dark zone, sometimes by using baits. On-site records were taken of arthropod abundance and behavior. Behavioral traits observed were normal arthropod activity and response to stimuli (such as exposure to light, touch, wind, and moisture). New or little-known animals were captured and taken to the laboratory for further study. Samples of the two common slimes were examined microscopically by M. F. Stoner. He also analyzed these for their fungal content by the dilution-spread plate technique (Stoner TR 75). Cave animals were collected only when necessary because the population sizes are small. The primary objective was to study the cave ecosystem, not to alter it.

DATA ANALYSIS

A uniform method for analyzing the different organism groups studied in the Kīlauea rain forest and the Hawaiian lava tubes was sought for facilitating synthesis. The method consisted of three main steps:

1. Preparation of a species list, quantified as much as possible.
2. Conversion of the species on that list into groups with closely similar ecological roles or those with closely similar structures and functions (e.g., life form groups).

3. Diagrammatic representation of the species groups in form of life form spectra (sensu Mueller-Dombois and Ellenberg, 1974:145).

During the field study, all investigators were concerned initially with step 1. Following this or consecutively with the search for all species of his or her concern, each investigator made observations on the ecological roles of his or her species in the prescribed ecosystems. Depending on existing knowledge and the relative ease with which new information could be derived for the different organism groups under study, these observations were pursued to different levels of detail and quantification. Regardless of the different detail obtained, it was possible for most investigators to process their data through step 3. The plant life form diagram (figure 6-1) was used as a model for this purpose.

6

Structural Variation of Organism Groups Studied in the Kīlauea Forest

D. Mueller-Dombois, R. G. Cooray, J. E. Maka, G. Spatz, W. C. Gagné, F. G. Howarth, J. L. Gressitt, G. A. Samuelson, S. Conant, and P. Q. Tomich

FOREST LAYERS AND QUANTITATIVE COMPOSITION OF THE PLANT COMMUNITY*

The vegetation in the Kīlauea forest can be stratified into five forest layers on the basis of height of crown biomass and other structural characteristics:

1. The herbaceous plant layer (0–50 cm height), composed primarily of herbaceous plants but including also the seedlings of shrub and tree species.
2. The tree fern layer (over 50 cm–5 m height), dominated by the tree fern *(Cibotium glaucum)* but including shrubs and saplings of tree species.
3. The low-stature tree layer (over 5–10 m height). This layer is composed of tree species that cannot grow any taller and smaller individuals of the taller tree species *Metrosideros collina* subsp. *polymorpha* and *Acacia koa* var. *hawaiiensis*.
4. The intermediate-stature tree layer (over 10–15 m height) in which *Metrosideros* is the most abundant.
5. The emergent tree layer (more than 15 m height) in which *Acacia koa* (koa) is the dominant tree.

In addition to categories of rooted vegetation, all vascular epiphytes were enumerated in a separate category.

The quantitative relationships between the woody plant layers in the forest were evaluated for each species by two parameters, cover and density. The forest floor components were quantified in terms of cover, but for the epiphytes frequency was used as the criterion for quantification because of the difficulty in assessing cover. Shoot and crown cover in the different forest layers, stem

*By D. Mueller-Dombois, Ranjit G. Cooray, and Jean E. Maka

cover or basal area, and herbaceous shoot cover all express the degree of horizontal space utilization. Since the relative saturation of space is a factor determining niche availability, cover within layers is an important quantitative parameter for plant populations. Density provides an assessment of the structure in terms of the numerical relationships of the different species in the different layers. This is useful in understanding the maintenance trends of the forest.

Floristic Composition and Life Form Spectrum of Kīlauea Forest

Table 6-1 gives a checklist of all vascular plants found in the Kīlauea rain forest plot. Names of angiosperms follow the citations in the complete checklist of Hawaiian flowering plants by St. John (1973). The fern names were adjusted according to a mimeographed checklist by Lane (undated) held in the Department of Botany, University of Hawai'i. The species are listed in order of their quantitative importance in each of the five forest layers. The vascular epiphytes are listed as a separate group, following the ground-rooted herbaceous layer plants. All species are further classified into life form types according to the well-known Raunkiaerian system. A generalized, common, taxonomic life form term is added in parentheses for clarification.

Seventy-nine species were recorded in the herbaceous plant layer on the 80 ha forest plot. Of these, twenty-seven are fern species, eighteen are forbs, eleven are seedlings or suckers of tree species, ten are seedlings of shrub species, four are vines, three are grasses, two sedges, two rushes, and two club mosses. Introduced or exotic species are twelve of the eighteen forbs, two of the nine shrubs, and all of the grasses and rushes.

All herb layer species combined covered only 7.7 percent of the forest floor. According to general abundance, they can be divided into three groups: species with cover values of 0.1 percent or greater (moderately abundant), species present with insignificant cover on the sample transects (occasional), and species outside the sample transects but inside the 80 ha plot (locally rare). Among the moderately abundant species are two native sedges (*Carex alligata* and *C. macloviana*), three native ferns with potentially upright woody trunks (*Dryopteris paleacea, Cibotium glaucum,* and *Sadleria pallida*), four other native herbaceous ferns, seedlings of two native tree species (*Metrosideros collina* and *Cheirodendron trigynum*), seedlings of three native shrub species (*Broussaisia arguta, Rubus hawaiiensis,* and *Vaccinium calycinum*), and only one native forb *(Nertera granadensis)* but four exotic forbs (*Veronica serpyllifolia, Ludwigia octivalvis, Hydrocotyle sibthorpioides,* and *Hypericum mutilum*). The four exotic forbs are all small, matted plants that are shade tolerant and well established in this otherwise dominantly native forest. The other fifteen exotic herb layer species are only occasional or locally rare. Of these, the two *Juncus* species (*J. planifolius* and *J. tenuis*) are found in a few wet

TABLE 6-1. *Vascular Plant Species, Life Forms, and Quantities in Each Layer*

Life form	Species	Quantitative value
Herbaceous plant layer (0–50 cm height)[a]		
Ch(H) caesp (sedge)	*Carex alligata*	1.7%
Ch ros (fern)	*Dryopteris paleacea*	1.1
Ch(H) caesp (sedge)	*Carex macloviana*	0.8
Ch(H) rept (forb)	X*Veronica serpyllifolia*	0.5
G rhiz (fern)	*Adenophorus tamariscianum* var. *tripinnatifidum*	0.4
G rhiz (fern)	*Athyrium microphyllum*	0.4
(tree fern sporeling)	*Cibotium glaucum*	0.4
Ch(H) rept (forb)	X*Ludwigia octivalvis*	0.3
Ch(H) rept (forb)	*Nertera granadensis*	0.3
Ch ros (fern)	*Sadleria pallida*	0.3
G rhiz (fern)	*Athyrium sandwichianum*	0.2
Ch(H) rept (forb)	X*Hydrocotyle sibthorpioides*	0.2
(tree seedling)	*Metrosideros collina* subsp. *polymorpha*	0.2
(shrub seedling)	*Broussaisia arguta*	0.1
(tree seedling)	*Cheirodendron trigynum*	0.1
Ch(H) rept (forb)	X*Hypericum mutilum*	0.1
(shrub seedling)	*Rubus hawaiiensis*	0.1
(shrub seedling)	*Vaccinium calycinum*	0.1
Ch(H) rept (fern)	*Vandenboschia davallioides*	0.1
(tree seedling)	*Acacia koa* var. *hawaiiensis*	+
G rhiz (fern)	*Asplenium contiguum*	+
G rhiz (fern)	*Asplenium lobulatum*	+
G rhiz (fern)	*Asplenium normale*	+
Ch(H) caesp (forb)	*Astelia menziesiana*	+
(tree fern sporeling)	*Cibotium chamissoi*	+
(shrub seedling)	*Clermontia hawaiiensis*	+
(tree seedling)	*Coprosma rhynchocarpa*	+
(shrub seedling)	*Cyanea pilosa*	+
(shrub seedling)	*Cyrtandra lysiosepala*	+
G rhiz (fern)	*Dicranopteris emarginata*	+
G rhiz (fern)	*Dryopteris glabra*	+
G rhiz (fern)	*Dryopteris* sp.	+
G rhiz (fern)	*Elaphoglossum hirtum*	+
G rhiz (fern)	*Elaphoglossum wawrae*	+
Ch(H) rept (forb)	X*Epilobium cinereum*	+
T scap (forb)	X*Erechtites valerianaefolia*	+
Ch(H) ros (forb)	X*Gnaphalium purpureum*	+
G rhiz (fern)	*Grammitis hookeri*	+
Ch(H) caesp (grass)	X*Holcus lanatus*	+
(tree seedling)	*Ilex anomala*	+
Ch(H) caesp (rush)	X*Juncus planifolius*	+
Ch(H) rept (club moss)	*Lycopodium cernuum*	+
Ch(H) caesp (club moss)	*Lycopodium serratum*	+

TABLE 6-1. *(Continued)*

Life form	Species	Quantitative value
G rhiz (fern)	*Marattia douglasii*	+
(tree seedling)	*Myoporum sandwicense*	+
(tree seedling)	*Myrsine lessertiana*	+
(tree seedling)	*Pelea clusaefolia*	+
(tree seedling)	*Pelea volcanica*	+
Ch(H) rept (forb)	*Peperomia leptostachya*	+
(shrub seedling)	*Pipturus hawaiiensis*	+
G rhiz (fern)	*Pleopeltis thunbergiana*	+
G rhiz (fern)	*Polypodium pellucidum*	+
G rhiz (fern)	*Pteridium aquilinum* var. *decompositum*	+
(shrub seedling)	X*Rubus rosaefolius*	+
Ch(H) rept (fern)	*Sphaerocionium obtusum*	+
G rhiz (fern)	*Sphenomeris chusana*	+
PL suff (semiwoody vine)	*Stenogyne calaminthoides*	+
Ch(H) rept (forb)	X*Veronica plebeia*	+
G rhiz (fern)	*Xiphopteris saffordii*	+
(woody vine seedling)	*Alyxia olivaeformis*	*
Ch(H) caesp (grass)	X*Anthoxanthum odoratum*	*
Ch(H) rept (grass)	X*Axonopus affinis*	*
G rhiz (fern)	*Cyclosorus sandwicensis*	*
Ch suff (forb)	X*Eupatorium riparium*	*
(tree seedling)	*Gouldia terminalis*	*
Ch(H) ros (forb)	X*Hypochoeris radicata*	*
Ch(H) caesp (rush)	X*Juncus tenuis*	*
(tree seedling)	*Nothocestrum longifolium*	*
Ch(H) caesp (forb)	*Peperomia macraeana*	*
PL suff (semiwoody vine)	*Phyllostegia floribunda*	*
Ch suff (forb)	*Phytolacca sandwicensis*	*
G rhiz (fern)	*Pteris excelsa*	*
G rhiz (fern)	*Pteris irregularis*	*
(shrub seedling)	X*Rubus penetrans*	*
PL suff (forb)	*Rumex giganteus*	*
Ch(H) caesp (forb)	X*Senecio sylvaticus*	*
Ch suff (forb)	X*Solanum nigrum*	*
(shrub seedling)	*Styphelia tameiameiae*	*
PL herb (herbaceous vine)	*Vicia menziesii*	*
Vascular epiphytes (on tree ferns and trees)[a]		
P E (tree)	*Metrosideros collina* subsp. *polymorpha*	49.4
P E (tree)	*Cheirodendron trigynum*	46.3
Ch E (shrub)	*Vaccinium calycinum*	46.3
G E (fern)	*Grammitis hookeri*	31.9
Ch E (tree)	*Ilex anomala*	23.8
G E (fern)	*Athyrium microphyllum*	23.2

TABLE 6-1. *(Continued)*

Life form	Species	Quantitative value
G E (fern)	*Adenophorus tamariscianum* var. *tripinnatifidum*	18.8
G E (fern)	*Polypodium pellucidum*	18.8
Ch E (shrub)	*Rubus hawaiiensis*	18.8
G E (fern)	*Dryopteris glabra*	17.5
Ch(H) E (forb)	*Astelia menziesiana*	14.4
Ch E (tree)	*Coprosma rhynchocarpa*	12.5
G E (fern)	*Elaphoglossum hirtum*	11.9
Ch(H) E (forb)	*Peperomia leptostachya*	11.9
Ch E (fern)	*Dryopteris paleacea*	11.3
Ch E (tree)	*Pelea clusaefolia*	11.3
Ch(H) E (sedge)	*Carex macloviana*	10.0
Ch E (tree fern)	*Cibotium glaucum*	9.4
G E (fern)	*Asplenium lobulatum*	8.8
G E (fern)	*Xipthopteris saffordii*	8.0
G E (fern)	*Pleopeltis thunbergiana*	7.5
G E (fern)	*Asplenium contiguum*	6.9
Ch E (shrub)	*Broussaisia arguta*	6.9
Ch E (shrub)	*Cyanea pilosa*	6.9
G E (fern)	*Elaphoglossum wawrae*	6.7
Ch E (vine)	*Alyxia olivaeformis*	5.6
G E (fern)	*Athyrium sandwichianum*	5.0
Ch E (tree)	*Myrsine lessertiana*	4.4
Ch(H) E (forb)	*Nertera granadensis*	3.8
G E (fern)	*Vandenboschia davallioides*	3.1
Ch E (tree seedling)	*Acacia koa* var. *hawaiiensis*	2.5
G E (fern)	*Asplenium normale*	1.9
Ch(H) E (sedge)	*Carex alligata*	1.3
Ch E (tree fern)	*Cibotium chamissoi*	1.3
Ch E (fern)	*Sadleria pallida*	1.3
Ch(H) E (fern)	*Asplenium schizophyllum*	0.6
Ch(H) E (fern)	*Coniogramme pilosa*	0.6
Ch E (tree)	*Coprosma ochracea*	0.6
Ch(H) E (fern)	*Ctenitis rubiginosa*	0.6
Ch E (tree)	*Gouldia terminalis*	0.6
Ch(H) E (club moss)	*Lycopodium serratum*	0.6
G E (fern)	*Marratia douglasii*	0.6
Ch E (tree seedling)	*Myoporum sandwicense*	0.6
Ch E (shrub)	*Pipturus hawaiiensis*	0.6
Ch(H) E (fern)	*Sphaerocionium obtusum*	0.6
Ch E (forb)	*Stenogyne calaminthoides*	0.6
Ch(H) E (hemiparasite)	*Korthalsella complanata*	+
Tree Fern Layer (0.5–5 m height)[b]		
PMi ros (tree fern)	*Cibotium glaucum*	77.3
(tree sapling)	*Cheirodendron trigynum*	1.5
sPMi (shrub)	*Broussaisia arguta*	1.1

TABLE 6-1. *(Continued)*

Life form	Species	Quantitative value
mPMi (shrub)	*Rubus hawaiiensis*	0.9
(tree sapling)	*Coprosma rhynchocarpa*	0.8
(tree sapling)	*Metrosideros collina* subsp. *polymorpha*	0.5
mPN (shrub)	*Vaccinium calycinum*	0.5
(tree sapling)	*Ilex anomala*	0.4
(tree sapling)	*Acacia koa* var. *hawaiiensis*	0.2
mPMi (shrub)	*Cyrtandra lysiosepala*	0.2
(tree sapling)	*Myrsine lessertiana*	0.2
(tree sapling)	*Pelea clusaefolia*	0.1
PL (woody vine)	*Alyxia olivaeformis*	+
PMi ros (tree fern)	*Cibotium chamissoi*	**
mPMi (shrub)	*Clermontia hawaiiensis*	**
sPN (small tree)	*Gouldia terminalis*	**
(tree sapling)	*Pelea volcanica*	**
mPN (shrub)	X*Rubus rosaefolius*	**
mPN (shrub)	*Cyanea pilosa*	*
(tree sapling)	*Myoporum sandwicense*	*
sPMi (small tree)	*Nothocestrum longifolium*	*
mPMi (shrub)	*Pipturus hawaiiensis*	*
mPN (shrub)	X*Rubus penetrans*	*
sPN (shrub)	*Styphelia tameiameiae*	*
PL suff (semiwoody vine)	*Rumex giganteus*	*
PL suff (semiwoody vine)	*Stenogyne calminthoides*	*
PL suff (semiwoody vine)	*Phyllostegia floribunda*	*
PL herb (herbaceous vine)	*Vicia menziesii*	*
Low-stature tree layer (5–10 m height)[c]		
sPMa	*Metrosideros collina* subsp. *polymorpha*	12.9
mPMe	*Cheirodendron trigynum*	12.0
mPMe	*Myoporum sandwicense*	2.5
sPMe	*Ilex anomala*	1.2
mPMe	*Myrsine lessertiana*	0.9
mPMe	*Coprosma rhynchocarpa*	0.8
sPMe	*Pelea clusaefolia*	0.1
sPMe	*Pelea volcanica*	0.1
sPMa	*Acacia koa* var. *hawaiiensis*	+
PL frut	*Alyxia olivaeformis*	*
PL suff	*Rubus hawaiiensis*	*
Intermediate-stature tree layer (10–15 m height)[d]		
sPMa	*Metrosideros collina* subsp. *polymorpha*	5.1 m^2/ha
sPMa	*Acacia koa* var. *hawaiiensis*	0.1 m^2/ha

TABLE 6-1. *(Continued)*

Life form	Species	Quantitative value
Emergent tree layer (>15 m height)[d]		
sPMa	*Acacia koa* var. *hawaiiensis*	14.7 m^2/ha
	Acacia koa snags	5.4 m^2/ha
sPMa	*Metrosideros collina* subsp. *polymorpha*	5.0 m^2/ha

Notes: The life form symbols are explained in figure 6-2.
X: exotic species; +: present in sampled plots but with <0.1 percent cover or frequency; *: present in study area but outside sampled plots; **: present in sampled plots but not intercepted on the ten lines.

[a] Cover for herbaceous plant layer—7.7 percent—and frequency of vascular epiphytes based on estimated percentage values of 160 15 m^2 quadrats.

[b] Cover of tree fern layer—83.6 percent—and cover of individual species in this layer based on percentage interception on ten 100 m long lines.

[c] Cover for low-stature tree layer—30.5 percent—equals sum of percentage cover of individual species in forty 6 × 100 m plots. Cover for each individual species was computed using dbh-crown cover regression values (from Maka, TR 31).

[d] Quantitative value is stem cover by species within layer.

depressions, four others are shade tolerant *(Rubus rosaefolius, Rubus penetrans, Eupatorium riparium, Veronica plebeia),* and the remaining nine exotics are heliophytes that participate as minor components in locally open places in the forest, where they have only an early successional position.

The vascular epiphytes group contains forty-seven species that grew on upright trunks of tree ferns or on standing trees. Ferns are the best-represented subgroup, with twenty-three species. Only three species are not also found among the ground-rooted herb layer plants and thus can be considered obligatory epiphytes. These are *Coniogramme pilosa, Ctenitis rubiginosa,* and *Asplenium schizophyllum.* Included in the epiphyte group for convenience is one native vascular hemiparasite, *Korthalsella complanata,* which is quite rare in this forest. Another well-represented subgroup of the epiphytes is woody plants. These include ten species of trees or tree seedlings. Among these, *Metrosideros collina, Cheirodendron trigynum,* and *Ilex anomala* are the most frequent. Five shrub species were found growing as epiphytes also. Of these, *Vaccinium calycinum* and *Rubus hawaiiensis* were the most frequent. Also the two tree fern species *Cibotium glaucum* and *C. chamissoi* occurred as epiphytes.

None of the exotic herb layer species was found to grow epiphytically in this rain forest, while nearly all native herb layer species grew also as epiphytes. Thus the epiphytic life form can be considered a mechanism of survival for native herb layer species when ground conditions are repeatedly disturbed, such as in this case from pig activity. The fact is that most native herbs, shrubs, and trees have the capacity to grow on the soil and above it on trees. All tree and most shrub species that start epiphytically have the capacity to extend their roots into the soil at a later stage, and then they can grow to their normal size potential. However, relatively few individuals of woody plant species grow to maturity after an epiphytic start. Their growth depends to some extent on the height above the ground, the size and branching patterns of the supporting tree, and the species itself. Among the tree species, *Cheirodendron trigynum* and *Metrosideros collina* appear to have the greatest capacity to survive from a higher epiphytic start as high as 5 m above the ground.

The tree fern layer is quantitatively very prominent, with over 80 percent cover. One species, *Cibotium glaucum,* is dominant since it alone contributes 77.3 percent to the cover of this layer. All other species contribute 6.5 percent or less in cover. Yet the tree fern layer is relatively rich in species with twenty-eight. Of these, seventeen species reach their height-growth capacity within this layer (up to 5 m height). This group includes the second tree fern species, *Cibotium chamissoi* (which is locally rare), and several native arborescent shrubs (*Broussaisia arguta, Cyrtandra lysiosepala, Clermontia hawaiiensis, Cyanea pilosa,* and *Pipturus hawaiiensis*). Only two exotic species, *Rubus rosaefolius* and *R. penetrans,* participated in this layer. Both had insignificant cover. The layer also includes four native vine species that are locally rare.

The low-stature tree layer contains eleven native and no exotic species. One of these is a thin-branched woody vine *(Alyxia olivaeformis),* another a thin-branched shrub that changes into a vine in this layer *(Rubus hawaiiensis).* Two species are taller-growing trees (*Metrosideros collina* and *Acacia koa*). This leaves seven native tree species that find their height potential in this 5 to 10 m stratum. Among these, *Cheirodendron trigynum, Myoporum sandwicense,* and *Ilex anomala* are the quantitatively more important.

The intermediate and emergent tree layers, over 10 m in height, contain only two tree species in this forest. *Metrosideros collina* is equally well represented (with about 5 m^2 stem cover or basal area per ha) in both height strata (from 10 to 15 m and from 15 m on up). *Acacia koa* is by far the dominant of the two species in terms of stem cover (14.8 m^2 per ha) and height. Most of its population members are emergents (from 15 to 25 m tall).

Figure 6-1 is a plant life form spectrum of the Kīlauea rain forest.It was developed by summarizing the number of species and quantitative information given in table 6-1 and by classifying each species according to its Raunkiaerian life form type (Raunkiaer, 1918, 1934, 1937). Raunkiaer's subdivisions were further revised by Ellenberg and Mueller-Dombois (1967), and the use of international symbols follows that revision (Mueller-Dombois and Ellenberg, 1974:449). The system provides for a grouping of all plant species into structurally and functionally similar categories.

As shown at the bottom of figure 6-1, all species are divided into two main groups: structurally self-supporting plants and plants that grow by supporting themselves structurally on others. The second group includes two sorts of plants: *lianas* (symbol L, plants that germinate on the ground and remain rooted in the soil but that grow up by supporting their photosynthetic apparatus on other plants without investing much structural material in their own stem system) and *epiphytes* (symbol E, plants that grow nonparasitically on other plants). The lianas in the forest are of two kinds, woody and herbaceous.

The structurally self-supporting plants include the five basic Raunkiaerian life forms: *phanerophytes* (symbol P, plants that continue to grow over 50 cm tall, generally trees or arborescent shrubs); *chamaephytes* (symbol Ch, perennial plants less than 50 cm height, or those that die back periodically to about that height limit, generally low shrubs or semiwoody forbs); *hemicryptophytes* (symbol H, herbaceous perennials that die back periodically to a remnant shoot system near the ground); *geophytes* (symbol G, herbaceous perennials whose shoot system disappears from the ground periodically and who survive by subterranean storage organs); and *therophytes* (symbol T, plants that die after seed production and usually lack a mechanism for vegetative reproduction).

The traditional view about humid tropical rain forests is that they are dominated by phanerophytes and perhaps woody lianas and epiphytes, that chamaephytes play a minor role, and that hemicryptophytes, geophytes, and therophytes are either absent or rather insignificant. Our diagram shows a departure from that view.

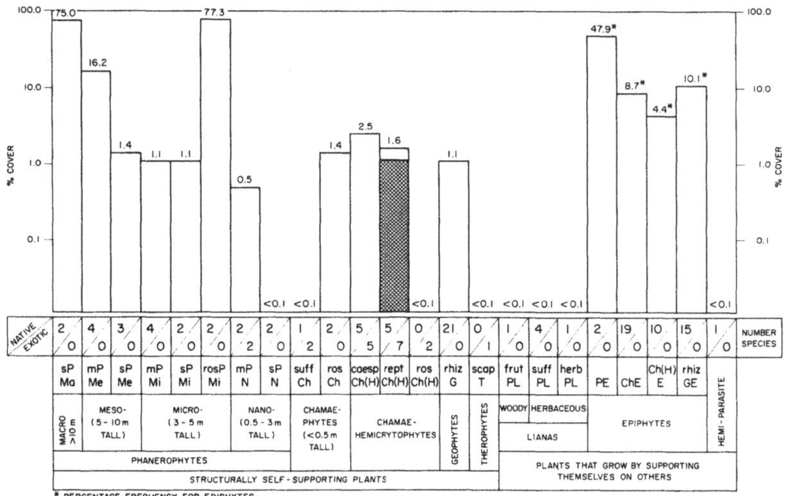

FIGURE 6-1. *Life form spectrum of vascular plant species in the Kīlauea forest. Cross-hatch = cover of exotic species >0.1%. Symbols: s = sclerophyllous; m = malacophyllous (soft-leaved); ros = leaves or fronds in rosettes; suff = suffrutescent (semiwoody); caesp = caespitose (branched from base or bunchy foliage); rept = reptate (creeping or matted), rhiz = rhizomatous (modified stem imbedded in soil or organic matter); scap = scapose (single stemmed, upright); frut = frutescent or woody; herb = herbaceous.*

True hemicryptophytes are indeed absent from the Kīlauea rain forest. Even introduced ones—for example, the grass *Holcus lanatus,* which behaves like a typical hemicryptophyte in the seasonal climate along the Mauna Loa transect—does not die back annually to a remnant shoot sytem in this humid environment. Instead only part of its shoot system dies at irregular or undefined intervals, while the other part remains green. In this caespitose (bunchlike) grass, the outer blades turn yellow and die, while the inner ones remain green or are replaced more or less continuously. For this reason, the term *chamaehemicryptophyte* was used for such herbaceous plants that are found occasionally with part of their shoot system dried up but still attached.

Another departure relates to the concept of geophyte, which here is broadened to include all herbaceous plants with subterranean storage organs, such as starchy rhizomes, corms, and bulbs. In temperate and arid climates, these plants lose their shoot system during the unfavorable season. In the tropical rain forest, where an unfavorable season is lacking, they may continue to grow. Since plants with subterranean storage organs do occur in these different climates, it would seem unwise to invent a new term for those growing in humid tropical climates. Their common characteristic in all climates is that they can spread subterraneously by vegetative reproduction (an important mechanism in plant competition) and that they are well equipped to cope with mechanical damage, such as caused by flash floods, ash fallout (Smathers and Mueller-Dombois, 1974), or herbivory (if not constantly repeated).

Apart from its well-known functional implication with respect to climate, the life form spectrum as presented here has further functional implications regarding plant matter production (Walter, 1971:28), plant-to-plant competition (Mueller-Dombois and Ellenberg, 1974:354), and recoverability from mechanical damage (Smathers and Mueller-Dombois, 1974). These four functional characteristics are further implied in the subdivisions of the seven general life form types shown in figure 6-1.

As is well known, individual members of the same species utilizing the same local habitat are also the strongest competitors (starting from a certain minimum population density) because of their similar physiological adaptation. The life form groups presented here contain species with closely similar structural and functional attributes. These species can thus also be considered as probable within-group competitors or as the members of a general niche, in the sense of the zoologist. The different life form groups or "synusiae" (Lippmaa, 1939) themselves represent the different general niches occupied by plant species in this forest. In terms of resource utilization of the habitat, the life form groups can be considered as complementary in general.

As figure 6-1 shows, the phanerophytes are subdivided into four size groups (macro-, meso-, micro-, and nanophanerophytes) according to their height growth potential. All are evergreen broad-leaved woody species. Within each group, they are once more subdivided into sclerophyllous (symbol s) and malacophyllous (symbol m = soft-leaved) woody species. The sclerophyllous evergreens are usually slower-growing woody species (Walter, 1971), but not too much can be said about their relative growth rates until they are studied further. A unique life form group is the two tree fern species, which are here recognized as microphanerophytes with large apical foliar rosettes (symbol PMiros); they are morphologically similar to palms and cycads. The implication is that species of such life forms, if introduced, may become serious competitors. Currently the tree ferns seem to have no serious competitor other than the species within the genus *Cibotium* themselves. For example, the local rareness of *C. chamissoi* could well be the result of competition by the much more successful *C. glaucum* in the Kīlauea Forest.

The presence of only two macrophanerophyte species in this forest implies a poorly filled general niche in this island forest. Thus there is a likely chance that a tropical rain forest tree, if introduced from another area, may become established rather easily if that species is preadapted to grow successfully in a montane humid tropical environment.

The chamaephytes are further subdivided into two groups: partially woody (suffrutescent) and woody plants with large apical leaf rosettes. The latter group includes the two native ferns *Dryopteris paleacea* and *Sadleria pallida,* which have short, woody trunks. These are the quantitatively more important chamaephytes, with a combined cover of 1.4 percent. The suffrutescent chamaephytes include two introduced (*Eupatorium riparium* and *Solanum nigrum*) and one native semiwoody forb *(Phytolacca sandwicensis).*

The chamae-hemicryptophytes are further subdivided into three groups: basally tufted (caespitose), matted or ground appressed, and creeping (reptate) and basal rosette forms. The caespitose chamae-hemicryptophytes are represented by ten species, which include the two native sedges (*Carex alligata* and *C. macloviana*), two native forbs (*Astelia menziesiana* and *Peperomia macraeana*) and one exotic forb *(Senecio sylvaticus),* two of the introduced grasses (*Holcus lanatus* and *Anthoxanthum odoratum*), the two introduced rushes (*Juncus planifolius* and *J. tenuis*), and an indigenous club moss *(Lycopodium serratum).* Thus the number of native and exotic species is equal, with five each in this life form group. The matted (reptate) form is represented with twelve species, most of them exotics. There are only five native species (two forbs, *Nertera granadensis* and *Peperomia leptostachya;* two ferns, *Vandenboschia davallioides* and *Sphaerocionium obtusum;* and a club moss, *Lycopodium cernuum*) in this group, but seven exotics (six herb species and one grass). All are relatively shade tolerant and often occupy places on the ground that were scarified by pigs several months previously. The third group of chamaehemicryptophytes, the basal rosette forms, include only two exotic composite weeds, *Hypochoeris radicata* and *Gnaphalium purpureum.* But these two species will remain only occasional members in this forest, since they are heliophytes growing only in open, disturbed places.

The lianas are represented by six native species in this forest. One of them is truly a woody (frutescent) liana *(Alyxia olivaeformis),* but it develops only thin-diameter (pencil-thick) stems that wind around the support trees. Four species are partly woody (suffrutescent) lianas (*Rubus hawaiiensis, Rumex giganteus, Stenogyne calminthoides,* and *Phyllostegia floribunda*) that prop their thin, semiwoody branches on other plants. Of these, only *Rubus hawaiiensis* was found to extend into the low-stature tree layer (up to about 8 m height). The others were seen to grow up to about 5 m height within the tree layer. The sixth species *(Vicia menziesii)* is a herbaceous liana that climbs with the aid of tendrils. Thus the forest is very poor in woody lianas. *Rubus hawaiiensis,* which is normally a typical shrub in forest openings, assumes the liana life form when it is closely surrounded by taller-growing trees, but there is no other high-growing liana except *Alyxia*.

The epiphytes are divided into four life-form groups: phanerophytes, chamaephytes, chamae-hemicryptophytes, and rhizomatous geophytes. The phanerophytic epiphytes include two tree species (*Metrosideros collina* and *Cheirodendron trigynum*), some of which grow into normal trees in certain situations after their seedlings germinated high off the ground (3 to 5 m). These could also be called hemi-epiphytes. All other woody species that grow as epiphytes in this forest were classified as chamaephytes. This group includes the two tree fern species, five native shrub, and eight other tree species (including *Acacia koa*). It could not be established what proportion of individuals of these other tree species will survive and grow into normal trees from such epiphytic beginnings. For this reason they were classified as chamaephytes. It was established, however, that all of these tree species can grow into normal trees from germinating on logs, sometimes (as in the case of an *Acacia koa* tree) from as high as 1.5 m off the mineral soil (Cooray, TR 44).

The chamae-hemicryptophytes among the epiphytes include three small caespitose fern species, four forbs, two sedges, and one club moss, all of them native species. The three caespitose ferns are the only obligatory epiphytes in this forest; that is, they were found only on trees or tree ferns but not on the ground. The fourth epiphytic life form is represented by rhizomatous geophytes. This group includes the fifteen fern species that were also recorded as rhizome geophytes growing on the forest floor. On the trees, their starchy rhizomes are usually imbedded in pockets of moist organic detritus. Occasionally their rhizomes can be found partially exposed. In that case they become stolons, technically speaking. This characteristic indicates a certain robustness of these rhizomatous ferns that explains their dual position as ground- and tree-rooted members in this forest community. It may also indicate an ability of these ferns to withstand a certain amount of ground disturbance by pigs, an animal influence that is very common throughout this forest. The physiological mechanism underlying this behavior is probably related to a superior capacity of these fern rhizomes to withstand a certain amount of soil-water stress.

One more life form is represented in the forest by *Korthalsella complanata,* a native hemi-parasite that superficially looks like an epiphyte.

In summary, the eighty-two vascular plant species in the Kīlauea rain forest are grouped into twenty-three life form types. These were defined by generalized structural and functional differences and similarities of the species. There are eight groups of phanerophytes differentiated on the basis of height-growth potential and leaf characteristics. The most dominant in cover are the sclerophyllous macrophanerophytes (*Acacia koa* and *Metrosideros collina*) and microphanerophytes with large apical leaf rosettes (the tree fern species *Cibotium glaucum*). However, the species-richest groups are the malacophyllous micro- and mesophanerophytes, with eight species. All phanerophytes, except two nanophanerophytes, are native. The other sixteen exotic invaders are found among the chamaephytes, chamae-hemicryptophytes, and therophytes. The last life form includes only one species, the composite weed *Erechtites valerianaefolia*. It is particularly the low mat-forming (reptate) herbaceous perennial plant life form that has gained a relatively stable position in the forest, while the bunch (caespitose) type is rather unimportant in cover. The native caespitose forms, however, are more dominant in cover. Species richness is particularly pronounced among the epiphytes, which are all native species, and except for three of these, the epiphytes occur also as ground-rooted plants.

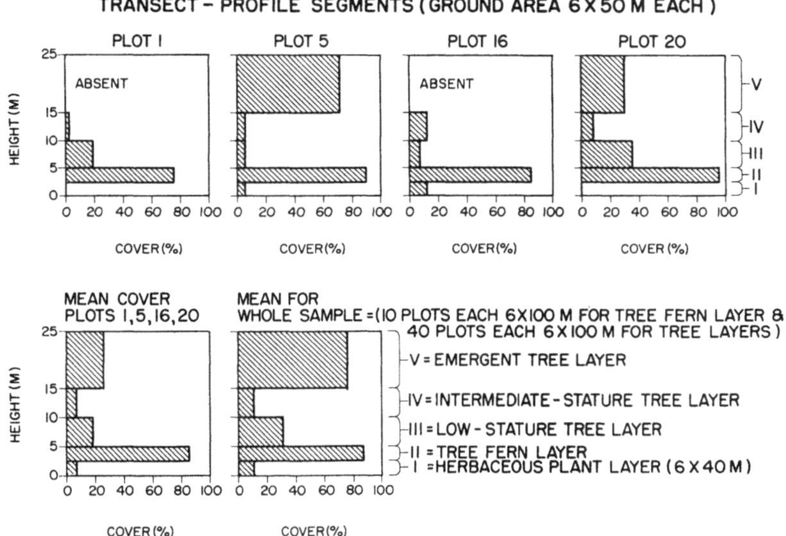

FIGURE 6-2. *Layer diagrams showing crown and shoot cover contribution in percentage of total area in Kīlauea rain forest. Plots 1, 5, 16, and 20 are samples of the cover variation throughout the forest. The lower two diagrams give mean values.*

Cover Variations Among Forest Layers

Figure 6-2 shows the cover contribution for the five forest layers on each of four 6 by 50 m transect profile segments. The plot numbers coincide with those on figure 5-3. The cover relationships in the forest are described by mean values for the five layers on the four transect segments and for the whole sample based on ten plots, 6 by 40 m each, for the herbaceous plant layer; on ten plots, 6 by 100 m each, for the tree fern layer; and on forty plots, 6 by 100 m each, for the tree layers.

The cover of the herbaceous plant layer is low on all four transect profile segments. It ranges from 2 percent in plot 1 to 11 percent in plot 16. For the whole sample it came to 8 percent.

Over 70 percent of the forest floor in the Kīlauea forest consists of exposed mineral soil, rich in humus (13 percent organic matter) but very acid (pH 4.1–4.6). Pig tracks, pig rooting, and wallows are found everywhere on the mineral soil, which explains in part the sparse cover of the herbaceous plants. Exposed rocks cover just over 4 percent of the forest floor. Nearly one-quarter of the forest floor is covered by rotting wood and logs, including trunks of tree ferns lying on the ground. The presence of great numbers of slowly decaying tree trunks on the forest floor is a characteristic feature of this forest. Mosses and liverworts cover 19 percent of the forest floor. These are concentrated on rotting wood and logs. A few occur on rocks.

The cover of the tree fern layer is consistently high on all four transect segments, ranging from 75 percent on plot 1 to 95 percent on plot 20.

The mean cover of the low-stature (5–10 m) tree layer is less than 20 percent on all four transect segments. The highest percentage cover (36 percent) for this layer is in plot 20, which here reflects advanced gap-phase regeneration by *Metrosideros* trees.* In plots 5 and 16, the low-stature tree cover was below 10 percent, indicating large openings in this layer.

The intermediate-stature tree layer had the smallest percentage cover. The highest value for this layer was 12 percent in plot 16. In the other three plots, this layer had values less than 10 percent. The cover of this layer, which is primarily from *Metrosideros,* is less than 10 percent for the whole sample. Thus its cover is similar to that of the herbaceous layer. These two layers show the poorest space utilization.

Emergent trees were present only on two of the profile segments, in plots 5 and 20. The mean cover of emergent trees on the four plots was 26 percent. The cover of the emergents for the whole forest was estimated to be 70 percent, about the same as for plot 5. The crown cover values of *Acacia koa* emergents derived from the aerial photograph revealed no significant difference between transect 1 (70 percent) and transect 4 (65 percent).

---

*Tree regeneration occurring in distinct openings of an otherwise closed-canopy forest.

Density Relations of Woody Species in the Different Forest Layers

The density (number of individuals per ha) of woody plant species in the five forest layers is shown in figure 6-3. If a line were drawn connecting the midpoints of the bars in the histogram, it would form almost a straight line sloping from left to right. There are nearly 21,600 individuals per ha of woody plants in the herbaceous plant layer and just over 30 individuals per ha in the emergent tree layer. The herbaceous plant layer (0–0.5 m tall) contained 19

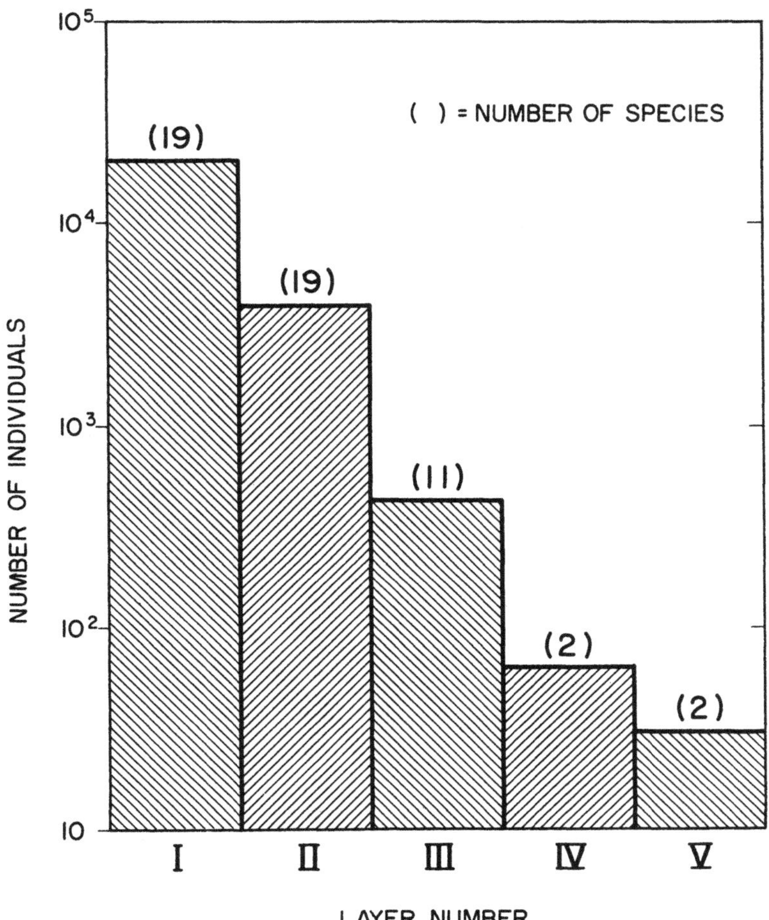

FIGURE 6-3. *Density of woody plants in the five forest layers on the basis of number per ha. I = herbaceous plant layer (0–0.5 m tall); II = tree fern layer (>0.5–5 m tall); III = low-stature tree layer (>5–10 m tall); IV = intermediate-stature tree layer (>10–15 m tall); V = emergent layer (>15 m tall).*

woody plant species and so did the tree fern layer (over 0.5–5 m tall). There were 11 plant species in the low-stature tree layer (5–10 m tall). Only two species, *Metrosideros collina* and *Acacia koa,* were found in the intermediate-stature (over 10–15 m tall) and the emergent tree layer (over 15 m tall).

The decrease in the number of species from the lower to the upper forest layers is related to the different height-diameter growth potentials of the species. That is, a number of small-sized woody plant species (for example, *Vaccinium calycinum* and *Broussaisia arguta*) will never grow into the tree layer and the low-stature tree species (such as *Cheirodendron trigynum* and *Ilex anomala*) will never grow into the tall tree layers because they lack the height-growth potential. This tree species pattern is generally true of tropical forests and is reflected in the forest stratification, which shows a decreasing number of species with increasing height strata (Richards, 1952; Whitmore, 1975).

HORIZONTAL PATTERN ANALYSIS WITHIN FOREST LAYERS*

The objective of this study was to determine the structural components of the plant community by resolving spatial arrangements and defining, where possible, species groups present within the superficially homogeneous forest.

Two possible spatial arrangements are considered to exist within a horizontally extending layer: homogeneous (random, without a specific pattern) and heterogeneous (nonrandom). Two types of heterogeneous arrangements were recognized as possible: a mosaic of recurring groups of species (contagious) and a continuum (gradual change). If an arrangement was determined to consist of recurring groups, the groups could then be identified.

Plant species were combined into three structural groups according to layers, one consisting of herbaceous plants (layer I on figure 6-2), one of shrubs, small trees, and tree ferns (layer II), and one of upper-canopy and emergent trees (layers III, IV, and V combined). Another group included the epiphyte species, which do not actually form a layer themselves but which, for simplicity, will be referred to as such in this discussion.

Data Analysis

A mathematical approach was used to analyze the relationships among individuals and species. On a larger-area basis (600 m^2) plots were compared by their composition (Q-technique, Cattell, 1952). On smaller sample areas (15 to 480 m^2), species were compared for their associated presence in subplots of various sizes (R-technique).

---

*By Jean E. Maka

The data collected in the field consisted of species quantities for each quadrat. For the plot (Q) comparisons, average species values were considered to be the attributes of the individuals (plots). These data were processed by the classical ordination method of Bray and Curtis (1957), using the equation of Beals (1960, 1965). Since the plots appeared very similar qualitatively, a quantitative coefficient by Spatz (1970) was utilized to maximize the difference between plots (further detail in Maka, TR 31).

In the species (R) comparisons, species pairs were ordinated with respect to their distribution among the quadrats or subplots. Species were considered to be the individuals, and the particular subplots in which they were present were considered their attributes. The species' presence values in each subplot were used to ordinate the species. The simple matching coefficient by Sokal and Sneath (1963) was used to compare species pairs. In addition the relative distance coefficient of Orloci (1967) was used for the tall-tree layer. For reference points on the ordination axes, the criteria of Newsome and Dix (1968) were used.

Since the commonly used methods of pattern analysis are designed to handle only a single species at a time (Greig-Smith, 1964), a new method of recurring group (or heterogeneity) testing was devised with the consultation of Dr. L. Orloci. This method tests the species arrangements jointly. It is based on the assumption that if random samples are taken from the vegetation, the similarities between species in the samples will approximate the similarities between species in the vegetation. Hierarchies constructed from the sample data will approximate the hierarchical class structure of the groups. If repeated testing with new random samples shows that a single hierarchical class structure exists, the existence of inherent groups is implied. In that case, the sample hierarchies would be consistently similar or statistically indistinguishable. If no single hierarchy exists, implying a homogeneous or very gradual continuum arrangement of species, then the sample hierarchies would only reflect chance arrangements and would be dissimilar. The null hypothesis was that the sample dendrograms of species were consistently similar (that is, statistically indistinguishable).

Ten random samples were selected from the available subplots, and tests were made among equally sized subplots. Each sample consisted of 20 percent of the number of subplots. The same subplots could reappear in a new random sample. The species of each sample were arranged in a dendrogram hierarchy by the sum-of-squares clustering (Orloci, 1967) using the computer program by Goldstein and Grigal (1971). Additional subroutines were written to construct a topological distance matrix (Phipps, 1971) from each dendrogram by counting fusions. A comparison of the matrices by the minimum discrimination information statistic (mdis) (Kullback, Kupperman, and Ku, 1962) was made, which was compared with a chi-square table to test for significance. A low number of significant values (at the 0.05 level) would indicate that the dendrograms were sufficiently similar so that the null hypothesis could be accepted. If there were many (95 percent or more) significant values, then the hypothesis would be rejected. For a more detailed description of the method see Maka (TR 31).

Whenever recurring groups were indicated from the test, species data for all subplots for that layer were clustered by the sum-of-squares test (Orloci, 1967) to produce one hierarchical classification of the species.

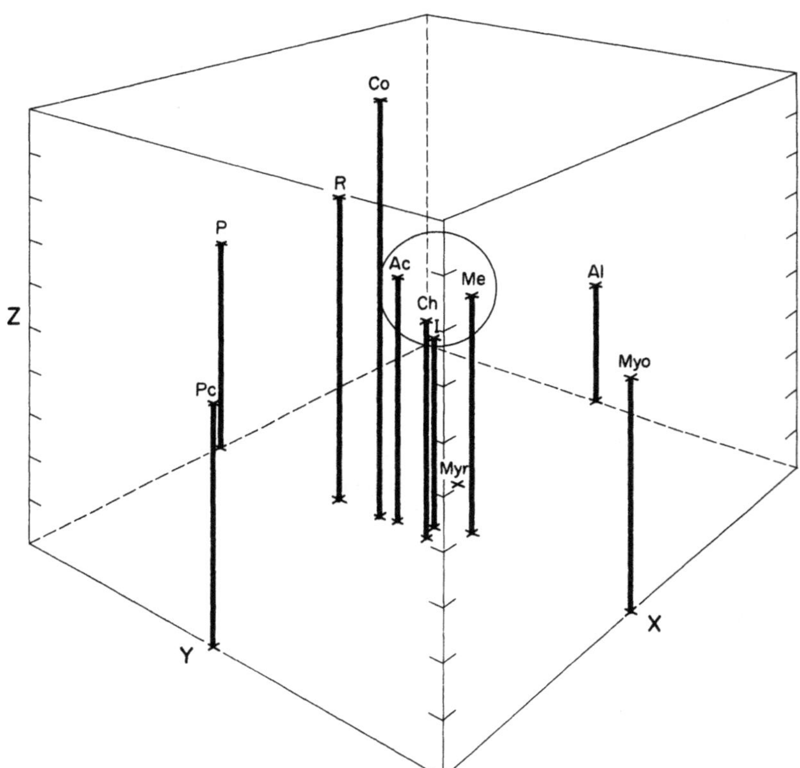

FIGURE 6-4. *Ordination of eleven tree layer species in three dimensions with relative distance coefficient. Dissimilarity is based on species counts in 200 subplots of 120 m² each. The following abbreviations for species names are used: Ac =* Acacia koa *var.* hawaiiensis; *Al =* Alyxia olivaeformis *(liana); CH =* Cheirodendron trigynum; *Co =* Coprosma rhynchocarpa; *I =* Ilex anomala; *Me =* Metrosideros collina *subsp.* polymorpha; *Myo =* Myoporum sandwicense; *Myr =* Myrsine lessertiana; *PC =* Pelea clusiaefolia; *P =* Pelea volcanica; *R =* Rubus hawaiiensis; *(tall shrub). One cluster of species is shown circled in the graph.*

Recurring Groups in Tree Layers

The plot ordinations showed greater quantitative-floristic similarities between plots on the same transect than among plots of different transects. In general the ordinations revealed a compositional change parallel to the contour lines of the study site. This change is reflected in each layer by either species presence or quantities. No distinct plot clusters are evident, and no individual plots stand out consistently from the rest.

In the species ordinations, epiphyte, herbaceous, and tree-fern layer species were scattered diagonally across the graphs for each of these layers. This is interpreted as a gradual change among the more common species in the subplots across the study site. The graph of the tree layers revealed a cluster of common species (encircled on figure 6-4). This is interpreted as one recurring group. Using the relative distance coefficient, a second graph of this layer was produced, which leads to the same interpretation: the spatial arrangement includes one recurring species group. The tree species grouped are *Acacia koa, Cheirodendron trigynum, Ilex anomala,* and *Metrosideros collina.*

The recurring group tests for the epiphyte, herbaceous, and tree-fern layer species at each subplot size all produced significant values (100 percent). This was interpreted as an indication that each of these layers consists of homogeneously arranged or somewhat continuously changing species. There was no difference in the results of different subplot sizes of these three layers. The tree layer species, however, produced fewer significant values at the subplot sizes tested. The size of 120 m^2 resulted in the lowest percentage of significant values, indicating the greatest heterogeneity or grouping tendency.

Since the tree layer showed recurring groups, the hierarchical class structure was calculated and portrayed in a dendrogram shown on figure 6-5. The species group consisting of the four species encircled in the ordination graph (figure 6-4) was isolated near the level of 1.0 (90 percent similarity). By drawing a cutoff line at 2, three other groups were also isolated at this level. However, these were not evident from the species ordination graphs, suggesting that these are perhaps overlapping and may not be distinct from the main group.

Discussion

The species ordinations gave results similar to the recurring group tests. From the ordinations, the three layers (epiphyte, herbaceous, and tree fern) appear to consist of species that merge spatially into one another in the form of a very gradual continuum. The test results show that these layers contain no recurring species groups. This confirms our initial assumption that the forest is essentially homogeneous. From the ordinations at subplot size 120 m^2, one associated species group recurs in the tree layer (*Acacia koa, Cheirodendron*

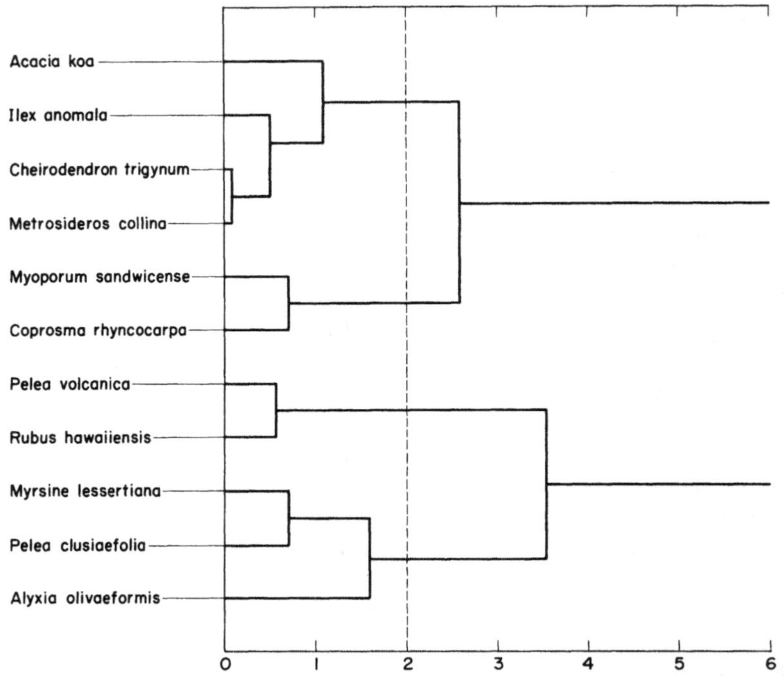

FIGURE 6-5. *Classification of tree species using qualitative data from one hundred quadrats of 240 m² each. The sum-of-squares method with relative distances was used to produce this dendrogram. The horizontal axis represents decreasing spatial similarity between species of a group, from left to right. The two large groups join at the 7.6 level (if the diagram were to be extended).*

trigynum, Ilex anomala, and *Metrosideros collina*). This agrees with the results of the recurring group test. They are the same four species that were revealed in the classification.

The presence of the recurring groups of trees may be interpreted as a relationship with certain recurring site conditions or as a stability in this particular species combination (cf. Poore, 1964). Goodall (1954) agrees that these regularly repeating patterns may be due to differences in the inanimate environment. The patterns can also arise through dynamic changes within the vegetation itself by regeneration and degeneration occurring throughout the area but not at the same time.

It is possible to interpret the recurring species groups as resulting in part from recurring site conditions which are due to the previous plant community by the presence of fallen logs. The patterns arising from these site conditions do so through dynamic changes by the degeneration of old trees and the regeneration of new trees in association with fallen logs.

PROFILE DIAGRAMS AND MAP*

The vertical and horizontal stratifications of a forest are difficult to visualize from numerical data, and photographs are inadequate for illustrating structural detail. Thus many authors have adopted the profile diagram method as a standard technique for illustrating forest stratification (Beard, 1946; Robbins, 1959; Chapman and White, 1970). One of the major problems in the construction of profile diagrams, the selection of truly representative forest segments, has led some workers (Holdridge, 1970; Holdridge et al., 1971) to illustrate forest structure by idealized profile diagrams, which are to represent the average condition. Because of the abstraction, an idealized profile diagram may omit important variations. Instead a spectrum of profile diagrams might portray the structural pattern of the stand more adequately.

The vertical structure and crown biomass layering in the Kīlauea forest was determined by profile diagrams of four systematically chosen 50 m long by 6 m wide transect segments in the forest. The four transect profiles were located systematically on the first 50 m of plots 1, 5, 16, and 20. Therefore they represent the four corners of the rectangular IBP study site.

Although all four profiles showed individual differences, only three are reproduced here because one was somewhat repetitive. All profiles were drafted in the field to a scale of 1:200.

Three Major Profile Patterns

The first profile segment (figure 6-6) depicts the situation for which the forest has been named, an *Acacia-Metrosideros-Cibotium* forest. The tree fern layer is dense and continuous. The tree ferns in this segment grow tall, reaching with their apices up to 5 m in height. Several low-stature tree groupings can be recognized. These are primarily formed by *Cheirodendron* and *Ilex,* but individuals of *Coprosma, Pelea,* and *Myrsine* participate in some groups. This open understory tree layer exhibits a greater species diversity than do the taller canopy and emergent trees. Emergent trees belong to only two species, *Acacia koa* var. *hawaiiensis* and *Metrosideros collina* subsp. *polymorpha*. These occupy the height stratum from 15 to 25 m with their crowns. Within this group, koa is clearly the dominant on this profile in terms of height, crown cover and diameter. In this profile the vertical space from 0.5 to 2 m is well utilized by foliar biomass. The 2 to 5 m height stratum is quite open, but the greatest number of tree species occurs in this stratum. The 5 to 10 m stratum is practically unoccupied in this profile. The intermediate-stature tree layer (from about 10 to 15 m) shows only a single clump of *Metrosideros* trees. The emergent tree layer (more than 15 m height) exhibits a more complete space utilization. In

*By Ranjit G. Cooray and D. Mueller-Dombois

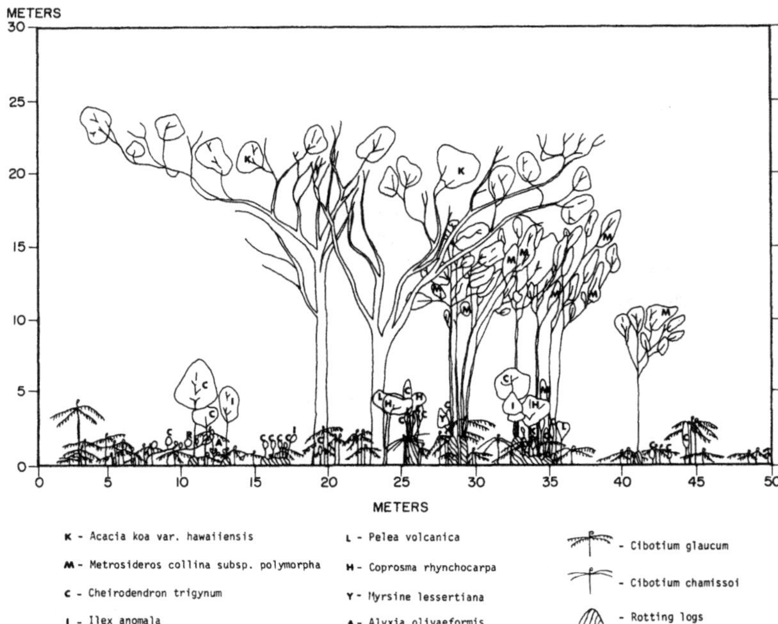

FIGURE 6-6. *Profile diagram of 6 by 50 m strip on transect 1, plot 5.*

summary, the profile can be typified as showing an emergent tree group with scattered low-stature trees and a closed tree fern understory.

On the second profile diagram (figure 6-7), tall emergents are absent. The tree fern layer is interrupted by gaps. Also the tree ferns themselves are shorter than on figure 6-6. Groups and individuals of low-stature trees reach above the tree ferns to heights of 5 to 10 m. One intermediate-stature (10–15 m) *Metrosideros* individual is present at 15 m. Two kinds of tree groups can be identified: a *Metrosideros-Cheirodendron* group at 5 to 7 m and 15 m along the profile and a *Cheirodendron-Ilex* group at 40 to 45 m. Thus along this stand segment, which lacks emergents, there are also fewer species among the low-stature trees than on the first profile. These low-stature tree groups are growing in fallen trees. In addition there are a few individually growing trees, notably *Myoporum sandwicense* at 19 m, and two smaller *Cheirodendron* individuals at 25 m and 27 m. These grow on mineral soil in an opening where tree ferns are absent. Such openings and segments without emergent koa trees occur quite frequently throughout the Kīlauea forest. The profile taken by itself is structurally quite similar to the more open *Metrosideros* rain forest that occupies a much wider territory in Hawaii Volcanoes National Park (as documented on the vegetation map of Mueller-Dombois and Fosberg, 1974). In summary, this second profile can be typified as showing tree fern patches with scattered low-stature tree groups.

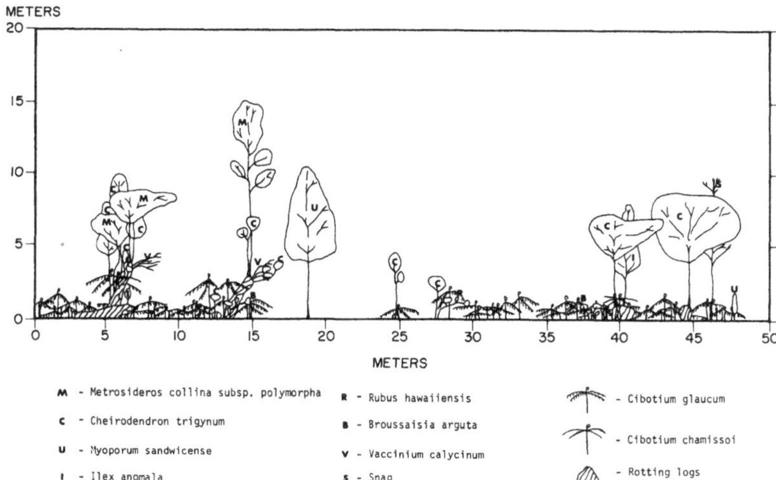

FIGURE 6-7. *Profile diagram of 6 by 50 m strip on transect 1, plot 1.*

The emergent koa makes the third profile (figure 6-8) appear to be similar to the first one. However, the low-stature tree layer is here well represented and composed of a relatively large group of evenly sized *Metrosideros* trees. Such closed *Metrosideros* tree groups occur at various places throughout the forest. Because of their uniform size, they can be assumed to be relatively even aged. The *Metrosideros* tree group is associated with fallen trees, which makes it probable that this group became established after a windfall of a group of trees that were formerly in the emergent layer. Thus the even-sized *Metrosideros* group probably represents one form of gap-phase replacement in the forest. From a structural viewpoint, the closed *Metrosideros* group is just one other variation of this forest, where low-stature trees make good use of the space that is not overshadowed by emergents. The tree fern layer is as dense and continuous as on figure 6-6. The third profile pattern may be typified as showing a closed, low-stature tree group with a koa emergent and a dense and continuous tree fern understory.

Four Major Map Units

The mathematical pattern analysis yielded recurring groups only among the trees that extended above the tree fern layer. The undergrowth layers exhibited a homogeneous pattern. However, there were two obvious undergrowth layer departures from the norm. One was caused by swampy depressions and the other by recently fallen trees that had disrupted the tree fern layer.

The swampy depressions were covered either with standing water and algae or with dense growth of *Carex alligata*. But these depressions were so

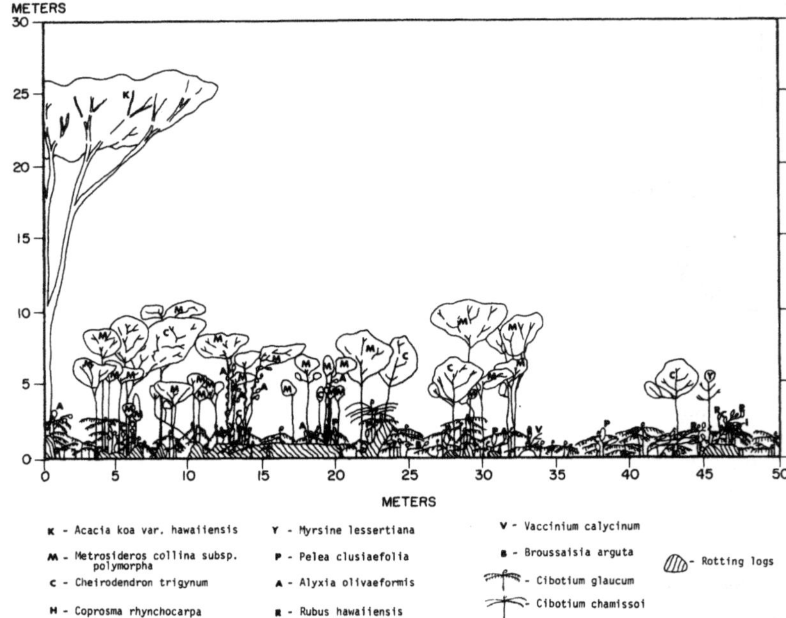

FIGURE 6-8. *Profile diagram of 6 by 50 m strip on transect 4, plot 20.*

few and small (maximum 10 by 20 m) that they were not detected by the systematically laid out transect plots used for the mathematical analysis.

These very few and small swampy depressions were the only significant substrate variations in the forest. Detailed soil depth analyses (Maka, TR 31) had shown that the dark-brown, organic-matter-enriched, uniformly textured silty clay ash soil varied from 20 cm to 60 cm depths. Thus throughout the Kīlauea forest plot, the soil is relatively shallow, overlying ‘a‘ā bedrock. Occasional small (3–5 m) rock outcrops occur, and the soil is distibuted not as a uniform blanket but rather in the form of a mosaic of deeper and shallower pockets. Most of the time the soil is at field capacity or wetter, kept moist by fog drip. Therefore there are no larger-area soil moisture regime variations, and the site can be considered homogeneous except for the few swampy spots.

A series of color slides was made by John I. Kjargaard from a low-flying aircraft at about 300 m elevation. The objective was to obtain a close aerial view of the forest for redetecting the patterns that were established on the ground through the mathematical and profile analyses. A number of 10 by 12 cm prints were made that covered particular sections and the entire forest plot at a large scale so that individual trees could be recognized. One of the photos taken centrally over the 800 by 1,000 m IBP study plot from about 500 m height could be enlarged without much distortion to a scale of 1:1,500. This photo served as the base for establishing the map, which is here shown much reduced as figure 6-9.

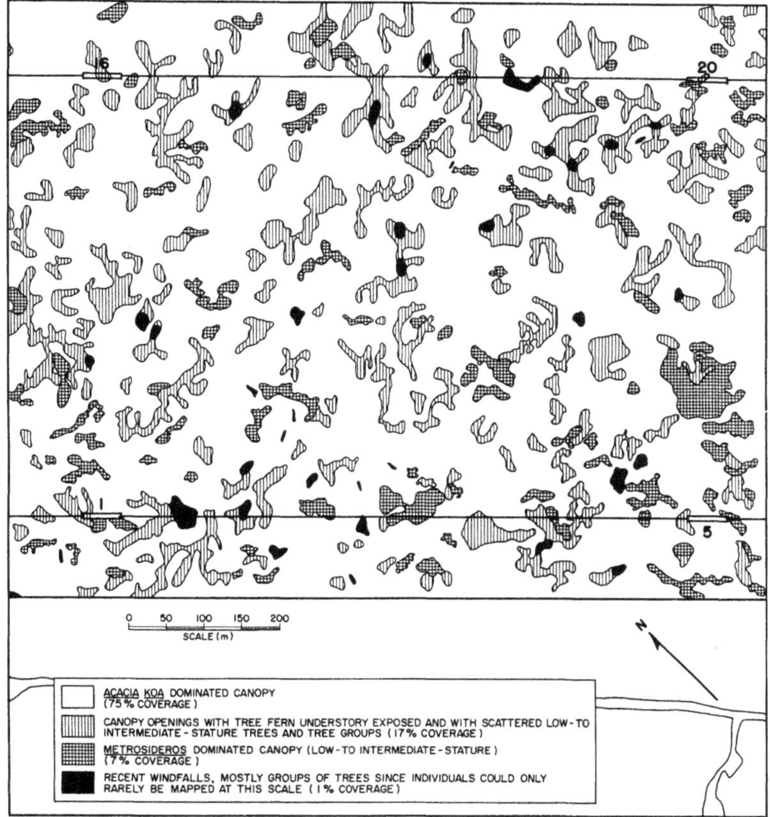

FIGURE 6-9. *Vegetation map of the 80 ha IBP study plot in the Kīlauea rain forest. Derived from a color air photograph taken July 1974. Original map scale before reduction was 1:1,500.*

It was possible to recognize the three major structural variations that were initially discovered as profile patterns:

1. An *Acacia koa*-dominated canopy corresponding to profile 1 (figure 6-6).
2. Canopy openings with tree fern understory exposed. These included, in the tree fern matrix, scattered groups of low- to intermediate-stature trees, a pattern that corresponded to profile 2 (figure 6-7).
3. *Metrosideros*-dominated canopy, showing mostly low- to intermediate-stature trees, a pattern that corresponded to profile 3 (figure 6-8).

A fourth spatially recurring pattern—that of recent windfalls—was recognizable on the photos. All horizontal patterns were checked once more on the ground with the air photos at hand. Figure 6-10 is an example air photograph

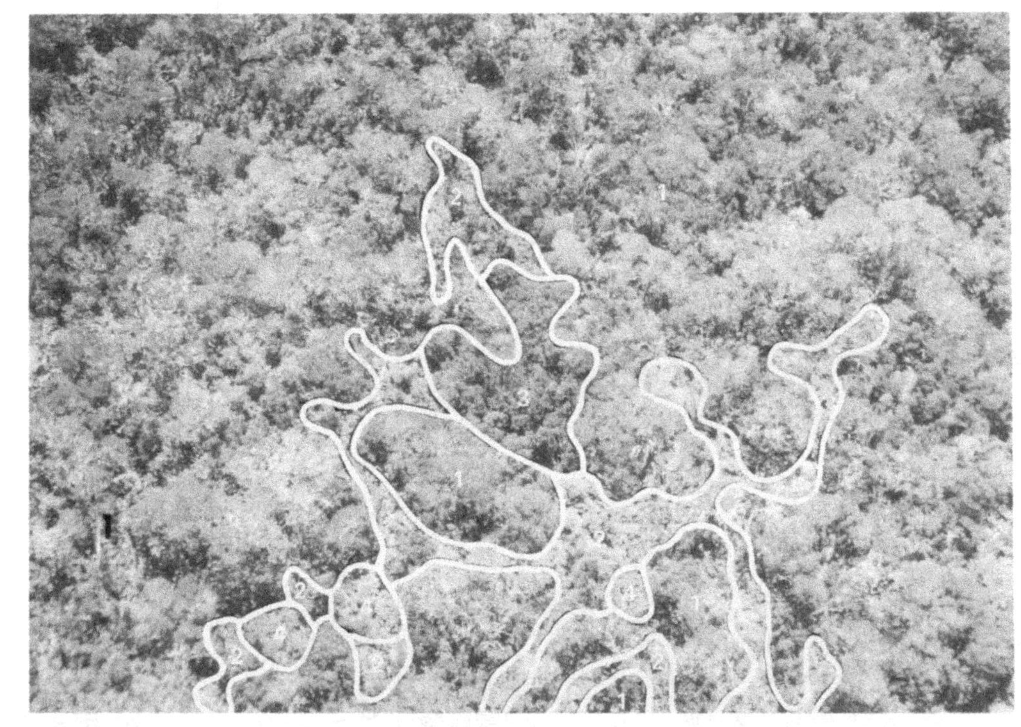

FIGURE 6-10. *Example air photograph showing a 150 by 225 m section of Kīlauea rain forest. Approximate scale 1:1,600.* 1=Acacia koa-*dominated canopy;* 2=*canopy openings with tree fern understory exposed;* 3=Metrosideros-*dominated canopy;* 4=*recent windfalls. Photograph by John Kjargaard.*

showing a section of Kīlauea rain forest at an approximate scale of 1:1,600. The four structural variations are recognizable on this photo and are outlined as such.

The recent windfalls relate invariably to tall and big-diameter emergents of *Acacia koa* that had fallen either as individuals or in groups of two to three. It is conceivable that not all windfalls were detected on the enlarged 1:1,500 air photo; some may be obscured by overtopping crowns. Also small-area mosaics of patterns 2 and 3 could not be mapped so the map appears somewhat biased in favor of map unit 1, the *Acacia koa*-dominated canopy. The smallest map unit accepted was approximately 4 mm wide (6 m in the field) which corresponded to the transect width.

Interpretation of the Map

The few swampy spots could not be mapped effectively because they were covered partially by overtopping tree crowns. The four recognizable map units, therefore, relate to structural variations in the forest that all occur on the same basic habitat. It can be concluded that the units must have a dynamic relationship to one another.

It is clear that the recent windfalls (map unit 4), which invariably have disrupted the otherwise closed tree fern undergrowth canopy, form new gaps in the forest. These gaps, which cover about 1 percent of the area, provide for the start of serules (Daubenmire, 1968), which are small-area secondary successions. The native raspberry *(Rubus hawaiiensis)* is usually an early important component in those gaps. The more advanced gaps are either those of map unit 2 or 3. Map unit 2, the canopy openings with tree-fern understory exposed and with scattered groups of trees of mixed species composition, is the spatially more extensive unit of the two, with 17 percent ground coverage. As shown on profile 2 (figure 6-7), the tree-fern layer may be interrupted also. Although the profile diagram itself did not show an *Acacia koa* sapling on the 6 m wide strip, it is in this unit that intermediate-sized *Acacia koa* trees can be found occasionally (figure 6-11). Also, many of the emergent root overturns showed advanced regeneration of *Acacia koa* (Cooray TR 44). The root overturns are found in this unit (map unit 2), and also in map unit 4 and a few in unit 1. The recurring tree groups discovered by the mathematical pattern analysis are also found in map unit 2 and also in map unit 1 (the *Acacia koa*-dominated canopy area). Map unit 2 thus can be interpreted as an advanced gap, in which *Metrosideros* did not become dominant and in which *Acacia koa* individuals have a chance to become new emergents. Map unit 3 represents another form of advanced gap vegetation in which *Metrosideros* has become dominant, almost to the exclusion of *Acacia koa*. Also the other tree species appear to be more or less excluded from these *Metrosideros*-filled gaps. This is also the reason why this unit was not discovered in the mathematical pattern analysis, which was sensitive only to recurring combinations of species, not to aggregations of individuals of the same species. However, since most of these species, with the exception of *Myoporum sandwicense,* appear to be rather shade tolerant, they

FIGURE 6-11. Acacia koa *sapling growing on root overturn of an emergent in a successionally advanced gap (map unit 2) in the Kīlauea rain forest. Dashed line outlines crown and stem of sapling and at the base, the overturned root. Ranjit G. Cooray in photograph.*

can be expected to join the *Metrosideros* patch stands later when the older *Metrosideros* individuals begin to break down.

Thus it appears that *Acacia koa* and *Metrosideros collina* compete for establishment in canopy gaps created by windfalls. Here *Metrosideros* has a certain advantage where its seedlings are already established, and it can take advantage of the shade release. *Acacia koa* has to reproduce nearly each year from new seedlings and root sprouts. It is more prone to depradation by pigs in the early seedling stage (Cooray, TR 44), but once established as a 2 m tall sapling, particularly when growing on the mineral soil of overturned roots, it seems to grow fast and has a good chance of survival.

POPULATION STRUCTURE OF WOODY SPECIES*

For the purpose of fully understanding the stability-fragility relations of the Kīlauea rain forest, it would be necessary to study the behavior of its species populations over a long period of time. These studies could be done in permanent plots, but time alone is not a sufficient criterion of ecosystem development or change. During that time of study, an ecosystem would need to be exposed to the ecological stress factors that are part of the periodically recurring regional perturbations, such as flash floods and storms in this particular environment. Because the time frame of the IBP project did not allow for such a permanent plot study, the alternative chosen was to deduce any dynamic relationship of forest development or change from a detailed structural analysis of its woody plant species and from signs of former perturbations. Periodic heavy rainstorms resulting in wind-thrown emergents, which in turn cause forest gaps, are the probable perturbations.

The enumeration of individuals in different size groups of the same species populations is a widely applied structural analysis technique (Daubenmire, 1968; Mueller-Dombois and Ellenberg, 1974; Whitmore, 1975), which can be used to predict developmental trends in forest stands. This section will present and synthesize such structural data for the twenty-one woody plant species encountered in the forest. The data are portrayed in the form of diagrams that allow grouping the woody species into structural similarity groups. According to their height-growth potential in the Kīlauea forest, the woody species can be discussed in four phanerophytic life form groups (after Ellenberg and Mueller-Dombois, 1967, slightly modified): nanophanerophytes (less than 3 m tall, five species), microphanerophytes (3 to 5 m tall, seven species), mesophanerophytes (5 to 10 m tall, seven species), and macrophanerophytes (over 10 m tall, two species).

---

*By Ranjit G. Cooray and D. Mueller-Dombois

Nanophanerophytes

Figure 6-12 shows the population structure data for the five low-growing woody plant species in the Kīlauea forest. Stem-length classes are interpreted as equivalent to tree heights. The reproduction classes are defined by a more detailed enumeration of the smaller individuals up to 1 m tall. *Gouldia terminalis* is a small native tree that rarely grows taller than 2 m; the others are typical shrubs. *Cyanea pilosa* is a native shrub, found only on Hawai'i in the montane rain forests. *Vaccinium calycinum* is one of the most common native shrubs in Hawaiian rain forests, and *Rubus rosaefolius* is the more abundant exotic woody species found in this otherwise native forest, which is also very common

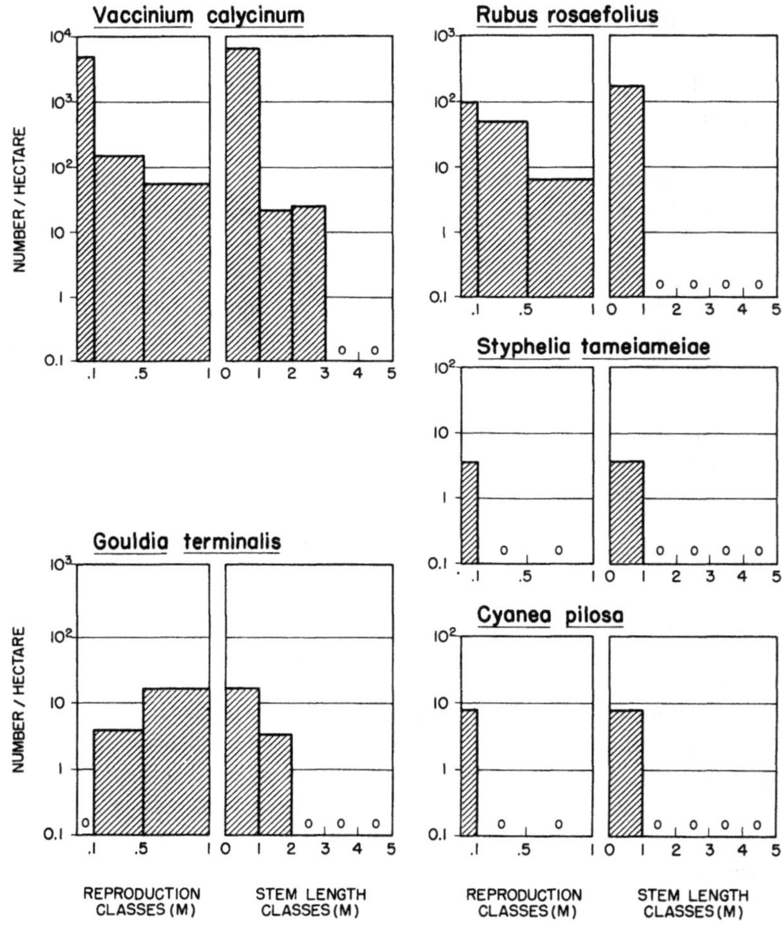

FIGURE 6-12. *Population structures of the five nanophanerophytes in Kīlauea rain forest.*

and widely naturalized in other Hawaiian rain forests. Few individuals of the other exotic woody species, *Rubus penetrans,* were found in the Kīlauea forest, all outside the transects. *Styphelia tameiameiae* is common in seasonally dry montane ecosystems—for example, along the Mauna Loa transect. In this rain forest, *Styphelia* is rather rare, with only five individuals per ha. *Cyanea* also is rare in this forest, with only nine individuals per ha. Of these two species, only seedlings under 10 cm height were encountered. However, mature fruiting individuals of both species occur in the forest outside the transect plots, where they were not enumerated. Thus the population structure of such sparingly distributed or locally rare species could not be presented adequately. In order to do so, we would have had to intensify the sampling by at least a factor of ten. The trends of numbers of individuals in size classes for *Vaccinium calycinum* and *Rubus rosaefolius* indicate active replacement of mature individuals by abundant reproduction. As the diagram shows, *Vaccinium* seedlings under 10 cm height are very abundant in the forest with about 5,000 individuals per ha, or about one seedling for every 2 square meters. There are about forty-five mature, fruiting individuals (from 1 to 3 m tall) per ha. The introduced *Rubus rosaefolius* is less abundant, with under ten individuals per ha in the mature category (here from 0.5 to 1 m tall). But both species appear to maintain a stable position in the forest on account of their good reproduction and adequate presence in their intermediate-size classes. The maintenance trend for the small native tree, *Gouldia terminalis,* is not so clear. There are about fifteen juvenile individuals (under 1 m tall) and fewer than five mature individuals (1–2 m tall) per hectare. While this indicates population stability in the stem-length classes, the trend in the reproduction classes shows no seedlings under 10 cm height. This may be caused by the relatively small enumeration in general (by sampling error) or by periodic germination (by seedling establishment every few years only).

Microphanerophytes

These are the more important, taller-growing members of the tree fern layer. They include seven species; the population structures are shown in figure 6-13. This group includes the two tree fern *(Cibotium)* species encountered in the forest. As the diagrams show, *Cibotium glaucum* is about fifty times more abundant than *Cibotium chamissoi.* There are approximately 5,000 *C. glaucum* individuals per ha, or one for every 2 square meters. The distribution of individuals by size classes (here trunk-length classes) indicates that both species are maintaining themselves in this numerical relationship, their populations are stable. Sporophytes under 10 cm height were not enumerated because they could not be properly identified. Subsequently Becker (1976) found a way to distinguish small tree ferns, which had developed from spores and gametophytes from those which had developed asexually from trunk buds. From his

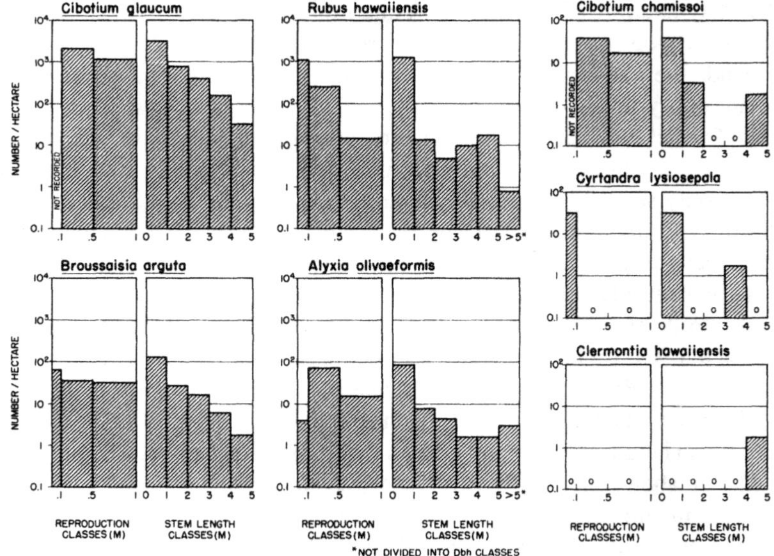

FIGURE 6-13. *Population structures of the seven microphanerophytes in Kīlauea rain forest.*

enumeration he estimated for this forest the former to amount to 1,750 per ha, while those of bud origin were estimated as 283 per ha. This implies that there was an abundant supply of small tree fern sporelings under 10 cm height in this forest. The two missing size classes (2 to 3 m and 3 to 4 m stem length) in the *Cibotium chamissoi* population can be attributed to the generally small population size of this species. Only two individuals per hectare were enumerated in the tallest (4 to 5 m) size class.

Next to the abundant *C. glaucum,* there are three moderately abundant species: *Rubus hawaiiensis, Broussaisia arguta,* and *Alyxia olivaeformis. Rubus hawaiiensis* is a native, tall-growing, thin-branched shrub, a major component of stand openings (gaps) that were caused by recent windfalls of emergents. In gaps with advanced growth, *R. hawaiiensis* often assumes a vinelike habit, climbing into the branches of taller-growing trees. Although a weedy disturbance species, this native *Rubus* is very prevalent and obviously reproducing vigorously in this forest. This provides another indirect indication of the gap-phase nature of the Kīlauea forest. *Alyxia olivaeformis* is a true woody vine, but it also has rather thin branches. The thick-branched native woody vine, *Freycinetia arborea,* was not encountered in the Kīlauea rain forest. It occurs usually at lower elevations in the Hawaiian rain forests. In contrast to *R. hawaiiensis, Alyxia* grows mostly on the taller trees in the more shaded sections of the forest. *Broussaisia arguta* is a common native arborescent shrub. These three woody plants indicate (as do the tree ferns) stable maintenance trends by

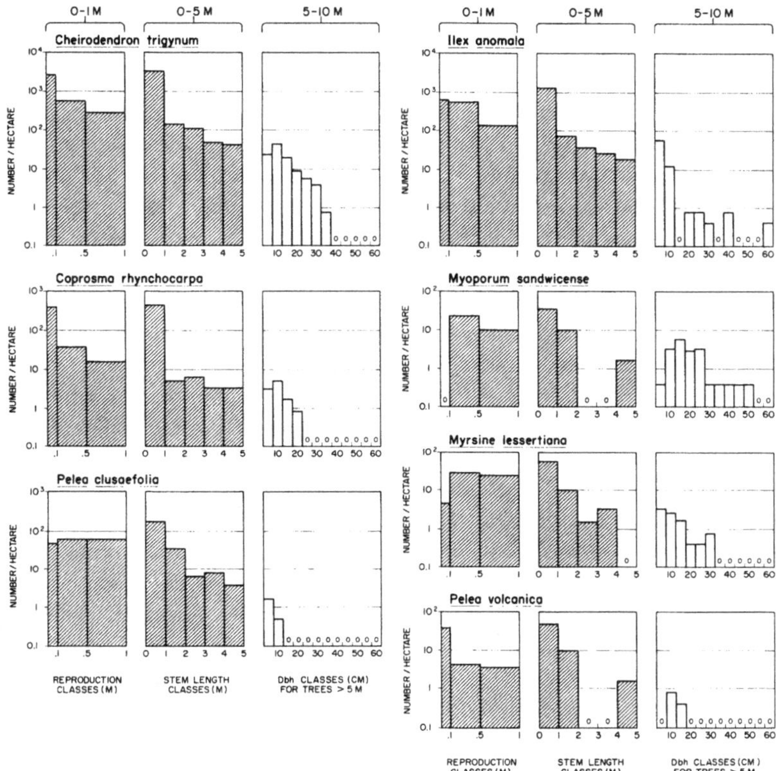

FIGURE 6-14. *Population structures of the seven mesophanerophytes in Kīlauea rain forest.*

their population structures. The smaller numbers of *Alyxia* seedlings under 10 cm height than in the next reproduction class may be attributed to fast, early, seedling growth in this species. The more sparingly represented *Cyrtandra lysiosepala* also seems to maintain itself in the forest, since a number of seedlings were recorded (about 35 per ha) in addition to the few (about 2 per ha) mature individuals. No seedlings were recorded for *Clermontia hawaiiensis*. However, this species was so sparingly represented along the transects (with only about 4 mature individuals per ha) that it is difficult to read a trend from its population diagram. Seedlings may occur elsewhere in the forest.

Mesophanerophytes

This group includes seven native tree species that are typically of low stature. In this forest they grow into the height range of 5 to 10 m. As indicated on figure 6-14, individuals of these species up to 5 m tall were measured in 1 m stem-length classes, and those exceeding 5 m in height were measured in 5 cm diameter classes. Heights of these taller individuals were measured with an Abney level for only a few sample trees (Maka, TR 31) to establish the general height-diameter relationship shown on the diagrams. Only *Ilex anomala, Myoporum sandwicense,* and *Myrsine lessertiana* may have a height-growth potential somewhat exceeding 10 m in other habitats. As the structural diagrams show, *Cheirodendron trigynum* and *Ilex anomala* are the most abundant low-stature trees in this forest. Both show stable maintenance trends by abundant reproduction and good replacement of mature individuals (those over 5 m tall) by juvenile individuals in the five stem-length classes. The absence of *Ilex* trees in some of the larger diameter classes can be attributed to the low sample number. While there are about 11 *Ilex* trees per ha in the 10 cm (4 in) diameter class, there are from 3 to 4 *Ilex* trees per ha that grow to bigger diameters. One *Ilex* per 2 ha may grow to a potential diameter at breast height (dbh) of 60 cm (or 2 feet). Stable population trends are also shown for the other five low-stature tree species. All show their greatest density in the reproduction class, which is less than 1 m stem-length class. Here *Coprosma rhynchocarpa* has about 500 seedlings per ha, *Pelea clusaefolia* about 200, and the other three species (*Myoporum sandwicense, Myrsine lessertiana,* and *Pelea volcanica*) about 50. As expected in stable tree populations, their numbers decline as the trees become bigger, but there is no radical decline with size increase. The gaps in the 2 to 4 m stem-length groups for *Myoporum* and *Pelea volcanica* can be attributed to sampling error because of the small number of individuals in the larger size classes. However, *Myoporum* exhibits a somewhat irregular pattern by its absence of seedlings in the under 10 cm class, by the gap in the 2 to 4 m stem-length classes, and by the population bulge in the 10 to 25 cm diameter classes. *Myoporum* was found to grow mostly on mineral soil (like the tree ferns), and it was clearly concentrated in the openings (or gaps) of the forest. It therefore appears to be a seral species that takes advantage of local disturbances by windfalls in the forest. The other six low-stature species appear to be typical shade-tolerant climax species. All of them were found to grow mostly on decaying logs. *Ilex anomala* and *Cheirodendron trigynum* had log-establishment indexes of over 90 percent (Cooray, TR 44:70).

Macrophanerophytes and Snags

Only two native species have a growth potential to become tall trees in this forest: *Metrosideros collina* subsp. *polymorpha* and *Acacia koa* var. *hawaiiensis*. Both can grow over 15 m tall. However, as indicated on figure 6-15, the majority of over 5 m tall *Metrosideros* trees are of low to intermediate stature—up to 15 m tall. Only about 6 trees per ha are found in the emergent tree stratum—those exceeding 15 m in height. In contrast, the majority of *Acacia koa* trees are over 15 m tall; few—about five trees per ha—are found in the 5 to 15 m height range. An interesting species difference is the dbh-to-height relationship of these species. A *Metrosideros* tree has to increase in diameter to about 65 cm (26 inches) before it grows taller than 15 m, while an *Acacia koa* tree only has to attain about 35 cm (14 inches) dbh to become recognizable as an emergent. *Acacia koa,* therefore, can be assumed to grow much faster in height than *Metrosideros* does. The population structure of *Metrosideros* is quite comparable to those of the low-stature climax species, such as *Cheirodendron trigynum*. However, there is a drastic reduction in the number of individuals in the stem-length classes of *Metrosideros*. When it grows from 1 to 2 m in height, the density of *Metrosideros* declines from about 8,000 to 8. Thus there is a tremendous seedling mortality of this species in the Kīlauea forest. Many of the *Metrosideros* seedlings grow in clumps on decaying logs in the shade of tree ferns. Here they seem to stagnate in shade-suppressed condition. Where tree ferns are slashed down from falling emergents, snags, or other trees, *Metrosideros* seedlings may be released to grow into the taller height classes. However, few of these succeed, a fact indicated also by the relatively few gaps with advanced growth of pure *Metrosideros* tree groups (figure 6-9). Since the *Metrosideros* seedlings are often found in clumps on logs, there must also be intensive intraspecific competition, and many individuals will probably die due to this cause. The regeneration pattern of *Metrosideros* in the Kīlauea forest

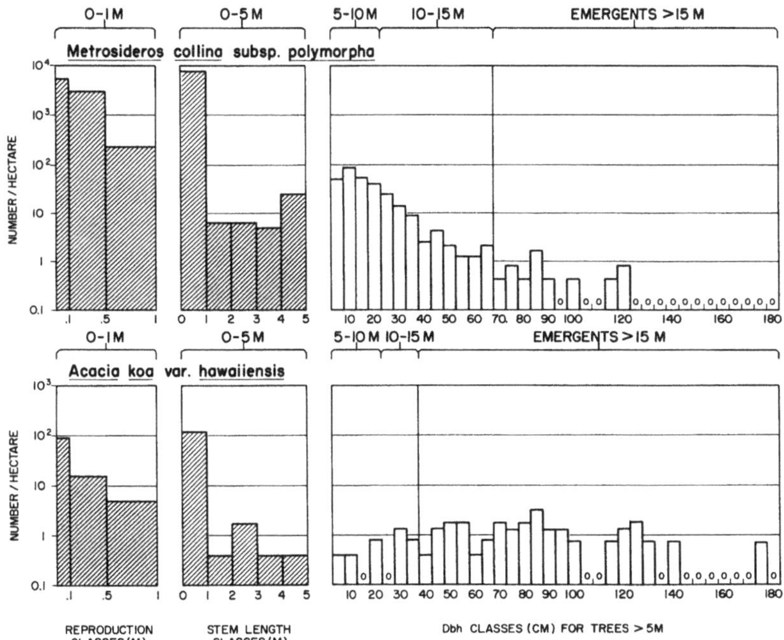

FIGURE 6-15. *Population structures of the two macrophanerophytes in Kīlauea rain forest.*

shows that it is not a typical climax species; rather it displays the characteristics of a species ready to take advantage of local disturbances in the forest.

In this regard, *Acacia koa* displays similar characteristics. Its seedling numbers also are very much reduced (from about 100 to less than 1 per ha) when growing from the 1 to 2 m height class. In addition, the numbers of seedlings, saplings, and taller juvenile individuals of *Acacia koa* are very much less than those of *Metrosideros*. Thus the maintenance of *Acacia koa* seems to be dependent entirely on disturbances of the canopy. But the structural diagram indicates that the species is not disappearing from the forest since individuals are present in all size classes. It was observed (Cooray, TR 44) that nearly all overturned roots of fallen *Acacia koa* emergents were stocked by *Acacia koa* seedlings or

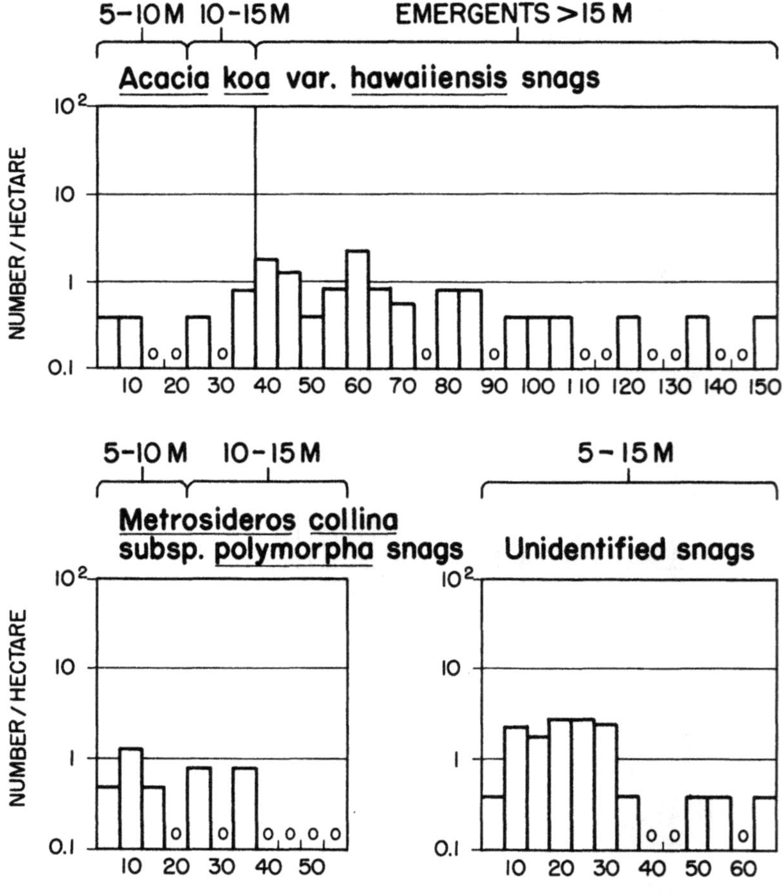

FIGURE 6-16. *Distribution of snags by size classes.*

saplings, with at least a one-to-one replacement. Of course, the density of *Acacia koa*, even in the emergent class with only about 24 individuals per ha, is far below that which one would expect of a productive *Acacia koa* forest. This is a natural forest, in which the tallest tree species is represented with low density, a situation typical for emergent tree species in tropical rain forests elsewhere (Aubréville, 1938; Davis and Richards, 1933; Richards, 1939; Jones, 1950; Whitmore, 1966, 1975).

Also typical for natural, unmanaged forests are dead-standing trees—the snags. Their number by size classes is shown on figure 6-16. There were about twelve snags in the emergent tree layer, all from *Acacia koa*. This represents almost one dead-standing emergent among three. In the low- to intermediate-stature tree group (5–15 m tall) there were about nineteen snags per ha—two from *Acacia koa*, four from *Metrosideros*, and thirteen unidentified snags. This supports the fact that a few trees die in all size classes due to competition and other causes. The same is shown indirectly for all woody plants in their population diagrams by the common trend of fewer individuals occurring in the next-larger size classes. Dead saplings and seedlings were not enumerated because few were found dead and still standing. Most of these smaller woody materials seem to decay relatively fast, but taller trees may remain standing for several years. The rate of decomposition of larger wood fragments appears to be slow in the relatively cool montane tropical rain forest, which is also indicated by the high number of decaying logs on the ground. In contrast to the tropical lowland areas on Hawai'i, termites are absent in the montane environment.

Conclusions

The population structure analyses indicate that all twenty-one species maintain themselves in this rain forest community under the prevailing environmental conditions. This can be concluded because all species are well represented by juvenile individuals. The only exception is the locally rare *Clermontia hawaiiensis*, whose seedlings were not encountered along the transect plots, probably because its species individuals are too sparingly distributed throughout the forest.

Most species show stable maintenance trends by only moderate reductions in numbers as individuals grow bigger from one size class to the next. The exceptions are the two macrophanerophytes, *Metrosideros collina* and *Acacia koa*, and one mesophanerophyte, *Myoporum sandwicense*. Both *Metrosideros* and *Acacia* show severe reductions in their population numbers when they grow from seedling to sapling stage. This high seedling mortality indicates that both are pioneer rather than climax species. The pioneering behavior of *Metrosideros*, the first tree on new lava flows, is well known (Smathers and Mueller-Dombois, 1974). The behavior of *Acacia koa* is less well known. It rarely occurs on rock outcrop soils such as created by new lava flows; instead it is usually associated with fine soil (Mueller-Dombois, 1967). Thus *Acacia koa* is not such an extreme pioneer. However, in the Kīlauea forest it seems to be dependent, even more so than *Metrosideros,* on periodic disturbances as created by windfalls. A devastating storm, such as a hurricane, would probably result in a substantial increase of its density. However, with the present rate of breakdown of emergent trees during annually recurring winter rainstorms (kona storms), *Acacia koa* seems to maintain itself more or less indefinitely in the forest.

Myoporum sandwicense is a gap phase species similar to *Acacia koa,* but *Myoporum* more so than *Acacia koa* appears to be dependent on mineral soil as a seedbed. Its log-establishment index was 11.6 (Cooray, TR 44), implying that 88.4 percent of *Myoporum* individuals were found growing on mineral soil. In *Acacia koa* only 37.5 percent of the trees were growing directly on mineral soil. The population-structural diagram of *Myoporum* indicates certain population fluctuations that are, however, small enough so that its presence in the Kīlauea forest will continue indefinitely under the prevailing natural conditions.

AGE STRUCTURE OF *ACACIA KOA**

Like other tropical trees, *Acacia koa* (koa) lacks growth rings, which are reliable for age determination. Efforts have been made to ascertain the age of koa trees from growth rings by staining (Wick, 1970) and microscopic examination (Burgan, 1970). Both methods were not satisfactory, particularly because the growth rings that are made discernible are not reliably correlated with age. This lack of confidence in age correlation with growth rings is caused by the absence of annual temperature seasons, which are found at higher latitudes. However, there is a relationship between size and age of trees, and particularly the trunk diameter is often correlated with age (Horn, 1971). This correlation is obscured by other factors, such as site quality, stand density, and biotic influences. Yet if these other factors are held constant by an appropriate selection of trees and if annual variations in climate are considered, tree age should be determinable from trunk diameter through diameter increment measurements from a sample of differently sized trees of the species population.

Sampling Scheme

Two sites with koa stands were selected for comparison, the Kīlauea rain forest plot and Kīpuka Puaulu, a mesic seasonal forest along the Mauna Loa transect (figure 2-1, site 4, and figure 2-2, segment 9). At each site at least two

*G. Spatz and D. Mueller-Dombois

trees were selected in each of 10 diameter classes ranging from 0 to 3.8 cm (0 to 1.5 in) dbh (class 1) to trees 114.3 cm (45 in) and larger (class 10). Fifty-six trees were included in the sample at the Kīlauea site and forty-three at the Kīpuka Puaulu site.

Tree diameters were measured once per month, from August 1971 through July 1972. Trees with diameters over 38 cm (15 in) were equipped with permanent diameter tapes. Smaller-diameter trees were remeasured with a steel diameter tape at exactly the same points marked with nails or paint. For obtaining information on other factors that influence diameter growth, measurements were made of height, crown height, crown diameter, and light or exposure angle. The light angle was obtained by sketching the crown of the sample tree in relation to other tree crowns. An angle was then superimposed with its vertex at the base of the crown, which showed how much direct light interference the sample tree was receiving from neighboring trees. The light angle may vary theoretically from zero (for completely overtopped trees) to 180° (for emergent or free-standing trees that are exposed to sky light from all sides for their full crown length). These additional parameters were obtained for explanatory reasons. They were not included in the data computation.

Calculation of the Growth Curve

The monthly diameter increments were added for the whole measurement period, and this total was converted into daily increments as a function of stem diameter. This relationship was expressed in a regression equation, in which the rate of daily increment was the dependent variable and the mean stem diameter the independent variable. The parameters for the equation depend on the specific growth curve of the species as well as the site. However, the site difference in the diameter-increment, stem-diameter pattern was found to be insignificant, so that the general relationship of diameter increment of koa to its stem diameter could be expressed for both sites by the following equation:

$$y = \frac{x}{ax^2 + bx + c}$$

where y = rate of diameter increment at breast height (in mm · day^{-1}), x = diameter at breast height, and a, b, c = regression coefficients. When the specific growth curve at a site is known for the species over a wide range of diameters, it is possible to convert this information into the age of differently sized trees.

Age Computation

For estimating the age of a tree of a given diameter, it is necessary to compute the cumulative rates of growth increments that are required of the tree to grow to that diameter. The daily rates at any diameter are obtained from the regression equation. The equation necessary for computing the age is given by the following integration:

$$A = \int_{Z_1}^{Z_2} \frac{dx}{y}, \tag{1}$$

where A = age of tree, Z_1 = minimum diameter measured, Z_2 = diameter of the tree for which age is to be determined, dx = an infinitely small diameter difference between two neighboring points on the curve, and y = current growth rate as defined by $x/(ax^2 + bx + c)$.

Equation 1 is integrated as follows:

$$A = \int_{Z_1}^{Z_2} \frac{(ax^2 + bx + c)\,dx}{x}$$

$$A = \frac{aZ_2{}^2}{2} + bZ_2 + c \cdot \log_e Z_2 - \frac{aZ_1{}^2}{2} - bZ_1 - c \cdot \log_e Z_1. \tag{2}$$

This equation gives the time it takes for a koa tree to grow from a diameter of Z_1 to a diameter of Z_2. The value for Z_1 is 1 cm because no smaller tree was included in the measurements. Therefore, the simplified equation can be written in this way:

$$A = aZ^2 + bZ + c \log_e Z - a - b, \tag{3}$$

where Z is now the final diameter.

Growth Curves for Koa

Figure 6-17 shows the regression curves of diameter growth rate as dependent on diameter for the koa populations at Kīlauea forest and Kīpuka Puaulu. In addition, the mean diameter increments are plotted for the fifty-six measured trees of the Kīlauea koa population. The same trends occurred at both sites. The initially very slow diameter growth increases rapidly with increasing diameter to an early maximum, and then the growth rate decreases more gradually with increasing diameter. The fastest growth rate occurs in the Kīlauea forest in trees of near 20 cm (8 in) diameter, and in Kīpuka Puaulu in trees of near 15 cm (6 in) diameter. The fastest growth rate appears somewhat faster in the Kīlauea forest, but with increasing diameter the growth rate slows down more than in Kīpuka

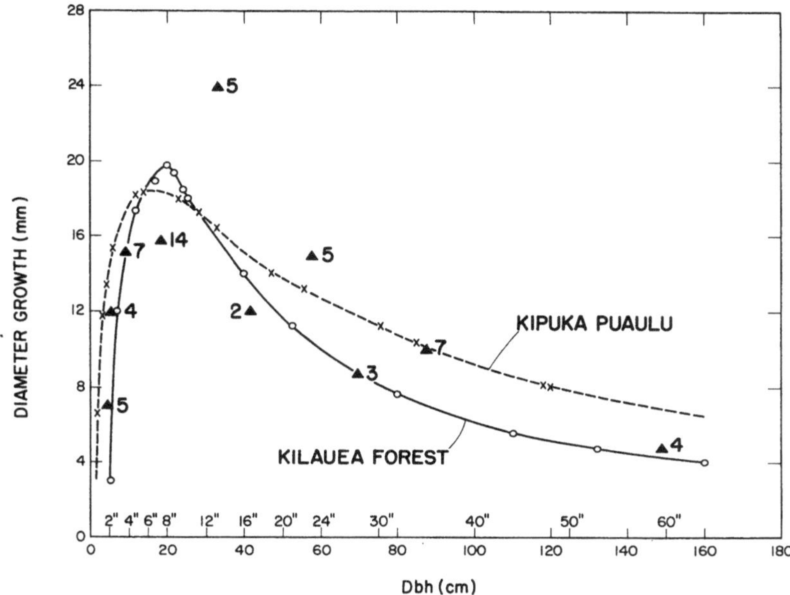

FIGURE 6-17. *One-year diameter growth in relation to dbh for* Acacia koa *populations at two sites expressed by least-squares fit curves. ▲5 = actual mean growth rate and number of trees per dbh class in Kīlauea rain forest.*

TABLE 6-2. *Regression Parameters for Growth Curves of Koa*

	a	*b*	*c*	r^2
Kīlauea forest	0.0088	−0.1240	5.8016	0.46
Kīpuka Puaulu	0.0037	0.3191	1.6217	0.45

Puaulu koa population. The values for the regression equations are shown in table 6-2.

The deeper ash soil (over 2 m, Mueller-Dombois and Lamoureux, 1967) and the warmer climate of Kīpuka Puaulu (16.2°C mean annual temperature for 1971–1972 versus 13.8°C for Kīlauea forest) may contribute to a somewhat faster diameter growth rate. Also the stocking density of woody plants of all species is greater in the Kīlauea forest than in the more open, savanna-like koa forest of Kīpuka Puaulu.

The plot of the grouped mean diameter growth rates of the fifty-six measured trees of the Kīlauea koa population gives an indication of the fit of the calculated curve (see figure 6-17). The curve estimates the annual increments somewhat conservatively.

The exact age of an individual tree cannot be determined reliably from its diameter, but it will be possible to determine the average age of a sample of several trees with similar diameters. Thus the method is adequate for the intended purpose of determining the age structure of the quantitatively enumerated koa population in the Kīlauea forest.

Age Curves for Koa

The age of the two koa populations was derived from the regression curves, which have had the diameter increments integrated for a range of sizes. This makes the age estimation over that range easier than calculating each case

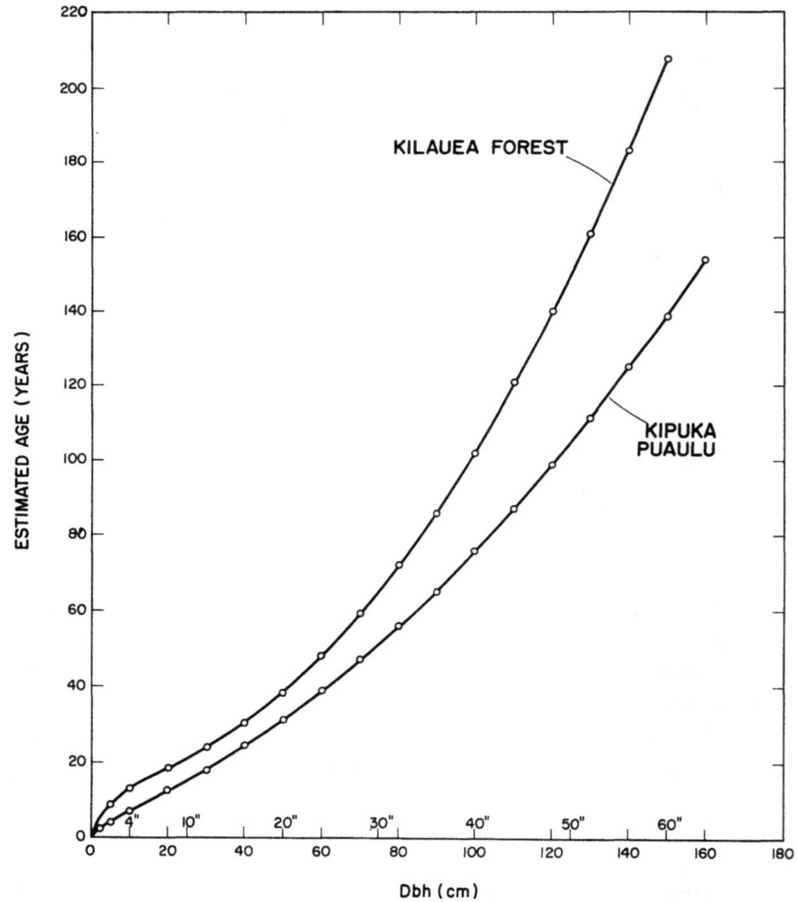

FIGURE 6-18. *Curves for determining approximate age of koa trees from dbh measurements at two sites.*

from the equation. The age curves are shown on figure 6-18. The curves are used as follows: a koa tree with a dbh of 100 cm (40 in) is 102 years old in the Kīlauea forest and 76 years old in Kīpuka Puaulu. To this age must be added the time required for a tree to grow to a dbh of 1 cm. This value was determined by direct measurements and can be estimated as two years for both sites. With this correction, therefore, a tree with a dbh of 100 cm would be estimated as 104 years old in the Kīlauea forest and 78 years old in Kīpuka Puaulu. The result applies only to an average tree of 100 cm dbh. An individual may deviate in age considerably from this value because of the scatter of growth-increment values obtained for the individual sample trees.

Age Structure of the Kīlauea Koa Population

The age curves make it possible to evaluate the estimated age composition of the currently existing koa population in the Kīlauea forest. This was done by assigning an age scale to the different size groups that were established from the population structure analysis. The outcome is shown on figure 6-19.

The emergents, which exceed 15 m in height and are generally between 20 and 25 m tall, range in age from approximately 30 to about 330 years. The maximum age of 330 years was obtained by extrapolating the calculated curve to the largest diameter tree. There are at least two generations among the emergents, indicated by the fact that certain big-diameter individuals were found growing on fallen logs of other big-diameter koa trees. The approximate age range of 300 years among the current stand of emergents puts to rest the earlier belief that the tall emergents are a one-generation stand that resulted from a major disturbance. The broad age range does not necessarily allow the opposite conclusion either—that emergents are replaced continuously by subcanopy trees. However, it seems rather probable now that the maintenance

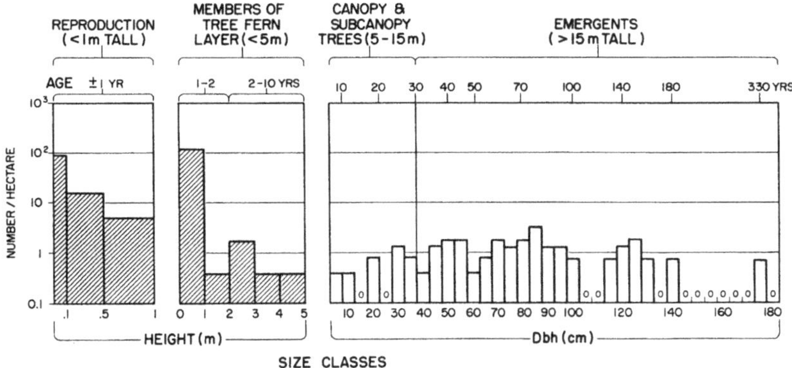

FIGURE 6-19. *Estimated age structure of* Acacia koa *var.* hawaiiensis *tree population in Kīlauea rain forest.*

mechanism of koa in the Kīlauea forest is by gap-phase replacement. When the koa emergents attain a diameter of 100 cm (40 in), which corresponds to an approximate age of 100 years, they have also reached their approximate maximum height of 25 m. At this stage the trees seem to become vulnerable to occasional storms, and they tend to fall down as individuals or in small groups. The fallen trees then produce sufficient openings for new koa individuals to grow from seedling stage through the intermediate size classes in 30 years, at which time they are recognizable as emergents again.

The current reproduction of a little more than 100 seedlings per ha is cut down by over 90 percent during the first year. Only about one tree per ha survives the first year, and from then on we find only one or two trees per ha that are two years old. Similarly rare are the consecutively older trees; for instance, there are only two to three trees between two and ten years and about four to six trees between ten and thirty years of age.

Conclusions

The causes of the early (over 90 percent mortality of koa individuals have not yet been fully clarified. But four factors have been recognized: shade, pig damage, poor seedbed, and insect damage.

Many of the small seedlings germinate in shaded locations rather than in the gaps. It was shown experimentally (Spatz, TR 17) that koa seeds need no light for germination, but they grow very poorly at low light intensities (less than 10 percent of sky light) and low air temperatures (less than 20°C, such as found in the Kīlauea forest). Spatz's experiments on light intensity were carried out for only ten weeks. Thus they are not entirely conclusive, but they indicate that koa seedlings grown in less than 10 percent of sky light have a small chance of surviving.

Practically all koa seedlings that germinate on mineral soil are destroyed in the first year by pig activity, mostly by trampling (Cooray, TR 44). However, even if pigs were removed from the forest, koa seedlings on mineral soil would not survive in shaded positions.

Because of the log-strewn forest floor, koa seeds also germinate on logs. In these positions they are usually not damaged by pigs, but they often die of starvation (Gagné, unpublished data). The starvation may occur either in form of substrate drying or nutrient deficiency. Both factors relate only to insufficiently decayed logs. Survival on the more decomposed logs seems to be good in well-illuminated positions (Maka, TR 33).

The larvae of a native moth (*Enarmonia walsinghami*, Fam. Tortricidae) were found to feed on koa shoots, causing terminal dieback and branching or death. This damage was noted particularly where koa seedlings were aggregated in clumps. The moth, therefore, may have a deaggregation effect on koa reproduction. It did not appear to be an important mortality factor among the naturally regenerating koa seedlings in the Kīlauea forest (Gagné, personal communication).

We can conclude that the rareness of koa individuals in the sapling and pole-size stage in the Kīlauea forest is caused by two factors: the high rate of seedling mortality during the year of germination (as a function of seedling position in the forest) and the rapid growth rate of the few surviving individuals in the gaps. Since it takes only thirty years for a koa tree in a gap to become an emergent, few low- or intermediate-stature trees can be found in the forest. Once koa seedlings have grown into 2 to 3 m tall saplings (in two to four years after germination), their survival chances appear to be very good. Even during the first year, their survival chances are improved in such gaps because the fallen koa trees often act as barriers or exclosures to pig activity. Thus the presence of scattered emergents of koa in the Kīlauea forest appears to continue indefinitely in spite of the pig activity in the forest. The high rate of germinating seedlings per year, however, indicates that *Acacia koa* var. *hawaiiensis* is not a typical climax tree species. Such a species would be expected to grow successfully in shade and to be more conservative in its seed production. Instead koa can be considered an opportunist that is well adapted to maintaining itself in the Kīlauea forest within the existing species composition and under the existing environmental conditions. The latter includes an occasional storm that results in blow-downs of emergents and canopy gaps. So far koa lacks a serious competitor, which would be a tall-growing shade-adapted tree species.

ARTHROPODS ASSOCIATED WITH FOLIAR CROWNS OF STRUCTURAL DOMINANTS*

The arthropod fauna associated with native Hawaiian trees has received considerable attention in the literature (Zimmermann, 1948–present; Swezey, 1954), but most of these studies were concerned with selected taxa only. A comprehensive community-ecological approach was called for according to our IBP objectives. Attention therefore was given to the whole spectrum of arthropod taxa associated with the canopies of the two dominant native trees in the Kīlauea forest, *Metrosideros collina* subsp. *polymorpha* (ʻōhiʻa lehua) and *Acacia koa* var. *hawaiiensis* (koa).

Beyond identifying the crown-associated arthropod fauna, the specific objectives were to find out what spectrum of general niches they occupy on these trees, in what quantities they occur, and to what extent, if any, exotic arthropods had invaded this relatively undisturbed native rain forest. In addition, data were obtained to find out what sort of temporal variation patterns occur in the arthropod fauna of this year-round humid forest.

*By W. C. Gagné and F. G. Howarth

Sampling Procedure

The four Kīlauea forest transects were each sampled in their central sections, near plots 3, 8, 13, and 18 (figure 5-3). Here representative trees of both species were sampled by the poison-fogging method with synergized pyrethrum with large canvas collecting funnels. Although this technique gives an unbiased sampling of a broad spectrum of arthropods, it is not suitable for arthropods that are sessile, nonmobile, internally situated in the host, strong flyers, resistant to the fog, or temporally active (nocturnal insects well hidden during the day). Representative trees of *Metrosideros collina* were those with exposed canopies, described as advanced gap-phase pattern. Among *Acacia koa* trees, typical emergents were selected. On these, the 10 to 25 m height interval was fogged (figure 6-6). Canopies with a minimum of standing dead branches, twigs, detritus, and epiphytes were selected in order to concentrate the sampling on the arthropods associated with the living canopy.

Sampling was done approximately every second month from March 1971 through September 1973. Each time two replicate samples were taken at each site. These were combined into one sample for this analysis. *Metrosideros collina* trees were sampled twenty-one times and *Acacia koa* trees nineteen times. An effort was made not to sample the same trees more than once a year.

Arthropod Communities and Their Composition

A list of the canopy arthropod taxa collected is given in table 6-3. Nymphs of delphacid planthoppers, *Oceanides* spp. and *Orthotylus* spp., could not be identified to species, and several other arthropods could be identified only as to their family relationship. However, most taxa could be identified as to their way of life or life form (guild membership) on the two trees, and these are also shown in the table. In some cases, numbers of species known to be present on these hosts in the forest in a given higher taxon and not separable in the samples are given on the list.

The four major life forms recognized on both trees were the phytophagous, the anthophagous (flower pollen and nectar feeders), the entomophagous (predators and parasites), and the saprophagous arthropods. A fifth group of transient arthropods was recognized; it included perchers and those whose ecological function or life form was unidentifiable. The quantitative values are based on the number of times a taxon was collected in the successively taken tree samples. These frequency values were converted into percentages to make the ‘ōhi‘a and koa samples comparable. The most important consistent contributors to the arthropod biomass in this study were the spiders (*Araneida* spp.) on both hosts, the true bugs (Hemiptera) on ‘ōhi‘a, and moth larvae (Lepidoptera, mostly *Scotorythra* spp.) on koa.

TABLE 6-3. *Canopy Arthropod Taxa, Life Forms, and Quantities on the Two Dominant Trees, Kīlauea Rain Forest*

Life form		Species or taxon	Frequency in samples on ʻōhiʻa	koa	Mauna Loa transect figure
Phytophagous					
De	(moth larvae)	Lepidoptera (predom. native *Scotorythra* spp.)	85.7%	94.7%	
Sa	(planthopper nymphs)	Delphacidae spp.	81.0	89.5	
Sa	(leaf hopper)	*Nesophrosyne* spp. (Cicadellidae)	47.6	47.4	
Sa	(mealybugs)	(X)Pseudococcidae spp.	42.9	42.1	
Tw	(beetle)	*Oodemas konanum* (Curculionidae)	28.6	31.6	3-11
Tw	(beetle)	*Proterhinus similis* (Aglycyderidae)	19.1	63.2	3-11
Sa	(planthopper adults)	*Nesodyne rubescens* (Delphacidae)	14.3	26.3	
Sa	(psyllids)	Psyllidae (3 *Trioza* spp., 2 *Kuwayama* spp.)	100	0	
Sa	(true bug)	*Sarona adonias* (Miridae)	90.5	47.4[u]	3-11
Tw	(beetle)	*Proterhinus tarsalis* (Aglycyderidae)	57.1	0	3-11
Se	(true bug)	*Oceanides vulcan* (Lygaeidae)	42.9	0	3-11
Se	(true bug)	*Oceanides pteridicola* (Lygaeidae)	19.1	0	3-11, 3-14
Sa	(planthopper adults)	*Leialoha nanicola* (Delphacidae)	14.3	0	
Tw	(beetle)	*Proterhinus* sp. (p.) (Aglycideridae)	14.3	0	
Se	(true bug nymphs)	*Oceanides* sp. (p.) (Lygaeidae)	9.5	0	
Sa	(planthopper adults)	*Nesodyne pseudorubescens* (Delphacidae)	4.8	0	
Tw	(beetle)	*Proterhinus ferrugineus* (Aglycideridae)	4.8	0	3-11
Sa	(planthopper adults)	*Nesodyne anceps* (Delphacidae)	4.8	0	
L	(fly)	Agromyzidae sp.	4.8[u]	15.8	
Sa	(psyllids)	X*Psylla uncatoides* (Psyllidae)	0	84.2	
Sa	(true bug nymphs)	*Orthotylus* sp. (p.) (Miridae)	0	15.8	
Sa	(planthopper)	X*Siphanta acuta* (Flatidae)	0	15.8	3-12
Sa	(aphids)	XAphidae sp.	0	15.8	
De	(beetle)	X*Pantomerus cervinus* (Curculionidae)	0	10.5	3-12

TABLE 6-3. *(Continued)*

Life form		Species or taxon	Frequency in samples on 'ōhi'a	koa	Mauna Loa transect figure
Tw	(beetle)	*Plagithmysus varians* (Cerambycidae)	0	10.5	
Sa	(planthopper)	*Nesodyne koae* (Delphacidae)	0	5.3	
Anthophagous					
Fl	(thrips)	*Neurisothrips antennatus* (Thripidae)	14.3	15.8	3-11, 3-14
Fl	(thrips)	X*Neurisothrips multispinus* (Thripidae)	4.8	15.8	3-12
Fl	(thrips)	*Neurisothrips* sp. (Thripidae)	4.8	10.5	
Fl	(honey bee)	X*Apis mellifera* (Apidae)	4.8	0	
Fl	(thrips)	X*Thrips hawaiiensis* (Thripidae)	0	5.3	
Fl	(bee)	*Nesoprosopis* sp. (Colletidae)	0	5.3	3-14
Entomophagous					
Pr	(spiders)	(X)Araneida (10 or more spp.)	100	89.5	
Pr	(lacewings)	(X)Chrysopidae (3 native, 1 exotic spp.)	90.5	79.0	
Pr	(true bug)	*Nabis oscillans* (Nabidae)	90.5	73.7	3-13
Pr	(beetles)	(X)Staphylinidae spp.	81.0	84.2	
Pa	(wasps)	(X)Chalcidoidea (10 or more spp., most of them native)	81.0	73.7	
Pr	(beetles)	Carabidae spp.	71.4	84.2	
Pr	(fly)	*Campsicnemus* sp. (Dolichopodidae)	52.4	26.3	
Pr	(fly)	*Campsicnemus setiger* or *modicus* (Dolichopodidae)	52.4	15.8	
Pa	(wasps)	(X)Diapriidae spp.	47.7	10.5	
Pa	(wasps)	*Seriola* sp. (p.) (Bethylidae)	42.9	68.4	
Pa	(fly)	*Pipunculus* sp. (p.) (Pipunculidae)	33.3	31.6	
Pa	(wasps)	(X)Scelionidae spp.	33.3	26.3	
Pr	(fly)	*Campsicnemus crispatus* (Dolichopodidae)	28.6	26.3	
Pr	(true bug)	*Koanoa hawaiiensis* (Miridae)	28.6	21.1	
Pr	(fly)	*Eurynogaster* sp. (p.) (Dolichopodidae)	28.6	10.5	

Pr	(true bug)	*Oechalia* sp. nr. *acuta* (Pentatomidae)	19.0	36.8	
Pr	(true bug)	**Orthotylus azalais* (Miridae)	19.0	10.5	
Pr	(true bug)	*Pseudoclerada morai* (Miridae)	9.5	5.3	
Pa	(wasps)	Dryinidae spp.	9.5	5.3	
Pa	(wasps)	Cynipidae spp.	4.8	21.1	
Pr	(ant)	XFormicidae sp.	4.8	10.5	
Pr	(beetle)	*Acritus minor* (Histeridae)	4.8	5.3	
Pr	(fly)	*Campsicnemus loxothrix* (Dolichopodidae)	38.1	0	
Pr	(fly)	*Eurynogaster kauaiensis* (Dolichopodidae)	33.3	0	
Pr	(lacewings)	(X)Hemerobiidae spp.	9.5	0	
Pr	(fly)	*Lispocephala* sp. (p.) (Muscidae)	9.5	0	
Pr	(wasp)	Vespidae sp.	4.8	0	
Pr	(true bug)	*Psallus sharpianus* (Miridae)	0	47.4	3-12
Pr	(beetle)	Cucujidae (mostly *Parandrita* spp.)	0	26.3	
Pa	(stylopid)	*Elenchus* sp. (Strepsiptera)	0	15.8	
Pa	(wasps)	(X)Ichneumonidae spp.	0	15.8	
Pr	(false scorpion)	Pseudoscorpionida sp.	0	10.5	
Pa	(wasps)	(X)Braconidae spp.	0	5.3	
Pr	(true bug)	*Psallus luteus* (Miridae)	0	5.3	3-12
Pr	(beetle)	X*Rhizobius ventralis* (Coccinellidae)	0	5.3	3-12
Pr	(beetle)	X*Scymnus bilucernarius* (Coccinellidae)	0	5.3	
Pr	(true bug)	*Lasiochilus denigratus* (Anthocoridae)	0	5.3	
Pr	(thrips)	X*Karnyothrips flavipes* (Phlaeothripidae)	0	5.3	3-12
Saprophagous					
D	(barklice)	Psocoptera spp.	100	100	
D	(mites)	(X)Acari spp.	100	73.7	
D	(fly)	*Forcipomyia* 2 spp.	71.4	73.7	
D	(springtails	(X)Collembola 10 spp.	71.4	47.4	
F	(fly)	*Bradysia* sp. #8 (Sciaridae)	66.7	68.4	3-16
F	(fly)	*Ctenosciara hawaiiensis* (Sciaridae)	61.9	68.4	3-16
F	(fly)	*Hyperlasion magnisensoria* (Sciaridae)	38.1	42.1	3-16

TABLE 6-3. *(Continued)*

Life form		Species or taxon	Frequency in samples on ʻōhiʻa	koa	Mauna Loa transect figure
F	(beetle)	X*Lathridius nodifer* (Lathridiidae)	33.3	26.3	
F	(beetle)	*Cis procatus* (Ciidae)	28.6	47.4	
F	(fly)	*Lycoriella hoyti* (Sciaridae)	28.6	36.8	3-16
F	(thrips)	***Haplothrips davisi* (Phlaeothripidae)	23.8	42.1	
D	(isopod)	XIsopoda sp.	19.1	84.2	3-13
F	(beetle)	*Cis* sp. #786 (Ciidae)	19.1	52.6	
F	(fly)	*Sciara prominens* (Sciaridae)	19.1	26.3	3-16
D	(cricket)	*Paratrigonidium* sp. (p.) (Gryllidae)	19.1	5.3	
F	(beetle)	*Cis setarius* (Ciidae)	14.3	47.4	
D	(fly)	*Psychoda* spp. (Psychodidae)	14.3	10.5	
D	(beetle)	Nitidulidae spp.	4.8	68.4	
F	(thrips)	***Haplothrips rosai* (Phlaeothripidae)	4.8	5.3	3-14
F	(thrips)	*Neurisothrips carteri* (Thripidae)	4.8	5.3	3-12
D	(fly)	*Limonia* spp. (Tipulidae)	33.3	0	
D	(bristle tail)	*Machiloides heteropus* (Machilidae)	4.8	0	
F	(thrips)	*Haplothrips laticomis* (Phlaeothripidae)	4.8	0	
F	(beetle)	*Cis cognatissimus* (Ciidae)	0	31.6	
F	(fly)	*Bradysia setigera* (Sciaridae)	0	23.8	3-16
D	(fly)	(X)Phoridae spp.	0	21.1	
F	(beetle)	*Cis signatus* (Ciidae)	0	10.5	
D	(millipede)	(X)Diplopoda spp. (1 native, 1 exotic sp.)	0	10.5	
F	(thrips)	*Conocephalothrips tricolor* (Urothripidae)	0	5.3	
F	(thrips)	*Haplothrips* sp. (Phlaeothripidae)	0	5.3	
F	(thrips)	*Haplothrips mauiensis* (Phlaeothripidae)	0	5.3	
W	(beetle)	XBostrichidae sp.	0	5.3	

F	(beetle)	*Cis serius* (Ciidae)	0	5.3
Perching or ecological role undetermined (transients)				
P	(fly)	*Drosophila & Scaptomyza* spp. (Drosophilidae)	76.2	79.0
P	(thrips)	*Neurisothrips* sp. (Thripidae)	42.9	47.4
U	(fly)	(X)Calliphoridae spp. (2 exotic, 1 native sp.)	15.8	47.4
U	(fly)	*Asteia apicalis* (Asteiidae)	15.8	15.8
U	(fly)	Anthomyidae spp.	9.5	10.5
U	(fly)	Cecidomyidae spp.	9.5	5.3
P	(true bug)	*Nesiomiris* sp. (Miridae)	4.8	15.8
P	(true bug)	*Orthotylus kanakanus* (Miridae)	4.8	10.5
U	(beetle)	Ptiliidae sp.	0	21.1
U	(fly)	Chironomidae spp. (prob. 2 species)	0	15.8
U	(beetle)	Orthoperidae sp.	0	5.3
U	(beetle)	Elateridae spp.	0	5.3
U	(fly)	X*Leptocera mirabilis* (Sphaeroceridae)	0	5.3

Note: Symbols are explained in figure 6-20.
X = exotic species, (X) = probably includes some exotic species, * = sometimes also phytophagous, ** = may also be predaceous, (p.) = may include more than one species, u = present with no apparent function on this tree (transient), P = percher, U = ecological role undetermined. These last two groups are collectively referred to as transients on figure 6-20.

Canopy arthropods of the Kīlauea forest, which were also recorded in the distribution analysis along the Mauna Loa transect, are identified in table 6-3 by the figure number on which their distributions are diagrammed in chapter 2 to permit an assessment of the similarities and differences between the two IBP sampling areas.

Twenty-four phytophagous arthropod taxa were collected in the Kīlauea rain forest. The nymphs of the Delphacidae and *Oceanides* species are excluded from this enumeration because their respective adults are listed also. Sixteen of the phytophages could be identified to the species level. The others may contain one or more species in their higher taxon. This complicates the interpretation of table 6-3 somewhat, but the data represent the best estimate that can be obtained to date. Overall seven taxa (29 percent) were common to both host trees. The phytophagous arthropods in the two host tree communities therefore have a relatively low similarity. If one compares the identified native species only, the similarity decreases further. Fourteen or fifteen native species were unique to ʻōhiʻa and only three or four were unique to koa, indicating perhaps that ʻōhiʻa is the evolutionarily older native tree species in the islands. Also, koa is inhabited with more exotic phytophages than ʻōhiʻa.

The Lepidoptera larvae or caterpillars are an important phytophagous group, and, if identified as to the species level, the majority of them would be host specific. This primarily defoliating group might include other life forms, such as leafminers, twig borers, detritivores, bud and inflorescence devourers, and even raptorial predators. Two of the taxa common to both trees have no apparent function on the alternate host tree. They are identified as such in the table.

Two native taxa unique to ʻōhiʻa were very abundant. These are the Psyllidae (three species of the gall-making *Trioza* and two species of non-gall-making *Kuwayama*) and *Sarona adonias*. Unique to ʻōhiʻa and rather frequently found on them were also the endemics *Proterhinus tarsalis* and *Oceanides vulcan*. Several others, including another endemic *Proterhinus* borer *(P. ferrugineus),* were found only on ʻōhiʻa, but with low frequency (under 10 percent).

Lepidoptera larvae make up the bulk of the arthropod biomass of the koa samples. These are predominantly *Scotorythra* species, two or three species of which are restricted to koa in the area. These are not shown separately on the list because they could not be identified with certainty. In contrast to ʻōhiʻa, four phytophagous arthropods on koa are exotics, including the very frequent *Psylla uncatoides,* which only recently was inadvertently introduced from Australia. Therefore ʻōhiʻa appears to harbor more host-specific endemics of the phytophagous arthropods than does koa in this forest, and koa also shows a greater number of exotic taxa (six) than does ʻōhiʻa (two only).

The anthophagous group includes only the flower-pollen and nectar-gathering species. It is a general niche analogous to the nectar-harvesting niche in the avian community. The group occupying this niche does not include flower consumers. Except for the thrips, most anthophagous arthropods are fast fliers or temporally active species, both of which would be missed in this spray collecting technique. However, the adaptable exotic honey bee, *Apis mellifera,* was taken once on ‘ōhi‘a, and a native solitary bee (*Nesoprosopis* sp.) was taken once on koa. Three species of *Neurisothrips* were found to occur relatively commonly on both trees; two of these were native.

The entomophagous (mostly predaceous) arthropod group was taxonomically the richest among the tree-crown associated fauna, with thirty-eight taxa. The taxonomic richness would be even greater if all taxa were identified to species level. More than half of the listed taxa (twenty-two) occurred on both host trees, which renders the entomophagous arthropod life form groups on both host trees very similar (58 percent). Their similarity increases to 73 percent if one considers the native taxa alone (21/29), but the undetermined taxa may inflate that ratio somewhat. A high similarity is generally expected of this secondary consumer group since their host relationship is only indirect. The weak flying dolichopodid flies were notably more abundant on ‘ōhi‘a than on koa, and two species of these, *Campsicnemus loxothrix* and *Eurynogaster kauaiensis,* were apparently restricted to ‘ōhi‘a. Possibly the weak flyers were found more abundantly on ‘ōhi‘a because of its tighter foliage, which may be more amenable for their form of prey stalking than is the open foliage of koa.

The saprophagous arthropod life form group is taxonomically also relatively rich, with thirty-three taxa identified in this study. Their communities on both host trees were also very similar (60 percent). This is not surprising either, since these arthropods feed mostly on dead organic materials or on fungi.

Nine taxa could not be ecologically identified as to their life form on the two trees (those with symbol U in the last group on table 6-3), including the endemic fly *Asteia apicalis*. These transients were much more frequent on koa than on ‘ōhi‘a. Five taxa were unique to koa, whereas no transient species was unique to ‘ōhi‘a.

Niche Differentiation in the Foliar Arthropod Communities

Within their four major life form groups, the tree-associated arthropod fauna could be assigned to ten general niches, ecological roles, or more specialized life form types. These were the defoliators, sapsuckers, seed predators, twig borers, flower pollen and nectar feeders, parasites, predators, fungivores, and two kinds of detritivores, one feeding on the outer surface of dead host material (or their epiphytes) and one feeding mostly inside dead wood. In addition, a category of transients, temporary or accidental members of the canopy arthropod communities, has been recognized.

Figure 6-20 portrays the life form spectra of the two canopy arthropod communities, one on *Metrosideros collina* (‘ōhi‘a) and the other on *Acacia koa*. For ‘ōhi‘a, nine general niches or specialized life form types are shown.

Koa was found to harbor ten. Koa lacked a seed predator group but supported two groups not found on ʻōhiʻa, the leaf miner (L), Agromyzidae, and the introduced detritivore (W), Bostrichidae, which operate in dead wood of koa crowns. The number of species or taxa for each general niche is recorded above the life form symbols on the diagram. The number is shown separately for native and exotic taxa, their sum being the total for the group.

A taxon identified with an X on the list was counted in the taxon enumeration as an exotic, but it was treated for the quantitative evaluation on the life form diagram, according to our best estimate, as half-native and half-exotic. In terms of taxon numbers, therefore, exotics are weighted more heavily than natives, but in terms of their population quantities, their proportion relative to the native forms is portrayed as realistically as possible in the bar graph on the life form diagram.

To obtain the relative quantities for each life form type, the frequency values given in table 6-3 were added. This is biologically not quite correct because in some cases species and higher-level taxa (which may contain one or more species) were added. A taxon currently identified as one entity may thus double when a second species with the same frequency is identified. This bias

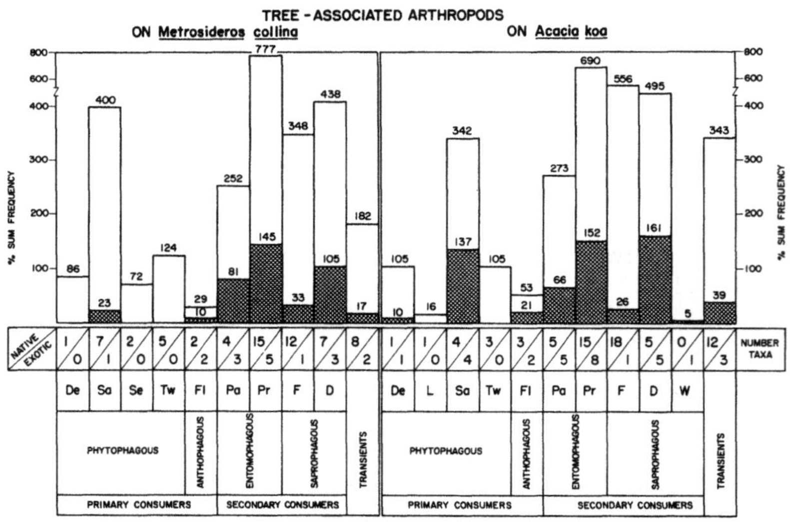

FIGURE 6-20. *Life form spectra of two canopy arthropod communities in the Kīlauea rain forest. Symbols: De = defoliators, Sa = sapsuckers, Se = seed predators, Tw = twig borers, Fl = flower pollen and nectar feeders, Pa = parasites, Pr = predators, F = fungivores, D = detrivores of plant and animal matter on host, W = detrivores in dead wood, L = leaf miners. Cross-hatch = combined frequency of exotics within respective life form group. Transients include perching arthropods and those with undetermined ecological roles.*

in the data must be taken into account, especially in some groups, such as the Lepidoptera, Araneida, Chalcidoidea, Acari, and Collembola, where more than ten species are included in each taxon. However, it is not expected that the relative magnitudes among the life form types shown would change very much with further taxonomic clarification.

The diagram shows that two life form types or general niches are occupied entirely by native species or taxa on each tree. These are in order of increasing magnitude on ‘ōhi‘a, the seed predators (Se) and the twig borers (Tw); on koa they are the leaf miners (L) and the twig borers (Tw). Thus koa apparently lacks native (as well as exotic) seed predators. Sampling bias may account for this, however, since seeds are in pods on koa and seed predators thus would be immune to spray. In contrast to koa, ‘ōhi‘a lacks native (as well as exotic) leaf miners (L), while both tree species are associated with several native twig borers. The fungivore niche (F) is almost entirely native on both host trees, with only one exotic species present and a good number of native species (12 on ‘ōhi‘a and 18 on koa). On both trees, the greatest number of exotic arthropod taxa is found among the predators (Pr) and detritivores (D). Together with the native forms, these represent the most diverse life form types. Other life form types with a high influx of exotics are the entomophagous parasites on ‘ōhi‘a and the phytophagous sapsuckers on koa. The detritivore niche in dead branches (W) was not searched for in this study because canopies to be sampled were selected for those with a minimum of standing dead branches, and this niche is internal, so that most of the life stages were missed. Nevertheless the niche was represented by an infrequent exotic Bostrichidae in koa. The number of species in the transient category appears to be very low.

Discussion and Conclusions

About 115 taxa were collected from the canopies of the two larger-sized native trees in the Kīlauea forest study site, of which sixty-eight were recognized at the species level. The faunal list would at least triple in length if all of the taxa were identified. This would refine the analysis and allow subdivision of some of the general niches, but we do not think it would change the gross pattern of figure 6-20. About ten general niches were found in the canopy arthropod community on each tree species.

Two niches appear to be vacant on ‘ōhi‘a: the leaf miner niche and the detritivore niche inside dead branches. However, both of these are internal and therefore were missed by this sampling method. Moreover, the leaf miner niche was also missed possibly because the Lepidoptera, which are a major group of native leaf miners, remain unidentified. Further, the sampling was designed to exclude canopies with dead branches, thus excluding the latter niche. Swezey (1954) lists two species of leaf-mining *Philodoria* restricted to ‘ōhi‘a. One of these is found on the island of Hawai‘i. In addition, the niche may be filled partially by the very successful native leaf gall makers (*Trioza* spp.), which we identified as sapsuckers in this study. Although not shown on the diagram, the detritivore niche inside dead branches on ‘ōhi‘a is at least partially filled by the native *Proterhinus* weevils, which may feed on both living and dead branches, by certain Lepidoptera, and by a native *Plagithmysus* borer.

On koa the one vacant niche was that of the arthropod seed predators. This, too, may be a failing of the sampling method to obtain arthropods feeding internally in the pods of this legume. However, this is an excellent example of resource partitioning and niche filling. ‘Ōhi‘a, with its many tiny (3–5 mm long) seeds packed in a capsule, seems admirably suited to a foraging, sapsucking seed predator, and indeed such species were found in this study. That specific niche does not exist on koa, a legume tree with a few very large seeds buried in a big pod, ideally suited to internal larvae, which may eat the store of seeds. The niche must be defined as the insect or potential colonizer would find it. It is known that this niche has been filled by native and exotic arthropods on koa trees outside this habitat—for example, along the Mauna Loa transect (Lanner, 1965). This observation correlates with the fact that along the Mauna Loa transect koa reproduction is almost never from seed (Spatz, TR 17), whereas reproduction from seed is the most prevalent regeneration mechanism of koa in this rain forest (Cooray, TR 44).

Some other niches that might occur in the canopy but that we did not encounter or recognize separately in our data are flower feeders (inflorescence devourers) represented by at least one Lepidoptera species on each tree; kleptoparasites, possibly represented by one native bee that usurps provisions of other native bee species; gall makers (three species occur on leaves, stems, and buds of ‘ōhi‘a only); the arthropod life forms associated with epiphytes; perchers, such as bark-resembling moths, that hide in the canopy for protection; and those occupied by mites. The number of general niches could be increased by a finer breakdown of the predator and parasite niches, once their ecological roles are better known. In addition, fleshy fruit feeders occur as a separate niche or life form group in the lower canopy on other tree species.

The ten or so general canopy arthropod niches are very differently filled in terms of the number of taxa represented. Even after correcting for unidentified groups and excluding some exotic species, there is a preponderance of entomophagous arthropod species, especially predators, on the two trees when compared to the number of phytophagous species. There are several interrelated reasons for this.

First, phytophagous species are often restricted to a given host, and for these species their entire niche is circumscribed by the canopy habitat. We collected large numbers of immatures of these species. Second, and most important, the general niche of the secondary consumers is often much larger than the canopies. These organisms range throughout the forest community according to their habitat, substrate, or prey-host preferences. Many spend most of their life cycles elsewhere in the forest—for example, many dolicopodids in the leaf litter or as epiphyte associates. The stage of the secondary consumers most often collected by our technique was the adult or dispersal stage, as the animals are moving from one substrate to another. Some of these are transients. With regard to the secondary consumers, we effectively sampled a much larger habitat than the tree canopy community. Third, the defoliator niche appears to be poorly filled. According to our data, it is represented only by the Lepidoptera, which include possibly fifteen to twenty native and one exotic species. This still relatively low arthropod diversity may be related to the low diversity of the canopy habitat; that is, only two tree species were sampled. Other defoliators would be found if the native trees in the lower canopy were included in the sample.

With regard to the invasion of exotics, the study has shown that the Kīlauea rain forest is still dominantly composed of native arthropod taxa. On ʻōhiʻa the influx of exotic arthropods was roughly 18 percent, on koa 27 percent. As one would expect, the primary consumer (or herbivore) niches on both trees are still more dominantly occupied by native arthropods (17/20 on ʻōhiʻa, 12/19 on koa) than are the secondary consumer niches (38/50 on ʻōhiʻa, 43/63 on koa).

The greater invasion of the secondary consumer niches by exotic arthropods can be related to the following reasons. Undoubtedly many of the intentionally introduced modern arthropod propagules have been predators and parasites because of the extensive and continuing efforts in biological control in Hawaiʻi. Moreover, secondary consumers generally have broader niches than primary consumers among the arthropods and thus have a better chance of becoming established. Also, because of the broader niche size of these species, we sampled them from a much larger community than is circumscribed by the tree canopy.

Surprising in this regard is the dominance and high diversity of the native arthropod taxa in the fungivore niche. One would have expected this general niche to be invaded by exotic taxa because of the cosmopolitan origin of most tree-associated fungi, which provide a less specialized food substrate. This is in contrast to the Mauna Loa transect, where three exotic sciarids have invaded this niche.

In comparison to continental tropical rain forests, the number of arthropod niches in our native island forest seems rather small. For example, our forest seems to lack a large tree-hole community, which elsewhere may house vertebrates and a specialized arthropod fauna. We also seem to lack fungus-culturing species and the insect-plant mutualists, such as the *Acacia*-ant community. With many additional tree species in the upper forest canopy as found typically in continental tropical forests, one would expect also a greater diversification among the defoliators than was possible to determine in our forest. Life forms such as leaf tiers, leaf skeletizers, leaf rollers, and leaf cutters could then be expected. Also, nonphytophagous arthropods, which gather leaves for nests and other nonfood purposes, did not occur in our island forest. Thus two factors result in the lower number of foliar-arthropod niches here: the lower tree species diversity and the absence of some specialized dominants among forest-tree arthropods.

CERAMBYCID TREE BORERS*

The tree borer niche in the Kīlauea rain forest was found rather sparsely occupied by several kinds of cerambycid beetles, all of which were endemic species belonging to the genus *Plagithmysus*. Two additional endemic cerambycids, *Parandra puncticeps* and *Megopis reflexa,* are known mainly from lower altitudes in the Hawaiian rain forest. Neither species was sampled, even though they both may reach at least the lower limit of the Kīlauea forest. They are secondary borers, boring only in dead, and usually decaying, wood. No exotic species of Cerambycidae were encountered.

A thorough search was made repeatedly along all four transects of the Kīlauea rain forest site. The search included inspection of trunks, foliage, fallen trees, and observations of beetles in flight. Trunks and branches were investigated by cutting through bark at suggestive scars, by observing the seepage of fluids, and by dissecting a representative number of twigs, branches, and logs of living and dead trees. A log-baiting experiment was carried out in which equivalent sections of living branches or stems (mostly 60 cm long and 8 to 10 cm in diameter) were cut from common trees in the forest, primarily from *Acacia koa, Metrosideros collina, Myoporum sandwicense,* and *Cheirodendron trigynum*. These were hung by wires in equal samples at intervals along transect 1 and inspected periodically. Traps were placed around some of the samples. However, these experiments gave negative results for primary borers. Fungus and other decay quickly made the bait logs unattractive to the beetles. Only some secondary scavenging lepidopterous larvae were detected.

Extensive Malaise trapping was done jointly with Dr. Steffan. This gave results, but these were very meager until a Malaise trap was elevated above the *Cibotium* cover.

Borer Abundance and Activity Patterns

Only five species of plagithmysines were found in the Kīlauea rain forest. Their names, hosts, and general quantitative importances are given in table 6-4.

The most common borers were *Plagithmysus varians* and *P. claviger,* which are associated with *Acacia koa*. On this tree species, most intact dead

---

*By J. L. Gressitt and G. A. Samuelson

TABLE 6-4. *Cerambycid beetles of the Kīlauea Rain Forest*

Borer species	Host-Plant species	Host importance in forest		Borer abundance
		Layer	Cover	
Plagithmysus vitticollis	*Rubus hawaiiensis*	II	0.9%	Occasional
P. vitticollis	*Vaccinium calycinum*	II	0.5	Scarce
P. perkinsi	*Myoporum sandwicense*	III	2.5	Moderately common
P. bishopi	*Pelea volcanica*	III	0.1	Very scarce
P. bishopi	*Pelea clusiaefolia*	III	0.1	Very scarce
P. varians	*Acacia koa*	V	68	Common
P. claviger	*Acacia koa*	V	68	Common

Note: II = tree fern layer, 0.5–5 m height, total cover 83.6 percent, III = low-stature layer, 5–10 m height, total cover 30.5 percent, V = emergent tree layer, over 15 m height, total cover 75 percent.

and dying branches in the live canopy probably contained borer galleries, with live borer larvae and pupae associated with some of the galleries. Even though these borers were rated as common in the Kīlauea forest, they averaged only about 20 percent as abundant as along the Mauna Loa transect. *P. varians* is the larger of the two borers, with a body length of 8 to 16 mm. It usually bores in the larger branches (from 2 cm diameter upward) and tree trunks. *P. claviger* is the smaller one, ranging from 5.6 to 6.5 mm in length. It works mostly in the thinner terminal branches and twigs of koa. *P. varians,* with its relationship to thicker wood, is more common in the lower part of the crown and the trunk area of the koa trees. Thus there is a rather well-defined resource partitioning among the two borers in the same host tree species. Large fallen branches attracted adults of *P. varians* in open koa colonies on the Mauna Loa transect, but in the closed Kīlauea rain forest such fallen branches were less effectively exploited by the borer.

Next in abundance was *P. perkinsi,* which occurred only on the low-stature tree *Myoporum sandwicense. P. perkinsi* is one of the largest of the plagithmysines known, with a length of 22 mm. It bores in the lower tree trunk in the cambium of living trees. This borer obviously has a considerable impact on the tree species in this forest, resulting occasionally in the dieback of portions of a tree.

Plagithmysus vitticollis was found occasionally on the tall native raspberry, *Rubus hawaiiensis,* which grows in locally disturbed places in the forest, often where a large koa tree has fallen. It bores in the living stems and kills portions of the host. This beetle also attacks *Vaccinium calycinum,* another common shrub in the Kīlauea forest. From other areas it is also known to attack the native trees *Perrottetia* and *Bobea. P. vitticollis* is larger (13.5 mm) than the average species of this genus. Among the five plagithmysines treated, it is the most generalistic in terms of host relationships.

Both *P. perkinsi* on its single host, *Myoporum sandwicense,* and *P. vitticollis* on its main host, *Rubus hawaiiensis,* in the Kīlauea forest were more common at the fringe of the forest in disturbed places than in the interior. These two host species were described as successional rather than climax species.

The rarest borer encountered was *P. bishopi*. It occurred on *Pelea,* of which there were two species in this forest (*P. volcanica* and *P. clusiaefolia*), and both constituted a very low cover (0.1 percent).

The absence of *P. bilineatus* in samples of 'ōhi'a in the Kīlauea forest is notable. This borer species is known to be host specific on *Metrosideros collina* subsp. *polymorpha,* the second most dominant tree in this forest.

Discussion and Conclusions

In general, the abundance of borer species, except *P. bilineatus,* correlated closely with the abundance of their hosts in the Kīlauea forest. Abundance levels of borers, however, tended to be lower on a given host in denser forest situations and higher at edges of the forest or in areas where disturbances had occurred *(P. perkinsi, P. vitticollis)*. This was also reflected in the abundance of the larger koa borer *(P. varians),* which had a greater ability to bore in large, fallen branches in open, drier forests on Mauna Loa transect and a lesser ability to do so in the closed Kīlauea forest. Such logs were usually more shaded in the Kīlauea forest, where it was also moister near the forest floor and where daytime temperatures were cooler. The frequency of adults of both koa borers *(P. claviger, P. varians)* was significantly less in shaded locations as shown in Malaise trap catches from open and closed forests, as well as in catches from different trap heights above the ground in closed rain forest.

The scarcity of adult plagithmysines in general near ground level in the Kīlauea forest, with its denser shade and consistently higher humidity, was enhanced by a tree fern stratum, which possibly also served to obstruct their flight paths. By contrast, these beetles were somewhat more common in kīpukas that lacked a tree fern stratum.

The absence of the 'ōhi'a borer *(P. bilineatus)* in the Kīlauea forest samples is not easily explained. It may be related to the pristine condition of the forest and to the absence of factors that have triggered explosive population increases of the borer elsewhere on the island (Papp et al., 1979).

In spite of their general scarcity, the plagithmysine borers are important to native Hawaiian forests in which they may hasten plant succession. They are the earliest borers to attack trees when these become physiologically weakened from old age, competition, or other causes. All other borers are secondary to these beetles. The plagithmysine borers have not been found to have a similar importance in exotic tree plantations.

NICHE DIFFERENTIATION IN THE AVIAN COMMUNITY*

For Hawaiian conditions, the Kīlauea rain forest contains a rather rich bird fauna. In a first census, Berger (TR 11) recorded nine endemic bird species in 1972: six honeycreepers (Drepanididae) and three other endemic birds. The six honeycreepers were the ʻapapane *(Himatione sanguinea)*, the ʻiʻiwi *(Vestiaria coccinea)*, the ʻamakihi *(Loxops virens virens)*, the ʻākepa *(Loxops coccinea coccinea)*, the creeper *(Loxops maculata mana)*, and the ʻakiapōlāʻau *(Hemignathus wilsoni)*. The three other endemics were the ʻio or Hawaiian hawk *(Buteo solitarius)*, the ʻelepaio *(Chasiempis sandwichensis sandwichensis)*, and the ʻōmaʻo or Hawaiian thrush *(Phaeornis obscurus obscurus)*. Berger (TR 11) also noted two exotic birds as occurring in relatively large numbers, the Japanese white-eye *(Zosterops japonica)* and the red-billed leiothrix *(Leiothrix lutea)*.

Subsequent studies confirmed these eleven species as well established in the forest. In addition, four other exotic bird species were observed in small numbers in the forest: the house finch *(Carpodacus mexicanus)*, the cardinal *(Cardinalis cardinalis)*, the melodious laughing-thrush *(Garrulax canorus)*, and the spotted dove *(Streptopelia chinensis)*. Historic, taxonomic, distributional, and general behavioral information on all of these fifteen species are given in Berger (1972).

Much has been said about adaptive shifts in the Hawaiian honeycreepers, which are believed to have evolved from a single ancestral species. Such information has been deduced primarily from the occurrence of members of the endemic honeycreeper family in a wide range of Hawaiian forest habitats and from their differentiation of bill shapes. These factors suggest considerable specialization and a high degree of niche differentiation among the endemic bird fauna.

With the decline of natural forests and the introduction of exotic birds, an increased species packing could be expected in certain areas, such as the Kīlauea rain forest. Beyond determining bird population densities, the objective of this study was to record the foraging and spatial use pattens for the avian community in an attempt to explain how the bird species relate to each other and to the forest as a whole.

Sampling Procedure

Monthly censuses were made along alternate transects in the Kīlauea study site from December 1972 to July 1973 and again from August 1974 to March

---

*By Sheila Conant

1975. The method was the same as used at the other IBP sites. In addition, the birds' foraging behavior was observed during many hours of separate walks and stationary observations along the transects and the unused logging road reaching into the study site. Only ten of the fifteen bird species were abundant enough to record their feeding trends with sufficient reliability. For the others, best estimates were made from observations elsewhere. Nine more or less self-explanatory foraging patterns were recognized in this study: aerial hunting (extrapolated from observations elsewhere for the Hawaiian hawk), bark drilling (observed for the 'akiapōlā'au), bark gleaning (observed for several species), foliage gleaning (one of the more common insect-feeding activities observed), fruit harvesting (another common activity observed for several species), nectar harvesting (observed many times for several honeycreepers), seed harvesting (extrapolated only, for certain exotic species), sallying (insect hunting in the air observed only for the 'elepaio), and ground gleaning (in this case arthropod and mollusc feeding, observed also only for the 'elepaio).

Approximately one to two hours of observation time was required for each of the ten more common bird species to obtain twenty minutes of foraging data, the minimum considered necessary for providing a reliable indication of a bird's typical feeding activity in this forest. The actual time of feeding records taken for each species varied from twenty-one to seventy-five minutes.

For determining the spatial utilization of the forest by the bird species, the feeding data were recorded in three height strata or canopy levels: below 7 m, from 7 to 18 m, and above 18 m. Another spatial segregation recognizable and recorded for the birds' feeding areas was the crown perimeter of a tree or its outer crown area, the inner area of a tree near the trunk, and the middle crown section between these two areas. A special effort was made to determine the tree and shrub species on which the birds were found to feed at each time for detecting possible bird-plant associations.

Bird Population Quantities and Activity Patterns

Table 6-5 lists the fifteen bird species and their estimated population numbers for the 80 ha study plot as computed from the repeated censuses. The birds are arranged in five vertical distribution groups according to their major area of activity in the Kīlauea forest canopy. A glance at their population numbers shows that 'apapane *(Himatione sanguinea)* is by far the most abundant forest bird. This endemic honeycreeper outnumbers two other endemic birds, the 'ōma'o or Hawaiian thrush *(Phaeornis o. obscurus)* and the 'elepaio *(Chasiempis s. sandwichensis)*. Two of the exotic birds, the Japanese white-eye *(Zosterops japonica)* and the red-billed leiothrix *(Leiothrix lutea)* are of comparable density (300 to 400 birds per 80 ha) to the second most important endemics. Two other honeycreepers rank third in quantitative importance by being quite common (100 to 300 birds per 80 ha), the 'i'iwi *(Vestiaria coccinea)*

TABLE 6-5. *Bird Species, Their Life Forms, and Quantities by Canopy Distribution Group*

Life Form	Species	Density (birds/80ha)
Restricted to upper canopy		
Carnivore, AH	*Buteo solitarius* Hawaiian hawk, ʻio (Accipitridae)	2
Ranging through upper and middle canopy		
Nectarivore-insectivore, NH, FG	*Vestiaria coccinea* ʼiʼiwi (Drepanididae)	262
Nectarivore-insectivore, NH, FG	*Himatione sanguinea* ʼapapane (Drepanididae)	1724
Seedeater-insectivore, SH, FG	X*Carpodacus mexicanus* house finch (Fringillidae)	2
Occupying full canopy range		
Insectivore, FG	*Loxops c. coccinea* ʻākepa (Drepanididae)	34
Insectivore, BG	*Loxops maculata mana* creeper (Drepanididae)	40
Nectarivore-insectivore, NH, FG	*Loxops v. virens* ʻamakihi (drepanididae)	172
Insectivore, BD, BG	*Hemignathus wilsoni* ʻakiapōlāʻau (Drepanididae)	28
Frugivore-insectivore, FH, FG	*Phaeornis o. obscurus* Hawaiian thrush, ʻomaʻo (Turdidae)	376
Ranging through lower and middle canopy		
Insectivore, FG, BG, GG, S	*Chasiempis s. sandwichensis* ʻelepaio (Muscicapidae)	386
Nectarivore-insectivore-frugivore, FG, NH, BG, FH	X*Zosterops japonica* Japanese white-eye (Zosteropidae)	352
Seedeater-insectivore, SH, FG	X*Cardinalis cardinalis* cardinal (Fringillidae)	2
Restricted to lower canopy		
Seedeater, SH	X*Streptopelia chinensis* spotted dove (Columbidae)	<.5
Frugivore-insectivore, FH, FG	X*Leiothrix lutea* red-billed leiothrix (Timaliidae)	304
Frugivore-insectivore, FH, FG	X*Garrulax canorus* melodious laughing-thrush (Timaliidae)	2

Note: Foraging patterns: AH = aerial hunting, BD = bark drilling, BG = bark gleaning, FG = foliage gleaning, FH = fruit harvesting, NH = nectar harvesting, SH = seed harvesting, GG = ground gleaning, S = sallying, X = exotic species.

and the ‘amakihi *(Loxops v. virens)*. The remaining three honeycreepers—the ‘ākepa *(L. c. coccinea)*, the creeper *(L. maculata mana)*, and the ‘akiapōlā‘au *(Hemignathus wilsoni)*—were less common. Six species can be considered rare or sporadic. These had population densities of fewer than ten for the study site. Among these are the other five exotics and one endemic bird, the ‘io or Hawaiian hawk *(Buteo solitarius)*.

Table 6-5 shows the major life forms of each bird species, which are based on the five predominant types of food material taken by the bird in question:

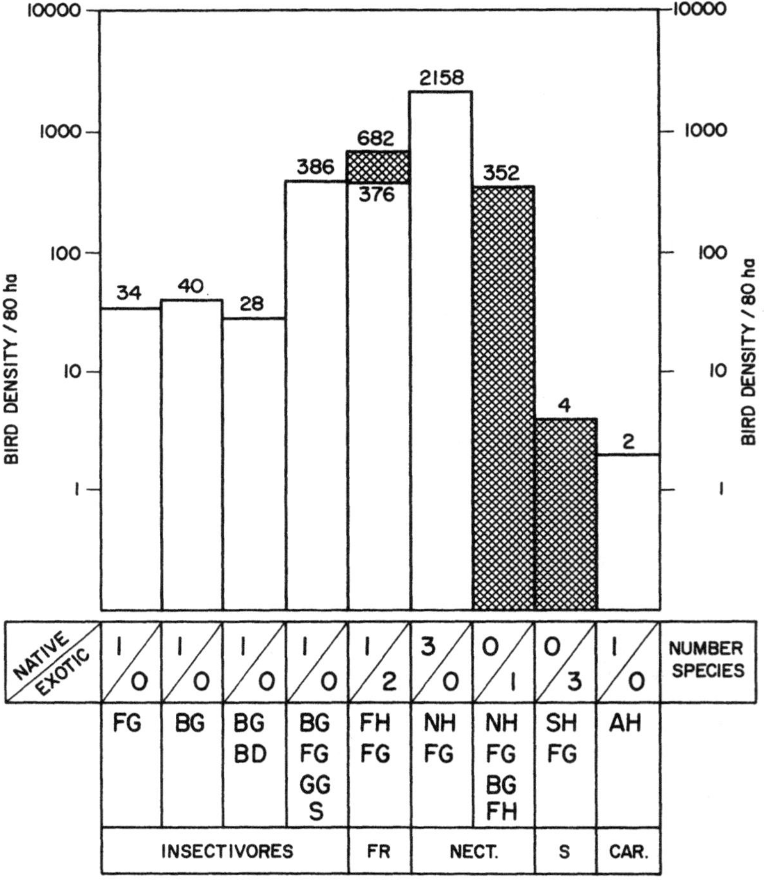

FIGURE 6-21. *Life form spectrum of avian community in Kīlauea rain forest. Cross-hatched marking refers to exotic birds and their densities.* Foraging patterns: *FG = foliage gleaning, BG = bark gleaning, BD = bark drilling, GG = ground gleaning, S = sallying, FH = fruit harvesting, NH = nectar harvesting, SH = seed harvesting, AH = aerial hunting*. Food relations: *Fr = frugivore-insectivore, Nect. = nectarivore-insectivore, S = seedeater and seedeater-insectivore, Car. = carnivore.*

flower nectar, fruits, seeds, insects, and flesh. The order in the name combination indicates which type of food material is the more prevalent for birds that feed on several kinds of food. Added to these major life forms are the foraging patterns (in symbols) as observed for the common birds or estimated for the sporadic birds in this study.

Based on this table, a life form spectrum of the Kīlauea avian community has been prepared (figure 6-21). The spectrum reflects nine feeding niches as determined in this study from combining the five basic food relations and the nine foraging patterns. The four endemic, pure insectivores do not overlap much in their feeding niches because their foraging patterns differ. Included in this group are the three less common honeycreepers (the ‘ākepa, the creeper, and the ‘akiapōlā‘au) and the relatively abundant and versatile ‘elepaio. A feeding niche overlap is seen in the frugivore-insectivore niche, which includes one endemic bird, the Hawaiian thrush, and two exotic birds, the red-billed leiothrix and the melodious laughing-thrush. The laughing-thrush is as yet so sporadic in this forest that its competition for the same feeding niche as the Hawaiian thrush can be considered negligible. Considerable overlap in their feeding niches is indicated among the nectarivore-insectivores, which include the three more common endemic honeycreepers—the ‘apapane, the ‘i‘iwi, and the ‘amakihi—and the exotic Japanese white-eye. The Japanese white-eye displays a more versatile foraging pattern, which may partially explain its success as an exotic invader although it must face competition for flower nectar with three well-represented native species. In addition, the seedeater niche appears to be a formerly vacant one that now is occupied by three exotic bird species, although all three are as yet very rare in the Kīlauea forest. Among them, the cardinal and the house finch may have the potential to become more important. The last feeding niche on the diagram is occupied by a carnivore, the Hawaiian hawk, apparently the only native vertebrate carnivore in the forest. The carnivore niche is now occupied also by exotic mammals.

Spatial Segregation of Bird Species in the Forest

While feeding niche overlap is clearly indicated for several bird species on the life form diagram, spatial segregation can be used as another parameter to elucidate their competitive relationships. Five spatial distribution groups were recognized on the basis of where in the canopy the birds spend most of their feeding time: birds more or less restricted to the upper canopy in their feeding, birds feeding primarily in the upper and middle canopies, full-range canopy users, lower to middle canopy users, and birds restricted to the lower canopy. Presence symbols are shown in table 6-6 for the five sporadic bird species, for which no feeding time was recorded. The Hawaiian hawk can be expected to use the full range of the canopy, including the forest floor, for its hunting ground, but it was listed here as restricted to the upper canopy because of its

TABLE 6-6. *Vertical Distribution of Bird Activity Areas, Kīlauea Rain Forest*

Birds	Canopy levels (%) Lower (< 7 m)	Middle (7–18 m)	Upper (> 18 m)	Observation time (mins)
Restricted to upper canopy				
Buteo solitarius	—	—	+	nr
Primarily in upper to middle canopy				
Vestiaria coccinea	—	11	89	37
Himatione sanguinea	1	45	54	38
X*Carpodacus mexicanus*	—	+	+	nr
Full-range canopy users				
Loxops c. coccinea	10	21	69	72
Loxops maculata mana	4	41	45	29
Hemignathus wilsoni	39	47	14	43
Phaeornis o. obscurus	32	63	5	22
Loxops v. virens	38	57	5	41
Lower to middle canopy users				
Chasiempis s. sandwichensis	95	5		32
X*Zosterops japonica*	74	26		25
X*Cardinalis cardinalis*	+	+		nr
Restricted to lower canopy				
Leiothrix lutea	100			21
X*Streptopelia chinensis*	+			nr
X*Garrulax canorus*	+			nr

Note: X = exotic birds, nr = no feeding time recorded, + = presence observed repeatedly.

tendency to carry its prey to that level for feeding. As table 6-6 shows, there is a rather clear segregation for the red-billed leiothrix, which was found to feed only in the shrub and low-stature tree canopy up to 7 m height from ground level. This indicates that this well-established exotic bird does not seem to interfere much with the important native frugivore-insectivore, the Hawaiian thrush *(Phaeornis o. obscurus),* which uses the full range of the canopy. The other quantitatively important exotic species, the Japanese white-eye *(Zosterops japonica),* also feeds more commonly in the lower canopy area and thus does not seem to form a significant threat to the native nectarivore-insectivores, which are the three honeycreepers, the ʻapapane *(Himatione sanguinea),* the ʻiʻiwi *(Vestiaria coccinea),* and the ʻamakihi *(Loxops v. virens).* The ʻiʻiwi feeds mostly in the upper canopy; the ʻapapane overlaps spatially with the ʻiʻiwi, but uses more of the middle canopy layer than the ʻiʻiwi, and the ʻamakihi uses the full canopy range and therefore is the endemic bird species that overlaps most strongly in its feeding niche with the Japanese white-eye.

The horizontal within-crown analysis shows that the ten more common bird species can be divided into three spatial groups (table 6-7): those feeding

TABLE 6-7. *Horizontal Distribution of Bird Activity Areas within Tree Crowns*

Birds	Inner	Crown area (%) Middle	Outer	Observation time (mins)
Crown-perimeter users				
Himatione sanguinea	4	3	93*	38
Loxops c. coccinea	4	4	92*	72
Vestiaria coccinea	10	15	75*	37
Zosterops japonica	4	12	84*	24
Loxops v. virens	19	20	61*	41
Crown-interior users				
Loxops maculata mana	82*	9	9	29
Leiothrix lutea	75*	14	11	21
Hemignathus wilsoni	58*	27	15	43
Full-range users				
Chasiempis s. sandwichensis	63*	2	35*	32
Phaeornis o. obscurus	36*	40*	24*	22

Note: Values with an asterisk emphasize areas of concentration.

predominantly at the crown perimeter of trees, those using mostly the crown interior, and those using the full crown space. None of the birds was restricted to the middle part of the crown. Two of the crown-interior users are pure insectivores and endemic drepanidids, the creeper and the ‘akiapōlā‘au. The third is the exotic red-billed leiothrix, which feeds on fruits and insects. The crown perimeter users are all the nectarivore-insectivores (the ‘apapane, the ‘i‘iwi, and the ‘amakihi) and the Japanese white-eye. The fifth outer-crown user is the ‘ākepa, a pure insectivore. The full-range users include two endemics, the ‘elepaio and the Hawaiian thrush. The ‘elepaio is the most versatile endemic insectivore. It employs sallying and ground gleaning among other feeding modes. The Hawaiian thrush, the only endemic frugivore-insectivore, is also a more generalistic feeder.

In general, the spatial segregation analysis shows a rather full complementation in the use of the canopy by the Kīlauea forest birds. However, there is considerable spatial overlap, particularly among the endemic forest birds.

Bird-Plant Associations and Foraging Patterns

The outcome of the bird-plant association analysis is given in table 6-8. Accordingly the ten more common Kīlauea forest bird species can be classified into five groups.

TABLE 6-8. *Percentage of Foraging Time Spent on Different Plants by Bird Species in Kīlauea Forest Plot*

Birds	Macro-phanero-phytes[a] 1	2	Mesophanerophytes[b] 3	4	5	6	Micro- and Nanophanero-phytes[c] 7	8	9	Observation time (mins)
Foraging exclusively on ʻōhiʻa										
Himatione sanguinea	100									38
Zosterops japonica[d]	100									24
Spending much time on koa										
Hemignathus wilsoni		62	34		4					43
Loxops c. coccinea	71	26	2		1					72
Loxops v. virens	50	22	9	2	4	5				37
Foraging primarily on ʻōhiʻa and *Myoporum*										
Vestiaria coccinea	82		11	2		5				37
Loxops maculata mana	71	2	27							29
Chasiempis s. sandwichensis	39		46		2	6	2		5	32
Foraging primarily on *Myoporum* and *Cheirodendron*										
Phaeornis o. obscurus	3	1	55	23					17	21
Primarily on *Rubus hawaiiensis* and other low, woody plants										
Leiothrix lutea[d]			13				14	66	7	21

[a]1 = *Metrosideros collina* subsp. *polymorpha* (ʻōhiʻa), 2 = *Acacia koa* (koa). [b]3 = *Myoporum sandwicense*, 4 = *Cheirodendron trigynum*, 5 = *Coprosma rhynchocarpa*, 6 = Other mesophanerophytes (*Ilex anomala*, *Myrsine lessertiana*, *Pelea* spp.). [c]7 = *Cibotium* spp. (tree ferns), 8 = *Rubus hawaiiensis*, 9 = Others (*Vaccinium calycinum*, *Rumex giganteus*).
[d] = Exotic bird species.

The first group includes two species found to forage exclusively on ʻōhiʻa. Among these is the most abundant honeycreeper, the ʻapapane, and the most successful exotic bird, the Japanese white-eye. Both species exploit the nectar of the ʻōhiʻa flower, which is probably the most abundant and nearly always available flower in this forest. Thus bird abundance and tree-flower abundance are positively correlated here. The spatial analysis showed that the ʻapapane uses mostly the upper- and middle-canopy areas of the trees, while the Japanese white-eye uses mostly the lower to middle canopies. Therefore significant competitive overlap occurs only in the middle range of the ʻōhiʻa canopies. Both bird species use mostly the crown perimeter area, where the flowers are located. Occasionally both species have been observed foraging on plant species other than ʻōhiʻa.

The second group of birds spends much time on the tall emergent tree species, *Acacia koa*. This group includes only native birds, the three honeycreepers: ʻakiapōlāʻau, ʻākepa, and ʻamakihi. Only the ʻakiapōlāʻau of these three was not seen to feed also on ‘ōhi‘a, while the other two spend more time feeding on ‘ōhi‘a than on koa. The greatest generalist among the three was the ‘amakihi, which, as a nectar and insect feeder, was seen to forage on several other tree and shrub species (table 6-8).

The third group includes three other endemic bird species—the ‘i‘iwi, the creeper, and the ‘elepaio—which were found to forage primarily on ‘ōhi‘a and *Myoporum sandwicense* (the naio tree). The ‘elepaio is the most versatile endemic insectivore in this forest.

The fourth group includes only one species, the Hawaiian thrush, which was found to spend most of its foraging time on two native trees of secondary quantitative importance, *Myoporum sandwicense* and *Cheirodendron trigynum* (the ōlapalapa tree). The Hawaiian thrush, the only endemic frugivore in the Kīlauea forest bird community, seems to be the main eater of the usually abundant fleshy fruits (berries) of the ōlapalapa tree.

The fifth group also includes only one species, the exotic red-billed leiothrix, which was restricted in its foraging to the lower canopy, where it fed mostly on the juicy Hawaiian raspberries and associated insects of the tall shrub, *Rubus hawaiiensis*. Thus the two important frugivores are well separated in their foraging behavior. They overlap only somewhat in their use of the naio tree *(Myoporum sandwicense)*. This overlap is probably most important when *Rubus* is not in fruit (September to May).

Discussion and Conclusions

This study on the niche differentiation in the Kīlauea forest bird community has shown that there is probably little direct competition among the native insectivores, which include three honeycreepers and one fly catcher (the ‘elepaio). These four species have rather diverse foraging patterns. They range throughout most of the forest canopy, but the ‘akiapōlā‘au employs bark drilling as an alternate insect-feeding technique to bark gleaning, the ‘ākepa uses foliage gleaning at the crown perimeters mostly, the creeper uses bark gleaning in the crown interiors mostly, and the ‘elepaio catches insects in the air and employs ground gleaning as an alternate insect-feeding technique. No exotic species has yet obviously disturbed this foraging pattern of the native insectivores.

The situation is very different among the nectarivores-insectivores. This life form group includes the three more abundant native honeycreepers—the ‘apapane, the ‘i‘iwi, and the ‘amakihi—and the exotic Japanese white-eye. There is only limited spatial segregation in their use of the ‘ōhi‘a flower nectar insofar as the ‘i‘iwi tends to exploit the upper flowers most, the abundant ‘apapane overlaps considerably with the ‘i‘iwi and uses more of the mid-canopy flowers, and the ‘amakihi overlaps heavily at the mid-canopy and exploits the lower-canopy flowers. In this vertical canopy segment, the ‘amakihi overlaps heavily with the Japanese white-eye. There is little doubt that these four species compete directly for 'ōhi'a flower nectar, and the limited spatial segregation pattern of the three endemic honeycreepers appears to be merely a current dynamic competitive equilibrium rather than a pattern of evolutionary specialization. This has become evident also from other observations.

Carpenter and MacMillan (1976a, 1976b) and Carpenter (1976) have demonstrated that the 'i'iwi *(V. coccinea)* will defend a feeding territory against other members of its species, as well as the 'apapane *(H. sanguinea)* and the 'amakihi *(L. v. virens)*. Several times I have observed an 'i'iwi chase an 'apapane and also an 'amakihi from a feeding tree even when a feeding territory was not established. 'Apapane sometimes chase 'amakihi from feeding areas under the same conditions. I have also observed certain shifts in foraging patterns and resource use by these species when they are feeding in close proximity. In the presence of the dominating species, the subordinate species may shift from harvesting nectar in the outer canopy where flowers occur, to gleaning insects from foliage, stems, and large branches in the inner and center portions of the canopy. However, none of these species have been observed to defend feeding areas against the Japanese white-eye *(Z. japonica)*. Van Riper (1976) reported this species to dominate the 'amakihi at sugar-water feeders, suggesting that the Japanese white-eye may compete in nature with the 'amakihi. More detailed field and experimental studies are needed to clarify whether the Japanese white-eye has a negative effect on the population density of the 'amakihi. It is also clear that the competitive relationships among the four nectarivores depend on the varying seasonal and year-to-year availability of the 'ōhi'a flowers.

Although the frugivore-insectivore life form group has been increased by two exotic bird species, there seems to be little interference among them. The only native frugivore, the Hawaiian thrush, exploits mostly *Myoporum* and *Cheirodendron* fruits, while the only successful exotic frugivore, the red-billed leiothrix, seems to exploit primarily the fruits of the Hawaiian raspberries in season. Only year-round monitoring of frugivore relationships can determine whether there is any interference among them. The second exotic invader in this life form group, the melodious laughing-thrush, is as yet too sporadic to be considered as a competitor with the Hawaiian thrush. The future role of the melodious laughing-thrush is unpredictable.

The exotic seedeaters apparently have begun to fill a vacant niche. Native seedeaters are known only from dry forests, and none seems to have occupied the rain forests in the past (Berger, 1972).

The carnivore niche, formerly occupied only by the Hawaiian hawk, has become invaded by introduced mammals, notably among them the mongoose *(Herpestes auropunctatus)*. However, this may have had little effect on the Hawaiian hawk, since mice and rats were introduced also, forming additional prey for the hawk. A strong effect of the mongoose and the roof rat *(Rattus rattus)* on the behavior and survival pattern of the other forest birds was expected.

COMMUNITY STRUCTURE OF INTRODUCED RODENTS AND CARNIVORES*

The greatest biological modification of the Kīlauea rain forest has been brought about by the unintentional introduction (or escape into the forest) of several mammals, none of which were part of this island ecosystem during its evolutionary history. Seven species of mammals were found to roam in this forest: the pig *(Sus scrofa),* the small Indian mongoose *(Herpestes auropunctatus),* the roof rat *(Rattus rattus),* the Norway rat *(R. norvegicus),* the house mouse *(Mus musculus),* the feral cat *(Felis catus),* and the feral dog *(Canis familiaris).* The presence of these mammal species was more or less expected from prior knowledge (Tomich, 1969a), but little was known about their relative abundance, their activity patterns, and their residence status in the forest. The activity pattern of the feral pig will be treated later. This study was concentrated on the other six species, which occupy two general niches in the community, the carnivore niche and the rodent niche.

Sampling Methods

Sampling consisted of systematic trapping for the small mammals on the four major transects from June 1971 to May 1973. Traplines were labeled with the same numbers as the respective transects (see figure 5-3). Three styles of traps were used, with spacing of sixty traps per transect at a standard 15 m, making lines 885 m in length extending into the study area from its northwestern end at 1,615 m elevation. Line I was an arrangement of forty rat live traps and twenty rat snap traps, with each third trap in the line a snap trap. Line II was similar, but with each third trap a mongoose live trap took the place of the snap trap. Rat traps were baited with fresh coconut; meat scrap (largely tallow) was used as bait in the mongoose traps. All mammals captured on line I were removed for autopsy in the laboratory and those from line II were examined alive (mongooses under ether anesthesia), tagged, and released at the site of capture. Lines III and IV were replicates of lines I and II. Each line was operated for three nights on a bimonthly schedule so that monthly samples came alternately from transects I and II or III and IV. Biological data taken in the laboratory were standard linear measurements, body weight, sex, age class, reproductive state, and in the case of *R. rattus,* coat color. Stomachs were preserved in 95 percent alcohol for analysis of contents. Released animals were weighed, age class was determined, reproductive state was estimated, and body color was noted in *R. rattus.*

---

*By P. Q. Tomich

Species Abundance and Activity Patterns

Table 6-9 gives an overview of the six mammal species with regard to their life form and relative abundance in the Kīlauea rain forest.

Carnivores

Among the three introduced carnivores, the small Indian mongoose, *Herpestes auropunctatus,* is the only nonferal species. It is a moderately common resident that is well established in the Kīlauea forest. Our sampling for this species with only sixty trap nights per month was adequate to gain some evidence of its place in the forest ecosystem. Mongooses seldom entered the rat live traps, and only three were captured in these traps in the entire study. This was on lines I and III, which had no mongoose live traps. Of thirty-one mongooses marked and released, thirteen (41.9 percent) were recaptured at least once (mean 2.6 captures). Movement data correspond to those obtained in long-term studies at lowland sites in Hawai'i (Tomich, 1969b). Six animals repeated capture in a single three-day trapping period, and thirteen were recaptured from two to four times one month or longer after the first capture. For

TABLE 6-9. *Introduced Rodents and Carnivores in Kīlauea Rain Forest*

Life form	Species	Quantitative evaluation[a]
Rodent	*Rattus rattus* (roof rat)	Abundant (460 individuals caught)
Rodent	*Rattus norvegicus* (Norway rat)	Extremely sparse (4 individuals caught)
Small rodent	*Mus musculus* (house mouse)	Sporadic (12 individuals caught)
Carnivore	*Canis familiaris* (feral dog)	Occasional (1 live sighting, 7 fecal scats/6 mo., occasional tracks)
Carnivore-omnivore	*Herpestes auropunctatus* (small Indian mongoose)	Moderately common (31 individuals caught)
Carnivore	*Felis catus* (feral cat)	Occasional (2 individuals caught in mongoose traps)

Note: In phylogenetic sequence.
[a]Quantities refer to the two-year study period and trapping system described in the text, unless otherwise stated.

thirteen records of movement by males, the mean was 354 m (extremes 90 to 675 m), and for six females the figure was 180 m (45 to 495 m). The mean time elapsed between captures was 5.4 months (extremes 1 to 18). Mongooses ranged freely in the study area, along lines II and IV where they had been marked, as well as on lines I and III.

A place for *H. auropunctatus* is well provided in the Kīlauea forest by numerous refuges among fallen trees and possibly in occasional crevices in the substrate and under the roots of trees, for secure shelter. Food sources—rats, birds, invertebrates, and fruits—appear to be adequate for a carnivore with omnivore tendencies. We were unable to describe details of food preferences of the mongoose, and can merely suggest that it thrives in the forest as it does in other regions of Hawai'i. Generally wherever rodents are abundant, they are commonly found in the food residues of the mongoose (Kami, 1964). Kīlauea may support an unusually dense population of the mongoose for such a high elevation. The present consensus is that it generally does not range higher than 2,200 m in Hawai'i. Mongooses occasionally molested live rats caught in cage traps, but we had no record of a dead rat being removed by a mongoose from a snap trap.

The other two carnivore species in the forest have escaped domestication. The feral cat, *Felis catus,* is widely distributed on Hawai'i, where it subsists on a large variety of animal foods (Tomich, 1969a). We made an occasional sighting of this species along the nine-mile road to the forest during the course of the study. Two cats were captured in traps baited for the mongoose, one in October 1971 and one in June 1972, in the interior of the study area on lines I and II. Both were adult females. The first cat was tagged and released, but the second was killed and examined in the laboratory. This animal weighed 1,850 g and was parous lactating. The stomach was full, packed with the remains of three successive meals. One item was an adult roof rat *(Rattus rattus)* represented by pieces of skin with black pelage and segments of the tail and feet. This rat was apparently taken as carrion because the remains were intermixed with numerous blowfly larvae (*Calliphora* sp.). A second item was a few feathers, trunk and legs, and some chunks of muscle tissue of a Hawaiian thrush *(Phaeornis o. obscurus)*. We could not determine whether the bird was a fledgling or an adult. Finally the cat had eaten a lump of tallow used as bait in the trap. Included also in the contents of the stomach were bits of detritus and two mature proglottids of a tapeworm, probably from the intestine of the rat, and seventy-eight nematode stomach worms, certainly as a parasite of the cat. The colon of the cat contained a fecal mass composed largely of hair and bone fragments of a rat (probably *Rattus rattus*) and some miscellaneous detritus.

The feral dog, *Canis familiaris,* ranges through Kīlauea Forest Reserve, probably with some regularity. The most frequent evidence of the dog was in the form of tracks and fecal droppings (scats) along the entrance trail to the four research transects. Between February and September 1972 we collected seven scat groups, each deposited characteristically on clumps of a native sedge

(Carex macloviana), which occurred in colonies along the trail. The scat specimens were preserved in 95 percent alcohol and later were dissected in the laboratory. One dog was observed by Ranjit Cooray on an afternoon in June 1972, deep in the forest along transect III. The dog maintained a brief curious interest in Cooray and then fled.

The roof rat *(Rattus rattus)* is consistently eaten by the feral dog and is its primary food in the forest. Each scat represented a primary meal of one or more *R. rattus*. The bulk of the remains was matted or felted rat fur. There was digestion of the fur, but fur was predominantly in fluffy tufts when the preserved scats were dried and dissected. In part of one scat, the fur remained attached to pieces of skin. In another, digestion was so advanced that the fur was reduced to a claylike fibrous material in which entire hairs were not prominent. There was obviously a gradation in digestion of the resistant parts of the rat, apparently as a result of the time and rate of digestive action. Data on rat remains are shown in table 6-10.

Determination of rat numbers was through enumeration of incisor teeth, which are particularly resistant to digestion. Ends and other fragments of long bones, vertebrae, scapulae, ribs, phalanges and nails, and skull parts were also detected. Each scat contained some fifteen to forty easily identifiable bones. Two eye lenses were found. One scat contained seven quills of pinfeathers, possibly from a fledgling or molting passerine bird. Elytra of one beetle were found, along with a few bits of plant detritus. Tomich (1969a) reports the roof rat as a food of the feral dog in other parts of the island.

Rodents

Among the three rodent species, the roof rat, *Rattus rattus,* was by far the most dominant. For this reason it was possible to obtain some detail on its population structure. The other two species, though sporadic or rare in occurrence, are also considered established residents in the Kīlauea rain forest.

Only four individuals of the Norway rat, *R. norvegicus,* were caught during the entire trapping effort—one in June 1971, two in March 1973, and a

TABLE 6-10. *Rat Remains in the Feces of Feral Dogs*

Collection date	Sample Number	Number and Age Class of Rats
February 1972	1	1 adult
May 1972	1	2 adult, 1 young
	2	1 adult
	3	1 young
	4	1 adult
June 1972	1	1 adult
September 1972	1	3 adult

final one on the last day of fieldwork in May 1973. Three were adult males weighing 253 g, 262 g, and 288 g, respectively; the other was an immature female weighing only 88 g. One specimen was found near the beginning of line II, and the other three in the mid-section of line III. The last three were caught over a distance of only 60 m, between March and May of 1973, but the first specimen, caught in June 1971, was at least 200 m apart from the rest. These data suggest local aggregations within this overall very small population of rats, but it is impossible to make any prediction about the dispersal pattern of such *R. norvegicus* aggregates in this rain forest.

The Norway rat is only marginally adaptable to a noncommensal existence in Hawai‘i except in some lowland areas, the wetter sugar-producing regions in particular. The range of the species in this regard has shrunk considerably in the past fifteen to twenty years—for example, on the Hāmākua (windward) coast of Hawai‘i Island. The elevation of the Kīlauea forest at 1,630 m is near the upper limit of this rat's distribution range in Hawai‘i. The extreme record for the state is at 1,770 m in the Hāmākua district, where the Norway rat occurred near water catchment facilities on the sparsely forested grazing lands of Mauna Kea (Tomich, 1969a).

The house mouse, *Mus musculus,* can also be considered a resident of the Kīlauea rain forest, though its occurrence is sporadic and then only in very small numbers. During June, July, and October 1971, we trapped eight individuals, and in August and September 1972 another four were captured. The pattern corresponds to a trend for peaks in populations and occasional plagues of the mouse in Hawai‘i during the summer periods (from July into November). The montane rain forest is not an especially suitable environment for the house mouse, which is generally associated with the dry to wet grasslands, shrublands, or parklands. The factor of altitude is apparently of little direct significance to the distribution of *Mus musculus* in Hawai‘i (Tomich, 1969a).

Body weights of seven adult males ranged from 10.0 to 18.6 g ($\bar{x}$ = 14.4) and that of five adult females ranged from 11.0 to 15.0 g ($\bar{x}$ = 12.4). All specimens were adult. The seven males were in breeding condition; two females were parous lactating and three were parous inactive. The mice tended to be clustered in the forest, and the twelve individuals were caught in only eight different traps on lines I, II and III (figure 5-3).

Population Structure of the Dominant Rodent

The roof rat, *Rattus rattus,* is abundant in the forest and is distributed throughout the study area. The rat snap trap proved to be the most efficient means of catching rats (table 6-11). This type of trap produced an overall catch of 9.1 rats per 100 trap nights. The rat live trap had an overall catch of 5.3 rats per 100 trap nights, but only 4.2 per 100 in lines where rats were removed compared to 6.4 per 100 in lines where rats were marked, released, and recap-

TABLE 6-11. *Summary of Two Years of Captures of Roof Rats in Kīlauea Rain Forest, by Trap Type*

Trap type	Trap nights	Removed lines I & III	Released lines II & IV	All
Rodent live	2,880	121 (4.2)[a]	183 (6.4)	304 (5.3)
Rodent snap	1,440	131 (9.1)	—	131 (9.1)
Mongoose live	1,440	—	25 (1.7)	25 (1.7)
Overall total				460
Year 1 total				233
Year 2 total				227

[a]Figures in parentheses represent catch of rats per 100 trap nights.

tured. The mongoose traps were not designed specifically for rats, and smaller animals could escape through the 2.5 cm mesh, but these traps did capture rats at a rate of 1.7 per 100 trap nights. Analyses are made with these differentials in mind.

In spite of removal of all rats from lines I and III, there was no depression of these local populations as the study progressed. As in lines II and IV where no rats were removed, the yearly totals of rodents taken remained stable, well within usual limits of variation that might be expected for the population. Overall captures—233 rats in year 1 and 227 in year 2—suggest a remarkable stability in the population of *R. rattus* and excellent adaptation to the ecological conditions at the site of the Kīlauea rain forest.

Mobility

The roof rat tends to be resident at a particular site and may generally confine its activities to a distinct home range, which is scent marked by adult males in particular, as a territory. Social structure, however, is not well understood in free-living populations.

We tagged and released 103 rats, of which 43 (41.7 percent) were recaptured at least once. Table 6-12 summarizes determinations of range, frequency of capture, and longevity. Rats caught successively in the same trap were regarded as having moved zero distance ($D = 0$), and this was treated as a frequency statistic. Females tend to be sedentary compared to males, as is observed from capture records extended in time and from frequent capture of a female in the trap of previous capture.

Of six rats that ranged the distance of 200 m or more between traps in adjacent traplines, all were males; the longest recorded move for a female was 90 m. The measure D (distance between successive captures) and the measure ARL (adjusted range length) are used as adapted by Tomich (1970).

TABLE 6-12. *Movements and Longevity of* R. rattus *as determined from Rats Marked and Released in Lines II and IV*

Repeats during 3-day trapping period				
Sex	*n*	*D = 0*	*Av. D (m)*	*Extremes*
M	9	3 (33.3%)	47.5	15–90
F	11	3 (27.3%)	37.5	15–60
Returns 1 month or longer between captures				
Sex	*n*	*D = 0*	*Av. D (m)*	*Extremes*
M	22	2 (9.1%)	92.3	15–300
F	22	7 (31.8%)	35.0	15–90
Adjusted range length (ARL)				
Sex	*n*		*ARL (m)*	*Extremes*
M	18		123	30–378
F	18		44	15–105
Number of captures				
Sex	*n*		*Mean captures*	*Extremes*
M	22		2.4	2–5
F	20		2.7	2–6
Months between captures				
Sex	*n*		*Mean months*	*Extremes*
M	22		3.9	2–12
F	22		3.8	2–10
Observed longevity				
Sex	*n*		*Mean months*	*Extremes*
M	18		4.8	2–18
F	18		4.7	2–10

Sex and Age Characteristics

Of 460 captures, including repeats and returns, we recorded 234 males and 226 females (50.9 percent males). Ratios in subsections of the data showed a similar slight excess of males. On lines I and III, where all rats were removed, the ratio was 133 to 120 (52.6 percent males), and on lines II and IV, where all rats were released, the ratio was 54 to 49 (52.4 percent males) among 103 individual rats tagged. Jackson (1962) has discussed this aspect of *R. rattus* in the Pacific and reports that either sex may be slightly in excess in various populations. Differentials in mortality by age class may alter sex ratios in time, but also differentials in behavior toward traps and in size of home range by sex may affect the accuracy of determining actual sex ratios in populations of wild rats. No evidence suggests that *R. rattus* in the Kīlauea rain forest is likely to exhibit other than approximate equality in numbers of males and females.

Male rats were predominantly adult. Of 133 taken in the two years and examined by autopsy for evidence of maturity (presence of motile sperm in the cauda epididymides), 119 (89 percent) were adult. There was no significant difference between year 1 and year 2. Of the 14 immatures identified, 10 occurred in September to December and four in April to July. For females, of 120 examined in the laboratory, 94 (78.3 percent), were adult as determined by past or current evidence of sexual activity, including estrus through a parous-inactive condition. The 26 immatures were distributed in all months except February to March, with a maximum of 9 in December.

Pregnancy and Litter Size

During year 1, 76.1 percent of females were adult ($n = 45$) and in year 2, 80.3 percent were adult ($n = 49$), as determined by autopsy of rats from lines I and III. Two periods of pregnancy occurred: August to November with 7 of 32 pregnant (21.9 percent) and February to June with 11 of 44 pregnant (25.0 percent). The annual prevalence of pregnancy was 19.1 percent of 94 adult females examined in the two years of study. The patterns of reproduction were similar in both years and do not bear separate analysis.

Litter size was determined from counts of embryos and from occasional new embryo scars that indicated recent births. Resorption of embryos was negligible. In year 1, records of 10 females indicated a litter size of 3.8 and in year 2, it was 4.25 as recorded from eight specimens. The mean for all was 4.0 embryos per litter, and the range was 2 to 6.

Discussion and Conclusions

Mammals form predator-prey relationships previously lacking in this rain forest ecosystem. In the case of the moderately common and most important carnivore, the small Indian mongoose, a firm predator-prey relationship was established, with the roof rat as the prey species when the mongoose spread into the forest less than one hundred years ago. This rat itself has been present for less than two hundred years.

The question often arises in Hawaiian studies whether the mongoose exerts an effective control on any of the rodent species or whether it merely skims the surplus animals resulting from a natural rapid turnover in rodent populations. The stable abundance of the roof rat at Kīlauea forest suggests that the mongoose is indeed a harvester of excess rats rather than a strong depressant on their numbers. Poor adaptation of both the Norway rat and house mouse to the particular conditions at Kīlauea forest explains their sparse and sporadic occurrence. The density and stability of the mongoose populations in this forest is primarily a function of its major prey, the roof rat, which was found to be abundant and stable as a population.

Baldwin et al. (1952) have discussed the status of the mongoose as a highly adaptable carnivore-omnivore and suggest that it has helped to check rodent populations in natural communities. The mongoose is not a skilled climber and therefore is not expected to prey significantly on native birds. On the forest floor it shares food resources, such as earthworms and arthropods, with the omnivorous pig. The pig takes the greater share of this food resource, making the mongoose more dependent on the rats.

The roof rat feeds primarily on the seasonally abundant fruits of forest plants. As a facultative predator, the roof rat appears to be significant in its ability to climb well into the tree canopy to raid nests of passerine birds. As the important prey species of the mongoose in this forest, tree climbing of the roof rat can have two purposes, escape from the predator and search for food. A new dynamic equilibrium with the surviving native birds can be expected from this relationship.

The feral cat may combine in part the predator roles of both the mongoose and the roof rat. It feeds on the roof rat and is an expert climber with a propensity for raiding birds' nests. Such a heavy mammal is not especially adaptable to reaching nests suspended on small branches, so its effectiveness may be limited as a predator of passerine birds in canopy nests at Kīlauea forest. Notably high populations of cats have accompanied excessively high numbers of the house mouse in drier regions of Hawai'i.

Whereas the feral dog is classically a predator on domestic stock and feral sheep, in the Kīlauea forest it probably feeds primarily on the roof rat. A general conclusion is that the dog and other carnivores would have much less support, and in some regions be unable to exist at all, in the absence of such an abundant widely distributed rodent as the roof rat. In the Kīlauea rain forest this rat is a key prey species for three diverse carnivores.

FERAL PIG ACTIVITY*

Pigs of Asian stock *(Sus scrofa)* were introduced to the Hawaiian Islands by early Polynesian settlers, perhaps a thousand years ago. After Cook's first visit to the islands in 1778, many introductions (Tomich, 1969a) of European pigs (also *Sus scrofa*) have followed. All introduced pigs were of domestic stock, but during the course of time many escaped into the wild and became feral. It is not well known whether the Polynesian pigs became feral after their introduction by the Hawaiians. It has been observed that Polynesian pigs in Samoa have not become feral to any significant degree (Lamoureux, verbal communication). This puts some doubt on the feral status of the Polynesian pig in Hawai'i and gives strength to the proposition that the pigs, like the goats (Spatz and Mueller-Dombois, 1973), present a new stress factor in the native ecosystems in Hawai'i.

---

*By Ranjit G. Cooray and D. Mueller-Dombois

Today feral pigs are found on all of the high islands of the Hawaiian chain except on the island of Lāna'i, which has been almost completely cleared for pineapple growing. Large populations of feral pigs are found on the island of Hawai'i, where two types of pigs, the mountain pig and the forest pig, have been recognized by Nichols (1962). The mountain pigs are confined to the remote and isolated sections of the island, particularly to the slopes of Mauna Kea. Nichols (1962) considered the mountain pigs to resemble the wild pigs of Asiatic or European origin in most of their physical characteristics, such as pelage color, hair structure, and body configuration. He believed that the pigs in these remote areas stopped interbreeding with domestic stock a long time ago. As a result, they reverted in form and now resemble their ancestors, the wild *Sus scrofa*. The forest pigs are considered more similar to domestic pigs and are believed by Nichols to have interbred continually with domestic stock that escaped or were intentionally released. The pigs found in most of the rain forests of Hawai'i, including the Kīlauea forest, are of the latter kind. Most have a black pelage.

A large concentration of feral pigs on the island of Hawai'i is found in the rain forest region (Nichols, 1962; Giffin, 1972). The dense vegetation cover, the abundance of food materials and water, the absence of an animal predator, and the ease with which pigs can escape hunters into inaccessible forest sections contribute to making the Hawaiian rain forest an ideal habitat for the feral pig.

In the Kīlauea forest, the effect of feral pigs is very noticeable. Although live pigs are not seen very often, some manifestations of pig activity are very striking even to casual observers. Among the two most striking manifestations are extensive ground digging and feeding on tree fern trunks. A recognizable survey of pig activity manifestations through a wider rain forest territory on Mauna Loa indicated that pig activity in the Kīlauea forest was moderately high. The rain forest area northeast of the Kīlauea forest shows still much greater feral pig activity effects than does the Kīlauea forest. This heavily disturbed forest section is relatively inaccessible to hunters.

Problems in Assessing Feral Pig Activity in Hawai'i's Rain Forests

The impact of feral pigs on the native rain forest vegetation of Hawai'i has been the subject of much concern (Lamb, 1938; Berger, TR 6). However, this aspect did not receive any systematic study before the IBP efforts.

The assessment of feral pig activity in these dense tropical vegetation types is rather complicated. It is difficult to follow pigs for behavioral studies. Therefore they cannot be observed for long periods of time. In such dense vegetation, the use of indirect evidence for estimating the kind and intensity of animal activity may serve as the primary source of information. Indirect evidence of an animal's behavior as shown by modifications to its habitat has been used in other areas to estimate population density (Eisenberg et al., 1970) and animal activity patterns (Mueller-Dombois, 1972). Such indications can be qualitatively and quantitatively evaluated by repeated recording of selected parameters of animal activity within an appropriately designed set of sample plots. The objectives of this study were to adapt an animal activity survey method developed and used in other tropical vegetation types (Mueller-Dombois, 1972) to the Hawaiian rain forest and then to use this modified procedure to assess pig activity patterns in the IBP study site.

Important in this context was a search for a set of pig activity parameters that would give reproduceable results in repeated surveys of the same area and in comparative surveys of other forest habitats. Pigs were observed for certain periods of time until certain manifestations were clearly verified as caused by their activity. Subsequently an intensive study was made of such pig activity signs, which could be quantified and which were sensitive enough for detecting periodic differences of pig activity in the same area.

Sampling Procedure

After some familiarity with the pig's use of the vegetation and habitat, eight major pig activity indexes were recognized as worthy of census:

1. Digging: This included any pronounced scarification or ploughing of the soil by pigs. The total area dug was estimated as a percentage of the area of a 15 m^2 subplot, of which a large number (eight hundred) were surveyed each time at annual intervals. In addition, significantly fresh or recent digging was recorded as either present or absent in each subplot.
2. Tracks: The presence or absence of pig tracks was recorded in each subplot.
3. Plant feeding: Foraging activity of pigs on several plant species was observed. These included primarily tree ferns *(Cibotium glaucum)* but also other ferns, forbs, and woody plants. Enumerations were made of the number of individuals of each species of plant partially eaten by pigs in each subplot. For all species except tree ferns, only recent feeding was recorded. Tree fern feeding was subdivided into six categories: (a) frond pulling and chewing (fronds pulled and stripped by chewing); (b) trunks with leaf trace holes (one or more fronds pulled out and broken at the leaf trace, and the starchy core exposed); (c) trunk scarification (tree fern trunks damaged on the surface by chewing and/or scarification with the snout); (d) trough feeding with apex undamaged (trunk cracked open below the apex and the starchy core eaten by pigs from the apical end to the basal end forming a trough); (e) trough feeding with apex damaged (similar to d but apex was eaten or damaged

and the tree fern was killed); and (f) apical feeding (apex of plant eaten by pigs, but trunk intact, not fed near base; tree fern killed also).

4. Rubbing: The number of rubbing places on tree trunks, prop roots, and tree ferns was enumerated. Tusk-marked trees were also included in this category.
5. Feces: The number of feces clusters was enumerated in two categories. Feces with a coating of slime and malodorous were recorded as fresh. Those with a dry, hard surface were recorded as old.
6. Resting: Resting places were enumerated as wet wallows and as dry resting places. Wallows occurred usually in depressional and open-canopy situations, and dry resting places in raised and sheltered positions under trees or shrubs.
7. Live pigs: The number of pigs sighted during each of the surveys was recorded.
8. Dead pigs or remains: The presence of bones, skin, and other remains was recorded.

These eight indexes with their subcategories were recorded systematically for each 3 by 5 m subplot along two of the 1,000-m belt transects (numbers 1 and 4) in the Kīlauea forest plot (figure 5-3). The data were thus obtained for eight hundred subplots and a sample area of 12,000 m^2 at a time. Four surveys were made, each during the month of August, in 1971, 1972, 1973 and 1976.

Stomach analysis was used as a second method to evaluate feeding patterns. Seven pigs were shot within the Kīlauea forest in 1971 and 1972, and their stomach contents were analyzed. The stomach contents were emptied into the top tray of a stack of sieves with decreasing mesh size from top to the bottom. The contents were then washed down the stack of sieves with running water. All identifiable materials were picked out, and percentage estimates were made of the total volume of materials present.

Activity Patterns Observed

For the analysis, the recorded pig activity indexes were converted to activity indexes, shown on table 6-13.

The areal extent of pig digging was assessed by the proportion of pig-scarified forest floor and by the dispersal of pig-digging activity over the whole sample area. Pig activity index 1a was computed from the percentage area estimate of pig digging recorded in each of the eight hundred subplots. Excluded from the total area was the log-strewn or log-covered area, which was estimated as 24 percent of the floor surface in this forest. Also excluded was the surface area in rocks, which was 4 percent. Thus 72 percent of the forest floor area was considered mineral soil surface. This was taken as the total surface area available to the pigs, and the proportions of mineral soil surface dug by pigs relate to this reduced total area. The data in the table show that this value

TABLE 6-13. *Pig Activity Indexes*

Activity Indexes	August 1971	August 1972	August 1973	August 1976
1. Areal extent of pig digging				
a. Mineral soil surface dug up by pigs	14.1%	32.8%	37.8%	23.8%
b. Digging frequency (800 subplots, 3 × 5 m)	49.3%	83.5%	92.0%	79.9%
2. Tree fern damage by pigs				
a. Trunk feeding damage	2.5%	1.5%	1.3%	1.3%
b. Proportion of tree ferns killed	8.3%	12.4%	11.5%	7.6%
3. Feeding on other plants[a]				
Sadleria pallida	nr	c	c	m
Dryopteris paleacea	nr	+	c	+
Athyrium microphyllum and *A. sandwichianum*	nr	+	+	+
Carex alligata	nr	+	+	+
Acacia koa root sprouts	c	c	m	+
4. Track frequency (800 subplots, 3 × 5 m)	68.6%	96.5%	96.6%	94.1%
5. Feces counts				
a. No. of feces clusters/ha	100.0	112.5	66.6	62.5
b. Proportion of recent feces clusters	17.5%	6.7%	15.0%	4.0%
6. Resting places				
a. No. of wet wallows/ha	1.6	1.6	2.5	4.2
b. No. of dry resting shelters/ha	4.1	4.1	5.0	5.0
7. Rubbing places (trees/ha)	18.3	26.4	17.5	20.0
8. Live pigs seen (no./ha)	1	2	4	0

Note: The indexes were developed from the activity manifestations recorded in four repeated surveys of two 1,000 m long transects.

[a]+ = sporadic feeding, m = moderately common, c = common feeding, nr = not recorded.

never reached 50 percent. In two of the four sampling years, pig-digging activity involved over 30 percent of the available soil surface, and in the other two years, the values remained under 25 percent. This would indicate that August 1972 and 1973 showed greater pig activity than either August 1971 or August 1976. Digging frequency (index 1b on the table) gives an indication of the distribution or dispersal of pig digging throughout the forest as a whole. This was under 50 percent in August 1971, and 80 percent or greater in the other three survey years. Thus the pigs made much greater use of the forest floor area as a whole during the later years, particularly during August 1973, when 92 percent of the eight hundred subplots showed pig-digging activity.

Tree fern damage by the pigs was assessed by feeding on the starchy core of the trunks and by feeding on the apical meristem. Apex feeding was done only on leaning or ground-trailing tree ferns, which always resulted in death of the tree fern individuals. Therefore this feeding damage is shown on the table as proportion of tree ferns killed (index 2b). Because of the time factor involved, the total number of tree ferns (with over 0.5 m stem length) was counted only once. The density was 2,487 per ha. The proportion of tree ferns killed is related to this total. In the two years with the greater pig-digging activity, the proportions of tree ferns killed were over 10 percent; in the two other years, they were under 10 percent. Thus there is a general correlation between ground disturbance by pigs and tree fern deaths due to pigs. However, the tree fern death values fluctuated only between 7.6 and 12.4 percent, which indicates a somewhat stable influence in this forest. An even less fluctuating feeding mode is indicated by the trunk-feeding index (2a) on the table, which varied only from 1.3 to 2.5 percent. This included a count of all tree fern trunks on which the pigs had opened the starchy core to feed on. Most of these tree ferns were still alive, and many will probably survive this feeding damage. The feeding does not necessarily kill the ferns, but certainly it weakens them physiologically. Both tree fern damage indexes could be made more precise by simultaneous counting of all tree ferns (over 0.5 m stem length) at each resurvey. However, the survey time effort was very great in situations where the damage was only moderate or low. The other tree fern feeding signs (frond chewing, opening of leaf-trace holes, and trunk scarification) were much less commonly recorded and thus did not lend themselves to a quantitative analysis. The same applied to pig feeding on other plants in the forest. The table shows that six other plant species were currently used for feeding by pigs.

The remaining five activity indexes (numbers 4 to 8 on the table) give further evidence of a more or less stable and continuous use of the Kīlauea forest by pigs. Track frequency was lowest, with almost 70 percent in August 1971. During the other years, pigs criss-crossed the forest floor nearly everywhere; about 95 percent of the eight hundred subplots showed pig tracks. Feces counts and resting and rubbing places also document a steady use. The last index, particularly the resting places, may be used for an indirect estimate of pig density in the Kīlauea forest and similar rain forests elsewhere in Hawai'i.

Giffin (personal communication, State Wildlife Biologist) estimated the pig population in the Kīlauea forest by a method involving dogs at approximately 1 pig per 1 to 4 ha during the winter of 1971 and spring of 1972. This means that an estimated population of 20 to 80 pigs was probably present on our study site (of 80 ha). Converted to the estimate of resting places given on table 6-13, this would imply that an average number of 50 pigs on 80 ha would create perhaps 128 wallows ($1.6 \times 80 = 128$) and 328 dry resting shelters (4.1×80). It would seem, however, that the population estimate of 20 to 80 pigs for such an area was rather conservative judging from the number of pig resting places in the study area.

Stomach Content Analysis

The estimated volume percentages of food materials found in the stomachs of seven pigs are listed on table 6-14. The bulk of material was made up of

TABLE 6-14. *Percentage Volume Estimates of Food Materials in Stomach Contents of Seven Pigs Killed in Kīlauea Forest, October 1971–August 1972.*

Food Material	1 Oct. 1971	2 Nov. 1971	3 Nov. 1971	4 Aug. 1972	5 Aug. 1972	6 Aug. 1972	7 Aug. 1972
Ferns							
Cibotium glaucum (starch pieces)	95	79	88	70	70	94	92
Cibotium glaucum (fronds)	1	+	+	+	+	1	2
Sadleria pallida (fronds)	+	+	+	+	+	+	+
Dryopteris paleacea (fronds)	+	+	0	+	0	+	0
Athyrium microphyllum (fronds)	+	0	+	+	+	+	+
Athyrium sandwichianum (fronds)	+	+	+	+	+	0	+
Sedges and herbs							
Carex alligata (leaves)	2	1	1	1	+	2	2
Carex alligata (seeds)	+	0	+	1	2	1	1
Astelia menziesiana (leaves)	+	0	0	0	0	0	0
Woody plants							
Acacia koa (shoots)	+	0	+	0	2	1	0
Ilex anomala (leaves)	1	+	+	1	1	0	1
Ilex anomala (berries)	+	0	0	0	0	0	0
Animals							
Earthworms	1	+	1	1	1	1	1
Insect larvae	+	+	+	+	+	+	+
Amorphous	+	20	10	25	24	+	+

Note: + = present in traces, with less than 1 percent of volume.

pieces of starch, which were obviously from the starchy core found in tree fern trunks. The pieces varied in size from about sugar-cube (1 cm^3) size to coarse-grained sand. Even in the amorphous, more digested portion, most of the material appeared to be from finely broken down starch as tested with iodine. A small remnant in this amorphous material was from mud, including perhaps a few other indistinguishable food materials. Thus it would seem reasonable to add the volume percentages of this amorphous material to the values for tree fern starch listed in the first row on the table. This then shows that the greater bulk of the food material—over 90 percent—was from tree fern starch in all seven animals analyzed from the Kīlauea forest.

The other materials give more of a qualitative rather than quantitative picture of the food materials eaten by the pigs. The diversity of food materials is not great but includes the seven plants noted also in the activity surveys. In addition, two more plant species were detected: the often low-growing epiphyte *Astelia menziesiana* and leaves and berries of a fairly common native tree, *Ilex anomala*.

Of interest also is the repeated occurrence of *Acacia koa* shoots in the pig stomachs. These were from root suckers, which had been noted repeatedly as chewed during the surveys. An occasional *A. koa* seedling may also be eaten by the pigs, but such feeding would leave no trace in the field. From the stomach analysis and the frequency with which *A. koa* root suckers were seen chewed in the field, it can be concluded that *A. koa* seedlings belong to the preferred food items in the Kīlauea forest. From the regularity with which the other listed fern species were found in the different pig stomachs, it can be concluded that these also belong to the regular diet of the pigs in this forest. The same applies to the sedge *Carex alligata,* which is the most abundant ground cover plant in this forest (table 6-1).

Earthworms and insect larvae are certainly a preferred food item. This can be concluded from both the constancy of detection in the pigs' stomachs and from their extensive digging activity. Digging for plant roots was not detected in this study, and no root fragments were found in the pigs' stomachs.

Discussion and Conclusions

The pig activity surveys and stomach analyses have provided new insights into the dynamic interaction of the feral pig with the Hawaiian rain forest vegetation. Significant interactions relate to three important forest components: the ground-cover vegetation, the tree seedlings, and the tree ferns.

There is little doubt that the widespread pig digging in the Kīlauea forest has been a major factor in reducing the native ground-cover vegetation. The present total cover was estimated as only 7.7 percent (table 6-1). It is possible, though not documented in this study, that a few of the rarer species were eliminated from pig digging over the years. Elimination of ground cover species by digging activity of feral pigs has been reported from the Smoky Mountain National Park (Bratton, 1974, 1975). However, the Kīlauea forest study site still contained *Marattia douglasii* (table 6-1), a fern well known to be a favored food item of feral pigs. This fern was found with only a few specimens between fallen logs that formed natural exclosures for the pig. It also occurred as a low-growing epiphyte in a few places. The capacity of most native ground-rooted ferns to grow as epiphytes can be considered an important factor in the stability of the native plant species composition.

Pig digging undoubtedly has an effect on the invasion of exotic plant species. Several exotic grasses (*Holcus lanatus, Anthoxanthum odoratum,* and *Axonopis affinis*) and composite weeds (*Erechtites valerianaefolia, Hypochaeris radicata, Gnaphalium purpureum,* and *Senecio sylvaticus*) occurred sporadically in forest openings where the ground was disturbed by pigs. The effect of pig digging on *Holcus lanatus* invasion has been experimentally documented for the mountain parkland along the Mauna Loa transect (Spatz and Mueller-Dombois, 1975). In this rain forest, however, another group of exotic weeds is more important. These are the relatively shade-tolerant mat-forming weeds, such as *Veronica serpyllifolia, Ludwigia octivalvis, Hydroco-*

tyle sibthorpioides, and *Hypericum mutilum,* which form the major exotic plant cover (see figure 6-1). These species seem to invade pig-disturbed ground in more shaded areas, where pig digging in the same places recurs at intervals exceeding a year. At the present moderate level of pig activity in the Kīlauea rain forest, these mat-forming species may be considered permanent new members of the forest vegetation.

Another important ground-disturbance effect is related to the tree seedlings. As the stomach analyses show, the small *Acacia koa* leaves seem to be a regular component of the diet. The observed chewing of root sprouts is probably not important, since *A. koa* reproduction is mostly from seedlings in this forest (Spatz, TR 17). However, the pigs probably will not distinguish between seedlings and root sprouts, and some of the *A. koa* seedlings are undoubtedly eaten by pigs. No tree seedlings of other species were found in the stomach analyses. Yet it was established (Cooray, TR 44) that most tree seedlings, except *Myoporum sandwicense,* grow on logs. This high log association in tree seedling establishment is probably also a reflection of the high level of ground disturbance by pigs.

The quantitative interaction of the feral pig is very important with regard to the tree ferns. The pig exploits the tree ferns for the starch contained in their trunks. However, tree fern killing is not very common in the Kīlauea forest at this time. It occurs only on leaning or ground-trailing tree ferns, whose apices are within the feeding range of the pig. Whether leaning or ground trailing of tree fern trunks is caused by the pigs has not yet been established. The pigs do not seem to knock tree ferns over. It is possible, however, that trunk feeding of the starchy core in erect tree ferns, which causes some physiological weakening, may lead to a gradual felling of tree fern individuals.

Pig activity in the Kīlauea forest continued to be moderate and relatively constant throughout the study period. Major disruptions of the forest ecosystem were not noted, but various alterations were recorded. An increase in pig density could, however, result in further retraction and perhaps elimination of the native ground cover. Increased pig density also has the potential of disrupting the reproduction cycle of *Acacia koa* and may affect the overall stability of the rain forest through reducing the number of tree ferns (Becker, 1976). If this happens, other exotic plant species, including trees, may invade readily and in quantity. Moreover the carrying capacity for feral pigs may reduce rapidly because it seems to depend largely on the tree ferns.

Since pig density is such an important factor, knowing this parameter becomes critical for rain forest management in Hawai'i. Direct censuses are extremely difficult in this type of vegetation, although the enumeration of pig resting places shows some promise as an indirect estimate of pig density. Further precision of this method could be worked out by case examples of simultaneous pig censusing and resting place enumerations.

7

Community Structure and Niche Differentiation in Hawaiian Lava Tubes

F. G. Howarth

Of the four environmental zones found in lava tubes, the true cave community is restricted to the dark zone. The troglobites, obligatory cave animals, have become specialized to live in the rigorous environment of the dark zone. Few other animals can exploit this habitat, and most of those that wander in fall prey to specialized hunters or die from physiological stress.

The lava tube ecosystem is not closed. It is connected to the surface and soil ecosystems by its energy sources, which therefore are allochthonous. As an ecosystem, the lava tube does have functional integrity. A suite of organisms, many of them uniquely adapted to the lava tube environment, carry out the many diverse functions of energy transfer and nutrient cycling occurring within the system.

The three main energy sources are plant roots, which penetrate the roof and dangle into the caverns, fine organic materials, which are washed in by percolating groundwater and deposited in the slimes, and accidentals, surface and soil animals that blunder into the cave but cannot survive there.

The specific objectives developed to test the initial hypotheses of this IBP component study were to survey, identify, and quantify the cave fauna of the Hawaiian Islands; to describe the ecosystem; to study the ecological roles or general niches in the cave community; and to determine to what extent exotics have invaded the ecosystem and to evaluate their impact.

SAMPLING PROCEDURE

Initially, as many different caves were surveyed as possible to collect and list the cave organisms present. Potentially rich caves were visited repeatedly for more thorough investigations. Later the Kazumura lava tube was selected for detailed studies of the ecosystem and development of a life-form spectrum analysis. This cave was the largest lava tube found to date. It was easily accessible, relatively little disturbed, and contained the most diverse troglobitic fauna found during the survey. Located in the Puna district between the city of Hilo and Hawai'i Volcanoes National Park on Kīlauea volcano between 300 m and 450 m elevation, it is exemplary of low- to mid-elevation caves on Hawai'i island.

Nineteen visits were made to Kazumura lava tube. The fieldwork represents more than 500 man-hours underground. Sampling, done between July 1971 and December 1976, was concentrated in a 2 km long dark-zone section near 400 m elevation. Both the upper and lower ends of the lava tube were sampled also. During each trip, observations were made on relative abundance, behavior, and feeding habits of the arthropods. A few animals were taken alive to the laboratory for further biological studies. Eighteen species, 58 percent of the cavernicoles found in this cave, proved to be new to science. These, and other new cavernicoles, are being described in a series of papers written by collaborating specialists and will appear in *Pacific Insects*.

DESCRIPTION OF THE ENVIRONMENT

Kazumura lava tube has been described, including a preliminary interpretation of the food chain (Howarth, 1973). Further detail was given by Gagné and Howarth (1975a). One of the largest known unbroken lava tubes in the world, it has a surveyed length of more than 10 km. The temperature in the cave ranges between 18° and 19°C at the upper end and 20° to 21°C at the lower end, depending on season. The warmer temperature in the lower section is probably due to the 150 m altitudinal range of the cave.

The forest over the lava tube has recently been altered by grazing, fire, and cutting for a subdivision and is now a swampy savanna with scattered *Metrosideros* trees and a ground cover of staghorn fern *(Dicranopteris linearis)* and introduced grasses. Many of the tree roots are dead but still hang from the ceiling (figure 7-1). The high rainfall (3,800 mm per year) and the boggy nature of the ground surface supply a constant drip from the ceiling. This maintains a saturated atmosphere and a damp or wet surface in the cave. Occasional puddles occur on the glazed surfaces of the floor, but in most places the water percolates rapidly into the cracks and spongelike vesicles of the lava tube floor.

The cave appears young, with little organic detritus. It is certainly under 20,000 years old, probably much younger. Much of the cave is solid, with few cracks and no roots. Such sections of the passage are nearly sterile, except for occasional patches of white slime. Other sections with tree roots support a varied fauna, but passages in and around breakdown piles support the highest populations of many arthropods, especially *Caconemobius* crickets and lycosid spiders. Such breakdown piles are usually associated with roots and organic debris because they are under areas with more cracks in the ceiling. The pile itself may afford more hiding places and a greater accumulation of moisture and food than does the adjoining cave passage.

FIGURE 7-1. *Passage in the dark zone of Kazumura lava tube. Jim Jacobi inspecting the long string-like roots of native ʻōhiʻa trees* (Metrosideros collina *var.* polymorpha). *Photo by F. G. Howarth.*

SLIMES*

Two distinct types of slimes, a white and a brown, commonly occur in the cave; they support a number of small scavenging arthropods. A sample of each was analyzed for fungi by M. F. Stoner (TR 75 for technique).

*By M. F. Stoner and F. G. Howarth

The white slime is by far the more common. It has a white, powdery surface and consists of a clear or milky-white, gelatinous, slimy material up to 3 mm in thickness. Patches of this white slime cover many hundreds of square meters of rock surface. The slime has a slight actinomycete soil odor and a moldy taste. Some patches are gritty and mineral-like. White slime also occurs as small or circular patches 0.5 to 10 mm in diameter. Near the entrance, the powdery surface of the same type of slime is often colored a bright orange or yellow. The origin of the slime remains a mystery, but it appears to be formed or deposited in part by percolating water.

A sample of the white slime, upon drying, was relatively odorless, opaque, and had the consistency of a moist, compacted, light powder. This friable material appeared microscopically to be composed mostly of inorganic substances, possibly of gypsum. The organic content of the sample was low. It was amorphous and noticeably devoid of roots or root fragments. Fungal hyphae and spores were present but sparse.

Three species of fungi were isolated: *Doratomyces stemonitis, Isaria* sp., and *Trichoderma viride*. Based on colony counts in dilution plates and distribution in the sample, *Doratomyces stemonitis* was most common. This species, which was also found in soil of the A_1 horizon near site 4 along the Mauna Loa transect, is frequently associated with decomposing plant material and animal excrement. It is cellulolytic and capable of utilizing xylan (Domsch and Gams, 1972). The isolates of *Isaria* sp. were all similar and produced irregular-clavate, dry, dull, flesh-colored coremia. *Isaria* fungi are frequently found as parasites or saprophytes on insects (Barron, 1968). This may explain their presence in the lava tube ecosystem. The same *Isaria* sp. was found also in the A_1 horizon of soils near site 4 of the Mauna Loa transect. *Trichoderma viride* is noted frequently as an inhabitant of the rhizosphere in soils.

Trichoderma and *Doratomyces* are known to be physiologically versatile fungi that are widespread in many different environments (Domsch and Gams, 1972). *Trichoderma* and *Isaria* species are known to utilize efficiently inorganic sources of minerals such as nitrogen. Thus they would be expected to be adapted for growth in this largely inorganic white slime.

In contrast, the distinctive brown slime is a semitranslucent, rubbery, dark brown gel, up to 2 cm thick in places. It is composed largely of organic matter, mostly dead roots and their decomposition products. It also contains fungal hyphae and very fine mineral colloids leached from the soil.

Microscopic examination of the brown slime sample revealed sloughed cells and tissues from dead roots, living roots, and fungal hyphae and spores. Macroconidia of *Fusarium* and oospores of *Pythium* were observed, as well as spores that resembled those of mycorrhizal Endogonaceae. Except for *Penicillium ochro-chloron,* all of the species found are frequently associated with roots (Clarke, 1949, Katnelson, 1965)—for example, *Cylindrocarpon theobromicola, Doratomyces stemonitis, Fusarium oxysporum,* and *Pythium irregulare*. *Cylindrocarpon, Fusarium,* and *Pythium* are external root inhabitants and also could be root parasites. Aseptate fungal hyphae and oogonia-like structures (probably *Pythium*) were observed within intact roots. The isolates of *Doratomyces, Fusarium,* and *Penicillium* were cellulolytic. Most fungal species that were found in the brown slime are known to prefer an organic source of nitrogen, and all but *Penicillium ochro-chloron* and *Doratomyces stemonitis* require the organically rich environment of the rhizosphere. *Doratomyces stemonitis, Penicillium ochro-chloron,* and *Fusarium oxysporum* were detected also in the A_1 horizon of the soil near site 4 (Stoner, 1974).

Aerobic bacteria were present in both white and brown slime. Both types of slime represent incipient soil formation in the young substrate and contain the agents of decomposition and nutrient cycling, particularly of inorganic nitrogen into organic forms. Slimes also capture minerals and nutrients, which are dissolved from the young rock by water. They hold these in place, where they become available for recycling.

Much of the fauna and flora associated with the brown slime are rhizosphere species. Many of the animals are considered to be accidentals, which migrate into the deep cave along roots but are unable to complete their development or survive for long. White slime, though organically poorer, covers large areas of the cave and serves as a habitat for some cave endemics. The white slime may not contain as much fungal biomass, but all three fungi associated with this slime produce large numbers of spores per thallus and, hence, might serve as important food sources for microbes or specialized animals. *Isaria* sp. may have a specialized role in recycling nutrients from insect bodies.

COMPOSITION OF THE DARK ZONE COMMUNITY OF KAZUMURA LAVA TUBE

A list of the cave macrofauna collected in Kazumura lava tube is given in table 7-1. The animals have been separated to the species level for all but two taxa. The species are listed under the three major categories (troglobites, troglophiles, accidentals) of the lava tube ecosystem. The mites (Acari) and oligochaetes have not been identified.

The quantitative values are derived from records of the animal's relative abundance in the cave on each of nineteen visits:

10 = Common throughout the cave, seen on every visit with this abundance.
9 = Widespread in cave but less common; also seen on every visit.
8 = Locally common with patchy distribution; seen on nearly all visits.
7 = Widespread in cave but rather sparse; seen on nearly all visits.
6 = Attracted to head lamp and then common throughout the cave; seen on nearly all visits.
5 = Scattered throughout the cave but locally sparse; seen on most visits.
4 = Adults sparse, but immature states common throughout the cave; seen on half of the visits.

TABLE 7-1. *Macrofauna, Living and Dead, Kazumura Lava Tube*

Life form	Species or taxon	Abundance scale value
Troglobites		
Rhiz. sa (planthopper)	*Oliarus polyphemus* (Cixiid)	10
Omni. (cricket)	*Caconemobius* sp. (Gryllid)	9
Rhiz. chd (millepede)	*Dimerogonus* sp. (Cambalid)	8.5
Zooph. sc (true bug)	*Speovelia aaa* (Mesoveliid)	8
Zooph. pr (spider)	*Lycosa howarthi* (Lycosid)	5
Rhiz. ch (moth)	*Schrankia* sp. (Noctuid)	4
Zooph. pr (earwig)	*Anisolabis* sp. (Labidurid)	0.5
Zooph. pr (centipede)	*Lithobius* sp. (Lithobiid)	0.5
Zooph. pr (spider)	*Oonops*? sp. (Oonopid)	0.2
Zooph. prs (spider)	*Erigone stygius (Linyphiid)*	< 0.1
Troglophiles		
Sap. m (fly)	*Forcipomyia* sp. (Ceratopogonid)	6
Sap. m (fly)	*Limonia* cf. *jacobus* (Tipulid)	3
Sap. m (springtail)	X*Sinella yosiia* (Collembola)	3
Rhiz. chd (millepede)	X*Oxidus gracilis* (Diplopoda)	3
Zooph. sc (fly)	X*Diploneura peregrina* (Phorid)	2.5
Zooph. prs (spider)	X*Nesticus mogera* (Nesticid)	2
Rhiz. ch (moth)	*Schrankia:* 2 spp. (Noctuid)	2
Zooph. pr (springtail)	X*Protanura hawaiiensis* (Collembola)	1.5
Sap. f (mites)	(X)Cryptostigmata	1
Zooph. pr (beetle)	X*Tachys arcanicola* (Carabid)	1
Sap. m (springtail)	X*Folsomia candida* (Collembola)	1
Sap. f (fly)	*Phytosciara volcanata* (Sciarid)	1
Sap. m (mites)	XAstigmata	1
Zooph. prs (spider)	X*Scytodes longipes* (Scytotdid)	1
Sap. m (mites)	XMesostigmata	1
Rhiz. ch (silverfish)	X*Nicoletia meinerti* (Lepismatid)	0.5
Zooph. prs (fly)	Mycetophilidae: 1 sp. (Mycetophilid)	0.4
Rhiz. sa (mealy bug)	X*Geococcus coffeae* (Pseudococcid)	0.2
Rhiz. chd (pillbugs)	XIsopoda	0.2
Sap. m (springtail)	X*Sinella caeca* (Collembola)	0.1
Sap. m (springtail)	X*Hawinella lava* (Collembola)	0.1
Accidentals		
Acc. (worms)	XOligochaetes	2
Acc. (fly)	ϕ X*Calliphora vomitoria* (Calliphorid)	1
Acc. (planthopper)	*Oliarus* sp. (Cixiid)	1
Acc. (roof rat)	ϕ X*Rattus rattus* (skeleton & dung)	0.3
Acc. (centipede)	X*Scolopendra subspinipes* (Chilopod)	0.2
Acc. (mite)	XProstigmata: 1 sp.	< 0.1
Acc. (springtail)	*Salina maculata* (Collembola)	< 0.1
Acc. (fly)	X*Limonia perkinsi* (Tipulid)	< 0.1
Acc. (true bug)	X*Rhytidoporus indentatus* (Cydnid)	< 0.1
Acc. (japygid)	X*Japyx* sp. (Japygid)	< 0.1
Acc. (beetle)	XStaphylinidae: 1 sp.	< 0.1
Acc. (mosquito)	X*Culex quinquefasciatus* (Culicid)	< 0.1

TABLE 7-1. *(Continued)*

Life form	Species or taxon	Abundance scale value
Acc. (moth)	?Microlepidoptera: 1 sp.	< 0.1
Acc. (wasp)	XMicrohymenoptera: 1 sp.	< 0.1
Acc. (ant)	X*Technomyrmex albipes* (Formicid)	< 0.1
Acc. (cockroach)	ϕ X*Allacta similis* (Blattoid)	< 0.1
Acc. (sandhopper)	X*Talitroides topitotum* (Talitrid)	< 0.1
Acc. (spider)	X*Bavia aericeps* (Salticid)	< 0.1
Acc. (flatworm)	X*Geoplana septemlineata*? (Planarian)	< 0.1
Acc. (snail)	ϕ X*Euglandina rosea* (shells only) (Mollusc)	< 0.1
Acc. (snail)	ϕ XUnknown snail (shell only) (Mollusc)	< 0.1
Acc. (bullfrog)	X*Rana catesbiana* (skeleton only)	< 0.1
Acc. (mongoose)	ϕ X*Herpestes auropunctatus* (skeleton only)	< 0.1

Note: X = exotic species or taxa, (X) = group may contain native species, ? = exotic or native status unknown, ϕ dead specimens found only. Abundance scale from 1 = rare to 10 = most common. See explanation in text. Rhiz. = rhizophage, Omni. = omnivore, Zooph. = zoophage, Sap. = saprophage.

3 = Adults sparse, immature stages moderately common or locally distributed and of variable abundance; seen on half of the visits.
2 = Widespread but of variable abundance (moderately common to sparse or absent); seen on less than half of the visits.
1 = Usually rare, sometimes locally common; seen on few visits only.
< 1 = One to a few individuals seen on one or two visits only.

Interpolations between these values were made in the table, where it was considered appropriate.

Thirty-one taxa of cavernicolous arthropods have been collected in Kazumura lava tube. Of these, ten (32 percent) are troglobites. All of the troglobites are endemic, and, except as noted, their eyes are greatly reduced or lacking. The troglobites are pale, and most have appendages that are thinner and more elongate than related surface species. Most have congeneric surface relatives. The troglobites are restricted to the dark zone of lava tubes and, except for *Dimerogonus,* will desiccate and die if the humidity drops to 95 percent. Many of these animals show an ability to enlarge their bodies and consume a prodigious amount of food at once. For example, adult *Speovelia aaa* can survive more than two weeks between meals.

Only five of the eighteen (28 percent) identified species of troglophiles are native. The mites, Cryptostigmata, may contain native species and are marked as such in table 7-1. The other mites are almost certainly exotic. Most of these smaller animals can survive equally well in the soil, and some of these mites may occur as accidentals in caves.

Twenty-three taxa are listed under accidentals. Of these, all but two or three are exotic. This category is analogous to the canopy transients discussed earlier, except that many of the cave accidentals are lost in the lava tube, where they will be eaten by the cave predators or scavengers. Many flightless soil and surface species are trapped in the dark zone since the cues of light and humidity gradients are absent, and migration upward would lead only to the top of protuberances. Some of these accidentals may be troglophilic in lower elevation caves.

Skeletons and dung of the roof rat *(Rattus rattus)* and skeletons of the mongoose *(Herpestes auropunctatus)* and frog *(Rana catesbiana)* are occasionally found on the floor, far away from any known entrance. The three vertebrates usually only indirectly supply energy to the cave. Their carcasses are generally fed on or are decomposed by fungi and introduced mites, Collembola, and scavenging flies—by *Calliphora vomitoria* and the troglophilous phorid *Diploneura peregrina*—which in turn may be eaten by the cave predators and scavengers. Two fungi, *Penicillium claviforme* and *Doratomyces* sp. (identified by P. Dunning), have been isolated from rat dung in the Kazumura lava tube.

NICHE DIFFERENTIATION IN LAVA TUBE ARTHROPOD COMMUNITIES

Four major life forms are recognized among the cave arthropods: the rhizophagous, omnivorous, zoophagous, and saprophagous cavernicoles. Nine general niches were distinguishable within the four major life-form groups: sapsuckers, chewers of living roots, chewers of both living and dead roots, an omnivore (not further subdivided), predators without snares, predators with snares, carcass scavengers, and two types of saprophages, one feeding on fungi and one feeding on microorganisms in general. Many cavernicoles are generalists and overlap in their general niches, or they fulfill several functions in the food web.

The life-form spectrum of the animal community in Kazumura lava tube is diagrammed in figure 7-2. The height of each column was determined by adding the quantitative values of the taxa in each general niche category. This gives an indication of the relative importance of the animals in the lava tube ecosystem.

Of the nine general niches, one, the omnivore, is entirely native, while two rhizophagous niches, the sapsucker and the chewer on living roots, are predominantly native (more than 90 percent). Three niches—the carcass scavengers, the rhizophagous chewers, and the predators without snares—are moderately invaded by exotics (24 to 30 percent). The predator-with-snares niche is almost entirely exotic, with one species *(Nesticus mogera)* predominating. The native-to-exotic ratio for the two saprophagous niches is tentative and may

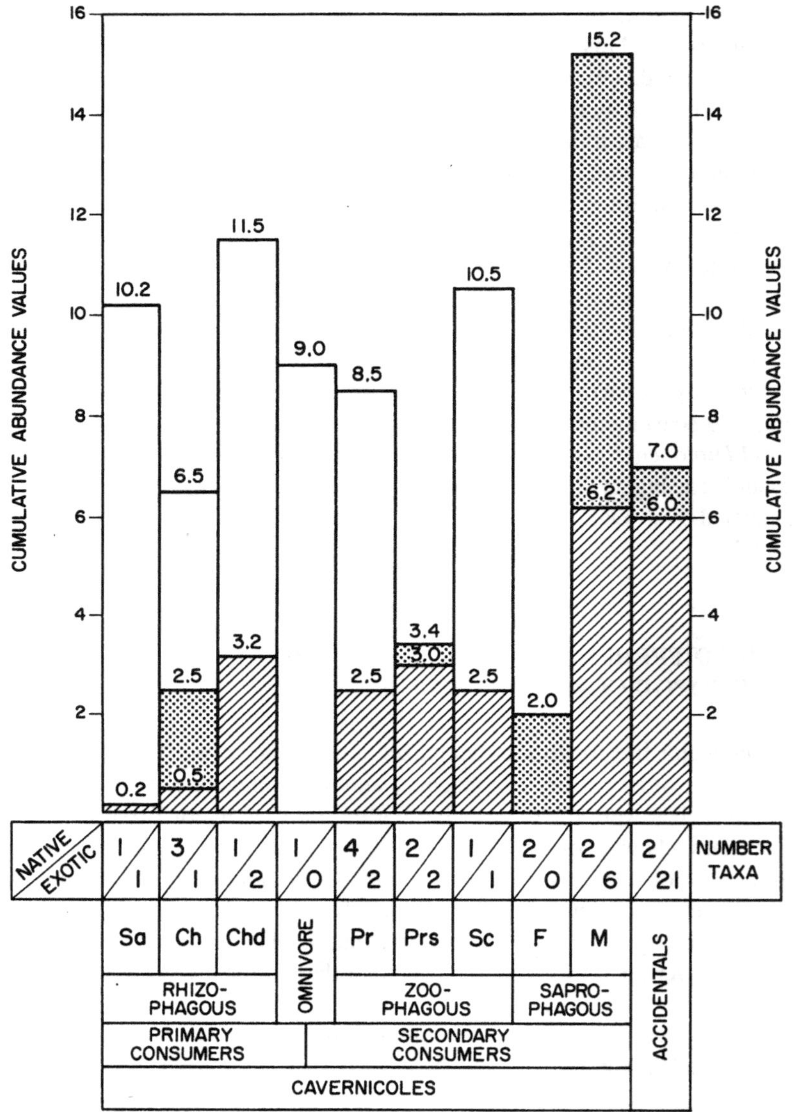

FIGURE 7-2. *Life-form spectrum of the animal community in Hawaiian caves. Sa = sapsucker, Ch = chewers on live roots only, Chd = chewers on live and dead roots, Pr = foraging predators, Prs = predators with snares, Sc = scavengers on carcasses, F = fungivores, M = scavengers on other microorganisms.* ☐ *Abundance of native troglobites* ▒ *Abundance of natives other than troglobites* ////// *Abundance of exotics*

change when the mites are identified and the tropical soil fauna becomes better known. The soil Collembola, mites, and other small arthropods are so poorly known in the tropics that it is difficult to make generalizations on the Hawaiian fauna.

The saprophagous column m, which designates the scavengers of microorganisms other than fungi, still includes an ecologically rather heterogeneous group. With further knowledge, this group can probably be subdivided into general niches more analogous to the others designated on figure 7-2.

Although the exotic cavernicoles comprise a large proportion (43 percent) of all recognized species, they represent a relatively small proportion of the population within the cave. The four main life-form groups are quantitatively well represented. In the rhizophagous niches there are five native and four exotic species, but the exotic species make up only 16 percent of the total group. In the zoophagous niches there are seven native and four exotic species, but here the exotic species make up nearly one-third of the whole. The saprophagous scavengers include four taxa (an unknown number of species) of natives and five or more exotics. The exotics appear to make up less than 20 percent of the group.

A native species feeding on or boring in large roots and bark is absent in Hawaiian lava tubes, although *Dimerogonus* occasionally feeds on root bark. This niche may have been filled when larger roots were more common, before humans destroyed much of the forest.

Rhizophages

The most common rhizophagous animal in Kazumura lava tube is the troglobitic planthopper (*Oliarus polyphemus,* figure 7-3), which in the nymphal stages sucks sap from older 'ōhi'a roots. The nymph makes an irregular wax filament cocoon in which it remains while feeding unless it is disturbed. When the final instar nymph has finished feeding, it moves to a protected area, makes another cocoon, and transforms to an adult in about one week. The adult, which is flightless, totally blind, and white, is the dispersal stage and can be found throughout the dark zone of the cave. It has abundant stored fat and probably does little feeding. The life cycle is unknown, but reproduction is asynchronous, as all stages usually can be found throughout the year. It is sometimes killed by an unidentified fungus. The exotic mealybug, another sapsucker that also constructs a cocoon but with much finer wax filaments, has invaded only the lower end of the cave. It is more common as a troglophile at lower elevations.

At least three species of noctuid cave moths of the genus *Schrankia* are known from Kazumura. One species, usually found nearer the entrance, has indistinctly patterned wings. In the two species found deep within the cave, the wings are uniformly pale gray without any pattern. They are reluctant to fly. One of these is considered troglobitic. The larvae are semiloopers and feed on

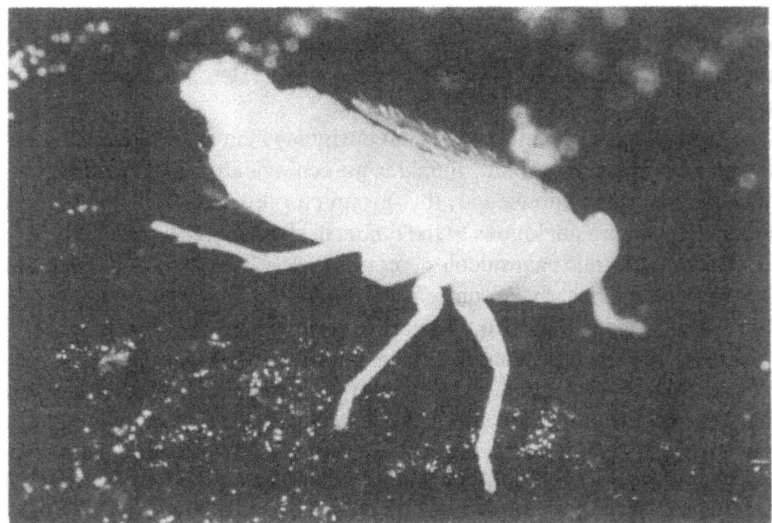

FIGURE 7-3. *Adult female of an endemic Hawaiian cave planthopper,* Oliarus polyphemus. *Photo by W. P. Mull.*

the succulent root flushes of ‘ōhi‘a. In the laboratory the larvae have been reared on sprouted wheat roots. They prepare a cocoon for pupation with pieces of tree roots spun together and tied to a dangling tree root. Both larvae and adults usually can be found on roots throughout the year. *Nicoletia meinerti,* also root chewing, is usually found only near large patches of roots.

One as yet undescribed blind, white species of the native cambalid millepede genus *Dimerogonus* feeds on tree roots and possibly also scavenges on rotting plant material in the dark zone. It has few enemies in the cave because it retains the chemical defense mechanism common to millepedes. In the laboratory it feeds on the living roots of sprouted wheat. It becomes inactive when the humidity drops below approximately 95 percent. However, some individuals can aestivate and survive desiccation in a moribund state for a week or more in the laboratory. They revive when water is added. This adaptation to survive temporary drying is unique among the Hawaiian cave animals so far studied. The exotic *Oxidus gracilis* is more common near entrances and dead and living individuals are scattered throughout the cave.

Zoophages

The cricket (*Caconemobius* sp.) is apterous, nearly blind, and restricted to the dark zone of lava tubes. An omnivore, it feeds on dead arthropods, plant roots, rat feces, and living prey.

There are ten predators; most, if not all, also scavenge on dead animals. Five are blind and considered troglobitic, one is a native troglophile, and four are exotic. The six native predators and the native omnivorous cricket differ in size and behavior and probably divide up the living and dead prey with little overlap. Four species construct snares, and six stalk their prey without snares.

Among the predators that hunt without snares is an unidentified blind, pale lithobiid centipede, which appears to be an endemic troglobite. It usually remains in cracks and only occasionally is found in larger passages. It preys on smaller organisms and also scavenges on dead insects. Another crack inhabitant of the same life-form group, only rarely found in the open, is an undescribed troglobitic earwig. This large endemic predator or scavenger is related to the native surface species, *Anisolabis perkinsi*. A tiny eyeless jumping spider (*Oonops*? sp.) is still known only from immatures but is considered troglobitic.

At the top of the food chain is the microopthalmic wolf spider *(Lycosa howarthi)*. This large troglobitic spider (nearly 2 cm in body length) does not construct a web but waits for prey to touch its outstretched forelegs or, when aroused, can stalk prey in total darkness with uncanny accuracy. It may scavenge on partially consumed previous kills.

Two minute exotic troglophilic predators, a springtail *(Protanura hawaiiensis)* and a carabid beetle *(Tachys arcanicola),* are sometimes found on the brown slime and are attracted to baits. They presumably prey on mites and springtails and may compete with the *Oonops* sp.

There are four snare-building predators, two exotic and two native species. The troglobitic spider *(Erigone stygius)* builds characteristic sheetlike webs up to 5 cm across in cracks. It remains almost exclusively in cracks and is difficult to study. An unidentified native mycetophilid fly is still known only from the larvae, which make slime tube snares in narrow cracks in the lava.

The exotic spider *(Nesticus mogera)* was introduced from Japan through commerce. Its sloppy webs are common in cracks and between protuberances. It is the same size as *E. stygius,* probably captures similar prey, and may be replacing *E. stygius* in some habitats. The other exotic web-spinning spider, *Scytodes longipes,* a large, common house spider in Hawai'i, is only marginally invading Kazumura. It is common in the entrance and twilight zones but generally drops out only a few hundred meters into the dark zone.

The troglobitic water treader *(Speovelia aaa)* is associated almost exclusively with the cave slimes. Notes on the biology of this interesting scavenger were given by Gagné and Howarth (1975a). Individuals walk very slowly and deliberately. They are blind but show keen sensitivity to slight wind and relative humidity changes. When touched or disturbed, they run nervously over the substrate; thus it is doubtful that *S. aaa* is able to feed on mobile prey. It feeds on freshly killed to partially decomposed insects, including dead of its own kind. An exotic troglophilic fly *(Diploneura peregrina)* breeds on rotting animals, including arthropods, and may compete with *S. aaa* for some food.

Saprophages

Two troglophilic groups are primarily fungivorous. The cryptostigmatic mites, still to be identified, are common but inconspicuous and are associated with the slimes. The native fungus gnat *(Phytosciara volcanata)* has been reared once from fungus on rat feces, and the adults occasionally were attracted to head lamps used during the search.

The most conspicuous saprophagous scavenger is the native troglophilic crane fly (*Limonia* cf. *jacobus*). Its larval trails are common in most larger patches of the white slime. The larva, up to 13 mm long, feeds by scraping the substrate in short, sweeping motions. It makes a proteinaceous slime tube, up to three times its body length, into which it can retreat. This tube is covered with white powder from the slime. The larva feeds partially extended out of its slime tube, then pupates within it. The volant adult emerges approximately five days later.

The adults of the native troglophilic biting midge (*Forcipomyia* sp.) are very common in the cave and are attracted to lights. The larvae are scavengers on the wet surface film on rotting roots and slimes.

Six species of springtails (Collembola) are known from Kazumura lava tube. None is considered troglobitic because soil Collembola are easily introduced with plant material. However, three species (*Protanura hawaiiensis, Hawinella lava,* and *Sinella yosiia*) are blind, white, and known only from Hawaiian caves. Two other species (*Folsomia candida* and *Sinella caeca*) are widespread troglophiles introduced to Hawai'i. The Collembola are probably all scavengers or fungivorous, except *P. hawaiiensis,* which was collected with a complete body of presumably *F. candida* in its midgut. *Salina maculata* is a common Hawaiian endemic surface species and is considered accidental in caves.

The mites (Acari) have not yet been identified. The immatures of one species of Astigmatic mite commonly ride on millepedes and crickets. The adults of this and other species are sometimes found scavenging on slimes, rotting roots, and dead insects in the cave.

FAUNA OF OTHER HAWAIIAN LAVA TUBES

Introduction and Dispersal

In lower-elevation caves on all of the islands, a higher proportion of introduced soil, leaf litter, and scavenging arthropods are found. These include isopods (listed in Schultz, 1973), spiders (listed in Gertsch, 1973), Collembola (listed in Bellinger and Christiansen, 1974) millepedes, cockroaches, symphylans, earwigs, termites, and others. This is the region most disrupted by humans, and these species are in large part adapted to either urban or agroecosystems, both human created.

Many of these taxa remain imperfectly known from the tropics, and some, especially the cryptic soil forms, are easily transported with soil. If close native relatives are unknown and the organism is known only from caves, it is difficult to tell if the cave animal is endemic, or introduced and not yet known from elsewhere. Two new monotypic genera of isopods and four new species of Collembola, all currently known only from Hawaiian caves, are considered to be introduced troglophiles.

That many troglobites on Hawai'i Island occur in widely separated caves, including caves less than one hundred years old, indicates that some intercave dispersal must occur. Both biological and geological evidence support the hypothesis that this dispersal can occur entirely underground. Young pāhoehoe basalt has numerous voids, such as gas vesicles, separated layers, small and large lava tubes, cooling fractures, faults, and tree molds. There is little soil; rainwater percolates through the lava almost as it falls. Lava tubes are a major land form on shield volcanoes. Since younger lava has crossed and covered older flows, the specialized fauna is able to invade newer caves from older caves. In Hawai'i the deep-rooted 'ōhi'a supplies food to the fauna living in crevices. Colonization and evolution are continuous if the basaltic lava flows are frequent enough to allow dispersal. The Mauna Loa and Kīlauea volcanoes on Hawai'i are at this stage, as are the lava fields on the lower slopes of Haleakalā on Maui.

The apparent ease of dispersal and specializations of some of these organisms in younger tubes suggest that they might be appropriately called lavacoles as well as cavernicoles. This intercave dispersal may be analogous to the dispersal of aquatic troglobites in temperate limestone caves (Culver, 1970).

Caves are windows to the specialized underground fauna. Some organisms, such as *Caconemobius* crickets, lycosid spiders, and cixiid planthoppers, freely come out of small cracks and spaces into the larger cave passages. Some, such as the linyphiid spiders, are especially adapted to live in cracks and smaller spaces and are found in larger passages only rarely. Others, such as the earwig, are thigmotactic; they live in tight places. Because of this behavior pattern, they only rarely enter large cave passages.

As soil formation progresses and erosion and siltation occur, the smaller cracks are filled, and these dispersal routes are closed to cavernicoles. The larger passages start to fill with silt and break down. Finally, breakdown continues, new entrances form, and cave passages shrink in size and fill with silt and breakdown materials. Erosion fills and dissects the passage until the cave is no longer able to support its specialized fauna, and these become extinct. Such is the case with the majority of caves visited on Kaua'i and O'ahu. Humans have modified the siltation processes by agriculture and development so that these degradations of the caves are now speeded up and the evolutionary history is obscured.

Hawai'i Island Lava Tubes

On Hawai'i Island the faunistically richest caves occur in areas of high rainfall and in the younger (one hundred years to a few thousand years old) flows of Mauna Loa and Kīlauea volcanoes, before significant amounts of soil have washed into the caves. However, there is more correlation of faunistic richness with size, shape, and environment of caves and degree of human disturbance, including alteration of the overlying vegetation, than with location of caves between climatic zones.

Most troglobites on Hawai'i Island are widely distributed. There is, however, a clear distinction between the fauna of low- to mid-elevation caves and those of higher elevations. A transition between the two regions occurs between 1,000 and 1,500 m. Many species common in lowland caves are rare or absent in higher caves, and a few other species are restricted to higher elevation caves.

Roots appear to be more important in higher caves, whereas slimes are relatively rarer; soil formation is much slower in higher montane regions. Significantly there are more rhizophagous troglobites in higher caves: *O. polyphemus* is rarer than in lower caves; *Dimerogonus* sp.; a different species of *Schrankia* with smaller eyes and a flightless female; the blind larvae of an unknown moth; and a troglobitic tree cricket (*Thaumatogryllus* sp.). Of the secondary consumers, *Caconemobius* is less common; the *Lithobius* is more common; *Lycosa howarthi* drops out above 1,200 m; and *Erigone stygius* is more common, as evidenced by its sheet webs.

Above 1,500 m a troglobitic thread-legged bug, *Nesidiolestes ana,* is the main predator. Its biology is given in Gagné and Howarth (1975b). It is a stationary predator, waiting in a suitable place for prey to come within striking distance of its raptorial forelegs. To date, it has not been found sympatrically with the spider, *Lycosa howarthi,* which may replace *Nesidiolestes ana* in lower-elevation caves. *L. howarthi* is clearly a much more efficient predator.

Maui Island Lava Tubes

The fauna of east Maui caves is exciting because it is closely similar to that in Hawai'i Island caves, yet the cave species have independently evolved their cavernicolous adaptations from separate ancestors. It is, so far as is known, a smaller fauna, possibly because of extinction from disturbance or age. The Maui species are much more restricted in range. To date at least seven groups have invaded caves independently on both islands. On Maui there is a different species each of cixiid (*Oliarus priola,* on 'ōhi'a roots), *Caconemobius, Thaumatogryllus, Schrankia,* linyphiid *(Meioneta gagnei), Dimerogonus,* and *Lithobius*.

Kaua'i Island Lava Tubes

The only lava tubes remaining on Kaua'i are in the low coastal dry zone and are remnants of the youngest lava flows, approximately six hundred thousand years old, in the Kōloa volcanic series. Most of the biota, so far as is known, is introduced and composed of a number of soil, leaf litter, and scavenging arthropods. The roots now entering these caves are all of recently introduced exotic plants, mostly of a *Ficus* sp. and presumably *Pithecellobium dulce* and *Eugenia cumini*. What animals were once in these caves when native forest overlay the area may never be known. Many cave species probably have become extinct with the advent of European contact and agriculture. Only two highly specialized cave animals survive: one a general scavenger, the terrestrial amphipod, *Spelaeorchestia koloana,* and the other a general predator, the wolf spider, *Adelocosa anops*. Each of these belongs to a monotypic genus, and both are among the most bizarre cave animals known. Bousfield and Howarth (1976) presented notes on the biology of the amphipod, and subsequent to its discovery a large complex of endemic amphipod species has been found in the Hawaiian Islands.

CAVE FAUNA ADAPTATIONS

Most groups appear to have invaded caves only once on each island. *Schrankia* may be an exception, with five species known from caves on Hawai'i Island, of which at least two species apparently are troglobitic. The close surface relative is still extant for many of Hawai'i's troglobites. It appears that Hawai'i Island troglobites have invaded caves from two very different surface habitats—one a barren rocky substrate represented by unvegetated lava flows and the other 'ōhi'a rain forests.

Of the species that may have invaded caves from barren rocky habitats, *Caconemobius* crickets are the best known, with two species endemic to Hawai'i Island, one cavernicolous species on Maui, and one surface species on Moloka'i. Based on genitalia, the two Hawai'i Island species are much more closely related to each other than either is to the Maui species. Yet in other characteristics, the two Hawai'i Island species differ strikingly. One is a surface species, apparently restricted to unvegetated lava flows near Kīlauea vent. It has functional eyes and is larger, shining black, active, saltatorial, and strongly nocturnal, whereas the cave species has greatly reduced, nearly functionless eyes and is pale straw-colored, much less active, more pedestrian, and lacking circadian rhythm. The lava cricket feeds on wind-borne debris, both plant and animal, whereas the cave species feeds on roots, plant debris, and living and dead arthropods. The two habitats join at cave entrances and in deep cracks in poorly vegetated flows. The Maui *Caconemobius* is more adapted to the cave environment. It is paler, has smaller eyes, and is less sensitive to light than is the Hawai'i cave species.

There are many parallels between the rain forest and the cave environment, the most conspicuous being the high relative humidity. Most of the Hawai'i Island cave species probably evolved from rain-forest-inhabiting species. Only two groups, however, are well known: the *Oliarus* planthoppers and *Nesidiolestes* bugs.

In kīpukas, islands of older substrate surrounded by new lava flows, the rain forest fauna becomes isolated and restricted in range. Animals that can adapt to the new surrounding cave environment have a much greater land area for colonizing. Although the surface of a young flow is usually desert-like, containing only a few scattered shrubs, the caves beneath usually have numerous roots. These new caves offer a wide territory for colonizing by a formerly isolated kīpuka fauna. Thus, the evolutionary process of adaptive radiation can be witnessed at various successional stages in the Hawaiian cave ecosystems.

Eleven species of *Oliarus* are endemic to Hawai'i Island. On the basis of the male genitalia, the rhizophagous cave planthopper *(Oliarus polyphemus)* belongs to the *O. inaequalis* species group (Fennah, 1973a), represented on Hawai'i by two species (*O. inaequalis* and *O. inconstans*). This relationship can be ascertained in spite of very different external appearances. *O. polyphemus* (figure 7-3) has no bodily pigment, no eyes, the tegmina are reduced to spiny thickened flaps, the hind wings are virtually absent, and the claws are highly modified for walking on rock. In contrast, *O. inaequalis* closely resembles other rain forest planthoppers on Hawai'i because it has similarly large protruding eyes, functional wings, deep brown and yellow pigments, and claws with pads adapted for walking on tree bark. This pair, then, reflects differences between soil and cave habitats, as immatures of both species feed on tree roots and the adults are the dispersers. In the rain forest species, however, the adult disperses above ground, whereas the cave species remains underground.

Significantly, *O. priola,* the Maui cave planthopper, closely resembles *O. polyphemus* externally. However, the internal characters of the male genitalia show that it is not at all related to it but is in its own species group, a striking example of convergent evolution.

Nesidiolestes is a highly evolved endemic Hawaiian genus of flightless thread-legged bugs with four described species and a few new ones. Based on male genitalia and spination of the legs, the cave thread-legged bug, *Nesidiolestes ana,* is most closely related to *N. selium,* recently rediscovered inhabiting mossy tree trunks and *Astelia* axils in the rain forests on Hawai'i Island. *N. selium,* in common with other surface species, is nocturnal, cryptic in behavior, and deeply pigmented in cryptic colors and patterns. It has small but functional eyes. In contrast, the cave species is pale white, displays no diurnal rhythm, and has vestigial unpigmented eyes.

The wolf spiders, family Lycosidae, are considered to be among the better-sighted spiders. They use some visual stimuli, although not exclusively, for prey and mate recognition. Therefore, it was surprising to discover two blind species in Hawaiian caves: *Lycosa howarthi* on Hawai'i Island and *Adelocosa anops* on Kaua'i Island. *A. anops* is so highly specialized that no hypothesis about its nearest relative is possible. Unfortunately, most of the native surface species of *Lycosa* remain unidentified so that the closest relative of *L. howarthi* also is unknown.

Preliminary observations have revealed a remarkably different prey-catching behavior between the two cave species. The microopthalmic species from Hawai'i rests with its first four legs held in front of its body. Prey touching these is captured. Some scavenging has been observed in the laboratory and may occur naturally. The active pursuit of prey has been observed, which presumably functions by sound. The eyeless species appears much more specialized. It rests high off the substrate on outstretched legs and actively stalks prey, sensing it by raising the front legs to a vertical position. It will pursue prey with uncanny accuracy. No scavenging has been observed in this species in spite of more observation of it.

The eyeless species lives in a lava tube at 25° to 26°C and is much less active than the intermediate species, which lives in lava tubes between 15° and 22°C. *A. anops* lays approximately twenty-five eggs, one-tenth the normal clutch size in the family, including native surface species. Clutch size in *L. howarthi* is unknown.

CONCLUSIONS

The detailed study of Kazumura cave has shown that the lava tube animals of the dark zone can be recognized as occupying nine general niches. These are three different rhizophagous niches, one omnivorous niche, two predator niches, one scavenger niche, and two saprophagous niches. In addition there is a noncavernicolous group of accidentals. These are important in the lava tube ecosystem because they provide a food source for some of the zoophagous cavernicoles.

The nine general niches are filled by ten troglobitic species and seven troglophiles, including six species and another taxon. As a result, there is a certain amount of niche overlap among these native organisms. This implies that these cave animals evolved into broad niches, a phenomenon probably related to their living in a rigorous environment, where food is difficult to find.

About half of the fifty-five species and taxa listed from Kazumura cave are cavernicolous. The fifteen exotic cavernicoles, all troglophiles, have with few exceptions not invaded this cave ecosystem to any great extent. Whether some of the exotic troglophiles have or will displace native troglobites is a question that can be answered only upon further research of the taxa in the different niche categories. Circumstantial evidence suggests that the common exotic spider, *Nesticus mogera*, competes with and may be replacing the rare troglobitic spider, *Erigone stygius*.

This niche analysis has shown that endemic troglobites are dominant in at least six of the nine niches: in three rhizophagous niches, in the omnivore niche, and in two of the three zoophagous niches. The third zoophagous niche, which relates to the snare-building predators, is the only cavernicolous niche dominated by exotics. The two saprophagous niches are also dominated by endemics but only by troglophiles. Thus the major exotic component occurs among the accidentals and the snare-building predators.

The nine niches are found also in other lava tubes on Hawai'i Island, but other species occupy some of the niches. For example, several more species occur in the rhizophagous niches in higher-elevation caves. Exotic troglophiles are much more in evidence in lowland caves, especially in those below 300 m elevation. Significantly this is also the region to which most of the successful surface arthropods have been introduced and become established. The majority of exotic plants grow in this region, and the roots penetrating into the lava tubes are mostly from exotic plants.

Only two troglobitic species still survive on the old island of Kaua'i. One is a general scavenger, the other a highly evolved general predator. East Maui caves are more depauperate in native species than those on Hawai'i Island. Although the Maui caves show almost identical niche differentiations to the latter, their fauna has independently evolved from separate ancestors. At least seven taxonomic groups have evolved troglobitic species independently on two islands, indicating that evolution of a lava tube fauna is a general process, not an exceptional case.

8

Niche and Life-Form Integration in Island Communities

D. Mueller-Dombois and F. G. Howarth

The information on the selected Hawaiian montane rain forest and the cave ecosystems was gathered with several objectives in mind. Foremost was the need to establish baseline data in relatively undisturbed native ecosystems. We hoped to gain insights into the maintenance patterns of these systems under natural conditions and to obtain a basis for comparing these with similar ecosystems elsewhere, which indicate breakdown or dieback problems. Another objective was the testing of our three niche hypotheses.

These objectives and hypotheses were approached through analyzing the community structure and the ecological roles of the more important organisms in the two selected native ecosystems. The first step was a detailed inventory of all species encountered in the different organism groups. This required repeated and differing methods of analysis for the organism groups involved. In conjunction with this basic inventory, each of the participating investigators attempted to quantify the species populations in his organism group (plants, birds, mammals, arthropods) and to analyze the ecological position of each species in the community. This resulted in a certain minimum of information that everyone was able to procure: a quantified species list and a life form spectrum for each organism group studied. Beyond this, individual investigators pursued population ecology studies to differing degrees of detail depending on prior knowledge and the relative ease with which such additional information could be obtained.

The concept of niche and life form integration, which provides the focus of this synthesis chapter, relates to a certain level of organization that can now be recognized among the community members. This level of organization can be interpreted from the life form spectra and other data tables, which were presented in chapters 6 and 7.

Principally there are two ways to look at these life form spectra. Horizontally, they serve as indicators of the general niche differentiation attained in each organism group and ecosystem. Vertically, they serve as indicators of the relative niche occupancy as achieved by the species with similar ecological roles. Identification of the proportion of introduced species in these life form spectra provides an added dimension, which is of particular relevance to our niche hypotheses.

LIFE FORM INTEGRATION IN AN ISLAND RAIN FOREST

Plant Life Form Spectrum

In contrast to the animal community, which lacked native terrestrial mammals in the Kīlauea rain forest, there was no obvious native life form paucity among the plant synusiae. Instead a rather rich spectrum of twenty-three different plant life form groups or plant synusiae could be recognized and portrayed in quantitative terms (figure 6-1).

In accordance with investigators of tropical rain forests in other parts of the world (Richards, 1952; Whitmore, 1975), plant life forms were separated into two general groups: structurally self-supporting plants and plants that grow by supporting themselves on others. The second group, which includes lianas, epiphytes, and hemiparasites in our forest, is typically found in tropical forests elsewhere. In temperate forests, lianas are not important, and there are few, if any, vascular epiphytes. The latter group, in particular, was well represented in our island forest, which underlines its tropical character.

Perhaps one plant life form could be said to be missing in our island forest: the heterotrophic vascular form. Such nongreen vascular parasites or saprophytes typically are found in some continental tropical rain forests, such as *Rafflesia* (Rafflesiaceae) in southeast Asia (Walter, 1971). An endemic member of the Loranthaceae, the hemiparasite *Korthalsella complanata,* occurred in our rain forest community, but its ecological importance was very small. In discussing the plant life form spectrum with regard to its ecological importance, it seems best to focus first on the phanerophyte synusiae and then on the others.

Phanerophyte Synusiae

Among the structurally self-supporting plants, fifteen synusiae were recognized. Of these, eight were phanerophytes, plants that grow over 0.5 m tall. There were no herbaceous phanerophytes (such as bananas and tall ginger plants), which are often found in tropical rain forests on the continents (Walter, 1971).

The eight phanerophyte synusiae resulted from grouping these woody plants into four height classes. These were based on each species' height-growth potential achieved in this montane rain forest. Subdivisions within height classes were made on leaf texture (leathery leaved or soft leaved) and on gross differences in leaf shape—large apical fronds (the tree ferns) or normal leaf blades (all other woody plants). Among the latter, two more categories of leaf structure could have been recognized: woody plants with phyllodes (in shape similar to eucalyptus leaves) and those with very small (microphyllous) leaves. However, there was only one species with each of these more extreme foliage characteristics in the plant community. For this reason they were not separated on the life form diagram. The phyllodial species was the endemic *Acacia koa* var. *hawaiiensis,* which also performs a foliage metamorphosis with height-growth, from a small (up to 2 or 3 m tall) sapling with typical feathery, legume-type, compound leaves to a taller (over 5 m tall) sapling and mature tree with sword-shaped phyllodes, which are the flattened and expanded leaf petioles. Thus the native koa has a unique life form in this forest. The same can be said for the microphyllous, native shrub, *Styphelia tameiameiae,* which, however, was found only sporadically in this montane rain forest. This shrub is a characteristic member of pioneer and seral communities on more recent volcanic surfaces throughout several climatic zones. Its sporadic presence in the rain forest seems to indicate a relict position from an earlier seral stage. This microphyllous (almost needle-leaved) heath-type shrub thus also represents a rather unique plant life form in this forest. One could argue that the twenty-one endemic phanerophyte species in the forest should have been grouped into seventeen rather than fifteen synusiae. Yet even the fifteen phanerophytic synusiae indicate a general niche spectrum for trees and shrubs, which is comparable to or just as rich as any continental tropical montane rain forest so far described (see reviews by Richard, 1952; Walter, 1971, 1973; Whitmore, 1975). However, what appears to be rather different is the general niche occupancy and biomass distribution among the phanerophytes.

Two tall-growing tree species (macrophytes over 10 m tall) occur in this native island forest: *Acacia koa* var. *hawaiiensis* and *Metrosideros collina* subsp. *polymorpha.* In most other Hawaiian rain forests, only the latter species occupies the upper tree synusia. Both species have been classified as sclerophyllous evergreens, a foliage characteristic that these trees have in common with the upper tree synusiae of montane tropical rain forests elsewhere. Both species are also shade-intolerant pioneer trees. Yet their differences make these species sufficiently distinct as to consider them in two separate general niches. *Acacia koa* is a relatively fast-growing tree, which requires only thirty years to reach a height in excess of 15 m, while *Metrosideros collina* is a relatively slow-growing tree (Porter, TR 27). Its height-growth potential is similar to that of *Acacia koa,* about 25 m, in this forest, but only few *Metrosideros* trees realized this potential in the Kīlauea rain forest. Most of the taller individuals were only 20 m tall, and these were far less numerous than the tall *Acacia koa* trees that reached up to 25 m.

In contrast to the species-poor upper-tree synusia (the macrophanerophytes), the lower tree synusiae were relatively rich in species. There were seven species of mesophanerophytes (trees with potential heights from 5 to 10 m) and eight species of microphanerophytes (trees or aborescent shrubs with 3 to 5 m height potential). Thus if all woody species with height-growth potentials exceeding 3 m are considered together, the species richness in this forest is not really so low, with seventeen species. It compares well with the relatively species-rich temperate deciduous forests in eastern North America (Curtis, 1959). However, while most tree species in these eastern deciduous forests compete for the upper tree stratum by having roughly similar height-growth potentials, the height-growth potential in this island rain forest is patterned in such a way that the number of woody species increases markedly in successively lower height strata. This characteristic, which separates our island forest from temperate forests in general, is shared with tropical rain forests elsewhere (Richards, 1952).

A big difference between continental tropical rain forests and our island forest undoubtedly would show up on a species-richness comparison among the various phanerophytic synusiae; our island forest has distinctly fewer species in these life form groups. In an idealized profile diagram, representing a similarly sized area of tropical rain forest on Puerto Rico, Holdridge (1970) shows thirteen species occupying the tree synusia between 10 and 25 m height (macrophanerophytes). Such richness in tall-growing tree species compares well to mainland tropical forests. This brings out the point that islands near continents (like Puerto Rico) may share more similarities with continental forests than with those on oceanic islands. Thus island size may be considered of lesser influence on tree-species richness than the degree of isolation. Puerto Rico and the Island of Hawai'i are of comparable size (approximately 10,000 km^2).

The nanophanerophytes in our island forest (woody plants with a height potential to 3 m) consist of six species. Two are introduced *Rubus* species. One of these, *R. rosaefolius,* is a widespread and naturalized but ecologically rather insignificant member of most Hawaiian rain forests. The other, *R. penetrans,* is a thicket-forming thorn shrub that has a potential for becoming a dominant member in this rain forest. However, at the time of our analysis, the shrub was found only with insignificant biomass in a few local spots. Larger openings and removal of tree ferns would undoubtedly encourage its spread.

These two *Rubus* species were the only exotic members among the phanerophytes. In spite of the fact that exotic phanerophytes with a high potential for invasion (such as the trees *Psidium cattleianum* and *Myrica faya*) are present in close proximity to the Kīlauea rain forest, no serious disturbance by foreign tree invasion had occurred as yet. Part of the reason can be found in the high competitive capacity of the tree ferns (Becker, 1976) and the excellent adaptation of the two tall-growing tree species populations.

Other Plant Synusiae

Three other broad life form categories were distinguished among the structurally self-supporting plants: chamaephytes, geophytes, and therophytes. The last group was ecologically insignificant. It was represented only by one species, an introduced tropical forb, *Erechtites valerianaefolia*. This weed, which belongs to the composite family, is found sporadically but widespread throughout Hawaiian rain forests, but it occupies only physically disturbed ground in open places, where it persists for short periods. It is shade intolerant and becomes displaced locally wherever it is overgrown by taller species.

The chamaephytes (plants generally under 0.5 m tall but growing perennially by investing their photosynthates mostly in shoot rather than underground storage organs) were grouped into five synusiae. Two of these contained primarily woody-stemmed plants, the others herbaceous plants. Among the woody chamaephytes, two ferns (*Dryopteris paleacea* and *Sadleria pallida*), which can be described as dwarf-tree ferns, assumed some quantitative importance. Suffrutescent (or semiwoody) forbs were present in insignificant quantity. Among the herbaceous chamaephytes, two synusiae stood out in quantitative importance: the caespitose (mostly graminoid) forbs and the reptant or mat-forming forbs. The two groups together contained twenty-two species, of which twelve were exotics. However, only among the mat-forming forbs did exotic species assume any quantitative importance. Seven species of exotic mat formers assumed a combined cover of about 1 percent. The more important of these were *Veronica serpyllifolia, Ludwigia octivalvis, Hydrocotyle sibthorpioides,* and *Hypericum mutilum.* This is the only ecologically significant exotic plant life form group that we observed in the Kīlauea rain forest. Their advance into the forest appeared to be directly related to pig scarification. Their ecological position is that of shade-tolerant pioneers or early seral species.

Another ecologically significant synusia of herbaceous plants was the rhizomatous geophytes, into which we grouped all ground-rooted rhizome-propagated ferns. This group included twenty-one species, all of them natives. Together they covered a little over 1 percent of the forest floor surface. This group was particularly interesting because the majority of them (fifteen species) were also represented as epiphytes. These rhizomatous native ferns, therefore, can be considered as having exceptional stability in this forest on account of their relative independence of the soil substrate. A few important representatives of this group were the ferns *Grammitis hookeri, Athyrium microphyllum, Adenophorus tamariscinianum* var. *tripinnatifidum,* and *Polypodium pellucidum*.

Three other epiphyte synusiae were recognized: herbaceous chamaephytes (ten species), woody chamaephytes (nineteen species), and phanerophytic epiphytes (with two species). It is ecologically significant that all epiphytes were native species and that most of these species occurred also as ground-rooted plants in this forest. Without exception, all ground-rooted endemic phanerophytes occurred also as epiphytes in this forest. Most of them, however, did not appear to achieve a height-growth potential exceeding 0.5 m when growing (above 2 m) on other trees. Therefore they were classified as chamaephytes. Two tree species, however, commonly grew taller and thus were designated as phanerophytic epiphytes. These were *Cheirodendron trigynum* and *Metrosideros collina* subsp. *polymorpha*.

The phenomenon that all phanerophytes can also grow as epiphytes is perhaps a characteristic unique to island rain forests. It brings out a stability characteristic, which has not yet been fully recognized. The two phanerophytic epiphytes can grow into normal trees once their roots have made contact with the soil. Their roots can descend down the trunks of their host trees over distances exceeding 3 m. All other native tree species in this forest, perhaps with the exception of *Myoporum sandwicense,* can also grow into normal trees if their epiphytic beginning is not too high off the ground (within 1 to 2 m). This observation does not apply to all individuals of these species, and there are interesting differences among the species in their survival capacities and growth development following an epiphytic beginning.

A quantitatively unimportant life form was the lianas, although it contained six native species. Only one of these was a woody liana *(Alyxia olivaeformis),* which forms pencil-thin stems that wind around the host tree. The others were semiwoody (or basally woody) and herbaceous forms that ascended only into the tree fern layer (up to 5 m height). Together they covered less than 1 percent of the forest floor area. This quantitative paucity of lianas is a characteristic held in common with montane tropical rain forests known from other areas (Richards, 1952).

Biomass Distribution and Structural Uniqueness

The biomass distribution in the plant community was quantified and portrayed in layer diagrams (figure 6-2). These layer diagrams show that the horizontally most continuous layer in this island forest was the tree fern layer, which was clearly dominated by only one species of tree fern, *Cibotium glaucum*. This layer, which filled out the second stratum above the herbaceous plant layer (a vertical height stratum from 0.5 to 5 m) is comparable to the shrub layers found in many forest plant communities elsewhere. However, in typically closed forests, particularly in the tropics, the shrub layer ties up relatively little foliar biomass, and this would be reflected by a low percentage of cover. In this island forest, however, this shrub layer contained most of the foliar biomass. A second concentration of foliar biomass was found in the emergent tree layer, which in this rain forest is also dominated largely by one species, *Acacia koa* var. *hawaiiensis*. It has a mean crown cover of about 75 percent. However, this emergent tree layer (extending from 10 to 25 m) was not uniformly distributed across the forest. Instead it showed many canopy gaps, which varied from about 100 m^2 to 4,000 m^2 (one acre) in size (figure 6-9). Moreover, the *Acacia koa* trees, having vertically oriented phyllodes, intercept a relatively small amount of light even where they are overtopping the lower tree strata. For this reason, one could have expected a more or less continuous subcanopy below the emergents. Instead the recognized subcanopy layers (an intermediate-stature tree layer from 10 to 15 m height and a low-stature tree layer from 5 to 10 m height) showed relatively little foliar biomass. The mean cover of the second tree layer was only about 10 percent, that of the third tree layer about 30 percent and that of the fourth layer (the tree-fern layer) about 85 percent. Thus below the variable emergent tree layer, the subcanopy showed many gaps. This indicates that the intermediate-stature trees (mostly *Metrosideros collina* subsp. *polymorpha*) and the low-stature trees (*Cheirodendron trigynum, Myoporum sandwicense, Ilex anomala, Myrsine lessertiana, Coprosma rhynchocarpa,* and six others), did not make efficient use of the available light beneath the emergent tree layer of *Acacia koa*. This apparent inefficiency, however, may be a reflection of the dynamic nature of this forest.

A high and relatively constant concentration of foliar biomass in the shrub layer (from 0.5 to 5 m height) appears to be an island-wide characteristic in the rain forests of the Hawaiian Islands. What varies more strongly from place to place is the tree overstory and the herbaceous understory. In the Kīlauea rain forest, the low cover (about 10 percent) of the herbaceous understory can be attributed to the deep shade under the tree ferns and, more important, the frequent rooting activity of feral pigs.

Three factors enter into the spatial constancy or continuity of the shrub layer and the variable nature of the tree overstory: site capacity, tree height-growth potential, and disturbance factors. Among the high Hawaiian Islands, Hawai'i is the only island that has relatively tall-growing forests, with trees reaching about 25 m in height. Our Kīlauea forest sample stand is one of these. The great height of *Acacia koa* here indicates a high site capacity, which can be attributed to the relatively recent (about 5,000 year old) volcanic ash soil, which still contains a good nutrient reserve. On the older islands, such as on O'ahu and Kaua'i, such tall-stature forests are absent, primarily because of the decreased site capacity due to soil aging in the same rainfall regime. Soil aging here implies latosolic leaching of soil nutrients. On these geologically older mountainous volcanic islands, which are from 3 million to 6 million years old (Macdonald and Abbott, 1970), one encounters only low-stature rain forests, which correspond in structure to the upper montane rain forests of tropical southeast Asia as described by Whitmore (1975). The low-stature rain forests on our older islands, however, occur at rather low elevation, from about 300 m to 1,500 m (the approximate upper altitudinal limit of Kaua'i). In tropical continental areas, these low-stature rain forests appear upslope at about 1,500 m elevation (Richards, 1952; Whitmore, 1975). This phenomenon has been attributed to the Massenerhebungseffekt, which implies that the upper montane forest boundary is raised to some extent wherever the mountainous land mass is increased. Our Kīlauea rain forest, although it occurs on the largest of the Hawaiian Islands and in an altitudinal range (1,200–1,800 m) that corresponds to the upper montane rain forests in continental mountains, is clearly more comparable to lower montane rain forests as described by Whitmore (1975).

Therefore, in extrapolating the characteristics of our forest example to other oceanic islands, island age (as reflected primarily in site nutrient capacity) has to be taken into consideration. Tropical island rain forests cannot be so easily generalized as tropical continental rain forests are generalized.

Another factor is the height-growth potential of the woody plant species involved. Tree ferns are often present in montane tropical rain forests elsewhere. But these as a rule are the taller-growing species of the genus *Cyathea,* while the Hawaiian tree ferns are all *Cibotium* species. The *Cibotium* species have a lesser height-growth potential than do the *Cyathea* tree ferns in general. The ecological success of the *Cibotium* tree fern genus on Hawai'i accounts for the high biomass representation in the 0.5 to 5 m height stratum. If *Cyathea* tree fern species had become established in a similarly successful way, the greater undergrowth biomass probably would be found in the intermediate- and low-stature tree strata of this forest rather than in the shrub layer. The layer distinctness might even disappear. Therefore tropical island forests elsewhere can be expected to differ considerably in structure, depending on the genetic growth potential of the tree species that became ecologically successful members of the island rain forest communities.

On the older high Hawaiian islands, tree ferns are quantitatively much less important in the shrub layer of rain forest stands, a phenomenon that may be related to the soil aging and nutrient loss. Also *Metrosideros collina* subsp. *polymorpha* trees are of low stature only, usually not exceeding 10 m in height. This size reduction in the same tree species can be a function of both the decreased site potential and a lower height-growth potential. The latter may be manifested in edaphic ecotypes.

A third important influence on island forest structure and biomass distribution, even in climax or steady-state forests, is the regional perturbation or disturbance factors.

Canopy Arthropods and Tree Borers

Direct comparison of our study with those of similar montane rain forests of continental regions is not possible. This is due to our limited taxonomic knowledge of both the Hawaiian arthropods and those of the continental tropics in general and also the absence of published reports of comparable studies (Janzen, 1973).

Janzen et al. (1976) listed 540 foliar insect species in a secondary growth field at 1,600 m in the Venezuelan Andes. Assuming a richer fauna in an undisturbed rain forest in that continental area, species diversity in the Kīlauea rain forest is low but not as low as is often assumed. Our list of 115 taxa would more than triple with all species identified. The lack of specific identities has also precluded the recognition of many niches.

Swezey (1954) listed more species of insects associated with koa and secondly 'ōhi'a than with any other native tree, Southwood (1960) and Strong et al. (1977) showed that this is more related to the dominance or area covered by these trees than to their evolutionary age.

Elton (1973) compared the insect fauna of the undergrowth vegetation (up to 2 m height) of neotropical rain forests in differing locales. Janzen et al. (1976) and Janzen (1973) compared the species diversity and abundance of foliar insects in several neotropical localities relative to a number of parameters. Both studies used other sampling techniques and different analyses so that only a few comparisons can be made with our results.

The numerical preponderance of predators is highly significant and according to Janzen (1973) is a characteristic of island faunas. However, Elton (1973) also found a large proportion of predators in his study of continental forests. Janzen (1973) speculated that these generalists at higher trophic levels become more easily established than the more host-specific primary consumers. We may add two other reasons: the general niche breadth of these secondary consumers, which is often much larger than the sampled habitat, and the purposeful introduction of parasites and predators to the islands for biocontrol, which has disproportionately added to these general niches.

This line of reasoning leads to the general corollary that the increase in species diversity within a habitat increases the chance for an exotic organism to become successfully established. The exotic organism is usually of the next higher trophic level but, as explained by Baker (1973), a greater plant diversity harbors a greater variety of pollinators and other commensals in the fauna, and thus an exotic plant has a greater chance of becoming established. Therefore the invasion of an exotic species does not always imply depauperate empty niches, as has often been assumed (Sailer, 1978). As Janzen (1977) pointed out, natural habitats are much closer to being full of species than empty. Unique perturbations, such as the introduction of exotic species, create empty niches that may be filled faster than is generally realized (Strong et al., 1977).

The number of species in a habitat is a function of habitat diversity, productivity, and harvestable resources (see Janzen, 1977). It is not valid to compare species packing or richness between different habitats without understanding that the niche must be defined as the organism perceives it. For example, the seed capsule of ‘ōhi‘a favors an external seed predator, while the koa seedpod is more suited to internal consumers.

The *Plagithmysus* borers are not as common in the Kīlauea rain forest as they are in more open situations such as along the Mauna Loa transect, due largely to their sun-loving propensity. The absence of *P. bilineatus* may be sampling error, but the species certainly is not as common as predicted from the dominance of ‘ōhi‘a, its host tree. The frequency of the other five species of *Plagithmysus* was correlated with the abundance of their host species.

Before humans, insects and birds were the dominant consumers. Zimmerman (1948) eloquently relates accounts by early entomologists of the past abundance of native moths in many habitats. However, Hawaiian rain forests, as exemplified by the Kīlauea Forest Reserve, are still predominantly native because the low species diversity of plants restricts the possibility of a new consumer finding a suitable niche. Also most continental rain forest insects are not adapted to travel with humans and the ports of entry are deserts to them. Most human introductions, including humans themselves, are adapted to lowland or open sunny habitats.

Bird Life Form and Activity Spectrum*

In comparison to structurally similar tropical rain forest ecosystems on continents and continental islands (Diamond and Terborgh, 1967; Karr and James, 1975; Terborgh and Diamond, 1970), the Kīlauea forest today has a somewhat lower diversity of avian life forms. A broad spectrum of guilds exists, but they are filled with very few species, particularly when the native Hawaiian species are considered. For example, there are no ground-feeding or seedeating forms at all and only one carnivore, the Hawaiian hawk *(Buteo solitarius)*. Similarly, the insectivore niche is occupied by only one native species, the ‘elepaio *(Chasiempis s. sandwichensis)*, and the only frugivore-insectivore species is the Hawaiian thrush *(Phaeornis o. obscurus)*. A ground-feeding niche may have been filled by the extinct Hawaiian rail *(Pennula sandwichensis)*. A second predator, the Hawaiian owl *(Asio flammeus sandwichensis)*, did not occur in the Kīlauea forest, though it is found in other Hawaiian rain forests. There is no evidence that the native seedeating forms were ever present in the rain forests (Berger, 1972). The frugivore niche has been rendered depauperate by the disappearance of *Corvus tropicus* (Corvidae) and *Psittirostra psittacea* (Drepanididae) from the Kīlauea forest area, though they occur elsewhere on Hawai‘i. Similarly five nectarivores or insectivores presumably found in this area previously (Melaphagidae: *Moho nobilis, Chaetoptila angustipluma;* Drepanididae: *Loxops sagittirostris, Hemignathus o. obscurus,* and *Drepanis pacifica*) have become extinct (Hawaii Audubon Society, 1978: 85–87).

Several species of introduced birds have invaded the Kīlauea forest with varying degrees of success. The nectarivore, insectivore, and frugivore niches are all exploited to some degree by the introduced generalist *Zosterops japonica* (table 6-5 and figure 6-21). *Leiothrix lutea* has become quite successful as lower-canopy frugivore-insectivore, and *Garrulax canorus,* which is common in similar ecosystems elsewhere in Hawai‘i, may begin to offer more serious competition to *Leiothrix* and *Phaeornis* in this niche if it invades the forest successfully. However, some degree of vertical separation (table 6-6) may lessen the effects of the potential competition among these species.

In the seedeating niche, two introduced species (*Cardinalis cardinalis* and *Carpodacus mexicanus*) have invaded along logging roads, and a third *(Strep-*

*By Sheila Conant

topelia chinensis) has been reported in the forest. Both of the finch species prefer edges and forest openings and are unlikely to become an important component of the Kīlauea forest avian community unless the forest is opened by clearing and logging.

A few of the species in the Kīlauea forest show a good degree of niche differentiation by vertical and horizontal foraging separation, species-specific foraging substrate preferences, and probably also in food preferences. However, the majority of the bird species inhabiting the forest belong to the nectarivore-insectivore group. Among these species, resource exploitation is accomplished primarily by a dynamic system of interspecific territoriality (Carpenter and MacMillen, 1976a, 1976b; Carpenter, 1976) and niche shifts (Carpenter, 1976, this study), although the other mechanisms of ecological separation are also important in this group. Although analyses of morphological differences among members of the avian community have not been made, previous studies (Karr and James, 1975) suggest that such differences may contribute significantly to the observed behavioral differences important in ecological niche differentiation in the Kīlauea forest community.

Even before the Kīlauea forest avifauna was reduced in diversity by extinction of species populations, its nectarivores, insectivores, and frugivores were limited to five passerine bird families: Corvidae, Melaphagidae, Muscicapidae, Turdidae, and Drepanididae. Now only three families (Muscicapidae, Turdidae, and Drepanididae) remain; however, the niche spectrum has probably changed little. There is no ground feeder now, but it is uncertain this niche was ever occupied. A new seedeating niche has been occupied with limited success (probably due to limited resources) by three exotic species, and one exotic frugivore is well established but offers little competition to the single native frugivore. The most successful exotic species *(Zosterops japonica)* shares the nectarivore-insectivore niche with three native species. The success of this exotic species probably can be explained by its versatility, especially when compared with its three native competitors. The Kīlauea forest differs from avian communities in other tropical rain forests of similar structural character (Diamond and Terborgh, 1967; Karr and James, 1975; Terborgh and Diamond, 1970) primarily in a lack of diversity at taxonomic levels above the genus, rather than a lack of ecological diversity in terms of niches.

Activity Patterns of Introduced Mammals*

Native terrestrial mammals are absent from Kīlauea forest and the Hawaiian Islands because of the extreme isolation of this island group. The single native forest mammal (not yet seen in the Kīlauea forest) is the volant hoary bat, *Lasiurus cinereus semotus*. It is admirably suited to colonization because it is a

---

*By P. Q. Tomich

strong flier. As a solitary tree-roosting species, it requires no specialized refuges such as caves. Small insects captured in flight comprise the diet. Most significant perhaps is that once arrived, the bat selectively suppressed a strong migratory instinct that may have prompted its dispersal to Hawai'i in the first place (Tomich, 1969a). The hoary bat is observed only occasionally in the rain forest, in edge environments used for feeding, and during movement or foraging in more open adjacent regions. The forest itself appears to serve adequately for roosting and may be of considerable significance in that regard and as a source of insect foods generated there. Niche interactions with the birds of Kīlauea forest may exist but have never been identified.

Until the arrival of humans and accompanying terrestrial exotic mammals, the forest remained empty of ground-dwelling forms. The present-day pig is a recent invader now selectively reverted to the wild from domestic breeds. If the small Polynesian pig did indeed form wild populations, these have been absorbed by large-bodied feral swine. Modern feral pigs are the greatest biotic agents of perturbation in the forest through their ploughing and digging and their feeding on tree ferns. Less than 25 percent of the forest floor is not subjected to disturbance by pigs. The long-term effect on species composition of the ground layer, if not of the successively higher layers, may be great. In the absence of control areas where pigs are absent, it is difficult to evaluate comparative trends in time.

The feral dog fills the niche of the roving carnivore-scavenger, and its daily range may be several kilometers in length, utilizing a much larger land area than the forested study site. The same foraging grounds are covered only sporadically. Such a pattern is suitable to the size of this mammal, its needs for food, and its social organization. Typically feral dogs forage in packs and frequently are active at night. That the roof rat is the major item of food taken by the feral dog in Kīlauea forest agrees with observations elsewhere on the island. This newly developed predator-prey relationship between exotic species is of considerable importance in explaining how the dog is adapted to living in the wild, where suitable and readily available food generally is in short supply. It provides another example of where an introduced species (rat) provided a niche for another introduced species.

It is unclear how the dog and pig interact in the forest, but it appears that predation by dogs on pigs is rare. Pigs may succumb temporarily when pursued by dogs, but pigs are obviously adapted to harassment of this sort.

Feral cats are elusive and nocturnal, and they may be more prominent in Kīlauea forest than our observations have revealed. Certainly this species is firmly established as a predator, with some tendencies to scavenge on carrion. There are indications that the cat, a highly adaptable predator of birds, has a depressing effect on bird populations at least of the lower-canopy layer, especially through destruction of nests (Berger, personal communication). The mongoose is the only carnivore present in larger numbers. This may be due to the small size of this species and its coincident omnivorous feeding niche. It ranges throughout the forest and may utilize the entire forest floor on a regular basis. Social organization in the mongoose permits extensive overlap of individual home ranges (Tomich, 1970). Nonselective feeding preferences permit a relatively high density for this species. The mongoose is obviously highly adapted to disturbance of the forest floor by pigs although it is unclear whether the total effect of rooting and wallowing by pigs is advantageous or detrimental to the mongoose.

Rodents of Kīlauea forest have not altered basic behavioral or activity patterns known in these species in continental environments. Although highly developed as commensals frequenting human abodes and storehouses, each species is admirably capable of existence in wild environments wherever suitable conditions of food, shelter, and climate exist.

Rodents invading Kīlauea forest, like the carnivores and pigs, encountered unfilled niches. The roof rat shows excellent adaptation and maintains a moderately abundant stable population based apparently on its ability to live on fruits of native plants. The starchy core of *Cibotium* is used by rats when exposed, as by pigs, and occasional wounds appear to have been initiated by rats. The role of rats as facultative predators on birds is suspected to be large, but quantitative information is lacking. Nests and contents appear to be most vulnerable to attack by rats. The brief nesting season is thus critical. Whatever the effect of rats on native bird populations, one can assume that by now a balance has been struck and a reasonable stability achieved. Bird species that have survived to this date are not likely to be annihilated by rats in the near future.

The Norway rat tends to be a ground-dwelling, burrowing species. This habit, coupled with its extreme scarcity, suggests that this species insignificantly affects birds, plants, or insects of the forest. Although this rat has a past record of abundance in some Hawaiian forests, such abundance is no longer known to occur. Instead the Norway rat retains its more typical habit of association with human habitations and aggregates where excess food supplies are available. Perhaps it is remarkable that it survives at all in remote natural forests.

The seedeating and insect-eating house mouse does not find the Hawaiian rain forest a particular attraction, yet it survives there. The tendency for seasonal rise in numbers, which leads to plagues of mice elsewhere, was not detected in Kīlauea forest. The reproductive niche of the mouse seems to be to provide a local excess of population, which emigrates to new centers of population wherever suitable habitat is encountered.

Management of Kīlauea forest directed at enhancing natural stability of the forest ecosystem, most obviously the plant and bird communities, might well be directed toward control or elimination of pigs and roof rats. Reduction in their numbers, or their elimination, might well lift depressant effects on perpetuation of the undergrowth plants, improve the survival of tree seedlings, and provide a considerable protection to certain of the canopy-nesting birds.

Forest Dynamics and Maintenance Trends

The forest plant community provides the structural matrix within which most of the activities and life functions of the associated biota occur. For this reason, it is useful to ask whether the structure of our rain forest can be expected to be maintained as it is now or whether it is changing. Although our study period was not long enough to observe any significant changes, we analyzed a number of structural attributes of the forest that can be used as indicators of the dynamic behavior of this forest. These were primarily the size-structure analyses of the woody plant populations, the detailed (very large-scale) air photograph and ground truth analyses resulting in the vegetation map and profile diagrams, the analysis of horizontal patterning in the different plant layers, the repeated pig-activity surveys, and the age-structure analysis of the *Acacia koa* tree population.

The analyses of bird-activity patterns, rodent population structures, and canopy-arthropod diversities gave some insight into the fine structure and dynamics occurring within the currently existing forest stand matrix. No major disrupting forces were detected among these biota groups. The possible interaction of the introduced rodent populations with the endemic forest bird populations has not yet been studied in detail. Some egg predation of bird nests by the dominant rodent, *Rattus rattus,* was expected. And some direct bird predation can be expected from feral cats. However, both of these suspected interactions appear as not significantly disruptive under present conditions. The three introduced seedeating bird species are quantitatively insignificant in the Kīlauea forest. It is also not expected that they will become more important, since they are marginal forest birds. Thus one may safely conclude that these new seedeaters do not play a significant role in reducing the seed crop of *Acacia koa*. The rodents may be of greater importance in this respect, but no detailed studies were done since koa maintenance was found to be regulated by more important factors.

Based on the information gained in this study, the maintenance of the forest plant community as a whole appears to depend largely on the physical environmental forces operative at the site and on the behavior of four dominant species: the three endemic woody plants *Acacia koa* (koa), *Metrosideros collina* (ʻōhiʻa), and *Cibotium glaucum* (tree fern) and the feral pig, *Sus scrofa*.

The structural analyses of the woody plant populations showed that all woody plant species, except the very rare *Clermontia hawaiiensis,** were represented by reproduction. This implies some degree of maintenance for all woody plant species in this forest. Most species displayed inverse J-shaped population curves, which indicates population stability under present condi-

*This species and a few other rare ones are of uncertain status because their enumeration in relation to the sample area was too low. Their survival behavior can be evaluated only by sampling a much larger area.

tions. The ecologically most successful native woody plant with an inverse J-shaped population trend was the tree fern, *Cibotium glaucum*. This species was also found to be the major food source of the feral pig. However, in spite of being the prime target of new herbivory pressure in this forest (new in the evolutionary sense), the tree ferns are holding their ground under present conditions. This fact was revealed through the repeated pig-activity surveys (1971 through 1976) and futher supported through a second detailed structural analysis (Becker, 1976). Observations on nearby areas of similar forest, where the tree ferns have been removed for commercial reasons, show that the tree ferns occupy a key position with regard to the stability of this native forest ecosystem. Large-scale removal of the tree ferns invites massive invasion of exotic plants. Such conditions can also be provided where feral pigs increase greatly in density. Examples of this are currently found in certain parts of the rain forest in Hawai'i Volcanoes National Park.

In our Kīlauea rain forest plot, feral pig activity was still relatively moderate. The question of whether pigs had any effect on decreasing the original species diversity of this forest cannot be answered from our analysis. Perhaps the most significant influence of feral pigs on altering the species composition of this forest was from the inroads made for the establishment of the exotic mat-forming herbs (the reptate chamaephytes, figure 6-1), such as *Veronica serpyllifolia, Ludwigia octivalvis, Hydrocotyle sibthorpioides, Hypericum mutilum,* and three others (table 6-1). These formed the only synusia in the Kīlauea forest plant community where exotic species had attained significant cover, and this amounted to only 1 percent of the total ground area.

In contrast to *Cibotium glaucum* and most other native tree species, koa and 'ōhi'a, the two tall-growing tree species, did not display typical inverse J-shaped population trends. Both species showed fewer saplings (1 to 5 m trees) than either mature trees or seedlings.

The number of seedlings (up to 1 m tall) was very high for 'ōhi'a, with over 10,000 per ha (more than one per square meter). But in the 1 to 2 m size class, there were fewer than 10 saplings per ha, indicating a very high mortality rate in early development. Approximately 10 trees per ha were found in successively larger height classes, indicating no further mortality after the initial numerical reduction from seedling to sapling. A population mode was found in the 5 to 10 m height class (about 200 per ha), but there were many (about 50 per ha) also in the 10 to 15 m class, and few (about 5 per ha) reached into the emergent tree layer (over 15 m tall). Since no age measurements were available for 'ōhi'a, this curve trend could mean both a one-generation population with a considerable spread in sizes or an uneven-aged population with wave generations as indicated by the population mode (of the 5 to 10 m trees) and the high current seedling crop. The second interpretation seems more likely.

The number of koa seedlings (up to 1 m tall) was only about 100 per ha. These were reduced to less than 1 per ha in the 1 to 2 m height class. About 1 sapling per ha was found in each of the successively larger height classes up to the emergent tree layer in which the number of trees was again much greater (about 20 per ha). This indicates both a high mortality rate in early development and a reduction to almost insignificant replacement. However, the age-structure analysis of koa indicated a very fast, early growth rate; koa trees in this forest require on average only about thirty years to grow into the emergent tree layer (greater than 15 m tall). Seedlings and saplings were often found growing on the root collars of overturned emergents, and there appears to be an approximate one-to-one replacement of new emergents to fallen snags under present conditions (Cooray, TR 44). This balance, however, can be easily upset.

A number of factors could disturb this balance. Two of these stand out as important. One factor is the feral pigs, which were found to feed on and destroy koa seedlings. Their ground-rooting activity, if severe, would result in the virtual elimination of any koa reproduction from the mineral soil surface, but there would probably still be survivors on fallen logs and especially on the mineral soil adhering to the big (often 2 m high) root collars that are turned upward during the breakdown of the overmature emergents. There is no lack of koa seeds as evidenced by the numerous seedlings that have come up in adjacent selectively logged areas of the same forest and elsewhere after logging in similar forests (Scowcroft and Nelson, 1976). Light is not necessary for germination of koa seeds (Spatz, TR 17). For this reason it is surprising not to find a much larger number of koa seedlings, similar to those of ʻōhiʻa in this forest. Recent evidence (Wirawan, 1978) indicates that koa seeds need exposure to oxygen for germination, in addition to favorable temperature and moisture conditions (Spatz, TR 17). Thus seeds do not seem to germinate when buried in poorly aerated surface soils. Instead they germinate as soon as they become properly aerated or exposed to the atmosphere. However, atmospheric exposure alone, which can be created by soil surface scarification, does not also provide survival conditions for the seedlings. Comparative observations have shown that koa seedlings are very intolerant of shade; they need a large amount of light. Therefore undergrowth competition, particularly from tree ferns, is another limiting factor for koa survival in this rain forest.

Wind throws, which occur frequently in this forest, are the important counterbalancing factor. They knock down tree ferns and other undergrowth vegetation and provide for a certain amount of ground disturbance. The early height-growth rate of koa seedlings, once they are established, is very rapid, about 1 m per year. As such, koa may often outstrip by rapid height growth the closing in of the tree fern fronds and associated undergrowth following a gap opening created by the falling of large-sized individuals and groups of trees.

ʻŌhiʻa, like koa, is a shade-intolerant pioneer species. It is the first tree species to appear on new lava flows in the montane forest environment (Smathers and Mueller-Dombois, 1974). Thus ʻōhiʻa also can grow up successfully only in forest gaps. ʻŌhiʻa seeds, in contrast to those of koa, are very small, like the achenes of some composite weeds. They germinate readily on moist substrates, and many seedlings are usually present in natural forests with open canopies (Mueller-Dombois et al., 1977), while 'ōhi'a seedlings are almost absent in forests with closed canopies. The high number of 'ōhi'a seedlings in the Kīlauea forest is a reflection of its gappy canopy. Thus tree-group breakdown, probably resulting from occasional Kona storms,* is an important maintenance mechanism for the two dominant tree species, which obviously compete for the canopy gaps in this forest.

The detailed structural map (figure 6-9) revealed a mosaic pattern of canopy gaps in this forest. The total gap area is 25 percent. Among the gaps, three types are recognized: recent windfalls (1 percent of the total gap area), *Metrosideros* ('ōhi'a)-dominated gaps (7 percent), and the tree-fern filled gaps which showed an occasional koa sapling and scattered groups of other native trees. This gap pattern can be interpreted as a highly dynamic chance pattern that moves across the habitat depending on the presence of tall trees and the occurrence of tropical storms.

Significant canopy gaps cannot be created unless there are tall trees, which break down individually, in groups, or in larger stand sections. A very heavy storm may mow down a large proportion of the tall emergents, resulting in a significant rejuvenation of the koa population. Such heavy storms of hurricane dimension are infrequent in Hawai'i. The current gap pattern does not reflect such a perturbation. However, the total map does because it shows a koa canopy dominance of 75 percent. Probably the present koa dominance resulted from such a major perturbation. Among the big koa emergents, we found a few trees growing on big-diameter koa logs, which were lying, still relatively undecayed, on the ground—clear evidence that the present crop of emergents is not comprised of merely one generation but of at least two. These may have been the results of such infrequent, perhaps 70 to 150 year interval, heavy storms of hurricane dimensions.

In contrast, the present gap pattern may reflect a series of moderate tropical storms that occurred at certain unspecified intervals through a few decades. It should be possible to specify the frequency of the perturbation intervals through within- and between-gap-type comparisons, a useful future research inquiry. Important from our results is the strong indication that both 'ōhi'a and koa tree populations are being maintained under present conditions, although the future position of koa is clearly more precarious (dependent on storm patterns) than the position of 'ōhi'a in this forest.

A significant slowdown in the storm-perturbation pattern would endanger the survival of first koa and eventually also 'ōhi'a. Both of these shade-intolerant, tall-growing species compete with each other to some extent. 'Ōhi'a has an advantage in that it can grow under the canopy of koa, whereas koa cannot grow under the canopy of 'ōhi'a as is apparent from their present distribution patterns. A displacement of both koa and 'ōhi'a from the rain forest is

*Tropical storms in Hawai'i that occur several times a year, mostly in the winter months, when the trade winds are interrupted.

currently exemplified in the ʻŌlaʻa forest, which occurs at a lower elevation (1,200 m) below the Kīlauea rain forest. Here in the ʻŌlaʻa forest the upper tree synusia is gradually giving way to a pure tree fern forest (Mueller-Dombois et al., 1977). The primary reason appears to be a somewhat less favorable storm pattern in this lower elevation forest, which occurs in a wetter and more continuously cloud-covered area. Associated with this also is a still more favorable tree fern climate and a deeper and thus more productive ash soil, which results in an ecological optimum for tree fern development. Thus the tree ferns, which can be considered the most stable native component of these young Hawaiian rain forests, present a competitive threat to both koa and ʻōhiʻa. However, a favorable wind perturbation pattern has maintained a dynamic equilibrium between these three dominant native woody species for at least a few thousand years in the Kīlauea forest, and there is no indication from our data that this will change for the next several centuries.

Two factors may interfere with this prediction, but both could be avoided by proper management. One of these factors would be a significantly higher pig population, which would disrupt the still stable maintenance pattern of the tree ferns. Such a disruption would invite massive invasion of exotic plant species. These in turn could seriously interfere with the reproduction cycle of the two shade-intolerant dominant native tree species, resulting eventually in a total stand conversion, a replacement by an exotic forest plant community.

The second factor, even more important, relates to the question of what would be an adequate size for establishing this koa-ʻōhiʻa-tree fern forest as a natural area or ecological reserve. Certainly a self-maintaining system of this forest is not as small as a hectare because some of the larger-sized gaps are already 0.5 to 1 ha in size. However, the forest seems to maintain its integrity, under current natural conditions, on minimal areas of approximately 80 to 100 ha. A forest segment of this size, but preferably much larger, should be considered if the objective were to set aside an example of this forest for safeguarding it in perpetuity. The consideration for a still much larger size is based on two additional factors (which were not included in our study): the minimal area required for the maintenance of the native bird community and the buffer zone required to prevent nonnative plant species from encroaching on the protected habitat. No doubt spatial relationships are an important factor, in addition to the specific site factors and the ecological properties of the species themselves. Therefore for a more adequate evaluation of the minimal size of this forest type for natural area purposes, all three factor complexes should be assessed.

LIFE-FORM INTEGRATION IN HAWAIIAN LAVA TUBES

Lava Tubes as Life-Support Systems

How and why an animal would lose its eyes, color pattern, and other characteristics and restrict itself perpetually to the rigorous environment of caves has long intrigued biologists. Much is known of the ecology of limestone caves in temperate regions (Vandel, 1965; Barr, 1968). The realization that lava tubes also harbor an analogous specialized fauna is more recent (Torii, 1960). In addition to Hawai‘i, significant lava tube faunas are now known from the Galapagos (Leleup, 1967, 1968), Japan (Ueno, 1971), and North America (Peck, 1973). In Hawai‘i, the fact that at least seven native groups have independently evolved troglobitic species on at least two islands indicates that adaptation to caves is a general process.

As is true generally, troglobites are strongly stenohygrobic and anemophobic and are restricted to the true dark zone of caves. The dark zone environment is similar to that described by Poulson and White (1969) for temperate limestone caves. It is a rigorous one, defined by the absence of light, the relative constancy of temperature at or near the average annual temperature of the region and of relative humidity above the physiological limits of most terrestrial animals, and the absence of many environmental cues. There is generally a rocky substrate, and this gives the impression of an absence of any food source.

Lava tubes are destroyed by erosion in a relatively brief geologic time. However, they are continually being created during volcanic eruptions, since oceanic volcanoes are characteristically built with vesicular basalts that often flow as pāhoehoe. Such flows produce lava tubes very frequently. Therefore lava tubes are a common land form on younger oceanic islands. It is important to realize that the voids in basaltic lavas probably offer some avenues for subterranean dispersal from older caves to younger ones. We can expect that a specialized cave fauna will develop wherever new basaltic lava flows are interlinked with older flows over a long enough period of time and wherever the climate allows the continuous colonization of caves.

In limestone caves new passages are formed by the dissolution of water as older passages collapse in a continuing long-term dynamic process. Limestone caves often are millions of years old, and deep, due to the downcutting and lowering of the water table with age. The overburden can be over 500 m, with 50 m being common.

Continental lava tubes are less studied but physically similar to those in Hawai‘i, except that often in continental lava tubes the cavernous basaltic flows partially cover relatively impermeable substrates such as andesitic lava. This can perch the water table, and vadose water can then be pirated by the lava tube, allowing the development of permanent, small, underground ponds and streams.

In Hawai‘i rainwater percolates rapidly into the young porous basalt. Only where the water table is near the surface is significant water found in the cave. This occurs near the coast. Here a remarkable aquatic fauna, including many troglobites, inhabits isolated coastal pools of brackish water (anchialine habitat) in young lava flows (Holthuis, 1973; Maciolek and Brock, 1974).

In younger lava tubes the substrate is usually barren lava rock, without any significant organic detritus. However, the rock substrate can vary considerably in texture, from a polished, glazed surface, to an irregular pile of broken blocks, to a highly vesicular porous lava, to ash rubble. Lava tube slimes, which affect incipient soil formation, often cover large areas. In the oldest caves, clay and soil have filled most smaller voids and cover the floor in many areas.

Limestone caves also have a rocky substrate, often as irregular blocks, but there is a greater variety of minerals, including crystals, biogenic minerals, and alluvial sediments. A fine residual silt from solution of limestone is characteristic and is closely associated with terrestrial troglobites (Barr and Kuehne, 1971).

The main energy sources in Hawaiian lava tubes are plant roots, especially those of *Metrosideros collina* subsp. *polymorpha,* slimes deposited by organically rich percolating groundwater, and accidentals, animals that blunder into the lava tube. In contrast, the main energy sources in continental limestone caves are trogloxenes, especially bats and crickets, and debris washed in with sinking streams, especially during floods. Additional energy is supplied by accidentals, percolating groundwater, autotrophic bacteria, and aerial plankton.

The greater overburden of limestone caves often precludes the importance of tree roots and, except for Paulian and Grjebine (1953) who discovered an intermediate cave-adapted cixiid in Madagascar (Synave, 1954), most biospeleologists have disregarded roots in their surveys. However, the discovery in Hawai'i of a cave cixiid (Howarth, 1972) stimulated other researchers to check for roots in caves. Recently Fennah (1973b) described three species of troglobitic fulgoroids—two from Mexico and one from Australia—and Peck (1975) listed an undescribed species from Jamaican caves.

The absence of native trogloxenes in Hawai'i may be related to the absence of winter and a need to hibernate and also to the fact that the continental trogloxenic groups did not colonize Hawai'i. Hawai'i's only bat and only native land mammal *(Lasiurus cinereus semotus)* is a forest species and is not known to enter caves.

Sinking streams are not important energy sources in Hawaiian lava tubes. It is unusual for a stream to enlarge a lava tube; rather it speeds the siltation and erosion processes. The few lava tubes visited that had captured a temporary stream were blocked with silt, had signs of periodic flooding, and had a poor fauna.

Niche and Life Form Integration of the Cave Fauna

Native cavernicolous animals are predominantly arthropods. In one lava tube, Kazumura, ten troglobitic and seven native troglophilic species occupy nine general niches: three primary consumer niches (feeding on living and dead roots), one omnivore, two predatory niches differing in strategies (with and without snares), one sarcophagous niche, and two saprophagous niches (one feeding on fungus and one generalist). A broad niche overlap is underscored by the omnivore niche occupied by the troglobitic cricket.

The roots that penetrate into Kazumura lava tube to any great extent belong to only one tree species, and five native and four exotic arthropods feed on them. The troglobitic cixiid, probably the most abundant arthropod in the cave, is a sapsucker. The living-root chewers are represented by three moths of the genus *Schrankia*. One of these is a weak flier and appears to be troglobitic. An undescribed troglobitic millepede occupies the living-and-dead-root-feeding niche.

At the top of the food chain is a large, striking, troglobitic wolf spider, which stalks its prey and does not build a web. The other five native predators are less well known and probably live in cracks, rarely entering larger passages. A troglobitic terrestrial water treader scavenges on dead arthropods. The two saprophagous niches are occupied by four troglophilic taxa.

To date no native organisms have been found boring into or specifically feeding on large diameter roots, and this is possibly an empty niche. However, in arthropod surveys it is difficult to generalize on negative evidence; one can only say that no borer has been found thus far, not that it does not exist. Moreover, with regard to arthropods, natural ecosystems may have few, if any, empty niches unless there is a major new disturbance, such as that created by an exotic species (Janzen, 1977).

Even though fifteen exotic species have colonized Kazumura lava tube, with one exception (snare building predator) they do not appear to have invaded existing niches to a great extent.

In the younger lava tubes on Hawai'i, troglobites have a wide distribution. Many lowland forms have not been found above 1,000 to 1,500 m, where other species often occur. The few lava tubes on Maui are significant because their fauna provides a control group for the Hawai'i studies. At least six native taxa have independently evolved troglobitic species on the two islands.

Although life form spectrum analyses have not been done for continental caves, some information is available to make comparisons. McClure et al. (1967) presented a food web analysis and listed over 150 arthropod and 23 vertebrate species living in Dark Cave, Batu Caves, Malaysia. The 0.8 km-long cavern is in a remnant outlier of limestone and is essentially a large bat roost without much of a dark zone. Nearly all the animals are troglophiles, a few are trogloxenes, and very few, if any, are troglobites. Trogloxenes, especially three species of bats, supply nearly all of the energy in the cave, although a number of accidental animals were noted also. The few small temporary and permanent streams described apparently do not bring in much organic material.

The rhizophagous niches were entirely absent in Dark Cave, but many more scavenging, predatory, parasitic, and detritivore niches were present than are found in Hawaiian caves. Some of the niches not found in Hawai'i caves included vertebrate and arthropod parasites, social predators, and exploiters of rotting fruit and insect remains in guano. A few of the troglophiles listed from Batu Caves have been introduced to Hawai'i. Most have not invaded Hawaiian lava tubes to any extent because their niche in guano or as general scavengers is not present. Notable exceptions are the phorid *Diploneura peregrina* and probably some of the Collembola.

Barr (1968) and Barr and Kuehne (1971) considered debris washed in with flooding, guano from cave crickets, and wood and litter from human activities as the most important energy sources in Mammoth Cave system, Kentucky, where about forty-five troglobites are known. Barr noted that the fresh stream-washed-in logs, sticks, and leaves often swarm with troglophiles and accidentals. A large number of saprophagous troglobites known from the cave prefer fine, well-decomposed detritus or the microorganisms in such material. Significantly no known terrestrial Hawaiian troglobites occupy such a detritivore niche. The other main niches of Mammoth Cave troglobites are predatory. These animals also show some specialized behavior and prey preferences.

Of the troglobitic taxa, only one family, Linyphiidae, is common to both Hawai'i and Mammoth caves, and the species are not closely related. With the exception of the spiders and amphipods, not a single order has troglobitic representatives in both regions. Both Vandel (1965) and Barr (1968) categorically excluded tree roots as an energy source because any organism that feeds on them is by definition a "deep soil organism, not a cavernicole."

Most continental troglobites are considered relicts of past climatic changes, especially glaciation and changes in sea level (Vandel, 1965; Barr, 1968; Mitchell, 1969). However, many of the Hawaiian troglobites have close surface relatives still extant. (Three such species pairs are *Caconemobius* spp., *Oliarius polyphemus-O. inaequalis,* and *Nesidiolestes ana-N. selium.*) This strongly suggests that troglobites are relicts only if the surface species have become extinct, not that they become cave adapted after extinction of the surface population. There are many examples of adaptive radiation among continental cave animals. Most are true relicts because of the more complex geological history of the continents.

Many continental troglobites maintain synchrony in reproduction, with egg laying occurring at a certain time of the year. In aquatic organisms this synchrony is usually an adaptation to the spring floods, either to exploit better the excess food at this time or, with some species, to have the life cycle at a stage resistant to the flood (Poulson, 1975).

There is no indication that Hawaiian cavernicoles display any seasonality in reproduction. Both immatures and adults of the more common species have been found throughout the year. An increase in population of the common species occurs in certain caves when seasonal rains and resultant moisture increases allow colonization of additional areas.

Data for most of the Hawaiian cave species suggest a much lower activity level, lower fecundity, and a different feeding behavior in the cave form when compared to its surface relative. This also is true for most continental species

(Vandel, 1965; Poulson and White, 1969). Because cavernicoles are poikilothermous animals and caves are generally cooler than the surface, especially in temperate regions, this activity may be temperature related. However, in at least one Hawaiian group, the Lycosidae, it is correlated to the degree of cave adaptation and not to temperature.

Cave Perturbations and Maintenance Trends

Continental caves are often viewed as islands, and their ecosystems share an apparent fragility in response to perturbations. Cave ecosystems on islands may be in double jeopardy, and several of the newly discovered arthropods are candidates for endangered-species status. What then is the future of this unique ecosystem, not even recognized before 1971? If perturbations had caused its demise sometime during the last two hundred years, biologists would have continued to believe no such fauna had ever existed in Hawai'i.

On Hawai'i Island there are still many avenues of dispersal between lava tubes, and continued new flows can be expected. Therefore one can expect the survival of the cave fauna, barring any major catastrophes. On the older islands of Maui and Kaua'i the caves are eroded remnants, many of the avenues of dispersal are closed through erosion, and the cave animals lead a tenuous, threatened, or endangered existence.

The major perturbations facing the Hawaiian cave ecosystem are destruction of the forest by grazing animals, fire, exotic plants, agriculture, and urbanization; creation of new entrances and increased siltation and filling of caves by the erosion resulting from the above activities; colonization by exotic animals; use of caves for refuse disposal; and direct disturbance by human visitation.

Since the main energy source is plant roots, the destruction of the overlying forest removes this energy source. The obligatory primary consumers die off rapidly, and the food web shrinks to a few scavengers and predators feeding on accidentals. A few primary consumers, notably *Schrankia* and the millepede, can feed on roots of some exotic species.

Troglobites are restricted to the true dark zone where relatively constant environmental conditions are maintained. New entrances introduce surface climatic influences, which destroy the cave fauna. Further, the erosion of the surface causes rapid filling of the voids in the lava and eventually the cave itself.

The introduction of exotic species, or biological pollution, is perhaps the most insiduous perturbation because it is usually irreversible, and it often pervades the ecosystem in unforeseen ways.

Using caves for refuse dumps hastens the filling of the cave and introduces large amounts of food that alter the cave food web in several ways. In relatively closed caves, it can foul the air and kill the inhabitants. By composting it can raise the cave temperature and thereby dry the cave. Most importantly, it often allows the colonization of the cave by opportunistic scavengers and predators. The high populations of these opportunists can destroy the endemic biota.

Kaua'i has few extant lava tubes, and the two known troglobites are among the most bizarre discoveries to date. Regrettably the fields with the largest caves known on Kaua'i were covered by five meters of sugarcane bagasse shortly before they could be analyzed scientifically. The caves are now gone, the fauna extinct, and no one will ever guess what that fauna might have been.

The entrance to Offal Cave, a relatively large lava tube on Haleakalā Volcano, Maui, was used as an offal pit by the local slaughterhouse, and the tallow, rotting bones, and other garbage are piled high near the entrance and scattered throughout the downslope portion of the cave. The cave is no longer used for this purpose, but even now the cave ecosystem approaches that reported for large bat caves in tropical continental areas, where a huge amount of animal matter is introduced into the cave. It now supports a large population of diverse troglophiles. These troglophiles are almost all exotic carrion feeders, scavengers, and predators, including such groups as earwigs, ants, moths, spiders, millepedes, and isopods. This is the only cave where many of these organisms have invaded any significant distance from the entrance, and it is assumed that the rich food supply allows them to colonize the cave environment. Only one endemic troglobite, the omnivorous cricket *Caconemobius* sp., occurs in this section of the cave.

Grazing has destroyed all native trees over Offal Cave, and no roots now penetrate into the deep cave. The influx of offal resulted in a second major perturbation, and it is thus difficult to predict whether other native troglobites would have survived the influx of offal and associated biota if the roots of the native trees had remained in the cave. We believe that most of these native troglobites would not have survived, as such a large amount of organic matter would have heated this cave and dried or otherwise affected the environment. Such a large influx of exotic predators, sustained by the high population of exotic scavengers, also would have preyed on any cave species, so that few endemics would have survived.

Caves are fragile ecosystems and, like some other discrete geologically defined habitats such as montane bogs and sand dunes, are easily disturbed by human visitation. Normal weathering processes are so changed and attenuated that even footprints can remain for centuries. In Hawai'i careless or destructive visitors kill or break tree roots, mark walls, litter the cave, and trample animals and their habitats. Tobacco smoke is a strong insecticide, and smoking in a closed cave may be lethal to the fauna. The heat from both the body and a torch, if used, can dry the cave. Any smoke also introduces a large number of condensation nuclei to the saturated cave atmosphere. Littering is related to the use of caves as refuse dumps. Cave visitation by the public should be discouraged until adequate protection of sample caves and ecosystems is assured.

CONCLUSIONS: TESTING OUR NICHE HYPOTHESES

We suggested three niche hypotheses:

1. That in island ecosystems we would encounter distinct differences in the numbers and kinds of general niches as compared to those in similar continental ecosystems.
2. That adaptive shifts may be common in island ecosystems in the sense that certain higher taxa assume functional roles there that they do not have in continental ecosystems.
3. That some general niches are not fully occupied in island ecosystems and as a result these niches are more easily invaded by exotic species.

In our analysis we were limited by necessity to only two island ecosystems: the very different montane rain forest and cave ecosystems. Moreover, our results apply to selected ecosystems in an oceanic island group, which was extremely isolated during the primary evolution of these ecosystems. We studied two original ecosystems, which, however, were not totally unaffected by the breakdown of isolation, which became very apparent with the advent of Europeans during the last two hundred years. Thus a certain influx of exotic biota had occurred in both island ecosystem examples. They are, of course, attributable to humans as the ultimate cause, but the more immediate cause relates to the mechanisms of interaction between the native and exotic species in given areas.

In attempting to verify the three hypotheses, we also have to keep in mind that we studied only four groups of biota in the montane rain forest: higher plants, tree arthropods, birds, and mammals. These may be considered among the most important, but there are others, such as the soil fungi, soil arthropods, fruit flies, and algae, which were considered in the gradient analysis. For verification of the three hypotheses, certain extrapolations can be made also for these other organism groups.

With regard to the higher plants, we did not encounter distinct differences in the numbers and kinds of general niches in our montane rain forest as compared to tropical montaine rain forests elsewhere. In this connection, we should clarify once more that we used plant life forms (in the sense of structurally and functionally similar species) as indicators of general niches. We found the same diverse spectrum of plant life forms in our evolutionarily isolated rain forest as is found in continental rain forests of the same kind.

However, a difference did become apparent, and it relates to the second hypothesis. The taxonomic composition of our forest plant community is no doubt extremely different at the species level. At the higher taxonomic level we could detect few differences. Legume trees (a higher taxon to which *Acacia koa* belongs) are common in the upper tree synusia in lower montane forests of tropical Asia, and myrtaceous trees (the taxonomic family to which *Metrosi-*

deros collina belongs) are common in upper montane rain forests of the same continental region. However, that we have only two tree species in the upper tree synusia is a significant departure from the typical tropical rain forests on the continents. Moreover, that both these species are shade-intolerant pioneer trees is another departure. In most continental rain forests studied thus far, the upper tree synusia is comprised of climax species, which can reproduce as populations under the shade of their own canopies. An adaptive shift can be recognized in the native ʻōhiʻa tree insofar as it has assumed the role of a climax tree in Hawaiʻi's rain forests. This tree species persists in the rain forest climates on all high Hawaiian islands, including Kauaʻi, which is estimated to have 5.6 million-year-old substrates. The tree species has adapted to a wide range of soil-moisture regimes, from excessively drained lava flows to permanently inundated boggy habitats. According to the latest information (Mueller-Dombois et al., 1977) it persists by periodic dieback in the form of canopy collapse and subsequent rejuvenation. Thus the second hypothesis can be positively verified to some extent, but it was not possible to do this without the above cited study, which was done after the IBP analysis.

With regard to the avian community, distinct differences in the numbers and kinds of general niches could not be detected. The bird life form spectrum appeared as complete as can be expected in relation to the native plant community. The dominance of the tree ferns (a spore bearer) in the understory may explain, in part, the absence of native seedeaters. The high life form diversity among insectivorous birds indicates a clear relationship to the rich, native, tree arthropod fauna. The three native nectarivores all overlapped to some degree in their exploitation of the major nectar-producing tree species, ʻōhiʻa. In the absence of large-fruit-producing trees, as are commonly found in tropical rain forests elsewhere, the frugivore niche was occupied by only one native bird species. Thus, the number and kinds of general niches occupied by the bird community must be viewed as a function of the structural diversity of the forest plant community. From our analysis all available bird niches appeared to be occupied, and thus the first hypothesis can be answered negatively.

Adaptive shifts are clearly evident in the avian community of the Kīlauea rain forest because the bird life form spectrum includes honeycreepers in all but three of the general niches. The three not occupied by honeycreepers were the frugivore, the seedeater, and the carnivore niche. According to our analysis, the six honeycreeper species in the Kīlauea forest occupied four different general niches.

The tree arthropod analysis did not indicate any distinct differences in numbers and kinds of general niches as compared to continental ecosystems of similar structure. Adaptive shifts were clearly indicated by the native tree borers, the *Plagithmysus* species. The five species had segregated into using six species of trees. In one case, where two borer species (*P. varians* and *P. claviger*) occurred on the same host tree *(Acacia koa),* they displayed considerable niche segregation in their use of different wood segments of the tree.

Thus our analysis resulted generally in negative evidence for the first hypothesis but in positive evidence for the second, although we found little information on the exact ecological role that the same higher taxa perform in continental ecosystems. This leaves us with an elucidation of the third hypothesis, which relates to the idea that some general niches are not fully occupied in island ecosystems, and, as a result, these niches are more easily invaded by exotic species. We were guided in this hypothesis by the general notion that high taxonomic diversity is positively related to community stability. Our results indicate that this is clearly not the case.

Relative niche occupancy relates to the number of species and their quantities in the various general niches as indicated by the life form groups (synusiae or guilds). As we found, life form groups with high numbers of native species are often occupied by high numbers of exotic species also. An example among the plants is the mat-forming herbaceous chamaephytes. Among the canopy arthropods there are several such examples, particularly among the entomophagous forms. Among the cave organisms, this applies particularly to the scavengers on microorganisms. Conversely there are a number of life form groups among the plants, birds, and tree and cave arthropods that are represented by rather few native species and that are not yet invaded by any exotics. Examples among the plants are the upper tree synusia (with two native species) and the tree fern synusia (with two native species). Among the birds we found four general niches among the insectivores, each occupied by one native species only, while in contrast, the nectarivore niche with three native honeycreepers was invaded by a rather successful exotic, the Japanese white-eye, a typical generalist. Among the canopy arthropods, we found purely native species assemblages only among the phytophagous life forms. Among the cave arthropods, the omnivore niche was filled by a single but quantitatively significant native species, a cricket (*Caconemobius* sp.). Thus there is evidence that, contrary to our earlier contention, native species diversity in a general niche has little to do with stability.

There is, however, a good possibility that the opposite relationship—low native species diversity in a given life form group (synusia or guild)—is correlated with high stability, particularly if this low native species diversity is combined with high population quantity. Good examples are the tree ferns, the phanerophytic epiphytes and the tall-growing native tree species among the plants, the insectivores among the birds, the native twig borers among the canopy arthropods, and the omnivorous cricket among the cave animals. The same relationship of community stability—low native species diversity per life form group combined with high native species quantity—holds for plant communities on new volcanic substrates (Smathers and Mueller-Dombois, 1974; Mueller-Dombois and Smathers, 1975) and for plant communities on old lava-derived soils (Gerrish, 1978).

This brings up another important point that is not included in the numerical relationship—the adaptation of native species to specific substrates, which often are too extreme for the normal exotic. Still another factor complex is the local environmental and species-interactive perturbations in given island ecosystems, which may prevent the successful establishment of the normal exotic. Only the extreme generalist among the exotics or the very similarly adapted specialist (which turns up rather rarely by chance) can be considered as potentially successful invaders in island ecosystems.

IV
Temporal Relationships of Island Biota

9

Introduction

K. W. Bridges

SHORT-TERM TEMPORAL PATTERNS

Examination of the temporal organization of ecosystems is a rich area for research. Studies range from those of paleoecology and evolution where vast time scales are examined through those focused on time periods of a few hours, such as in the organization of activity patterns within days. Intermediate scales include studies on successional problems, often with a time scale of decades. Seasonal studies focus on annual organization. Each of these time scales has some relevance for island ecosystems in Hawai'i, and studies performed as part of the IBP project have gathered evidence for each time scale. For example, insect (Gressitt, TR 71) and lava-tube studies (Howarth, TR 16) have gathered evidence on problems involving long time scales. Bird foraging behavior studies (Carpenter and MacMillen, TR 33, TR 63; Kamil, TR 68) have examined short-term processes. Successional studies of plant communities (Smathers and Mueller-Dombois, TR 10, 1974; Mueller-Dombois and Spatz, TR 13, 1975b; Spatz and Mueller-Dombois, TR 3, 1973, TR 15, 1975) involved intermediate time scales. The investigation of seasonality, emphasizing the annual pattern of ecosystem organization, is discussed in this part.

In a tropical area, especially one that is insular and has its climate moderated by oceanic influences, one would generally expect little if any pronounced seasonality. This is in contrast to continental regions with temperate climates where the seasonal responses in virtually all organism groups are so common and pronounced that even the most cursory examination reveals distinct phenological patterns. An examination of the degree of seasonality in the insular situation therefore seems appropriate if we are to test hypotheses relative to the role of temporal patterning. This should allow us to determine the generality of seasonality as an ecological process and the degree to which we can extend the utility of this concept to other geographic regions.

Where seasonal patterning exists, it provides an additional dimension in which plants and animals may be organized within an ecosystem. The importance of this organizational dimension has been widely discussed (Fretwell, 1972), particularly in relation to temperate climates where the temporal patterning is dramatic. Before examining whether such a phenomenon is of importance in the Hawaiian ecosystems examined as part of the IBP study, it is worth raising some questions and observations that tend to argue against seasonality organization in these environments.

The Hawaiian Islands exist in a tropical climate moderated by the surrounding ocean. The patterns of temperature change illustrate the degree of moderation. For example, an annual mean daily temperature range of 10.8°C was found for a typical transect site (1,600 m) for a typical year (1972). This value exceeds the temperature difference between the warmest and coolest mean monthly temperatures, 7.3°C, for the same site and period. This means that the organisms living at this site may experience a greater daily change in temperature than is exhibited by the mean annual trend. This may result in climatic cues that are either weak or obscured by climatic variability.

Even if there are sufficient climatic or other environmental cues to synchronize temporal responses, there are examples of atypical seasonal behavior. Mueller-Dombois (TR 4, 1972) found this to be the case for the relatively recently introduced perennial grass *Andropogon virginicus*. This grass becomes dormant during the wet but warm winter season on O'ahu, retaining a characteristic of temperate grasses. This seasonal behavior has implications relative to the water utilization in the soils in which it occurs and may be responsible for increased runoff and erosion in these areas. A number of other introduced grasses, however, do not show this seasonal behavior atypical to Hawai'i's climate patterns.

Other seasonal patterning requirements in ecosystems may relate, for example, to organism groups that temporarily utilize the resources of an ecosystem, such as birds feeding on flowers as they migrate. Since Hawai'i is isolated and not on many migration routes, there may have been less selective pressure for coevolution of such species relationships, and this in turn may have reduced the importance of seasonal behavior in these island ecosystems.

The results reported here are based on observations of the climate and the biota, primarily along the Mauna Loa transect. The rainfall and temperature data collected over a period of three or more years at the three transect climate stations form the basis for the climate-based seasonality analyses. Similar data are available for the climate station maintained at the National Park Service Headquarters. Records from this station extend back to the start of this century and allow for comparisons between the long-term record and the study period reported here.

Biological data collected during the same sample period from sites along the transect have been subjected to similar seasonality analyses. None of the studies of particular organism groups was designed specifically to study the temporal patterning, except the study of the phenological patterns of the vascular plants. In the other studies, data were collected with the primary purpose of examining the spatial distribution of the organisms or the genetic composition of the populations. The sampling design in some of these studies has allowed the dual use of the data. In the spatial-distribution analyses, the temporal variations in the measurements at a site were generally averaged. The emphasis in this part is focused on the temporal changes on a set of selected organisms at a few selected sites.

The importance of phenology and the analysis of seasonality were recognized early in the U.S.-IBP program with the establishment of the Phenology Committee (Lieth, 1974). The direct impact of this committee was not to initiate a separate phenology project but to provide a forum that emphasized the role of phenology as an important concept in the study of ecosystems. The results of this effort were seen in a number of IBP biome reports, including descriptions of biome models with distinct phenological elements (Cole, 1976). The Phenology Committee also provided a review of the status of phenological studies and a conceptual framework that can be applied to such studies. For example, seasonality was defined as "the occurrence of certain obvious biotic and abiotic events or groups of events within a definite limited period or periods of the astronomic (solar, calendar) year" (Lieth, 1974:5).

The emphasis here is placed not only on identifying whether biological events are seasonal but attempts to examine the correlation of these events with the climate patterns and the interrelations among separate events. As such, the results are concerned with phenology in the current sense of the use of this term (Lieth, 1974:4).

The analysis procedures used on the data collected in this program have blended the use of qualitative and quantitative techniques to demonstrate the existence or absence of seasonal patterns. The quantitative approaches have led to an objective testing of general (and simple) annual patterns. The qualitative analyses are less rigorous but have been more sensitive to revealing pattern complexities.

The application of some techniques used in these analyses, such as the use of climate diagrams, is well established. Other techniques are relatively new (Platt and Denman, 1975) in their application to ecological problems and are still subject to further evaluation. This emphasizes the relative infancy of this field of study and should be kept in mind while evaluating the results presented here.

The description of the climate of the Mauna Loa transect, based on analyses of rainfall and temperature data, is important to the establishment of whether environmental seasonality exists in the study area and if it does, what seasonality properties are shown. This is the primary subject of this chapter, which also presents brief descriptions of the techniques thought to have some utility in quantifying or describing seasonality patterns.

SEASONAL DISTRIBUTION OF RAINFALL AND TEMPERATURE

The general seasonality of the climate along the Mauna Loa transect is characterized here by the measurement of air temperature and rainfall. Measurements were made at three climate stations, located at IBP sites 4 (1,280 m, 4,200 ft), 6 (1,600 m, 5,250 ft), and 9 (2,040 m, 6,700 ft). The locations of these sites are shown in figure 2-1. The period of recording extended from early 1971 until mid-1975, with a small difference in the starting date of record for the three stations. The period considered in the analyses discussed in this chapter, however, will be restricted to the three calendar years with complete records for all three stations, 1972 through 1974. (Detailed listings of the climate measurements are available in TR 22, TR 38, TR 59, and TR 70.)

There are some historical records of air temperature and rainfall for the Hawai‘i Volcanoes National Park Headquarters site, located near the base of the Mauna Loa transect between IBP sites 1 and 2. This is called the HQ site in this chapter and is identified in government weather records as Hawai‘i Station 54. The historical climate records for the park have been summarized by Doty and Mueller-Dombois (1966). General comments on the climate of the State of Hawai‘i and some of the more specific aspects of the climate of the Island of Hawai‘i have been discussed in chapter 1 and can be found in the various publications cited there. Data listings with historical records have been summarized for this island in a report by the Department of Land and Natural Resources (State of Hawai‘i, 1970).

Rainfall and Air Temperature Measurements

Rain was collected in American Standard rain gauges, which were located in open areas at the recording sites. The gauges were read at approximately the same time each Monday morning. During weeks with heavy rainfall, the gauges were also recorded and emptied during the middle of the week. Since this was not always possible, some of the large rainfall values are underestimates of the actual rainfall due to the overflowing of the gauges.

The rainfall values used here do not incorporate the precipitation contributed by fog interception. In many places in Hawai‘i, this source of precipitation must be considered (TR 32). The fog belt on the Island of Hawai‘i is a well-defined band that occurs at elevations between 1,500 and 3,000 m. At about 2,500 m, the annual precipitation contributed by interception (as measured from standard fog interceptors) may be equivalent to 65 to 70 percent of the annual rainfall (Juvik and Ekern, 1978). Data collected along the Mauna Loa transect by these workers during nine months in the June 1974 to June 1975 period indicate that only at the Hawai‘i Volcanoes National Park Headquarters site was there an appreciable contribution to the total precipitation from this source. This was equivalent to approximately 20 percent of the measured rainfall for the period. These differences from expected higher values were attributed to the geographic position of the transect.

Air temperature was recorded on seven-day strip chart recording hygrothermographs located in Stevenson screen enclosures. The height of the recording shelters was approximately 1.5 m. Data from the strip charts were manually transcribed to data forms by reading the value at each two-hour interval. When the records were obviously in error, such as if a pen became stuck, the data values were not recorded. The transcribed data records were independently verified by comparing the data listings after keypunching to the original strip charts.

The monthly mean temperatures have been calculated from the sum of all the bihourly values for the month divided by the number of values included in this sum. There has been no correction for the bias that may have resulted from the missing values. An examination of this potential source of bias showed it to be negligible for the data analyzed here.

Temporal Patterns in Climate Data

The temporal scale chosen for examination is based on single years with the dominant pattern expected to be a single peak within the year. Obviously such a pattern applies only to those areas with relatively simple climates, such as are found throughout Hawai'i, and does not consider bimodal or more complex annual climatic changes. There has also been no attempt to consider the pattern of long-term change, although the long-term record is used here for comparisons with the short-term study period.

An analysis of the long-term rainfall and temperature patterns for the data available from the HQ site over a many-decade period has formed the basis of defining the climatic patterns typical for that site. These patterns can then be compared to the data for the three-year study period, both in terms of the mean annual values and the pattern of change within the year, to determine the amount of variation experienced during the study period relative to the long-term average. This leads to an evaluation of how representative the study period was, at least at the single site where the long-term record is available. The comparison of the three-year records over the same period for both the HQ and three IBP sites then allows an interpretation of the change along the elevational transect in the context of the historical record. This is the general procedure used in the following analyses of rainfall and air temperature.

Annual Rainfall Patterns

A plot of the annual total rainfall for the HQ site (1,210 m, 3,971 ft) is shown in figure 9-1 for the years 1899 through 1962. The mean annual rainfall for the sixty-three-year record, 2,365 mm, is shown as a horizontal line so that the year-to-year total rainfall can be evaluated relative to the average value. The frequency distribution of these yearly rainfall totals (figure 9-2) can be described as a normal curve, although it is skewed to the right and peaked, indicating that more years are drier than the mean rainfall. The deviations from the shape of a normal curve are not significant at the 1 percent level of significance, however. The standard deviation of the annual rainfall values is 633 mm.

The long-term record can be compared to the study period (1972–1974) to evaluate how typical these years were to an average year. The three-yearly total rainfall values, the three-year mean, and the long-term mean are shown in table 9-1 for the HQ and IBP sites. It is interesting to note that the HQ site recorded a

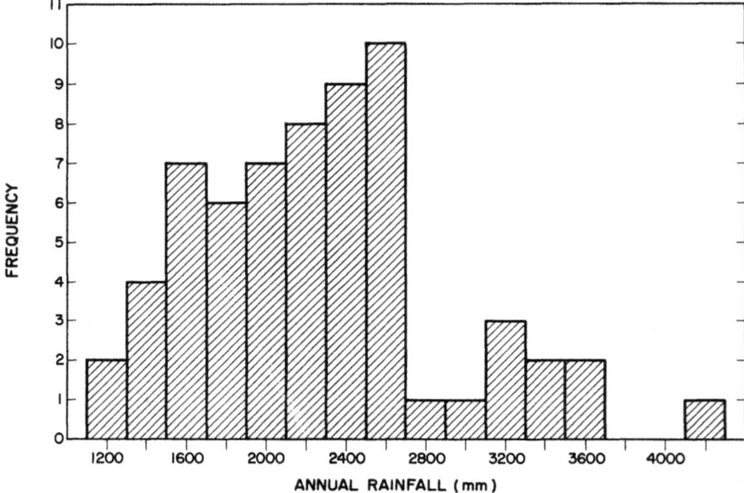

FIGURE 9-1. *Frequency distribution of annual rainfall at HQ station.*

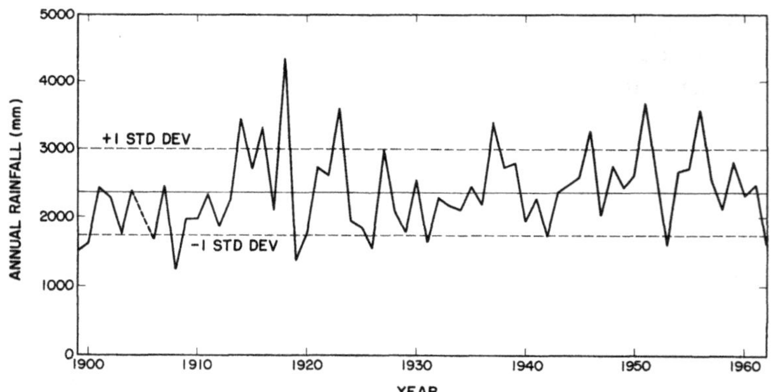

FIGURE 9-2. *Long-term pattern of annual rainfall at HQ site. Data are missing for the year 1905. The solid horizontal line is the mean annual rainfall (2,365 mm), and the dashed horizontal lines are the one-standard-deviation limits (1,732 and 2,998 mm).*

TABLE 9-1. *Annual Rainfall Totals for HQ and IBP Sites (in mm)*

	HQ site (1,210 m)	Site 4 (1,280 m)	Site 6 (1,600 m)	Site 9 (2,040 m)
1972	2,225	1,459	1,476	1,151
1973	2,343	1,391	1,696	1,657
1974	2,784	1,777	1,595	1,430
3-year mean	2,451	1,542	1,589	1,413
Long-term mean	2,365			

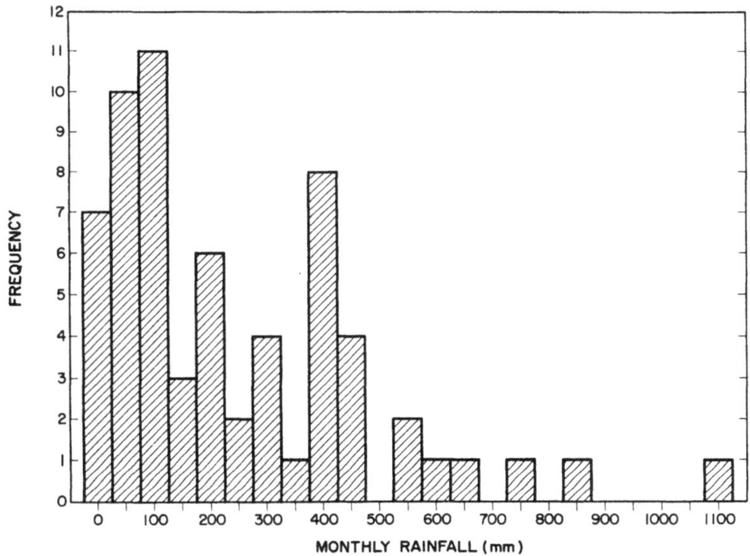

FIGURE 9-3. *Frequency distribution of January rainfall values at HQ station.*

nearly identical three-year mean as compared to the long-term mean. Based on this evidence, it appears that the study period had a reasonably typical annual amount of rainfall.

The monthly rainfall values examined for each month over the long-term record show a general tendency for skewed and peaked distributions. A typical month, January, is plotted as a frequency histogram in figure 9-3 to show how there is a dominance of years in which the January rainfall was lower than the mean (265 mm). In this case, 44 percent of the years had January totals less than 150 mm. At the other end of the distribution, several January totals had a very high value, with 11 percent of the years having over 500 mm rainfall and with one month having over 1,100 mm (42 in). Although the deviations from nor-

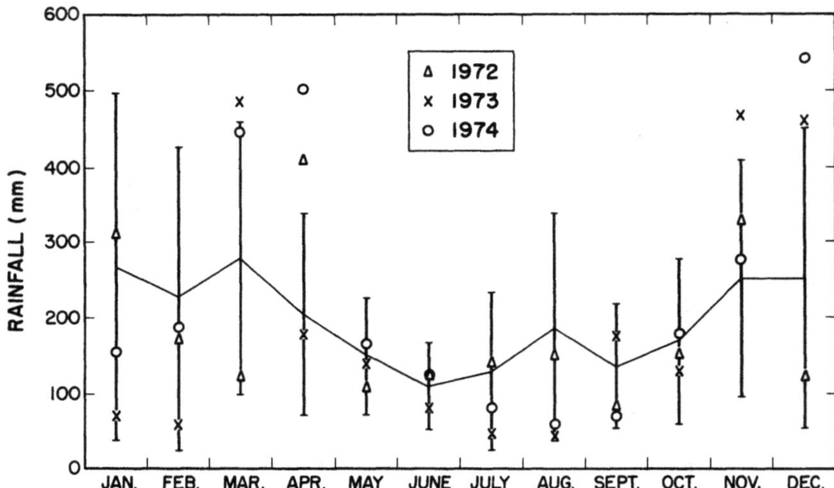

FIGURE 9-4. *Mean monthly rainfall and variances for long-term record at HQ site and actual monthly values for three-year study period (1972–1974). The vertical bars indicate the range for one standard deviation above and below the monthly mean.*

mality seen in the monthly distributions are generally significant, the standard deviation will still be used here as a general guide to the expected variation between years for a particular month's rainfall.

The monthly variance in rainfall is of interest because it is an expression of the deviation that might be expected from the average conditions. The monthly variance values have been plotted, based on the long-term record for the HQ site, with the monthly mean values, in figure 9-4. There is a large component of variability, and therefore a long-term record of rainfall should be used to obtain reliable estimates of the mean monthly rainfall values.

A more detailed evaluation of the study period is provided by an examination of the monthly rainfall values for the three-year study period as compared to the long-term monthly mean rainfall and deviations for the HQ site. This also is shown in figure 9-4. Only six of the thirty-six monthly records fell outside a confidence interval of one standard deviation above and below the mean value, less than might be expected, and thus giving further evidence of these being relatively average rainfall years for the HQ site.

The relationship of the HQ site to the higher elevation IBP climatic stations has been established by correlating the monthly rainfall values between the stations (table 9-2). While the IBP sites are closely correlated (with at least an *r* value of .95), there is considerably less correlation between the HQ site and each of the IBP sites. The correlation coefficient decreased slightly with increasing distance between the sites.

TABLE 9-2. *Correlation of Monthly Rainfall Values for Three-Year Study Period (r values)*

Site	Elevation	4 (1,280 m)	6 (1,600 m)	9 (2,040 m)
HQ	1,210 m	.68	.62	.60
4	1,220 m		.97	.95
6	1,600 m			.98

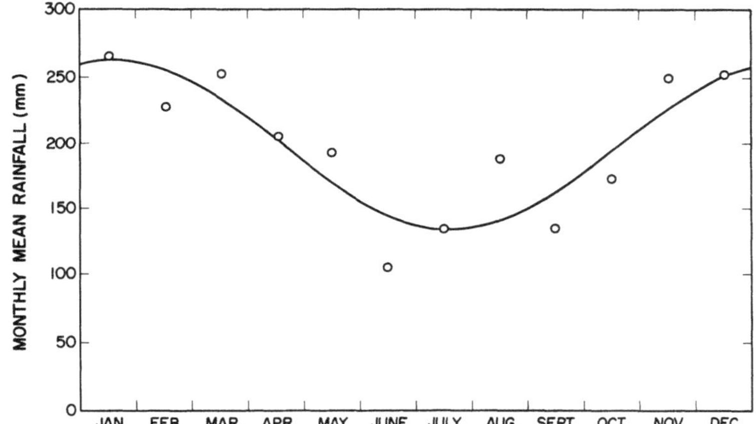

FIGURE 9-5. *Monthly mean rainfall values for long-term record at HQ site and least-squares fit curve using a sine-based function.*

The assumption underlying the hypothesis of seasonality is that this distribution is a simple annual pattern, so it is logical to test for this pattern by fitting data to a simple annual curve. If the fit is good, then a description of the climatic change at the site can be given as the parameters for this curve.

The typical distribution of the rainfall within a year has been obtained by an analysis of the monthly mean rainfall values for the HQ site for the long-term record. All of the values for each month were averaged, and then these mean values were fitted with a twelve-month-based sine function using a least-squares fitting procedure. The resulting curve has a correlation coefficient *(r)* of .88, indicating that the mean monthly rainfall values form a general trend that can be approximated by such a simple periodic function. This establishes, at least for the mean monthly rainfall values, that the sine-based annual pattern is a useful descriptive model for annual periodicity. Fitting values from shorter-term records to this model, therefore, can be done with some confidence.

The mean monthly rainfall values for the HQ site for the long-term record and the curve based on the least-squares fit to the sine-based function are shown in figure 9-5. The overall mean value for the fitted function is 199 mm monthly

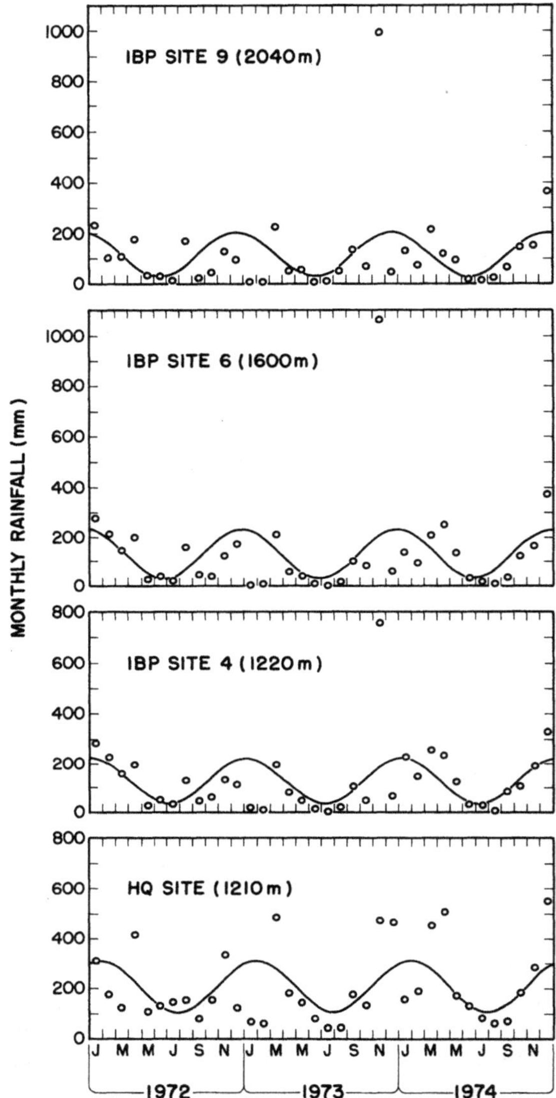

FIGURE 9-6. *Monthly rainfall totals for three-year study period at HQ and IBP sites and least-squares fitted curves using a sine-based function.*

rainfall. The calculated amplitude of the yearly change is 126 mm, predicting that there will be such a difference between the wettest month (262 mm) and the driest month (136 mm) during an average year. The rainfall peak is expected around mid-January, with the driest time coming six months later in mid-July.

The pattern of the rainfall over the three-year study period for the HQ and three IBP sites can be seen from the trend of monthly values, which have been fitted with the annual sine-based function (figure 9-6). The properties of these curves are given in table 9-3, along with the values for the fit of the long-term record at the HQ site. The monthly mean values show a generally decreasing trend as the elevation of the station increases. The three-year and long-term mean values for the HQ site were very similar. The three-year records, however, had a considerably larger annual amplitude than the long-term average. There was an interesting shift in the date of the maximum rainfall, becoming approximately a month earlier at the highest elevation (IBP site 9, 2040 m) as compared to the lowest elevation (HQ site, 1210 m).

Annual Air Temperature Patterns

The mean monthly air temperature data for the HQ site are shown in figure 9-7 for the fifty-year period extending from 1906 to 1956. The mean air tem-

TABLE 9-3. *Properties of Curves Fitted to Monthly Rainfall Values for Long-term HQ Site Record and Three-Year Study Period for HQ and IBP Sites*

	Long-term	Three-Year record			
	HQ site (1,210 m)	HQ site (1,210 m)	Site 4 (1,280 m)	Site 6 (1,600 m)	Site 9 (2,040 m)
Mean rainfall (mm)	199	204	129	132	118
Annual amplitude (mm)	126	203	185	200	177
Date of maximum rainfall	Jan. 18	Jan. 27	Jan. 6	Dec. 24	Dec. 15
r	.88	.49	.48	.39	.36

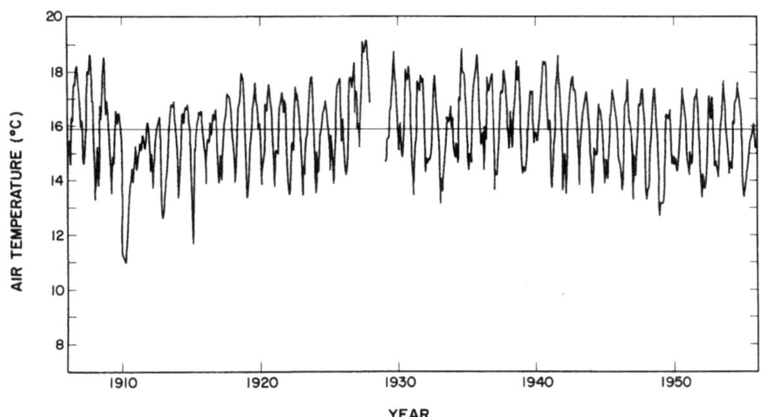

FIGURE 9-7. *Mean monthly air temperature for HQ site.*

perature for this period was 15.9°C and is shown in the figure as a horizontal line. The monthly variance in temperature, based on calculations from the long-term record at the HQ site, have been plotted with the mean monthly values in figure 9-8. The standard deviation is nearly identical each month, in contrast to that described earlier for the monthly rainfall where there was more variability expected in the winter months.

Comparisons of the long-term record to the three-year study period (1972–1974) are shown in table 9-4. The three-year mean air temperature at the HQ site was 0.2°C cooler than the long-term mean of 15.9°C. This is a close value, indicating that the study period was typical of the long-term value at this site.

The warmest temperatures were recorded at the HQ site. This is the lowest elevation site and is also located in the rain forest zone. It received approxi-

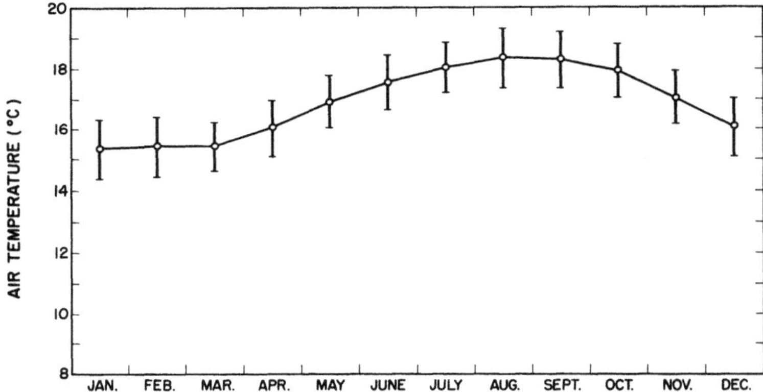

FIGURE 9-8. *Mean monthly air temperature values and variances based on long-term record at HQ site. The vertical bars indicate the range of one standard deviation above and below the mean.*

TABLE 9-4. *Comparison of Long-term Mean Annual Temperature for HQ Site, with Annual Mean Temperatures for HQ and IBP Sites*

	HQ site (1,210 m)	Site 4 (1,280 m)	Site 6 (1,600 m)	Site 9 (2,040 m)
1972	15.6	16.2	13.3	11.0
1973	15.4	13.3	13.3[a]	10.4
1974	16.0	13.9	12.9	10.4
3-year mean	15.7	14.5	13.2	10.6
Long-term mean	15.9			

Note: Temperatures are in °C.
[a]Only ten months (missing April and May) used in the calculation.

mately 60 percent more rain annually than did the higher elevation IBP sites. This difference may partially account for the steep lapse rate (1.7°C per 100 m) observed between the HQ and IBP-4 site, which is located in the mesophytic forest. A more moderate lapse rate was observed between the IBP sites (site 4 to 6, .4°C per 100 m; site 6 to 9, .6°C per 100 m).

The mean monthly air temperature values based on the long-term record for the HQ site have been analyzed for their annual pattern by fitting them with a sine-based function, following the same procedure described earlier for the long-term mean monthly rainfall values. Figure 9-9 shows the mean monthly values and the curve based on the least-squares fit to the sine-based function. The fit to this function is very good, with an *r* value of .99. The mean temperature was 15.9°C, and the annual amplitude was 3.2°C, indicating that an average year at the HQ site is expected to have a maximum mean monthly temperature of 17.5°C and a minimum mean monthly temperature of 14.3°C. The maximum temperature is predicted to occur on approximately August 18 and the minimum six months later on approximately February 18. There are no obvious periods of the year where the fit is any worse than in any other period, indicating that the sine-based curve also adequately describes the annual pattern of temperature change at this site.

The three-year air temperature values for the HQ and IBP sites have been analyzed using the annual sine-based function to determine how closely they fit an annual pattern. The plots of the monthly mean values and the least-squares fit curve are shown in figure 9-10. The properties of the curves are given in table 9-5.

The comparison of the pattern established by the long-term record for the HQ site is close to that of the three-year record for the mean temperature, annual amplitude, and the date of the maximum temperature. This indicates that the

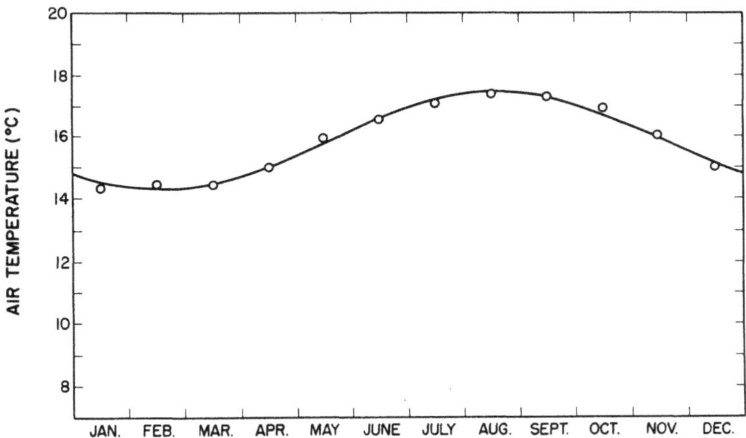

FIGURE 9-9. *Mean monthly air temperature values for long-term record at HQ site and least-squares fit using a sine-based function.*

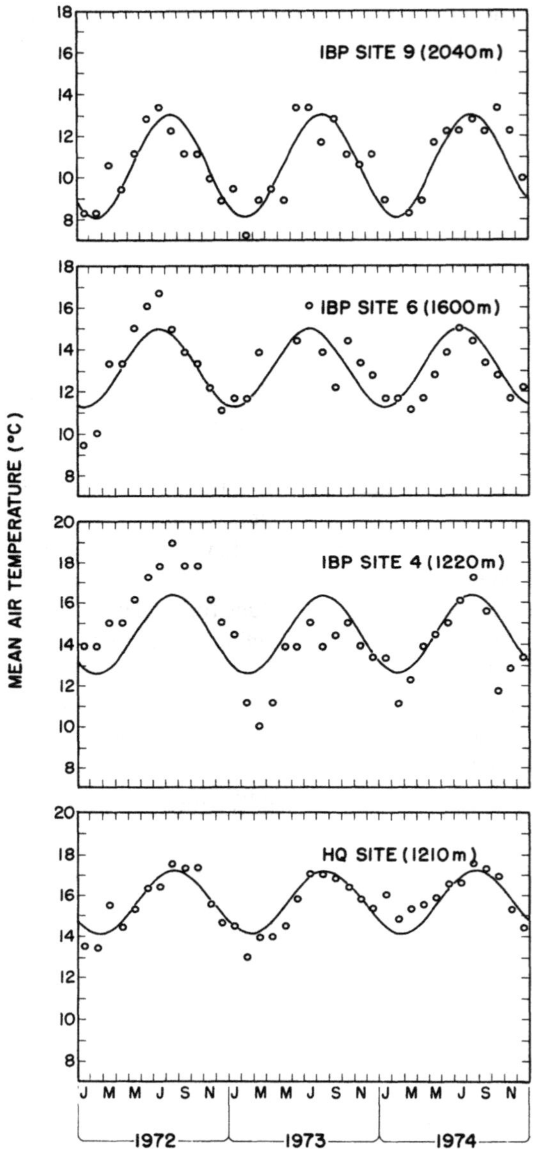

FIGURE 9-10. *Mean monthly air temperature for three-year study period at HQ and IBP sites and least-squares fitted curves using a sine-based function.*

TABLE 9-5. *Properties of Curves Fitted to Monthly Mean Air Temperature Values for Long-Term HQ Site Record and Three-Year Study period for HQ and IBP Sites*

	Long-term	Three-year Record			
	HQ site (1,210 m)	HQ site (1,210 m)	Site 4 (1,280 m)	Site 6 (1,600 m)	Site 9 (2,040 m)
Mean temperature (°C)	15.9	15.7	14.5	13.1	10.6
Annual amplitude (°C)	3.2	3.1	3.7	3.7	4.9
Date of maximum temperature	Aug. 18	Aug. 21	Aug. 15	July 15	Aug. 12
r	.98	.87	.64	.81	.80

annual pattern for the study period was typical of the long-term average pattern. The fit of the three-year data was good, with an *r* value of 0.87.

The patterns for the three-year records show the expected tendency for decreasing annual mean temperatures with increasing elevation. The values for the mean monthly temperatures were the same as determined by the least-squares curve-fitting procedure as given previously, except for rounding errors. The annual amplitude of the temperature change increased considerably with increasing elevation. The amplitude of IBP site 9 (2,040 m) was nearly 1.6 times greater than that of the HQ site (1,210 m). The date of the maximum temperature covered approximately a one-month period from mid-July to mid-August. Although it appears that the maximum temperature occurs earlier at higher elevations, the pattern is not clear based on these data. There was no systematic pattern to the fit of the data to the sine-based function. All of the sites except IBP site 4 have reasonably high *r* values. The *r* value at IBP site 4 was lower, although still high when compared to the *r* values obtained for rainfall.

SEASONALITY BASED ON CLIMATE DIAGRAM ANALYSES

Standard Climate Diagrams

The use of climate diagrams for the ecological characterization of the climate at a site was systematized into a model by Walter (1955, 1957). The method has been applied worldwide (Walter and Lieth, 1960–1967; Walter, Harnickell, and Mueller-Dombois, 1975). Climate diagrams for twenty-one sites on the Island of Hawai‘i were given in chapter 1 to show the spatial pattern of the climate across the entire island. The HQ site was included in the climate diagrams shown in figure 1-5.

An examination of the climate diagram for the HQ site indicates only a weak tendency toward seasonality. There is no arid period as would be indicated if the plotted rainfall curve were to fall below the air temperature curve. Indeed, the rainfall curve is generally in the section of the climate diagram indicating more than 100 mm rainfall per month, although this line does dip below this value at mid-year. The temperature curve is considerably below the rainfall curve throughout the year, indicating that there is no period of water stress to the vegetation.

This interpretation is based on the scaling of the climate diagram ordinates. On the left-side ordinate, 10°C corresponds to 20 mm precipitation on the right-side ordinate. In this way temperature values can be used as an index of drying. It is generally considered that this empirically derived relationship is useful because it considers the acclimatization of the plants. Other measures, which have a more direct focus on potential evapotranspiration, have been used as a means of determining the water balance of sites on the Island of Hawai'i, including the Mauna Loa transect (Juvik, personal communication). These studies have used continuous-level pan evaporimeter measurements as verification of the potential evapotranspiration relationships. In general, evaporation was observed to decrease with increasing elevation along the transect. There was a linear relationship, with 3.18 mm per day evaporation at site 4 (1,280 m), decreasing to 2.60 mm per day at site 9 (2,040 m). This resulted in a predicted evapotranspiration curve that came closer to approaching the rainfall curve during the summer than was seen when only a temperature curve was used. This change, however, is not sufficiently great to indicate a drought period at the HQ site, although it would be expected to indicate a drought at the higher elevations along the transect.

Although the climate diagrams show the average pattern of rainfall and temperature relative to each other, they do not give any idea of how much year-to-year variation can be expected in either the rainfall or temperature at a site. It is possible to use two slight modifications of the basic climate diagram technique to show the year-to-year variation. One modification is to plot the actual monthly values for a multiyear period in a climate diagram with an extended year axis to see the variation in the pattern; this is called a climatogram (Walter, 1971). The other modification is to plot the annual mean pattern, as in a standard climate diagram, but to add lines for the variances associated with the monthly values. Each of these modifications will be used to clarify further the climatic patterns of the study sites.

Climatograms

Climatograms for the HQ and three IBP sites are plotted in figure 9-11 for the three-year period examined in this chapter.

The climatogram for the HQ site does not show any arid months, which would be indicated by the rainfall line undercutting the temperature curve, for

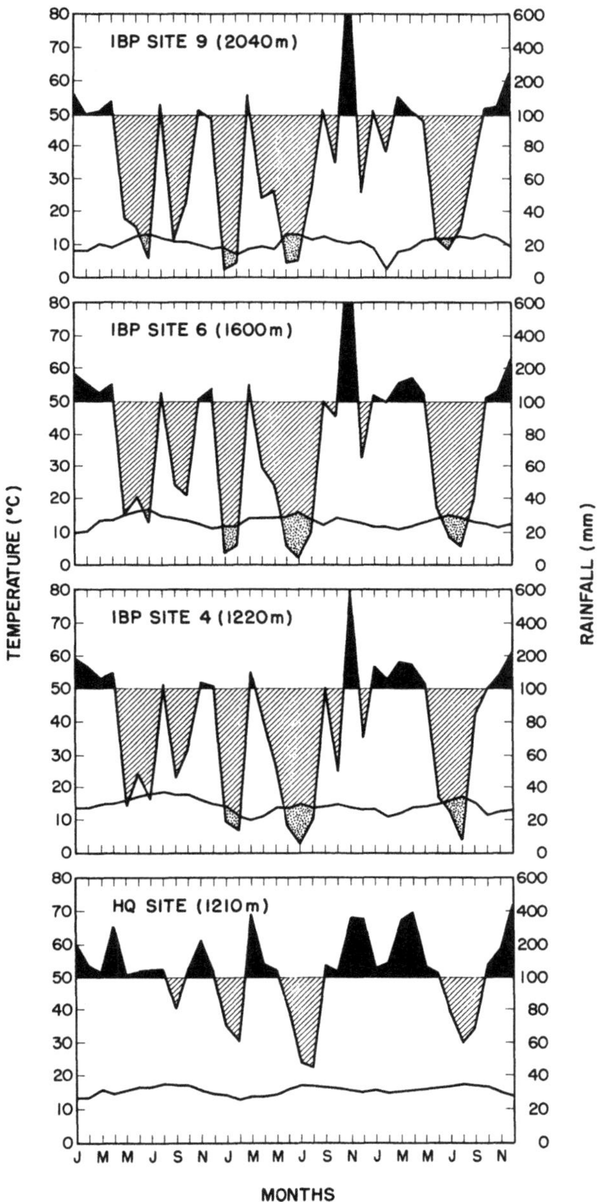

FIGURE 9-11. *Climatograms for three-year study period for HQ and IBP sites, from January 1972 through December 1974.*

the 1972–1974 period. In one period, July–August 1973, the rainfall curve approaches the temperature line, but it was not sufficiently dry to be considered as moisture deficit. The higher elevation IBP sites have climatogram patterns that show distinct arid periods at four times during the three years. Generally these arid periods fell in June and July. Each of the three years showed such a summer drought at all three IBP sites, although it was not pronounced at IBP site 4 in 1972. In addition, there was a drought period at all IBP sites during January–February 1973, indicating that winter dry periods do occur, even though winter is generally considered the wettest portion of the year and is so indicated on climate diagrams based on long-term mean values.

Climate Diagrams with Variance Lines

The use of climate diagrams in a standard twelve-month format can be enhanced by the addition of lines for the monthly variance in both temperature and rainfall. This allows a compact expression of the climatological variation for long-term records. To make such a plot, the standard deviations for each month's rainfall and temperature are calculated and then added to and subtracted from the mean values to obtain monthly high and low values for both rainfall and temperature. These six values for each month—the mean, high, and low for each of the two parameters—are plotted with lines connecting each value in the set using the standard climate diagram conventions for scaling.

A variance-enhanced climate diagram for the HQ site is shown in figure 9-12. The calculations are based on the sixty-three-year rainfall and fifty-year temperature records. This plot shows that the line for low rainfall distinctly undercuts the temperature curves for July and comes just to the mean temperature curve for February. This indicates that one, perhaps two, arid periods can be expected as a normal part of the climatological pattern for some years, even though this is not evident in a standard climate diagram for the same site. This confirms the patterns of arid periods that were seen in the three-year climatogram.

WET AND DRY MONTH ANALYSES

Alternating periods of wet and dry months have been suggested, in several different contexts (Blumenstock and Price, 1967), to be a measure of climatological seasonality. Mohr (Whitmore, 1975) has proposed an index that can be used to characterize the yearly wetness of a site based on the number of wet and dry months. Such an index is most useful for lowland tropics; however, it is discussed here to illustrate an alternative approach to defining climatological seasonality. In the case of lowland tropics, "wet" is defined as a month with more than 100 mm rainfall, with a dry month having less than 60 mm. The actual values for these limits vary from region to region. Janzen (1967) has suggested that 150 mm per month may be dry for an area with very high average wetness; thus the relative amounts in the rainfall are important.

Multiyear records can be analyzed by averaging the number of wet and dry months for all of the years and using the formula

$$\frac{\text{dry}}{\text{wet}} \times 100 = I_{\text{Mohr}} \cdot$$

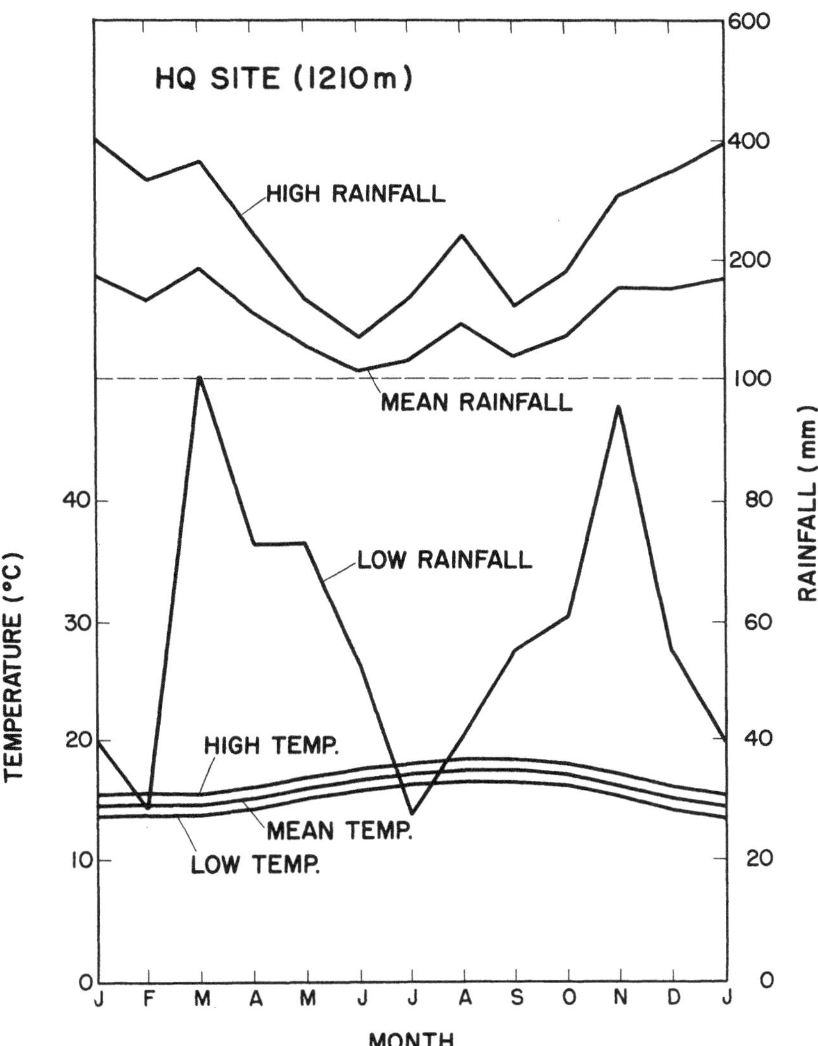

FIGURE 9-12. *Variance-enhanced climate diagram for HQ site. Dashed line at 100 mm indicates scale change for rainfall.*

A four-level classification scheme has been proposed (Whitmore, 1975) that groups sites by their index value into the ranges 0–14.3, 14.3–33.3, 33.3–100, and 100+. The lower the value, the wetter the site.

An examination of the general trend of the monthly values, based, for example, on the long-term monthly mean rainfall values such as were given in figure 9-4, indicates that no months are below the minimum wet value (60 mm) and that the index value, therefore, would be zero. This would imply that there is no climatological seasonality by this type of measure. Yet when each individual year is examined, it is found that there are averages of 1.38 dry and 8.13 wet months. This gives an index of 17 for the HQ site and places it within the next-to-wettest category, indicating the possiblity of some general wet-dry period seasonality. An average of each of the yearly indexes is 21.7, with a standard deviation of 28.8. Thirteen years of the sixty-three analyzed had an index value that would place them in the two driest categories (33.3 or greater). This shows that there are significant periods that are dry even at a site with a very high average wetness.

The tendency to have periods that are dry at a site with a high average wetness can also be seen from an examination of the yearly records to see how often the site is consistently wet. The measure that has been used is how often there are years in which there is no month with less than 100 mm rainfall. The long-term record for the HQ station shows only a single year in which this consistent wetness occurred, further indicating frequent dry periods at such a site.

An examination of the long-term record shows that there were eighty-four occurrences of dry months, or approximately 11 percent of the months of record. The length of these dry periods was usually a single month. The frequency of single months was sixty-one times (or 7.3 percent of the months of record). There were eight periods that had two consecutive dry months and one each for three and four consecutive dry months. The three-month dry period started in January, and the four-month dry period started in July. There was no consistent pattern to the months in which the dry periods occurred, with a nearly equal probability of this site experiencing a dry month in the winter as in the summer.

A comparison of the wet and dry months for the three-year study period (table 9-6) shows a generally increasing trend in the number of dry months (less than 60 mm per month) per year with increasing elevation. Few dry months were observed for this short-term period at the HQ site. The number of wet months (over 100 mm per month) shows the opposite pattern, with considerably more wet months recorded at the lower elevations. The three-year and long-term mean values for both the number of wet and dry months compare favorably at the HQ site, giving independent evidence that the three years being examined in this study were relatively typical for this measure of seasonality.

An examination of the Mohr index values for the four stations indicates a trend from wet to very dry as elevation increases. As a result, it is expected that

TABLE 9-6. *Wet and Dry Month Comparisons for Long-Term and Three-Year Periods*

	HQ site (1,210 m)	Site 4 (1,280 m)	Site 6 (1,600 m)	Site 9 (2,040 m)
1972	0/11	4/7	5/7	5/6
1973	2/7	7/3	6/3	8/3
1974	0/9	3/8	4/7	3/6
3-year mean	.7/9.0	4.7/6.0	5.0/5.7	5.3/5.0
Long-term mean	1.4/8.1			

Note: Wet months are defined as having less than 100 mm rainfall and dry months as less than 60 mm. Comparisons are given as number of dry months to number of wet months.

seasonality, which is based on alternating periods of wet and dry months, would be considerably more important at the three IBP sites than at the HQ site.

ANNUAL PATTERN ANALYSIS BY CURVE FITTING

The purpose of the temporal data analyses was to determine if there is any regular pattern to the changes in the organism groups sampled. In particular, the analyses were directed at determining whether the patterns showed annual periodicity. The analysis used here is a special case of the use of the spectral analysis technique with time-series data (Platt and Denman, 1975). This approach to the analysis of seasonality has been described by Waggoner (1974) and applied by Bliss (1970) to short-term patterns.

The analysis procedure used for the data collected in the IBP studies applied a least-squares curve-fitting technique to data transformed with a sine function. The more standard spectral analysis techniques were not used here due to the lack of uniformly spaced observations and since only one time frequency was of interest. The function being fit is described as

$$Y = a + b \cdot \sin((c + X) \cdot 2\ d^{-1}),$$

where a is the mean, b is the amplitude, c is the displacement of the curve, and d is the fundamental period of the function. Since we are restricting our consideration to an annual period, we can make the fundamental period, d, equivalent to one year.

The properties of this periodic function can be seen from plots of families of curves which differ in only a single parameter (figure 9-13). Note that three parameters are used to describe curves based on this function. Wherever this function is fitted, the displacement value, c, is expressed as a date rather than as its calculated equivalent numerical value.

Pearson's correlation coefficient is used as the measure of the fit of the observed data points to the sine-based function. The correlation coefficient, designated as r, has values ranging from 0 to 1, with 1 being the best fit.

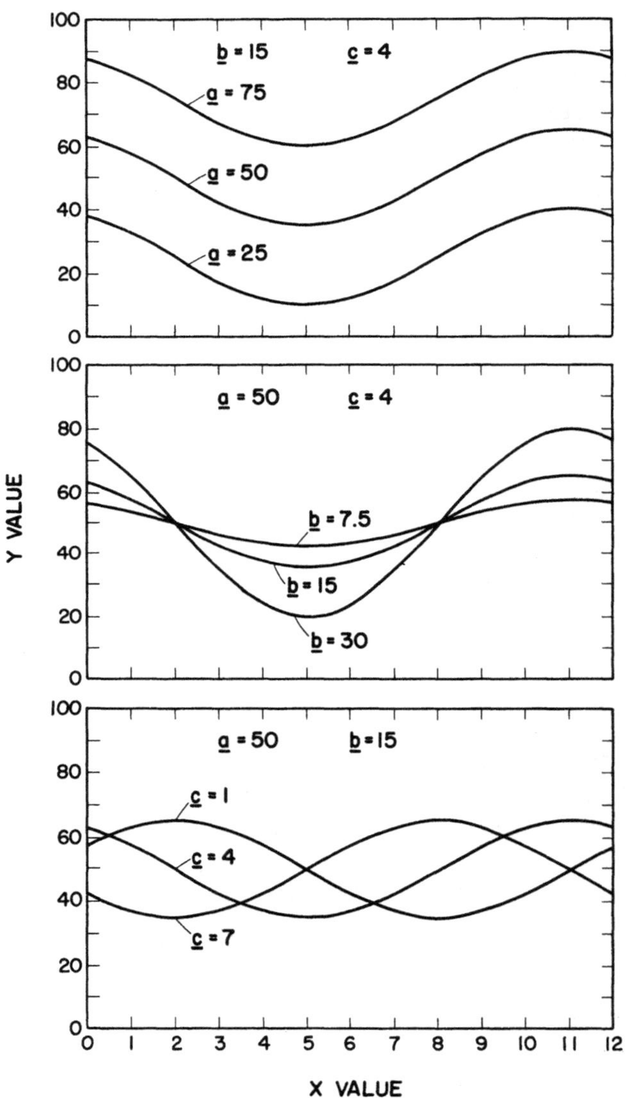

FIGURE 9-13. *Families of curves produced by using sine-based function and varying one parameter at a time.*

Negative values do not occur for this coefficient because of the manner in which the curve has been fitted to the data.

The utility of the sine-based function in analyzing climatic data has already been shown for the long-term records (figures 9-5 and 9-6) and three-year records (figures 9-9 and 9-10). The degree of fit in these analyses ranged from very high (.99) to low (.38). There is no indication that the variability encountered would be explained better by any other simple, functional form.

Several examples of the application of this curve-fitting procedure illustrate its utility for the analysis of biological data. The two sets of data have been chosen because one is a good fit and the other is a poor fit. Both curves fitted in figure 9-14 are for data on the flushing of *Sophora chrysophylla*. The data obtained at the higher elevation (2,040 m) fit the sine-based curve quite well, with an r value of .81. The data points generally occur near the curve, and the curve comes relatively close to meeting the range of the data points. Data for the *Sophora* flushing at the lower elevation (1,570 m) are quite a poor fit to this

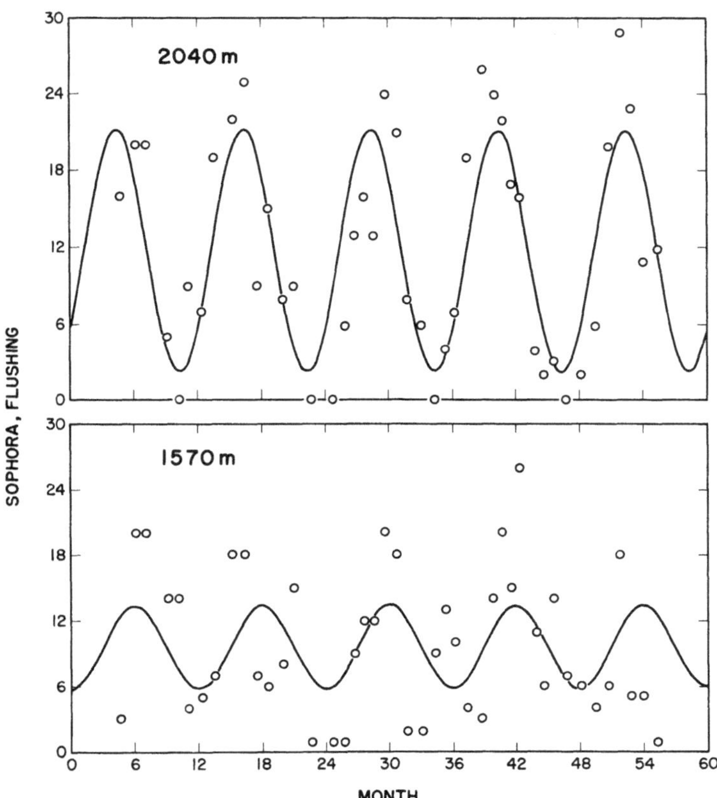

FIGURE 9-14. *Examples of sine-based curves that fit data reasonably well (2,040 m,* r = *.81) and quite poorly (1,570 m,* r = *.42).*

function, with an r value of .42. There is little pattern to the distribution of the data points relative to the curve, and the amplitude of the curve is considerably less than the range of the data points.

One of the problems in using this type of function is that it does not perform well when there are considerably different mean values for each annual cycle. For example, each annual period may have a very clear seasonal population pattern within the period but very different mean population levels for the successive annual periods, thus a poor fit can be expected. In these cases, mean values based on calendar months have been used to reduce this difficulty.

10

Temporal Variation of Organism Groups Studied

K. W. Bridges, C. H. Lamoureux, D. Mueller-Dombois, P. Q. Tomich, J. R. Leeper, J. W. Beardsley, W. A. Steffan, Y. K. Paik, and K. C. Sung

TREES*

Patterns of seasonality exist in the tree flora of the tropics just as they do in the temperate regions, but in different ways, and these patterns can be an important element in the maintenance mechanisms of these populations. Past studies have reported a general lack of information on tropical phenology, but this situation improved steadily throughout the 1970s.

Tropical trees occur in a variety of vegetation zones in what are generally termed tropical environments. Each of the vegetation zones, however, has its distinctive climate, which in turn partially determines the phenology shown in the communities that are acclimatized in these zones. Several recent studies point out the differences in seasonal behavior between wet and dry tropical forest zones. Frankie et al. (1974a,b) have done a systematic comparison between these two zones in Costa Rica. Their study documents the patterns of leaf fall, vegetative flush, flowering, and fruiting and relates these phenological patterns to both climatic triggers and plant-animal interactions. Burger (1974) has also provided comparisons between the climatically wet and dry forests in his study of the evergreen montane and deciduous forests of Ethiopia. He noted that seasonal differences in the occurrence of major phenological events is an important characteristic of these two types of forests.

The majority of studies on tropical tree phenology have involved forests in the dry tropics, that is, alternatively dry and wet seasonal environments. These tropical dry regions characterize the geographically larger area of the lowland tropics (Walter et al., 1975). Examples of such studies include earlier, classical descriptions on Java by Coster (1923, 1925, 1926, 1927, 1928), extensive phenological documentation in the wet and dry zones of Sri Lanka by Koel-

*By C. H. Lamoureux, D. Mueller-Dombois, and K. W. Bridges

meyer (1959, 1960), a botanical garden in Nigeria (Njoku, 1963), and more recent investigations of Zambezian woodlands (Malaisse, 1974) and Costa Rican semideciduous forests (Daubenmire, 1972; Janzen, 1967; Opler et al., 1976). Other studies have focused on wet tropical forests, such as on the New Hebrides (Baker and Baker, 1936) and in Singapore (Holttum, 1931; Koriba, 1958), while still others have examined forests that are probably transitional, such as in Trinidad (Snow, 1965). Another recent study (Wium-Anderson and Christensen, 1978) has been concerned with seasonality in a tropical mangrove forest, which can be considered an edaphic extreme.

The climatic range from tropical dry (seasonal) to wet is also included in this study of Hawaiian tree phenology. Sites showing such a range of climatic variation, except the edaphic extreme, were found on the Mauna Loa transect over the elevational gradient from 8 to 2,130 m. This brings in the additional dimension of temperature variation as a function of altitude in tropical environments.

Before the IBP studies, few phenological observations had been made on native or naturalized plants. Occasional notes on flowering or fruiting seasons have appeared in floristic works (Rock, 1913), and a few studies of the phenology of cultivated and ornamental plants have been made (Lanner, 1966; Neal, 1965; Pearsall, 1951; Rasid, 1963). Only two studies done prior to the IBP research dealt with native species growing under natural conditions. Baldwin (1953) studied the phenology of flowering in *Metrosideros collina* and *Sophora chrysophylla* in relationship to the seasonal cycles of Hawaiian honeycreepers on Kīlauea and Mauna Loa. Lanner (1965) did studies of flowering, fruiting, and shoot growth of *Acacia koa* on Mauna Loa. In addition to the work reported here, three studies relating specifically to plant phenology were done as part of the IBP research. Porter (TR 27) studied the phenology of *Metrosideros* on both O'ahu and Hawai'i. Leeper (TR 77) examined phenology of flushing in *Acacia koa* and *A. koaia* in relation to koa psyllid population fluctuation. Van Riper (TR 51) described flowering and fruiting phenology of *Sophora chrysophylla* and *Myoporum sandwicense* on Mauna Kea in relation to breeding cycles of Hawaiian honeycreepers. The studies of Porter and Leeper are treated in this chapter, since they also refer to the Mauna Loa transect.

Sampling Information

Phenological observations were made on trees growing in nine plots on the island of Hawai'i along the Mauna Loa transect and in the Kīlauea Forest Reserve. The locations and ecosystem types of all plots are given in the figures accompanying the results.

Measurements were made on twelve native tree species at monthly intervals over nearly a five-year period extending from February 1971 through August 1975. For each species at each sample site, ten individual trees were marked for observation. Trees were selected so that they would represent a nearly uniform distribution of sizes of mature trees available at each sample site.

The observations that were made include flowering (flower buds or flowers present, with indications of the quantity of each), vegetative flush (occurrence of conspicuous growth of new leaves and twigs, with indications of quantity), and circumference measurements. Other phenological measurements were made but are not reported here. These include fruiting, dispersal, leaf fall, and cambial activity.

Quantification of data on flowering and vegetative flush presents certain problems related to differences in morphology of the species studied. For certain species, such as *Erythrina sandwicensis* and *Sophora chrysophylla,* it was usually feasible to count accurately the number of inflorescences present on a tree. For other species, it was possible only to make estimates of the percentage of the branches on a tree that showed a particular phenophase. In still other species, the phenophase was determined as a qualitative class (such as absent, little, moderate, or heavy) for the plant as a whole.

Phenophase index values for flowering and flushing were established to allow for a comparison across all species. For this purpose, enumeration, percentage value, and qualitative data were reduced to a scale of four index values, ranging from 0 to 3. Count-based data were converted as follows: 0 = none, 1 = 1–19, 2 = 20–50, and 3 = more than 50 counts. In the percentage estimates, conversions imply the following: 0 = 0 percent, 1 = 1–5 percent, 2 = 6–25 percent, and 3 = more than 25 percent. Qualitative classes assigned were: 0 = none, 1 = little, 2 = moderate, and 3 = heavy. The index values determined for the individual trees were then summed over the ten plants of the same species in the sample population. This provided an index value for the population in the range between 0 and 30.

Growth changes in the trees were based on monthly circumference measurements. During 1971, measurements were made using a steel tape, which was replaced in the same position at each reading by using nails left in the tree as permanent alignment guides. In early 1972, simple dendrometers, made of aluminum tape held by small springs, were permanently placed on each tree. Growth values for each sample population were calculated as the change in circumference expressed as μm per day. These are used as the index values in the analyses.

The flowering phenology of an additional tree species, *Metrosideros collina* subsp. *polymorpha,* has been included in this synthesis based on data from Porter's one-year study. Sampling details are found in TR 27.

Patterns of Flowering

Flowering generally shows distinct annual patterns in each of the species populations. This is indicated by the relatively high average seasonality correlation value (r = .64). Moreover, 65 percent of the species populations have a correlation value of .6 or greater. This can be seen in figure 10-1 in the general smoothness and regularity of the curves and their tendency to have a single peak. This confirms that there was an overall annual flowering period among the species studied.

There were two periods in the year in which peak flowering occurred. The primary period came in the winter wet months of January through March and included the peak flowering periods for 80 percent of the high r value species populations. A secondary period occurred in August and September and coincided with the late summer dry, warm period. Only one high r value species *(Coprosma ochracea)* had its peak flowering outside these two peak flowering periods.

The constancy of the flowering is shown by the relatively infrequent occurrence of months with a flowering index of zero. Taken overall, the flowering appeared to be nearly continuous, with an average species population having 10.3 months of flowering each year. There are only three exceptions to this general flowering pattern.

Those species populations with the strongest annual patterning occurred in the coastal lowland, mesic forest (Kīpuka Kī), upper rain forest (Kīlauea rain forest), and upper mountain parkland. The lower rain forest (IBP site 1) and seasonal forest (Kīpuka Nene) had the least annual patterning. However, there is no statistical separation by r values into ecosystem types.

The flowering trends of *Acacia koa* populations at five different sites were generally quite synchronized. Flowering along the Mauna Loa transect in the montane seasonal environment began in late summer (September) or early fall (October) and continued through May. It ceased usually in July. During August there was no flowering. The approximately ten-month flowering of *Acacia koa* gradually reached a peak, which usually occurred in the second half of January or in February. The same annual trend shown for *Acacia koa* in the seasonal environment held also for the *Acacia koa* in the rain forest environment (Kīlauea Forest Reserve), except that the winter-flowering peak arose here more abruptly in December than on the Mauna Loa transect.

Since the flowering trends of *Acacia koa* were so similar on sites with different rainfall regimes, it can be assumed that they are largely controlled endogenously. Nevertheless some year-to-year shifts can be expected as indicated by the pattern parameters. The cessation of flowering in *Acacia koa* coincided with the driest and warmest part of the year.

The yearly flowering pattern of *Sophora chrysophylla* in the montane seasonal environment was very similar to that of *Acacia koa*. However, *Sophora* began to flower earlier, usually in August or early September. Once it started flowering, *Sophora* inflorescences usually became abundant more rapidly than in *Acacia koa*. The *Sophora* population at the highest elevation sampled (2,040 m) flowered through the year without interruption. The same was true for the *Sophora* population sampled at the lowest elevation (915 m).

However, at the highest elevation where flowering was most abundant among the *Sophora* populations sampled, there was a distinct trend of rapid development to a flowering maximum from late summer (September) to December through February, followed by a gradual decrease in flowering abundance to August. The overall quantitative flowering trend was similar to those *Sophora* populations with interrupted flowering. In contrast, the low-elevation *Sophora* population showed no such significant peaking and instead exhibited relatively uniform and moderately abundant flowering through the year.

Dodonaea viscosa flowering also followed the general trend of *Acacia koa* and *Sophora chrysophylla*. But *Dodonaea* flowered even more profusely and decidedly during the winter months. Its flowering peak was reached in the second part of January. During the summer there was little flowering, and the flowering period ended in July. There was a month-long period of cessation, and flowering began again, but gradually, in the second half of August.

The next three species (on figure 10-1)—*Santalum ellipticum, Sapindus saponaria,* and *Myoporum sandwicense*—showed flowering trends opposite to those just discussed. *Santalum, Sapindus,* and *Myoporum* had pronounced summer flowering peaks that occurred from the end of August through mid-September. Their flowering activity declined in the cooler and wetter part of the year. A low level of flowering continued through the year in *Santalum,* while in both *Sapindus* and *Myoporum* flowering ceased near the end of December. The new flowering period began in April or May for these last two species. The opposite flowering trends of the two species groups are particularly interesting because these species occur in the same environmental sections. For example, the summer-peaking *Santalum* and *Sapindus* occur in the same montane seasonal environment (section III) with the winter-peaking *Acacia, Sophora,* and *Dodonaea*. Likewise, the summer-peaking *Myoporum,* which was sampled in the upper rain forest section (IV), occurs there together with the winter-peaking *Acacia*.

A third group of species is represented by *Cheirodendron trigynum, Ilex anomala, Myrsine lessertiana,* and *Coprosma ochracea,* characteristic rain forest species that do not occur in the seasonal environmental sections. These four species have their individual patterns. Their flowering peaks, except for *Coprosma,* were not very pronounced and occurred at quite different times of the year—*Cheirodendron* in early July, *Ilex* in late September, *Myrsine* in March, and *Coprosma* in early May. These species not only had individually different flowering peaks and seasons, they were also out of phase with the two groups of summer-peaking and winter-peaking species. The four rain forest species followed a pattern that may be described as random peaking. Their flowering periods, however, were overlapping in such a way that the four rain forest species as a group formed a continuously flowering unit.

Finally, the two coastal lowland species, *Diospyros ferrea* and *Erythrina sandwicensis,* conform individually to the two first groups. *Diospyros* is a winter to early spring peaking tree, and *Erythrina* is a summer-peaking species. These two lowland species occur in a seasonal climate with winter rainfall and summer drought.

Thus, in summary, we can recognize three groups of species according to their flowering characteristics. There are summer-peaking and winter-peaking species. Both of these occur sympatrically in the same environments. More-

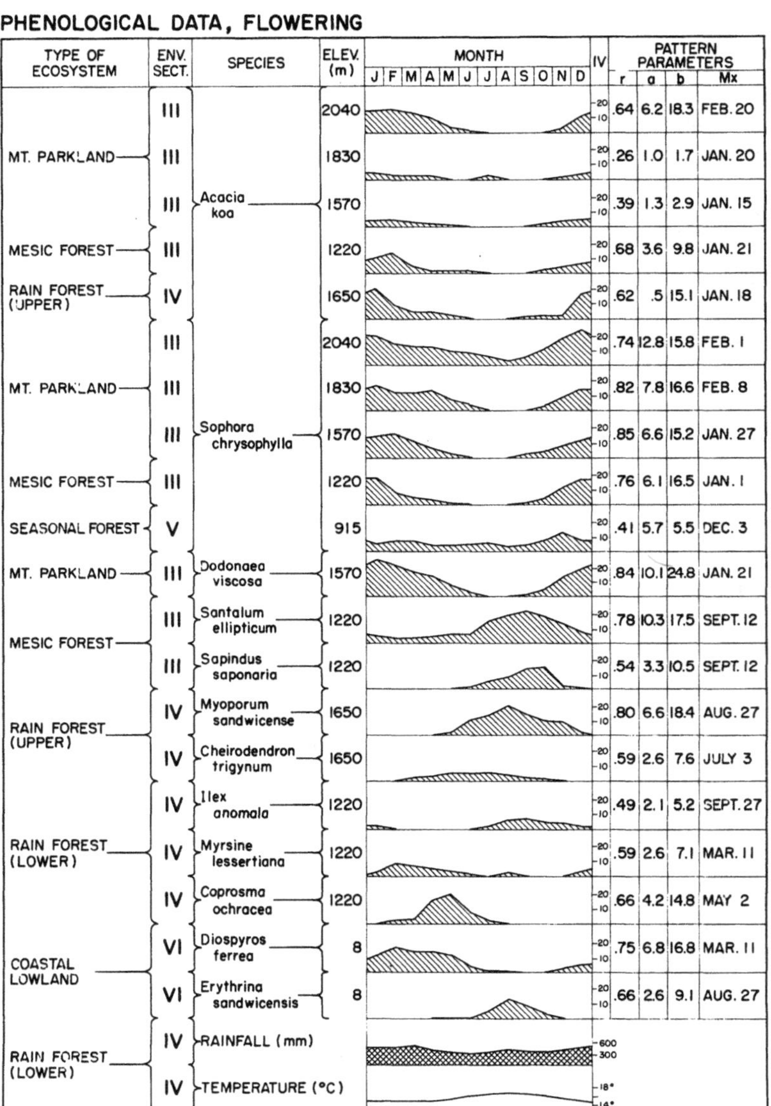

FIGURE 10-1. *Flowering phenology of twelve tree species along extended Mauna Loa transect.*

over, both groups are found characteristically in the seasonal environments over the whole altitudinal gradient, from the coastal lowland to the subalpine environment (at 2,040 m). A third group of species is characteristically restricted to the rain forest environment. They exhibit a pattern of flowering that can be described as random peaking. Most of these also had long periods of moderate or low flowering activity so that as a group they exhibited continuous flowering. This trait of continuous flowering was also characteristic for the tree groups in the seasonal environments, which showed peaks in either summer or winter with periods of moderate flowering during spring and fall.

Patterns of Flushing

The general tendency shown by these species populations was an intermediate expression of an annual seasonality pattern with a mean correlation value *(r)* of .5. Half of the species populations had an *r* value of .5 or greater (figure 10-2).

There was one general period of peak flushing that extended from December through July. The species populations with the highest *r* values had peak flushing in March through June, although there was still some spread throughout the entire peak period.

There was a high constancy of flushing, with all species populations showing some flushing 95 percent of the time, with an average of 11.4 months of flushing per year. There were several exceptions to this high constancy.

Except at the highest elevation sampled (2,040 m), *Acacia koa* leaf flushing was continuous through the year in the mountain parkland and mesic forest. At the highest elevation, the flushing peak occurred in the spring (end of May) shortly after flushing activity began in mid-February. Following the peak, flushing activity gradually decreased, ceasing in mid-October. A similar flushing pattern occurred at the rain forest site in *Acacia koa* with a May flushing peak, although here a little flushing activity continued through the year except in the December to February period. At the remaining three sites along the Mauna Loa transect, from 1,220 to 1,830 m, *Acacia koa* flushing activity was lowest in summer (July and August) and highest in winter (December through February). Flushing peaks for *Acacia koa* were not as pronounced as the flowering peaks, but there was a tendency for the vegetative flushing peak to follow (rather than coincide with or precede) the flowering peak.

A similar trend was shown for *Sophora chrysophylla*. Its vegetative flushing peaks were mostly in the spring to summer, and they followed the flowering peaks, which typically occurred in the winter. Here also, flushing peaks were not as pronounced as the flowering peaks. Flushing continued through the year in *Sophora,* while the flowering activity was usually interrupted in midsummer (July and August).

Dodonaea viscosa showed an alternating pattern of vegetative flushing and flowering, with vegetative-flushing peaks in the summer and flowering peaks in the winter. Flushing activity was continuous through the year, but with an indication of clear seasonality.

Santalum ellipticum and *Myoporum sandwicense* exhibited moderately high vegetative-flushing activity throughout the year. Their flushing peaks were not pronounced but occurred in the first half of the calendar year, while

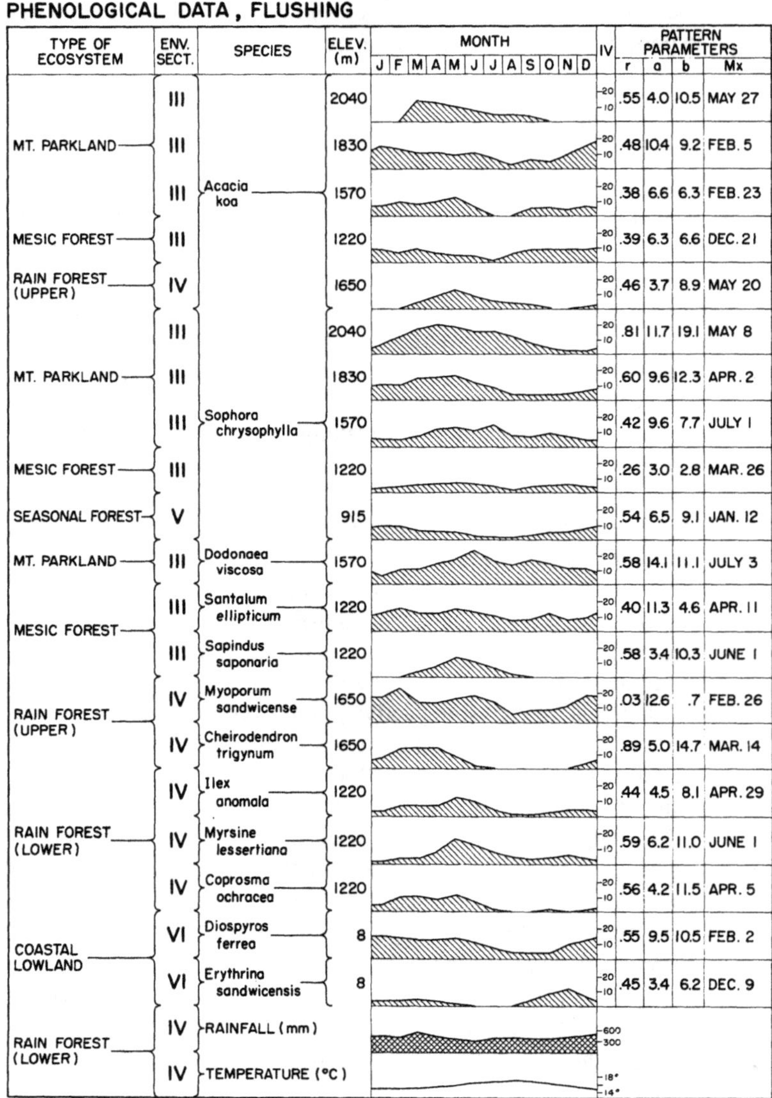

FIGURE 10-2. *Leaf-flushing phenology of twelve tree species along extended Mauna Loa transect.*

their flowering peaks occurred in late summer (late August to September). Thus in these summer-flowering species, flowering activity followed peak flushing activity by a few months. The third species with a late-summer flowering peak (mid-September), *Sapindus saponaria,* showed a pronounced vegetative flushing peak at the end of May and early June. In this species as well, flushing preceded flowering by a few (2.5) months. In terms of vegetative-flushing activity, *Sapindus* was unique in that there was no such activity from the end of September through mid-February. *Sapindus* is a winter-deciduous species. This is of some interest since winter rainfall is heavier where *Sapindus* grows, and there is a short but moderate dry season in the summer. Thus its observed flushing activity was opposite to what one might expect in relation to the climatic seasonality.

The four rain forest species—*Cheirodendron trigynum, Ilex anomala, Myrsine lessertiana,* and *Coprosma ochracea*—surprisingly exhibited a fairly correlated pattern of leaf flushing in that their vegetative activity was more pronounced in the spring and early summer. Two species, *Cheirodendron* and *Coprosma,* showed cessations in flushing activity during the second half of the calendar year, and all four rain forest species exhibited significantly lower flushing activity in the second half of the calendar year. Thus while their flowering peaks followed a random pattern, their vegetative flushing periods appeared to be well correlated. In the rain forest species, however, flushing and flowering did not seem to be as well correlated as in the species growing in seasonal environments.

The two lowland species *Diospyros ferrea* and *Erythrina sandwicensis* showed quite different flushing patterns. The flushing in *Diospyros* was continuous through the year, with heavier flushing in the cooler, wetter season from November through May. In this species, peak flushing activity coincided more or less with peak flowering activity. Both activities peaked during the spring months, with flushing in February and flowering in March. In *Erythrina,* vegetative flushing peaked in late fall. No flushing activity occurred during the height of the summer (end of June through August). This species is summer deciduous, and its canopy activity is clearly related to the dry-season, wet-season pattern of its coastal lowland environment. Flowering in *Erythrina* occurred during the dry season, when this tree is leafless. This species, therefore, exhibited seasonality with alternating flowering and flushing periods.

Patterns of Growth

The growth measurements fit the annual pattern curves least well of the three indexes (figure 10-3). The mean annual correlation value *(r)* is .45, with 45 percent of the species populations having an *r* value of .5 or greater.

The times of peak growth were spread irregularly throughout the year except that none of the peaks occurred in the summer period (May through

August). Some concentration of peaks was seen in the fall (October and November) for the species populations with high r values.

Growth appeared to be a nearly continuous phenomenon among the species studied, shown by the high constancy value (95 percent of the species populations showing growth at any time). The average growth rate (in circum-

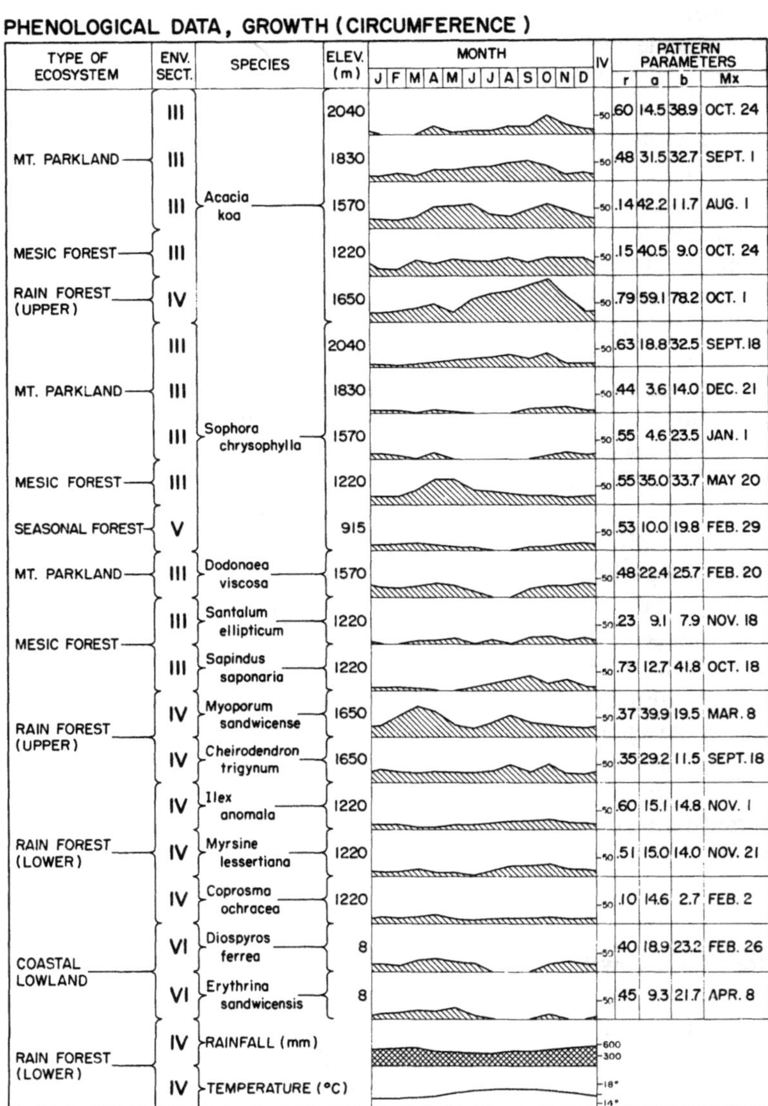

FIGURE 10-3. *Circumference growth of twelve tree species along extended Mauna Loa transect.*

ference) for all species populations was 22.3 μm per day (based on the means of the curves fitted to the population patterns), with a standard deviation of 14.8 μm per day.

Year-round continuous growth was expected for the four rain forest species—*Cheirodendron trigynum, Ilex anomala, Myrsine lessertiana,* and *Coprosma ochracea*—which also showed no real peak-growth activity. Instead growth was rather uniform through the year.

In *Acacia koa,* circumference growth was generally more pronounced in the fall. Its growth was clearly greatest at the rain forest site, and here it showed a decided peak in the fall (October). Thus for *Acacia koa,* flowering activity was greatest in the winter, flushing in the spring, and growth in the fall.

In *Sophora chrysophylla,* circumference growth was much less active than in *Acacia koa;* it was also commonly disrupted. The rate of growth of *Sophora* did not exhibit any clear peaking. An exception occurred at 1,220 m in the mesic forest, where maximum growth was in the spring (May). A cessation of growth in *Sophora* was common during the second half of July and the early part of August. This correlates with a lag effect from the short dry season in June, which at that time may have resulted in soil-moisture stress.

Dodonaea viscosa also showed this same interruption in growth during the second half of July and August, and there was moderate growth activity during the rest of the year, with no clear peaking at any time.

The three species with summer flowering peaks—*Santalum ellipticum, Sapindus saponaria,* and *Myoporum sandwicense*—each showed different growth patterns. There was no relation to their flushing peak, which occurred during the spring and was generally correlated with low or even interrupted growth activity in *Santalum* and *Sapindus*. The low growth activity in the spring was then followed by a somewhat increased growth activity in the fall, particularly in *Sapindus*. The data for *Myoporum* are unclear since repeated vandalism to the dendrometers terminated growth measurements after two years. The data do suggest, however, a peak growth in spring and perhaps another, less pronounced one, in August. In general, a relationship emerged in which vegetative-flushing activity with low growth was followed by increased circumference growth. This is not unexpected, however.

The two lowland species showed growth cessation during July and August, probably in response to the dry season in that environment. For the rest of the year, their growth activity was fairly uniform and at low levels, with slight peaks during the spring.

Thus we can conclude that all species in seasonal environments along the altitudinal gradient, with the exception of *Acacia koa,* showed growth cessation during or after the dry season, with a probable lag effect due to soil-moisture stress. Species in rain forest environments showed no growth disruptions. The continuous growth of *Acacia koa* along the Mauna Loa transect in montane seasonal environments is perhaps a species characteristic. *Acacia koa* exhibited the most efficient capacity for growth among the twelve native species sampled in this study.

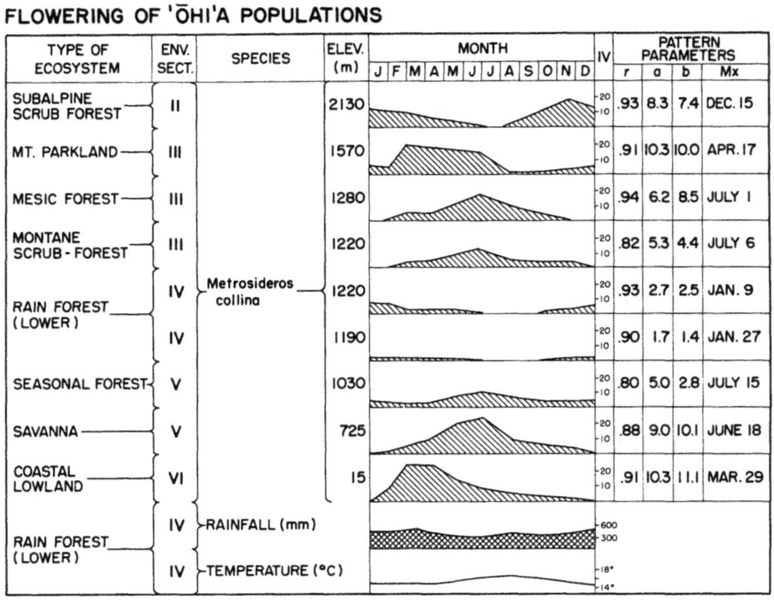

FLOWERING OF 'ŌHI'A POPULATIONS

TYPE OF ECOSYSTEM	ENV. SECT.	SPECIES	ELEV. (m)	MONTH J F M A M J J A S O N D	IV	PATTERN PARAMETERS r	a	b	Mx
SUBALPINE SCRUB FOREST	II	Metrosideros collina	2130		20 10	.93	8.3	7.4	DEC. 15
MT. PARKLAND	III		1570		20 10	.91	10.3	10.0	APR. 17
MESIC FOREST	III		1280		20 10	.94	6.2	8.5	JULY 1
MONTANE SCRUB-FOREST	III		1220		20 10	.82	5.3	4.4	JULY 6
RAIN FOREST (LOWER)	IV		1220		20 10	.93	2.7	2.5	JAN. 9
	IV		1190		20 10	.90	1.7	1.4	JAN. 27
SEASONAL FOREST	V		1030		20 10	.80	5.0	2.8	JULY 15
SAVANNA	V		725		20 10	.88	9.0	10.1	JUNE 18
COASTAL LOWLAND	VI		15		20 10	.91	10.3	11.1	MAR. 29
RAIN FOREST (LOWER)	IV	RAINFALL (mm)			600 300				
	IV	TEMPERATURE (°C)			18° 14°				

FIGURE 10-4. *Flowering patterns in* Metrosideros collina *along extended Mauna Loa transect.*

There was a slight indication that maximum growth in circumference followed the period of peak flushing in some species. However, the general year-round continuity of both of these processes does not permit a clear-cut interpretation of the interrelationship of these two processes. It should be noted that the species studied here are evergreen trees, except for the wet-season spring-deciduous *Sapindus saponaria* and the dry-season summer-deciduous *Erythrina sandwicensis*.

Flowering in *Metrosideros*

Only one year of data was taken for *Metrosideros collina* (Porter, TR 27), but its flowering phenology was studied over the same altitudinal range as that of the other native trees. *Metrosideros* is by far the most widely distributed tree species so it could be sampled over the whole altitudinal gradient, including the rain forest.

At the highest elevation (2,130 m), in the subalpine scrub forest, *Metrosideros* flowering coincided closely with that of *Acacia koa* and *Sophora chrysophylla*. Here *Metrosideros* showed its flowering peak in late fall and winter (figure 10-4). Flowering decreased during spring and ceased for a month during July. It started in August and increased steadily toward winter.

In the mountain parkland, near the IBP climate station at mid-level (1,570 m) on the Mauna Loa transect, *Metrosideros* flowered continuously through the year. Its peak flowering occurred in the spring, from March through June. Flowering activity was low for the rest of the year.

Lower down in the mesic forest (1,280 m) and in the montane scrub forest (1,220 m), *Metrosideros* showed a pronounced summer peak in flowering (end of June), but flowering began very gradually in late January and extended to early December.

In the lower rain forest (IBP sites 1 and 2), flowering was disrupted from early July through September. It continued at low levels without pronounced peaks through the rest of the year.

In the submontane seasonal forest and savanna, there was again a summer peak, and flowering continued for the rest of the year at low levels.

Near sea level, in the coastal woodland, flowering started in January and increased rapidly to a peak in March and April. Then it decreased gradually to a very low level of activity in winter.

In summary, *Metrosideros collina* showed several patterns. Cooler-season flowering peaks were shown at the highest elevations (from 1,500 to 2,130 m); summer-flowering peaks occurred in the mesic and montane scrub forests from 1,200 to 1,300 m and in the submontane seasonal environment (section IV); spring-flowering peaks appeared in the coastal lowland and mountain parkland; and summer-flowering cessation with low flowering activity during the remainder of the year was found in the rain forest.

The summer or winter peaking is typical for other species in the seasonal environments. *Metrosideros,* as a single species, seems to accomplish a phenological flexibility that otherwise resides only in a mixed-species stand. The spring peaking of *Metrosideros* in the coastal lowland habitat coincided with that of *Diospyros* in the same locality. However, the multiple pattern of flowering phenology does not indicate any correlation with climatic seasonality in *Metrosideros*. Conversely the disrupted but generally similar flowering pattern of *Metrosideros* in the rain forest habitat does not seem correlated with any environmental cue. Instead it seems that cohorts of *Metrosideros* exhibit an internally regulated flowering pattern, which over the entire area results in peak- or moderate-level flowering occurring at different localities and at any time during a year.

Discussion and Conclusions

The studies on thirteen native tree species have provided evidence on the timing, degree, and spatial distribution of phenological phenomena in Hawaiian ecosystems.

Flowering periods were very long for all thirteen species studied. Many species flowered from ten to eleven months per year.

Year-round flowering was observed in four species: *Acacia koa, Metrosideros collina* subsp. *polymorpha, Sophora chrysophylla,* and *Santalum ellipticum.* Such flowering was not a species characteristic per se; instead it was restricted to certain populations of these species only. These did not occur in the rain forest but in the seasonal environments along the transect.

With regard to peaks in the flowering period, we found three species-related patterns. First, peak flowering in the moister season of the year, which in Hawaii coincides with the slightly cooler winter season. This applied to *Acacia koa, Sophora chrysophylla, Dodonaea viscosa,* and *Diospyros ferrea.* Second, peak flowering in the drier season of the year, which in Hawaii coincides with the warmer summer season. This applied to *Santalum ellipticum, Sapindus saponaria, Myoporum sandwicense,* and *Erythrina sandwicensis.* Third, random peaking, i.e., peak flowering occurring at almost any time of the year. This applied to *Metrosideros collina* subsp. *polymorpha.*

In addition to the three species-related patterns, we found two ecosystem-related patterns: individual-species peaking and a special flowering pattern in the rain forest. This was displayed in the following way: *Myrsine lessertiana* peaking in early spring, *Coprosma ochracea* in late spring, *Cheirodendron trigynum* in midsummer, *Myoporum sandwicense* in midsummer to late summer, *Ilex anomala* in late summer, and *Acacia koa* in midwinter (during the rainier season).

Outside the rain forest in the more seasonal environment (regardless of elevation), we found two prevalent synchronized flowering peaks: either a dry-season peaking *(Erythrina, Sapindus, Santalum)* or moist-season peaking *(Diospyros, Dodonaea, Sophora, Acacia koa).*

Metrosideros collina subsp. *polymorpha,* the species with random peaking, fell out of line with the others. In the rain forest it displayed low-level flowering through most of the year, except in the warmest period (July, August, and September) when flowering ceased. Here it showed a slight winter peak coinciding with that of *Acacia koa.* Outside the rain forest, *Metrosideros* displayed spring peaking in the coastal lowland and mountain parkland, summer peaking in the submontane and montane seasonal environment (all over the mid-elevation range), and late-fall peaking in the subalpine environment. Thus, this species displayed staggered flowering peaks along the extended Mauna Loa transect similar to what several individual species did in the rain forest. Although flowering phenology of *Metrosideros* was observed for only one year, its universal lack of peaking during the warmest part of the year (August and September) may be of some significance.

Leaf flushing was even more continuous through the year than was flowering. A clear rest period, when flushing activity ceased, was observed only in three of the twelve species studied for this phenophase. These three were the summer-deciduous *Erythrina sandwicensis,* the winter- (or moist-season) deciduous *Sapindus saponaria,* and *Cheirodendron trigynum.* The last species showed a summer rest period in the rain forest, where one would have expected flushing activity. The higher-elevation (2,040 m) population of *Acacia koa* also displayed a distinct period of inactivity during the winter (November through January). This may be related to cool night temperatures during that time of the year, which may often approach freezing at this elevation.

Leaf-flushing peaks were not very clear or pronounced for most species except *Dodonaea viscosa* (in early summer). However, leaf-flushing activity was generally more apparent in the moister season of the year (from winter through spring), and less flushing occurred generally during late summer and fall.

Growth in circumference was similarly continuous through the year as was leaf flushing, but there is possibly a dry-season growth dormancy in the coastal lowland because the two coastal lowland species, (*Diospyros ferrea* and *Erythrina sandwicensis*) have opposite flowering and flushing peaks but showed a synchronized rest period in wood production. This may be treated as a new hypothesis.

In *Acacia koa* and some other species, there is a tendency for more circumference growth in the fall than in the spring and a broad pattern of alternating semiannual flushing in spring and circumference growth during fall. However, the growth peaking was generally weak, while growth continuity was strong.

The long-period flowering characteristic observed in the Hawaiian trees does not appear to be common throughout the tropics. For example, Janzen (1967), in a review of the tree-flowering patterns in the lowland tropics in Costa Rica, noted that only fifty-seven species of some 280 for which there were records showed continuous or intermittent flowering (six to twelve months per year). The remaining 223 species had a mean flowering period of approximately two months. By comparison, only one species in the study reported here had a flowering gap as long as six months. It is not clear whether the results of this study are in general disagreement with those reporting a dominance of short-flowering periods. Often reports of flowering period use only abundant flowering as the index to define the limits of flowering (Stiles, 1978).

There are several implications of a near-continuous flowering of species populations that relate both to the mechanisms for the maintenance of the tree populations and their interactions with the faunal elements in their communities. The year-round availability of seeds may play an important role in the establishment of some species in closed-canopy environments. For example, *Metrosideros* occurs in the upper rain forest where its seedlings have been observed to respond to randomly occurring openings in the upper canopy and thus fill the gaps resulting from large trees that have fallen. This gap-phase replacement would not be a general establishment mechanism for this species unless there was an abundant supply of seedlings throughout the year since *Metrosideros* seeds are small and they may not have sufficient reserves to last for long periods between gap openings.

Such general availability of seeds throughout the year may also be an important characteristic in an area like the Mauna Loa transect where new volcanic substrates frequently become available. Species with an ability to invade these areas quickly, such as *Metrosideros* with its small seeds which are easily transported by wind, probably would have an increased competitive ability. This topic has been considered in detail by Smathers and Mueller-Dombois (1974).

The timing relationships between the flowering of the plants and the pollinator species for other tropical areas have been examined by a number of workers. In particular, the studies on hummingbird pollinators (Janzen, 1967; Stiles, 1976, 1977, 1978) and bee pollinators (Janzen, 1967) have shown that intimate relationships exist. Hummingbirds are not present in Hawai'i. The native bird pollinator relationship to *Metrosideros* trees found near the Mauna Loa transect has been the subject of a critical study by Carpenter (TR 76, 1976). A general conclusion was that high levels of fruit set and outbreeding in *Metrosideros* resulted from the pollination activities of the endemic Hawaiian birds. Strong pollination segregation was found in this tree species. In general, however, there were indications of less specialized relationships between the plants and pollinators in Hawaiian forests than are reported for other tropical regions. There are many species of solitary bees native to Hawaii, but the honeybee was not introduced until 1857 (Krauss, 1978). Preliminary information (Corn, TR 2) indicates that birds were the primary floral visitors in the rain forest, but native bees (*Nesoprosopis* sp.) and wasps (*Polistes* sp.) became more prominent in the nearby drier areas. Our knowledge of the full role of pollinators in the development of flowering patterns awaits further study.

The environmental cues that act as phenological triggers were not as strong in any of the transect ecosystems as might be expected for many tropical areas. In many typical lowland dry-tropical areas, there is a very abrupt change from wet to dry and the reverse (Walter et al., 1975). A similar analysis of the climate for the transect shows a rather gradual change in the monthly rainfall values. This does not mean that intimate environmental relationships do not exist; rather it implies that the synchronization of short-period phenomena, an often-cited feature of tropical forests (Richards, 1952), would not be a regular feature of Hawaiian ecosystems. Indeed the level of coordination between individuals of a species, even in a limited area, was relatively weak. Only in the broader temporal and spatial context was there synchronization.

If the phenological responses of the plants were simply related to the climate of the area in which they exist, then it might be expected that sites with a strong annual patterning of their rainfall, for example, would show a similarly strong patterning in the plant phenology. There was no evidence in this study, however, that the degree of annual patterning (as measured by the r values) was related to any of the ecosystem types; sample sizes were possibly too small, however, to detect this relationship, and it should receive further attention.

An understanding of the phenological relationships in Hawaiian ecosystems is important to our unraveling the complexities of these systems. This continues to be a concern if we are to understand these ecosystems that have evolved in the absence of many selective pressures common to continental tropical regions.

RODENTS*

The two rodent species examined here, *Rattus rattus* and *Mus musculus,* have geographic ranges that include most tropical and temperate regions of the world. In general, these species occur in coastal regions where they are locally common up to mid-elevations. In Hawai'i, however, they range well up on the high mountains (Tomich, 1969).

In general, reproduction in *Rattus rattus* occurs throughout the year in Hawai'i. Although it is expected that there are seasonal peaks in the breeding cycle, past evidence has not been adequate to show this. *Mus musculus* has been shown to have a bimodal reproductive pattern in a coastal region on the island of Hawai'i, with peaks in midsummer and midwinter (Tomich, 1961).

Two measurements made of the rodent populations have been chosen for the analysis of seasonal change: population numbers and mean animal weight. Each of these can be expected to change through the year if there is annual seasonality and should demonstrate the integrated effects of reproduction, mortality, and growth. While there could be seasonal migration from one area to another that would also affect these two measures, this is not considered likely given the relatively small home ranges of these mammals. The study sites were selected along an elevational gradient so that a range of environmental conditions could be examined. Traplines at three elevation ranges have been selected for analysis. Each of these was run for a period of two years.

Sampling Information

The data collected for the spatial analysis study are also used for the temporal analyses described here. The traplines were sampled for four-day periods at three-month intervals over a two-year period, starting in October 1971. Traps were set in lines along the Mauna Loa transect, which were extended to lower elevations in the vicinity of the transect (TR 48). For the purposes of these analyses, separate traplines were lumped into three study areas, at high, mid-, and low elevations. The high-elevation traplines were located in the alpine to mid-montane seasonal environments at 2,740, 2,590,

---

*By P. Q. Tomich and K. W. Bridges

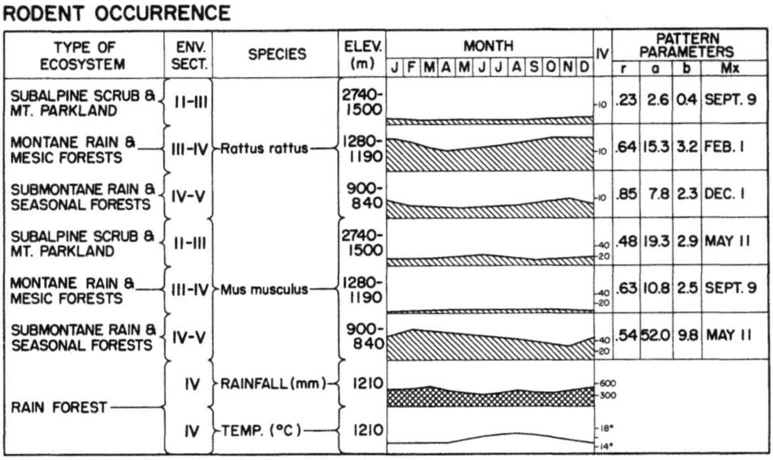

RODENT OCCURRENCE

TYPE OF ECOSYSTEM	ENV. SECT.	SPECIES	ELEV. (m)	MONTH J F M A M J J A S O N D	IV	PATTERN PARAMETERS r	a	b	Mx
SUBALPINE SCRUB & MT. PARKLAND	II-III	Rattus rattus	2740-1500		10	.23	2.6	0.4	SEPT. 9
MONTANE RAIN & MESIC FORESTS	III-IV	Rattus rattus	1280-1190		10	.64	15.3	3.2	FEB. 1
SUBMONTANE RAIN & SEASONAL FORESTS	IV-V	Rattus rattus	900-840		10	.85	7.8	2.3	DEC. 1
SUBALPINE SCRUB & MT. PARKLAND	II-III	Mus musculus	2740-1500		40 20	.48	19.3	2.9	MAY 11
MONTANE RAIN & MESIC FORESTS	III-IV	Mus musculus	1280-1190		40 20	.63	10.8	2.5	SEPT. 9
SUBMONTANE RAIN & SEASONAL FORESTS	IV-V	Mus musculus	900-840		40 20	.54	52.0	9.8	MAY 11
RAIN FOREST	IV	RAINFALL (mm)	1210		600 300				
RAIN FOREST	IV	TEMP. (°C)	1210		18° 14°				

FIGURE 10-5. *Seasonal variation in numbers of* Rattus rattus *and* Mus musculus *along extended Mauna Loa transect.*

2,130, 1,890, and 1,500 m. The mid-elevation traplines were located in the lower montane seasonal and rain-forest environments at 1,280, 1,220, and 1,190 m. The low-elevation traplines were located in the submontane seasonal environment at 900, 870, and 840 m. As a result, these trapline sets include an elevational range averaging 460 m and combine several types of ecosystems. The vegetation along each of the traplines has been described in TR 48. The range of trapline elevations was similar for the two years, but there were some trapline substitutions made in the second year, so the two years are not strictly comparable.

Annual Population Trends

The occurrences of *Rattus rattus* and *Mus musculus* along the elevational gradient for the monthly means of the two-year sample period are shown in figure 10-5. *Rattus rattus* caught at the high-elevation traplines had the lowest populations, with the largest number caught in a four-day period equal to the least caught on the other traplines. The mid-elevation traplines had the highest populations, and the low-elevation traplines had intermediate-size populations. The largest number of animals caught during a single trapping period on the mid-elevation traplines was approximately twice the highest value at the low-elevation lines. *Rattus rattus* populations at each of the elevations showed higher population levels from late fall into winter, although this was not pronounced at the high-elevation traplines, probably due to the small number of animals caught.

The *Mus musculus* populations were found to have an intermediate-size population at the high-elevation traplines, and these showed no obvious seasonal trend. Population sizes in the second year were slightly larger than in the first year. This difference in years is not apparent in the figure since the two-year values have been averaged; where relevant, however, such differences will be described to assist in the interpretation of the average patterns. The lowest population size occurred at the mid-elevation traplines where a spring to early summer peak was seen. This was indistinct, however, due to the low number of animals caught. There was no apparent difference in numbers caught during the two years. The highest populations, by a considerable margin, were at the low-elevation traplines. The population peak came in the winter to early spring period. There was a difference in the number of animals caught, with the second year catches being generally larger than the first.

Annual curves for the two species have been fitted. Since the mean catch for each of the two years was often quite different, each year's catch for both species was adjusted either up or down so that the mean of each year for a set of elevational transects would be the same and the annual curve would fit the average pattern.

Annual curves for *Rattus rattus* showed the strongest seasonality at the low-elevation traplines (r = .85) and a reasonably good fit at the mid-elevation traplines (r = .64). The fit was very poor, however, at the high-elevation traplines (r = .23). The strongest seasonality for *Mus musculus* was seen at the mid-elevation traplines (r = .63), where the fit was reasonably good. A poorer fit was seen at the higher (r = .48) and lower elevation traplines (r = .54).

Curves for both mammal species based on the two-year averages of mean animal weights are given in figure 10-6. The mean weight of *Rattus rattus*

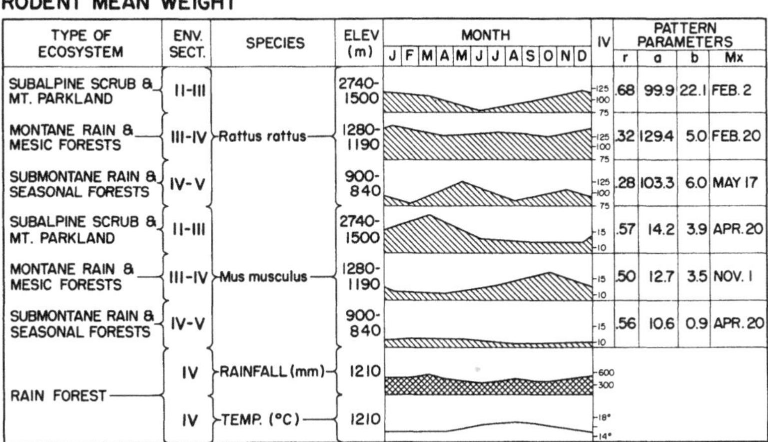

FIGURE 10-6. *Seasonal variation in mean weight of* Rattus rattus *and* Mus musculus *along extended Mauna Loa transect.*

observed at each of the trapline sets was similar, with the highest weight at the mid-elevation traplines. The high-elevation traplines showed a winter weight peak and had the greatest annual amplitude. The mid-elevation traplines did not have a discernible annual pattern. The low-elevation traplines showed considerable month-to-month variation, with no seasonal trend.

The *Mus musculus* populations had the highest mean weight at the high-elevation traplines, and this decreased downslope. At the high-elevation traplines also, a weight peak was seen in the winter to spring period. This set of traplines had the highest annual change in the mean weight. Most of the variation in weight came from the second-year catch. The mid-elevation traplines had a summer-to-fall peak in weight, with quite a high annual amplitude. The catch from both years showed similar variation. The low-elevation traplines had only a weak weight peak in the winter-to-spring period and very low annual weight changes. Virtually all of the weight changes came from the first-year catch.

The fit of the annual curves to the *Rattus rattus* data showed a reasonably good fit at the high-elevation traplines ($r = .68$) but a poor fit at the two other sets of traplines. The *Mus musculus* annual curves were similar in their fit at all three elevations but best at the high and low traplines.

A striking inverse similarity was seen in how well the annual curves fit an elevation-trapline set when comparing the occurrence and mean weight data. For example, the low-elevation *Rattus rattus* occurrence observations fit well ($r = .85$), but the mean weight fit poorly ($r = .28$). The meaning of this relationship is not clear. To see if the total biomass at each trapline might be a better expression, biomass values were calculated as the product of the occurrence and the mean weight from each transect sample. This has been plotted in figure 10-7.

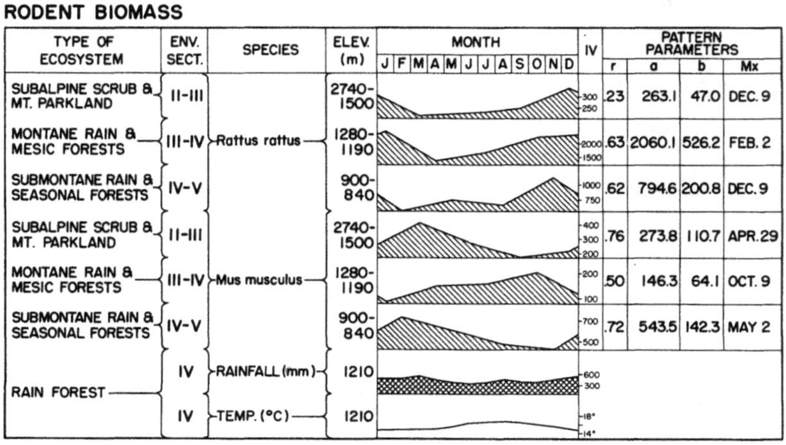

RODENT BIOMASS

TYPE OF ECOSYSTEM	ENV. SECT.	SPECIES	ELEV. (m)	MONTH (J F M A M J J A S O N D)	IV	PATTERN PARAMETERS r	a	b	Mx
SUBALPINE SCRUB & MT. PARKLAND	II-III	Rattus rattus	2740-1500		300, 250	.23	263.1	47.0	DEC. 9
MONTANE RAIN & MESIC FORESTS	III-IV	Rattus rattus	1280-1190		2000, 1500	.63	2060.1	526.2	FEB. 2
SUBMONTANE RAIN & SEASONAL FORESTS	IV-V	Rattus rattus	900-840		1000, 750	.62	794.6	200.8	DEC. 9
SUBALPINE SCRUB & MT. PARKLAND	II-III	Mus musculus	2740-1500		400, 300, 200	.76	273.8	110.7	APR. 29
MONTANE RAIN & MESIC FORESTS	III-IV	Mus musculus	1280-1190		200, 100	.50	146.3	64.1	OCT. 9
SUBMONTANE RAIN & SEASONAL FORESTS	IV-V	Mus musculus	900-840		700, 500	.72	543.5	142.3	MAY 2
RAIN FOREST	IV	RAINFALL (mm)	1210		600, 300				
RAIN FOREST	IV	TEMP. (°C)	1210		18°, 14°				

FIGURE 10-7. *Seasonal variation in total biomass of* Rattus rattus *and* Mus musculus *along extended Mauna Loa transect.*

The total biomass curves for *Rattus rattus* showed similar annual trends, with the seasonal peak occurring in the late fall or early winter at all three elevation traplines. The high-elevation traplines had the least biomass, with only about 13 percent of that found at the mid-elevation traplines with the highest biomass. The smallest annual amplitude (when measured as a percentage change from the mean biomass) was also found at the high-elevation traplines. The amplitudes at the mid- and low-elevation traplines were similar to each other, showing about a 25 percent annual change above and below the mean biomass. The low-elevation traplines had an intermediate biomass—approximately 39 percent as large as that found at the mid-elevation traplines. The fit to the annual curves was very poor at the high-elevation traplines ($r = .23$) and reasonably good at the other two lower-elevation trapline sets ($r = .63$ and $r = .62$).

Mus musculus total biomass curves showed two patterns of annual peaking. The high and low elevation traplines peaked early in the year and had low values in the fall. The mid-elevation traplines had the opposite pattern, with the peak in the fall and low values in the winter, slowly climbing to the fall peak. The low-elevation traplines showed an abrupt change from a low value in the fall to a high in the winter, while the high-elevation traplines had a more symmetrical change throughout the year. The average biomass comparisons had a pattern of highest biomass at the low elevation, intermediate biomass (50 percent of that at the low elevation) at the high elevation and the lowest biomass (27 percent of the low elevation) at the mid-elevation traplines. The amplitude of the annual change was highest at the mid-elevation, next at the high-elevation, and least at the low-elevation traplines. The curve fit to the annual pattern was good for the high and low elevations ($r = .76$ and $r = .72$) but fairly poor at the mid-elevation traplines ($r = .50$).

Discussion and Conclusions

The examination of the occurrence and mean weight of both the *Rattus rattus* and *Mus musculus* populations gave some indications of annual patterning along the elevational gradient studied, although there was quite a diverse set of patterns. The combination of these two measures into total biomass values reduced the observed variation in patterns into a few that appear to have greater consistency. The interpretation here is that the total biomass values may come closer to a definition of the carrying capacity of the areas in which the traplines were set. However, further examination of this aspect is required.

Rattus rattus showed a single biomass pattern for all elevations along the gradient, with the peak occurring in the late-fall to early-winter period, the time of increasing rainfall and decreasing temperature. *Mus musculus* had two biomass patterns. The high- and low-elevation populations had early-year peaks (during the wettest and coolest months), and the mid-elevation populations, which showed the lowest population sizes, had a fall peak during the relatively dry and warm period. This mid-elevation site includes rain forest ecosystems where the mice may be excluded generally by the high average wetness. Both species showed a relatively good fit to annual seasonality patterns of total biomass, except at the poorest environments for each species: high elevations for *Rattus rattus* and mid-elevations in the rain forest for *Mus musculus*.

AN EXOTIC PSYLLID ON *ACACIA KOA* *

Nine species of Homoptera are associated with native *Acacia* species on the island of Hawai‘i (Beardsley, 1971). Five of these homopteran species are endemic to the Hawaiian Islands and four have been introduced accidentally. *Psylla uncatoides* (Sternorhyncha: Psylloidea; Psyllidae), a recently introduced sapsucker, is the only psyllid presently known to attack *Acacia* species in Hawai‘i. However, it is exerting a profound influence on these plants. The results reported here are concerned with the population dynamics of this insect and its relationship to its host's phenology. A discussion of its biological control can be found in TR 23.

Psylla uncatoides was first collected on O‘ahu in a mosquito light trap near the Honolulu International Airport in March 1966 (Joyce, 1967). In April 1967, this psyllid was found established on introduced *Acacia confusa* at Pu‘u Ualakaa Wayside, Tantalus-Round Top, at the Pali Golf Course in Kaneohe, and on *Acacia confusa* and *Acacia koa* in Kalihi and Nu‘uanu valleys (Funasaki, 1968a). New island records for this psyllid were reported for Maui in June and Kaua‘i in July 1967 (Funasaki, 1968b,c). In July 1970, it was first reported on the island of Hawai‘i, although it had been observed there five months earlier (Davis and Kawamura, 1970). Gagné (1971) reported great numbers of this psyllid on the ground at the summit of Mauna Kea on July 28, 1970. Its presence at the summit was interpreted as an indication of the high populations on *Acacia koa* at lower elevations since the summit is devoid of vegetation and migration would have been necessary to reach this remote location. Beardsley (1971) found this psyllid on Lana‘i in 1971.

Sampling Information

Acacia koa is distributed between about 1,220 and 2,040 m along the Mauna Loa transect. Study sites at 1,280, 1,600, and 2,040 m (IBP sites 4, 6, and 9) were established along the transect in May 1971. Samples were taken at approx-

*By J. R. Leeper, J. W. Beardsley, and K. W. Bridges

imately monthly intervals until June 1974. Collections were made at each elevation on the same date.

Sampling methods were designed to obtain the best population estimates of the different life-cycle stages of *Psylla uncatoides* in the following four categories: eggs, small nymphs, large nymphs, and adults. Instars one through three were lumped together as small nymphs, and instars four and five were considered together as large nymphs.

Psylla uncatoides nymphs tend to cling to the foliage when disturbed, while the adults tend to jump and fly away. These differences in behavior necessitated the use of two sampling techniques. The counts of eggs and nymphs were made by taking ten 4-in (10 cm) samples of *Acacia koa* terminal branches at each study site. The samples were placed individually in plastic bags and chilled in a refrigerator until they could be observed under a dissecting microscope and the counts made. Adult populations were sampled by applying a D-VAC Model 24 vacuum collecting apparatus to the branch-foliage for three-minute periods. After collecting the D-VAC sample, the excess debris was removed and the arthropods were killed with ethyl-acetate and stored in 70 percent ethyl-alcohol until they could be sorted and counted.

Most psyllids breed on the new terminal growth, or flush, of their host plants. The percentage of total *Acacia koa* terminals in flush was estimated for each sample site at the time the psyllid samples were taken. This estimate was determined by averaging the percentage of the terminals with new growth for three counts of one hundred terminals each. This is a different technique from that used in the tree phenology study.

Annual Distribution Trends

The three-year sample period has been summarized by averaging the sample values for each calendar month. These mean monthly values for the flushing of *Acacia koa* and the abundance of each of the *Psylla uncatoides* life-cycle stages have been plotted in figure 10-8. A percentage flush value has been used as the *Acacia koa* flushing index value, and the index values for the *Psylla uncatoides* populations are the log (base *e*) values of the actual counts obtained in the samples. This figure also contains the seasonality parameters obtained by using the least-squares curve-fitting procedure. These curves are based on the use of the log values. The actual unaveraged values were used to calculate the seasonality parameters.

The *Acacia koa* flushing had a very broad, spring-peaking period, generally with some flushing throughout the year. Only at the lowest elevation site (1,280 m) was there any cessation of flushing, and then only for the August through September period.

The general pattern for all of the *Psylla uncatoides* life-cycle stages was for high populations in the midwinter through spring period. At the higher elevations, this period was longer than at lower elevations. Generally populations of all stages were found throughout the year except for the large nymphs, which were not found during the fall at the high-elevation sites, during June and July at the mid-elevation sites, and from July through October at the low-elevation sites. The peak periods were successively shorter and the abundances less for the three subadult stages at each of the lower elevations.

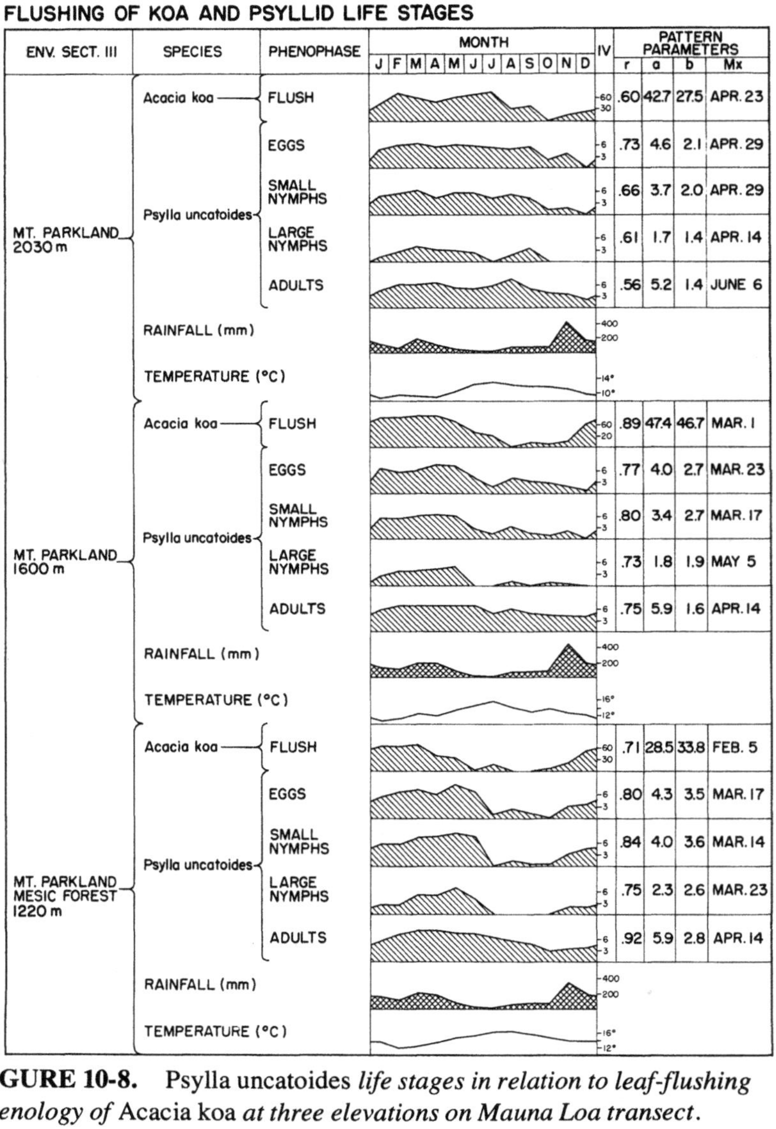

FLUSHING OF KOA AND PSYLLID LIFE STAGES

ENV. SECT. III	SPECIES	PHENOPHASE	MONTH (J F M A M J J A S O N D)	IV	PATTERN PARAMETERS r	a	b	Mx
MT. PARKLAND 2030 m	Acacia koa	FLUSH		60, 30	.60	42.7	27.5	APR. 23
	Psylla uncatoides	EGGS		6, 3	.73	4.6	2.1	APR. 29
		SMALL NYMPHS		6, 3	.66	3.7	2.0	APR. 29
		LARGE NYMPHS		6, 3	.61	1.7	1.4	APR. 14
		ADULTS		6, 3	.56	5.2	1.4	JUNE 6
	RAINFALL (mm)			400, 200				
	TEMPERATURE (°C)			14°, 10°				
MT. PARKLAND 1600 m	Acacia koa	FLUSH		60, 20	.89	47.4	46.7	MAR. 1
	Psylla uncatoides	EGGS		6, 3	.77	4.0	2.7	MAR. 23
		SMALL NYMPHS		6, 3	.80	3.4	2.7	MAR. 17
		LARGE NYMPHS		6, 3	.73	1.8	1.9	MAY 5
		ADULTS		6, 3	.75	5.9	1.6	APR. 14
	RAINFALL (mm)			400, 200				
	TEMPERATURE (°C)			16°, 12°				
MT. PARKLAND MESIC FOREST 1220 m	Acacia koa	FLUSH		60, 30	.71	28.5	33.8	FEB. 5
	Psylla uncatoides	EGGS		6, 3	.80	4.3	3.5	MAR. 17
		SMALL NYMPHS		6, 3	.84	4.0	3.6	MAR. 14
		LARGE NYMPHS		6, 3	.75	2.3	2.6	MAR. 23
		ADULTS		6, 3	.92	5.9	2.8	APR. 14
	RAINFALL (mm)			400, 200				
	TEMPERATURE (°C)			16°, 12°				

FIGURE 10-8. Psylla uncatoides *life stages in relation to leaf-flushing phenology of* Acacia koa *at three elevations on Mauna Loa transect.*

The seasonality patterning of *Psylla uncatoides* was strong at all three elevations studied, although there was a clear trend for all life-cycle stages to have a decreased fit to the annual pattern, as measured by the *r* values, with increasing elevation. The mean population levels and annual amplitude also showed a systematic decreasing value, with few exceptions, with increasing elevation.

The timing relationships among the maximum population levels of each life-cycle stage of *Psylla uncatoides* and the flushing of *Acacia koa* showed a generally expected pattern. The maximum egg numbers followed peak flushing by approximately twenty-three days, and the maximum adult populations followed the egg peak by approximately twenty-nine days. This latter value is very close to the approximately one-month developmental period known for this species.

The relationship between elevation and the timing of certain seasonal events revealed several interesting points. The period between the peak flushing of *Acacia koa* and maximum egg numbers decreased with increasing elevation, at the rate of about 44 days per 1,000 m elevation, so that at the high-elevation site (2,040 m), there was less than a week separating the two peaks, while more than six weeks' difference was found at the low-elevation site (1,280 m). The mean populations for the averaged *Psylla uncatoides* life-cycle stages peaked successively later in the year at the rate of approximately 54 days per 1,000 m elevation.

Discussion and Conclusions

The seasonality trend for the *Acacia koa* flushing observed in this study is in general agreement with that reported earlier. In this study, however, a more distinct annual trend for the *Acacia koa* flushing was seen at all elevations. The peak flushing periods are similar in both studies, with the same sites showing peak differences from one to six weeks apart.

Psylla uncatoides, a very recently introduced species, closely matches the flushing phenology of its host, *Acacia koa*. If the *Acacia koa* flushing occurred only during a short period, such a match would be expected if the psyllid were reproducing exclusively in this host. In the case of *Acacia koa* along the Mauna Loa transect, however, some flushing can be found in nearly every month of the year; it is only differences in the percentage of flushing that generally distinguish the phenological pattern. Therefore the *Psylla uncatoides* population patterns need not be coincident with those of *Acacia koa* for the persistence of the insect along the transect. The similarities in the patterns of the insect and its new host probably offer some increased probability of available reproductive sites if flushing were disrupted during unusually dry years. Whether this synchrony is due to preadaptation to similar phenological patterns in this insect's area of origin or its general timing flexibility is not known.

SELECTED DETRITOPHAGES*

Of the sixteen adult Sciaridae species found on the Mauna Loa transect, three have been chosen here for a more detailed examination of their temporal patterning. These are *Bradysia setigera, Bradysia* sp. 8, and *Ctenosciara hawaiiensis,* all from spatial group 2 identified in the transect analysis. The provisionally identified *Bradysia* sp. 8 has not yet been named since no males of this species have been found. Further field collection and larval rearing are required to resolve this taxonomic problem.

All three species are endemic and were found to extend along a considerable portion of the transect. They occur in the mountain parkland and open *Metrosideros* dry forest, which comprises the central portion of the transect. *Bradysia setigera* and *Ctenosciara hawaiiensis* are poorly represented in the rain forest portion at the lower end of the transect, and *Bradysia* sp. 8 continues into this zone at densities similar to those found at higher elevations.

The developmental rates of fourteen Sciaridae occurring in Hawaii have been determined from rearing studies using 20°C constant temperature cabinets (Steffan, 1974). The mean for adult emergence after oviposition ranged from 16.2 to 34.0 days, with the period for *Ctenosciara hawaiiensis* found to be 29.0 days. A preoviposition period of one to three days was also found. There was enough variation in this developmental period that the adults emerged over a five- to eight-day span from a single batch of eggs.

Some aspects of the ecology of *Ctenosciara hawaiiensis* have previously been reported (Steffan, 1973). Seasonal fluctuations in the adult populations were observed in the fifty-five-week period of study on sites in the Kīlauea Forest Reserve and at IBP site 6 on the Mauna Loa transect. Rainfall was an important correlate of population cycles, with the development of population peaks coming some three weeks after periods of high rainfall. The high moisture relationships may be associated with the fungal association thought to be important in the larval development of some Sciaridae species (Kennedy, 1974). *Ctenosciara hawaiiensis* occurs in confined, wind-protected areas with breeding sites in the top few centimeters of leaf litter.

The study reported here is an extension of the previously reported work of Steffan (1973).

Sampling Procedure

Insect trapping was done with Gressitt traps (Gressitt and Gressitt, 1962), a type of modified Malaise trap. Collections were made along the Mauna Loa transect at three IBP elevations (site 2 = 1,220 m, site 6 = 1,600 m, and site 9 = 2,040 m). Only the data from IBP site 6 will be reported here since this was

*By W. A. Steffan and K. W. Bridges

the longest-term record, extending over some two and one-half years, and includes eighty-one trap-weeks of samples. The collection period extended from January 20, 1971, to August 1973. During the first period, until July 5, 1971, insects were recovered from the net weekly. Thereafter, the net was left out for alternate weeks. These biweekly samples also represent the catch for week-long periods. Other methods of trapping these insects were employed but have not been used in the analyses reported here.

Temporal Distribution Trends

The abundances of the three species of Sciaridae are plotted in figure 10-9, along with the weekly rainfall values for the entire sample period. The primary characteristic of the three species is the dominance of the first-year populations as compared to those occurring in the remainder of the study. Another characteristic is the dramatic population changes all three species showed, with both upward and downward trends extending from the observed population peaks to very low levels between single sample periods.

Correlations were sought between rainfall from prior weeks and the weekly population levels of the insects. Earlier this was shown (Steffan, 1973) to be a useful predictor of abundance. That analysis, however, was based on a one-year sample and did not have information on the very small populations that were characteristic of the second-year samples. The best correlation values for the entire sample period being examined here were obtained with the rainfall that occurred five to six weeks prior to the observed population levels. The correlation coefficients were relatively low. Rather than reject a rainfall-population relationship, the evidence suggests that a more complex consideration of the microenvironment and multigeneration population buildup is needed to understand the fine-scale temporal patterning of these species. Such an examination is beyond the scope of this report.

The mean values for observations at each month for the three species are plotted in figure 10-10. This monthly averaged plot shows the general trends of the species for typical years. Rather abrupt population changes between successive months are also seen in this plot, although some general seasonal trends are apparent for each species. There are some similarities between species trends. A large population of *Bradysia* sp. 8 occurred in the six-month period between February and July, with a secondary, smaller peak in October. A generally similar pattern, except for the early-year peak ending a month sooner, was seen for the *Ctenosciara hawaiiensis* populations. The early-year peak was also seen in *Bradysia setigera* populations, except that it ended even a month earlier. This species did not show an October peak.

The monthly average data have been subjected to the sine-based curve analysis procedures. The best-fit parameters are included in the figure. The curves fit moderately well for all three species, with r values ranging from a low of .63 to a high of .69. The dates of the peak values fall at approximately monthly intervals: mid-March for *Ctenosciara hawaiiensis*, mid-April for *Bradysia setigera*, and mid-May for *Bradysia* sp. 8.

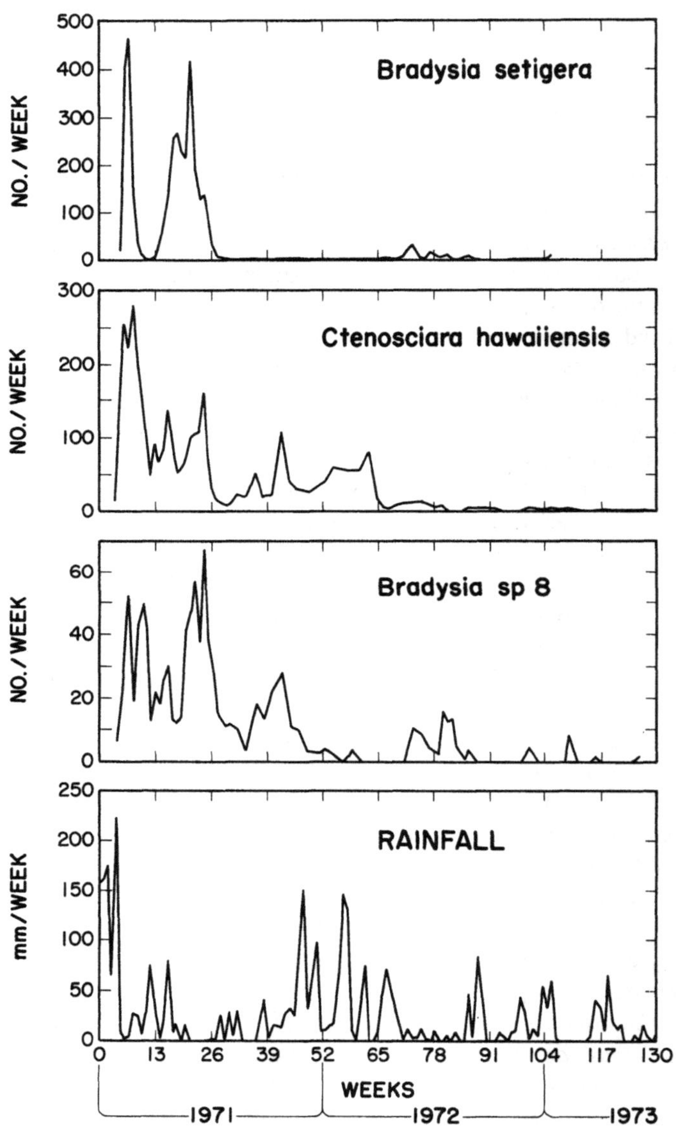

FIGURE 10-9. *Change in abundance of three sciarid species in relation to weekly rainfall, 1971–1973.*

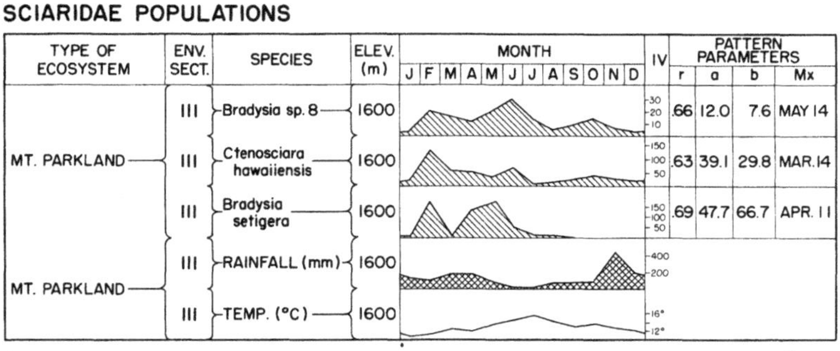

SCIARIDAE POPULATIONS

TYPE OF ECOSYSTEM	ENV. SECT.	SPECIES	ELEV. (m)	MONTH J F M A M J J A S O N D	IV	PATTERN PARAMETERS r	a	b	Mx
MT. PARKLAND	III	Bradysia sp. 8	1600		30 20 10	.66	12.0	7.6	MAY 14
	III	Ctenosciara hawaiiensis	1600		150 100 50	.63	39.1	29.8	MAR. 14
	III	Bradysia setigera	1600		150 100 50	.69	47.7	66.7	APR. 11
MT. PARKLAND	III	RAINFALL (mm)	1600		400 200				
	III	TEMP. (°C)	1600		16° 12°				

FIGURE 10-10. *Seasonal variation in abundances of three sciarid species.*

Discussion

The Sciaridae are ephemeral insects. Their relatively short developmental period (between one-half and one month) and their egg production capacity (between 120 and 175 eggs per female) allow a rapid population buildup from one or two reproductive periods. The effects of these population characteristics can be seen as the abrupt population increases observed in this study.

The specific environmental relationships that allow population changes are not completely clear. Microhabitat relations are important in the reproductive process, but what measurements are necessary to understand the processes for each species have not yet been established. The complexities can be seen in the difficulties of rearing *Ctenosciara hawaiiensis* larvae; the medium that was successful for all other species did not appear to be appropriate to *Ctenosciara hawaiiensis* (Steffan 1974), implying that optimal field conditions also differ between the species.

The Gressitt traps used in this study are thought to be relatively unbiased sampling devices for some Diptera (Julliet, 1963). However, there may have been some apparent population changes that could be explained by periods of windy weather. This factor is not examined here since wind records are not available for this site.

Both the detritophagous habit of the Sciaridae and their possible relationships with fungal populations indicate the environmental requirement for moist substrates. Such sites are expected to be generally available in the mountain parkland forest environment around IBP site 6. As was seen in the analysis of the climate relative to dry periods, however, more than 4.6 dry months (less than 60 mm rainfall per month) were found to occur at this site in an average year. These periods may be sufficiently dry to influence the reproduction of these species, especially since the important part of the environment is the upper litter layer, an area likely to become dry even with a short rainless period.

There are discernible seasonality trends in the three species of Sciaridae examined here. This is consistent with the similar population changes reported for lowland Sciaridae on Oahu (Steffan, 1974). The seasonality trends observed in this study were not strong, however.

TWO *DROSOPHILA* SPECIES*

The genus *Drosophila* is particularly well represented in Hawai'i with many endemic species, which have offered excellent material for studying the processes in microevolution. Collections of those drosophilids that could be captured by simple bait-trapping techniques were made along the Mauna Loa transect. The primary purpose for the collections was to obtain several exotic species so that the changes in their chromosomal patterns could be studied over spatial gradients as they changed during several seasons. The results of these studies are reported elsewhere in this synthesis as they relate to the chromosomal patterns and as the collections have been used for the spatial pattern analysis. The same collections are used here to provide a preliminary interpretation of the population changes of two of the dominant exotic *Drosophila* species at three elevations.

Sampling Information

Banana-baited traps were placed at different elevations along the Mauna Loa transect for twenty-four-hour periods. Three sites were selected for use in the analyses reported here: IBP site 2 (1,220 m in the open *Metrosideros* rain forest environment), site 6 (1,600 m in the mountain parkland ecosystem), and site 9 (2,040 m at the upper limit of the mountain parkland ecosystem). Collections were made in April and December 1971, October 1972 and April, July, August, September, October, and December 1973.

Two of the most abundant species, *Drosophila simulans* and *Drosophila immigrans,* were chosen for the analysis of the seasonal patterning. Because of the wide range of population values for the same species from different collections, all population values have been converted to their equivalent natural log (log base *e*) values. Where these values were averaged, they were first converted to the log value and then averaged.

Annual Distribution Trends

The mean monthly values for each of the two species have been plotted in figure 10-11. The gaps in this figure indicate those months for which there were no samples.

---

*By Y. K. Paik, K. C. Sung, and K. W. Bridges.

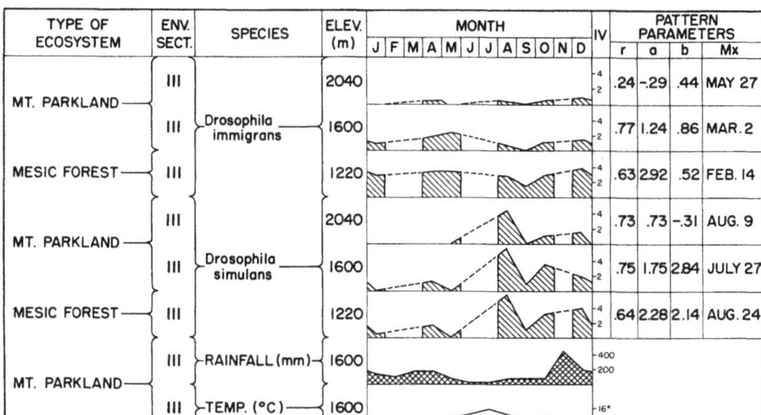

FIGURE 10-11. *Seasonal variation in abundance of* Drosophila immigrans *and* D. simulans *at three elevations on Mauna Loa transect.*

The *Drosophila immigrans* populations were lowest at the higher elevations. At the high-elevation site (2,040 m), there was no visible annual pattern, probably due to the small populations. At the mid-elevation site (1,600 m), there was a late-winter to spring population peak, with a distinct decrease in September. The annual amplitude of population change was greatest at this mid-elevation site, where it was more than 40 percent larger than either the higher or lower elevations. The high population levels found at the low-elevation site (1,220 m) were relatively constant throughout the year, but this site also showed a slight dip in the population levels in September. The predicted population peaks occurred later in the year at the higher elevations. The fit to the annual curve was relatively good at the two lower elevations but very poor at the high elevation.

The broader-ranging *Drosophila simulans* had a similar pattern of increasing abundance with lower elevation. Each of the three elevational sites had similar annual patterns. A summer peak was observed, followed by a decrease in September and a second slightly smaller peak in the fall. All of the populations peaked within a period of approximately one month. The seasonal pattern was relatively strong at all three elevations, as measured by the *r* values, although it was poorest at the lowest elevation.

Discussion and Conclusions

The environments along the Mauna Loa transect appear to be generally favorable for the two *Drosophila* species examined in this study. The patterns of response of the two species had both common and different elements. Both species had smaller populations at higher elevations, and there was a general synchrony of population trends within each species at all elevations. All species populations showed fairly clear annual seasonality patterns, except the *Drosophila immigrans* population at the upper limit (2,040 m) of its distribution along the transect, where few individuals were trapped.

The *Drosophila immigrans* populations had relatively little change in their population sizes at each site throughout the year, probably indicating that the environments utilized by this species along the transect are favorable throughout the year, but such environments are less abundant at the higher elevations. In contrast, *Drosophila simulans* showed marked population changes throughout the year, with its populations peaking in the warmer, drier seasons and decreasing during the cooler months with higher rainfall. This indicates a closer climatic relationship with this species. This differential annual trend suggests that the basic environmental relationships that influence the population processes of these two species are quite different. Further studies that might identify these differences seem to be warranted.

11

Short-term Temporal Patterns among Island Biota

K.W. Bridges

Considerable importance has been attributed to the phenomenon of seasonality in the tropics. Some of the earliest observers noted that without the constraints of temperature and drought, trees would be able to display their individual phenological patterns (Holttum, 1931). This possibility has many implications for tropical ecosystems since it should allow the coevolution of a variety of plant-pollinator and plant-disperser relationships and thereby play an important evolutionary role in the establishment of the timing of plant phenology (Stiles, 1978). Such time-based interaction is expected to be particularly prevalent in tropical regions where there are high numbers of species. In Hawai'i, where isolation has resulted in irregular numbers of species in its ecosystems, the importance of seasonality as an evolutionary factor can be questioned, and we can expect differences in temporal behavior as compared to other tropical regions.

Like most other studies on tropical environments, we found that the role of seasonality generally is not a problem of the survival of the species; low temperatures requiring some form of dormancy were encountered only at the highest elevations on the transect. More critical, especially for many of the insects, were the occurrences of dry periods, when either their food or reproductive sites might be disrupted. As a result, it is thought that temperature plays a secondary role because it controls the timing of the phenological behavior of the insects, their food plants, and the availability of reproductive habitat. The relationship of pollinator energetics (Heinrich and Raven, 1972) must not be overlooked in this regard, since the lower temperatures at the higher elevations may exclude some insect pollinators.

A number of other studies have attempted to define the extent, mechanisms, and implications of tropical seasonality. Some of the earliest studies sought out places with climates known to be very stable, such as the New Hebrides islands (Baker and Baker, 1936). Other workers have suggested that seasonality should be approached experimentally in controlled environments (Sweeney, 1969). The majority of studies, however, have been based on long-term observational records in climates that show varying degrees of temperature, rainfall, and photoperiodic change throughout a year.

In temperate areas, the seasonality expressed by the flowering of plants, for example, typically consists of a short period of blooming of a species, lasting from several weeks to a month. This occurs annually at a similar time and is synchronized by exposure to cold or changing day length or both (Evans, 1971) or is related to some thermal summation value (Jackson, 1966). In the tropics, various water relations are added to the possible factors that synchronize seasonality. Opler et al. (1976) reviewed the mechanisms that stimulate a break in dormancy for tropical species and proposed four categories of explanation: reduction in water stress, lowering of temperature, increasing photoperiod, and drought conditions.

Rainfall relationships are widely viewed in tropical environments as perhaps the most important factor involved in determining the flowering period. However, the role of rainfall probably is complex and not well understood, even in general terms. It appears, for example, that dry periods are necessary to break bud dormancy in coffee (Alvin 1960). Other workers have noted the strong correlation between rainfall that occurs during dry periods and flower induction (Opler et al., 1976).

Photoperiod, of course, varies considerably less in the tropics than in temperate regions. However, Njoku (1958) has shown that daily photoperiod differences in Nigeria are sufficient to induce flowering of some plants. Changes in temperature, especially those that are abrupt and may induce flowering, have not been systematically observed and analyzed relative to phenological timing. An example of this phenomenon, given by Walter (1971, 1973), is the epiphytic orchid, *Dendrobium crumenatum,* which occurs in the tropical rain forest of Java. This orchid has dormant buds that at certain times burst into flower simultaneously over areas of several square kilometers, probably following sudden chilling after a rain.

Explanations for phenological induction in plants generally are climate associated. Yet a sufficient number of detailed descriptive studies have not been carried out to document fully the climatological mechanisms involved in phenology. Data collection and analysis of even rudimentary climatic measurements over the long times required to establish patterns is a large undertaking. An attempt has been made in the studies reported in this part to analyze phenological phenomena and the associated climate along an elevational gradient that showed both temperature and rainfall variation so that greater variation (Jackson, 1966) could be observed in a shorter time. This elevation gradient has also allowed the study of both seasonal and nonseasonal forests, and two major types of environments found in the tropics.

An examination of the climatological seasonality was the first study reported here. It was important to establish not only the general pattern of the climate but to estimate the amount of variation that could be expected between years. This subject has been receiving increasing attention (Stiles, 1977, 1978).

The long-term air temperature record showed a regular annual patterning and relatively low variation between years. Because of the high diurnal temperature change relative to the annual mean temperature change, temperature is often cited as a relatively imprecise mechanism for behavioral synchronization in the tropics. At the higher elevations, however, there may be some possibility that temperature cues would be reliable. Not only were the mean temperatures lower, but there was a greater amplitude of summer-winter temperature change.

The typical pattern of the rainfall through the year, based on the long-term record, was seen to have a distinct wet winter period and a moderately dry summer period. This pattern also was seen for the study period at all sites along the transect. There was a decrease in rainfall, by more than one-third, as one moved upslope from the rain forest to the upper montane seasonal environments. Although not analyzed here, there was also a decrease downslope from the rain forest into the seasonal coastal lowland. There were considerable differences between the observed monthly rainfall and the long-term monthly averages. This variation indicated that even though a site may have a high average rainfall for a particular month, a dry period can occur unexpectedly. Unfortunately the analytical techniques that are generally used for climatic data often tend to obscure this relationship. Analysis difficulties include the skewed distribution of monthly rainfall values (reducing the utility of the standard deviation as a measure of variability), the tendency to use analyses based on long-term average conditions, and the neglect of some potentially important climatological factors such as nonrainfall precipitation. Since a relatively small difference in rainfall is sufficient to reclassify a rain forest climate to a seasonal climate, these problems should be given careful consideration.

The climatic data in this study were examined quantitatively for their rate of change by their fitting to a sine-based curve, an important consideration in comparing the results of this study with other areas of the tropics. In the ecosystems along the transect, both rainfall and temperature showed relatively gradual changes between months. In many other parts of the tropics, however, very abrupt changes may occur, especially in rainfall, between successive months. It is expected that this characteristic has resulted in more diffuse synchronization of the phenological behavior of the plants and animals along the transect than might otherwise be expected.

The general phenological flowering response shown by the thirteen trees studied on the transect was for relatively long periods of flowering. Species populations showed 10.3 months of flowering each year, with a range from 7 to 12 months, when monthly values were averaged between the five years of observation. The year-round pattern was characteristic of the seasonal environments along the transect. The species had peak-flowering behavior of three types: moist-cool period peaking, dry-warm period peaking, and random peaking, which could occur at almost any time of the year. Within the rain forest ecosystem, species showed individual patterns of peaking that were each relatively short but together spread throughout the entire year. In the more seasonal environments found along the transect both above and below the rain forest, species tended to group into those that had their flowering peaks in either the wet or the dry season.

Metrosideros collina, however, showed a complex pattern of more individual population peaking along its range on the transect. Leaf flushing was seen to occur even more often than flowering, with an average species population having 11.4 months of flushing each year. Patterns of flushing related to whether there was a period of flushing inactivity (as was shown by one-quarter of the species examined) and whether peak-flushing behavior was distinct. Only in *Dodonea viscosa* was there a pronounced peak, although flushing was generally greater for most species during the moist period of the year. Growth was seen to be as continuous a phenomenon through the year as flushing, with an average of 11.4 months of growth per year shown by the twelve species examined. The measurements in this study indicated some seasonal growth inactivity in the coastal lowland during the dry period of the year. *Acacia koa,* in particular, showed a tendency for more growth during the fall than in the spring. There were relatively weak growth peaks in most species. Of the three phenophases, flowering had the strongest annual patterning.

On first examination, the flowering patterns observed in the Hawaiian ecosystems seem to be different from those expected in other tropical environments. Most reports on plant flowering in the tropics, but certainly not all, indicate that the majority of species have short flowering seasons and considerable synchronization within a species. Extreme cases have been reported (Opler et al., 1976) such as with an entire population of *Casearia praecox* that flowered synchronously in a single day. This short-period flowering phenomenon is sufficiently well recognized that Gentry (1974) has classified tropical flowering patterns into categories ranging from those that describe single very short periods ("the big bang"), multiple occurrences of very short periods, short periods, and steady-state or near-continuous flowering.

One of the problems in comparing published results with the patterns found in Hawai'i has been the variety of recording and analysis techniques that have been used. Most studies reporting the length of flowering in tropical species select some taxomonic group, focus on a group with particular animal associations (such as in pollinator coevolution studies), or examine only a sample of the vegetation. Therefore we have only fragmentary evidence for comparisons between the flowering phenology in Hawai'i and other regions. These indicate that Hawai'i tends to have more longer-period and fewer short-period flowering plants than might generally be expected in the tropics in general but more characteristic of wet tropical regions. A few examples illustrate this difference. Snow (1965) examined species of *Micronia* in Trinidad and found that flowering lasted one to two months, depending on the species. Stiles (1978) has presented quantitative data on some 42 species of hummingbird food plants in Costa Rica. The mean length of the periods of good bloom was approximately three months but ranged to as long as six or seven months in a few species. Plots of flowering profiles of 25 representative species showed approximately one-third of the species to have near-continuous flowering. In an earlier review of the Costa Rica flowering phenology, Janzen (1967) reported that approximately 20 percent of the 223 species showed continuous or intermittent (six to twelve months per year) flowering, while the remaining species had a mean flowering period of approximately 2 months. The ten common species observed in the New Hebrides (Baker and Baker, 1936) had an average of approximately 5 months of flowering, and three of these species would be categorized by Janzen's criterion as continuous or intermittent flowering species. None of these studies had as long an average flowering period as was found along the Mauna Loa transect.

The fauna of the Mauna Loa transect showed a variety of seasonal patterns. Some patterns, such as those observed in the sapsucking homopteran *Psylla uncatoides,* were strongly synchronized with the flushing phenology of its host plant, *Acacia koa*. This synchronization extended throughout the montane seasonal forest. Other species showed patterns that appear to be related more to climatological seasonality. All three Sciaridae examined here were seen to have a generally midwinter through spring peaking period. This corresponds to the period following peak rainfall and ends at the start of the dry season in the montane seasonal environments where the populations were collected. While the first-year data showed a correlation between the populations and the rainfall that occurred several weeks prior, this relationship did not extend to the second year. This implies that a more complex climatological relationship probably exists.

The two *Drosophila* species that were examined had different seasonal patterns. The most consistent annual trend occurred in *Drosophila simulans* populations. The population peaks were seen in the late summer to fall period in both the seasonal montane and open rain forest environments in which samples were taken. The other species, *Drosophila immigrans,* showed relatively indistinct population patterning, with a general population decline in September, the wettest month of the year. This species may be relatively insensitive to annual climatological changes.

In contrast to the relatively short-lived insect species, the two rodent species examined have relatively long life cycles. Both showed systematic patterns of annual change, however, when the total biomass was analyzed for elevational ranges in both the seasonal and rain forest environments. *Rattus rattus* had a single pattern of peak biomass occurring in the late-fall to early-winter period in each environment. *Mus musculus* had two patterns of peak biomass, with an early-year peak in both the higher-and lower-elevation seasonal forests and a fall peak in the mid-elevation rain forest.

The general conclusion that can be reached from this examination of seasonality trends in the biota of the Mauna Loa transect is that a component of seasonality generally is found in the population processes of all of the species. It varies, of course, between species and in different environments, but much of this variation can be explained if the detailed life-cycle requirements of the species are considered.

V
Genetic Variation within Island Species

12

Introduction

H.L. Carson

ECOLOGICAL GENETICS

This term has been used by Ford (1971), Creed (1971), and others to specify a branch of biological science that deals with the dynamics of the evolutionary process. Evolution, simply described, is a process of descent with genetic change. The ultimate origin of such permanent change in organisms is the process of mutation. An extraordinary attribute of mutations is that they arise by a process that is random with respect to the requirements of the individual carrier in which the mutation occurs. Whether genic, chromosomal, or regulatory, mutations are cast by a chance process into the hereditary system of the organism. Without these intrinsic raw materials, no permanent change, no evolution at all, could occur. On the other hand, the directive forces of evolution (natural selection, for example) determine what, if any, of this raw genetic variability is retained and integrated into the genetic system. Natural selection and similar forces operate extrinsically on the organism through the environment. The ultimate source of permanent adaptive genetic change in populations, therefore, is determined by the constant delicate interplay between ecology and genetics. Many years ago, Turesson (1923) applied the apt term *genecology* to the study of this interactive process. Just as genetic change alone without selection is biologically meaningless for organisms in the long term, so also there can be no such thing as selection among individuals that are genetically identical.

Evolution results in the formation of new adaptations to the environment, the delicacy and intricacy of which have been a source of continual astonishment to biologists. Further, descent with change frequently results in new species—arrays of interbreeding individuals that are reproductively isolated from other such groups. Understanding how these processes operate in a state of nature requires the approach of the geneticist who has the techniques to catalog the basic genetic variations and separate them from noninherited somatic variations. It also requires close attention from ecologists who study the growing point of evolution, the local population or deme, as it comes under the influence of the local environmental conditions. At its basic level, this type of work is based on the premise that the large genetic differences frequently observed between species and between groups in geological time have arisen by a slow process of accumulation of individually small differences. In short, macroevolutionary change can be understood only by recourse to microevolution.

INTRASPECIFIC GENETIC VARIABILITY

Understanding and cataloging basic genetic variation as it occurs in natural populations is extraordinarily difficult. Darwin, for example, felt it necessary to abandon the study of variation in strictly natural populations and drew his most compelling examples from variation observed under domestication. Genetics was born as a laboratory science in Mendel's garden; it continued to be so during the explosion of basic genetic work on *Drosophila* flies and certain plants during the two decades following 1920. A handful of perceptive genetic biologists—Tschetverikov (1926), Dobzhansky (1933), Ford (1931), and Spencer (1932) in particular—attempted to assay genetic variability directly in natural populations. Methods of analysis of concealed genetic variability such as lethal genes, recessive visible mutants, and chromosome variants were laborious and had to be confined in actual practice to populations of such organisms as *Drosophila* flies, Lepidoptera, land snails, and certain selected plants.

This restriction of population genetics to genetically workable organisms was dramatically lifted in the mid-1960s by the exploitation of the new methods developed for the biochemical study of enzymes. Assays based on zone electrophoresis were introduced by Smithies (1959), subsequently developed by Ashton (1965), and applied to populations by Hubby and Lewontin (1966). The technique permits recognition of variant molecular forms of a large number of enzymes and other structural proteins. Tests with genetically suitable organisms have revealed that much of the variation observed is inherited in simple mendelian fashion. Application of these methods has revealed very high levels of genetic variability in natural populations of many organisms. Inference from earlier studies of chromosomal, lethal, and polygenic variants in natural populations had led to the conclusion that natural genetic variability was great, but with the new methods, quantitative studies of this variation became possible. What is even more important in the present connection, genetic tracking of the environment could be followed in detailed fashion. Sixty years after Turesson, genecology has come of age.

GENETICS OF ISLAND SPECIES

Organisms from insular ecosystems played a major role in the thinking of Darwin and Wallace, who pioneered the microevolutionary approach over a hundred years ago. In the present century, however, very few studies using modern genetic methods have made use of insular organisms. An exception is a large effort, begun in 1963, to study the evolutionary biology of Hawaiian drosophilids. This multidisciplinary team approach continued during the course of the US-IBP in Hawai'i; two reviews have appeared (Carson et al., 1970; Carson and Kaneshiro, 1976). The IBP and *Drosophila* research programs have been closely allied, and genetic data from the two sources have been integrated in this chapter.

The evolutionary process manifests itself in two ways: it produces adaptations and it produces species. One of the major goals of the work described here has been to specify the genetic changes that accompany adaptation and speciation. To this end, the approach has been to select certain organisms that are amenable to genetic analysis and to study these in detail. In order to obtain a dynamic view, it has been necessary to select populations in which new adaptations and species are apparently *in statu nascendi*. From this point of view, the selection of the ecosystems on Mauna Loa by the US-IBP was especially fortunate. The entire mountain and its forests are geologically new; therefore the evolutionary characteristics of the species found there may also show some novel features.

A striking aspect of evolution in the Hawaiian Islands is the fact that some groups of organisms have speciated extensively in the archipelago whereas others have not. This contrast between speciating and nonspeciating organisms is, of course, not an absolute one. Although the drosophilids, with over five hundred species, are generally considered examples of the speciating type, certain subgroups have relatively small numbers of species. This is true despite the fact that the amounts of geological time and ecological opportunity available appear to be similar for the two subgroups. Certain of the studies to be reported bear on this topic, which served as one of the major foci of the original island ecosystems proposal (see objective 1, p. 3).

The second aspect of microevolution to which attention has been given is the manner in which the genetic polymorphism carried within species tracks the environment. The detailed information on various ecosystems along the Mauna Loa transect, gathered by the island ecosystems workers, provides an extraordinary opportunity for the correlation of genetic variability with specific environmental parameters.

Finally, work has been done on genetic variability of several species of *Drosophila* that have been introduced by humans into the Hawaiian Islands. Organisms continue to be introduced into Hawai'i both inadvertently and purposefully. Each introduction is, in a sense, an experiment in evolution. In almost no case, however, has anything been known in detail about the genetics of the founders or what genetic changes may have ensued following the introduction. Many such exotic organisms not only are able to invade the island ecosystems but aggressively compete with and indeed oust the endemic biota. Some of the work to be reported bears on the genetics of this situation. Such competitive abilities may, for example, be inherent in the genetic system of the introduced organism or they may be evolved *in situ* following the introduction. An attempt has been made to investigate this matter by detailed genetic study of one such species, *Drosophila immigrans*.

GENETIC METHODS AND MATERIALS

Every character of biological importance shows some variance, part of it due to genetic and part to environmental causes. As expressed by the biometrical geneticist; $V_P = V_G + V_E$ where V_P is the variance of the phenotype and V_G and V_E are the variances due to genotypic and environmental factors, respectively. By definition, variance due to the environment is not inherited; accordingly the determination of the strictly genotypic component assumes paramount importance in evolution.

Prior to the development of the methods of mendelian analysis and quantitative genetics, no suitable estimates of hereditability of a character existed and only an educated guess was possible. Partitioning the variance of a metrical trait is a difficult task, even in laboratory strains that can be studied by growing them in various environments. For organisms in a state of nature, the task is virtually impossible, even with modern genetic methodology.

For these reasons, population geneticists have had to resort to the use of genetic variants, which have strongly discontinuous effects. Such variants have the advantage that environmental effects are relatively small so that the genetic variant may be easily scored in the phenotype. On the other hand such characters are disadvantageous because their immediate meanings, in terms of the adaptation of the organism to the environment, are difficult to specify.

Variations may be observable within or between species. From the dynamic evolutionary point of view, variation occurring within a species, called genetic polymorphism, is the most interesting. The two or more variants observed at a locus or chromosomal section may, of course, be held in a state of balance. Nevertheless the possibility always exists that directional change in time, leading toward the fixation of one or the other allelic condition, can occur. This latter process, as Haldane (1957) has said, is "the unit process of evolutionary change." It can occur only within a species population.

At first sight chromosomal inversions might seem an undesirable type of genetic variant to use in genecological and selection studies. Because it is indeed a character of the chromosome itself, however, no environmental influence blurs its recognition. In many cases, not only the heterozygote but both homozygotes may be recognized; such precision, however, is confined to organisms such as certain flies that display giant polytene chromosomes, wherein this type of change may be seen by microscopic study. An inversion involves a block of genes that under most circumstances are prevented, in heterozygotes, from undergoing effective meiotic recombination. Accordingly, inversions, by marking and rendering blocks of genetic material (supergenes) into the polymorphic state, enable workers to study not merely isolated gene loci but many associated loci as well.

Electrophoretically detected biochemical variants, like inversions, have the great advantage of being easily observed in many organisms taken directly from nature. Like chromosome inversions, this type of variant is unaffected by environmental conditions, and their inheritance in a mendelian fashion has been widely demonstrated (Lewontin, 1974). An even greater advantage in the use of these methods is their generally wide applicability to many types of organisms, some of which may indeed be rather unwieldy from the point of view of genetic manipulation or cytological analysis.

All of these methods are designed to read directly or infer the genetic condition as it exists in the wild parents. In this manner, gene frequencies in a state of nature can be determined. These may serve as a basis for comparison of both allopatric and allochronic populations. The method makes possible a direct measure of microevolutionary change.

Endemic and Exotic Insects as Objects for Genetic Study

Insects, especially Diptera, are particularly useful for studies of population genetics. The strong emphasis on the use of *Drosophila* for evolutionary studies stems from the fact that the genus, and the closely related genera, consists of many species of flies that may be reasonably easily collected in natural populations. Of the more than 1,200 described species of the genus *Drosophila* known in the world (Wheeler and Hamilton, 1972) only 17 belong to the category of cosmopolitan or colonizing species (Carson, 1965). Thus, most of the world's species are endemics, often narrowly confined to small areas of the continents or islands. Hawai‘i is most remarkable in this regard. It harbors 353 species of the genus *Drosophila,* more than one-fourth of those known in the world (Hardy, 1974). Only 22 of these have been introduced from faunas developing outside of the islands (Wheeler and Wheeler, 1972).

As part of the current genetic surveys, two introduced species, *D. immigrans* Sturtevant and *D. simulans* Sturtevant have been studied (Paik and Sung, TR 12, TR 65; Steiner, Sung, and Paik, TR 57). Among the endemic species, concentrated efforts have been made to study variation in *D. engyochracea* Hardy, *D. mimica* Hardy (Steiner and Carson, TR 50; Steiner, TR 52), and *D. silvestris* (Perkins) (Craddock and Johnson, TR 45). In practice, wild adult specimens are captured either with use of a net or by baiting with fermented banana or mushrooms in various localities and environments. The flies are then brought to the laboratory, where detailed analyses are carried out.

Chromosome Inversion Variability

If the sequences in the giant polytene chromosomes of the larvae of these flies are to be determined, each wild female is isolated in a separate culture tube and allowed to produce progeny that have been sired by an unknown wild male. These progeny are chromosomally examined by preparation of smears of the salivary gland cells of one to seven F_1 larvae, permitting inferences concerning the condition of the wild parents. In this manner, gene arrangement frequencies can be determined (Paik and Sung, TR 12; Craddock and Johnson, TR 45).

Chromosome variation within and between the various species of the Hawaiian members of flies of the genus *Telmatogeton* has also been studied (Newman, TR 56). In this case chromosome analysis is best done on larvae collected directly from the aquatic habitats in which these animals dwell. Material for this study was collected from freshwater streams in Kaua'i, O'ahu, and Maui, and Hawai'i. In addition, marine forms from island shores have been analyzed.

Protein Variability: Gel Electrophoresis

The technique most commonly used to detect genetic variation at the level of the protein is referred to as electrophoresis. The technique requires the establishment of an electric field in which the proteins, which are direct gene products, may be separated. This is done by using an electrophoresis apparatus consisting of two electrode chambers containing a buffered electrolyte solution, an electrical power source, and the support media, usually an acrylamide or starch gel, in which the protein solution is placed. The support medium also serves as a bridge between the electrode chambers in which the charged field is established. Within this field, the protein molecules separate according to the number of positively versus negatively charged amino acids present in the protein, the size and configuration of the molecule with respect to the pore size of the gel matrix, the ionic and steric forces created by the buffer system and its specific pH, and other unknown factors. Once separated, the molecules can be visualized by using histochemical techniques, many of them designed to demonstrate certain protein types.

A unique feature of the technique is that it can detect a minimum of 25 percent of all amino acid substitutions in any protein to which the technique can be applied. Thus mutations at the genic level can be investigated and their frequencies and the evolutionary implications of their occurrence calculated. Predicated upon the investigation of this variation are the molecular genetic studies described in the following chapters. For more detail concerning the techniques used in electrophoresis and for pertinent literature references, Steiner and Johnson (TR 30) may be consulted.

Analyses such as those described above have been done intensively on *Drosophila mimica* and *D. engyochracea* and to a lesser extent on *D. silvestris* and two introduced forms, *D. immigrans* and *D. simulans*.

Metrosideros collina subsp. *polymorpha*

The distribution of variability in populations of this species, which exists in various ecological situations, has been described by Corn and Hiesey (1973, TR 18), but it has been possible so far to deal with this variation only in terms of observed phenotypes. Accordingly, despite the fact that adjacent plants often differ in leaf shape, pubescence, and flower color, for example, proof of the nature of the genetic component of this variability has not been forthcoming primarily because of the difficulties involved in cloning and transplanting.

Furthermore, serious problems arose in the application of electrophoresis as an analytical tool to investigate genetic variation in this genus. Young leaf material was used, from which sample preparations were made. The sample material usually displays an unknown compound, believed to be a phenol or lipid by its characteristics after homogenization and extraction. This compound produced a molecular binding effect on the protein-enzyme systems. The effect was such that no separation of the protein systems was observed on starch gels after electrophoresis.

To circumvent this problem, a variety of types of vegetative material was used. The best result was obtained with freshly germinated seedlings about two to four weeks old. At this stage, the responsible phenolic compound had not reached levels inhibitory to meaningful study. However, two additional problems emerged. The first related to low concentrations of protein material and the second to effects due to the developmental stage. Because it was felt that more meaningful studies could be conducted within the time limits of the project, it was decided not to pursue and develop the technique needed to investigate *Metrosideros* further.

13

Genetic Studies of Natural Populations

H.L. Carson, E.M. Craddock, W.E. Johnson, L.J. Newman, Y.K. Paik, W.W.M. Steiner, and K.C. Sung

PROTEIN VARIABILITY IN TWO ENDEMIC *DROSOPHILA* SPECIES*

A primary goal of the Hawaii IBP was to determine what factors may have played a role in the extensive speciation observed to exist in different major taxa of the Hawaiian biota. A variety of hypotheses has been advanced to explain why, at least at a theoretical level, some groups of organisms have undergone extensive speciation while others have not. In developing ecosystems such as exist or have existed in Hawai'i, it is difficult to explain why this should be the case because arguments can be raised that support suggestions that abundant niches do exist, some of which offer little opportunity for competitive clashes. Thus parallel opportunities have existed for nonspeciating organisms to undergo diversification. The importance of genetic variability in species formation has also been pointed out; it is an integral although sometimes undefined part of any hypothesis that attempts to explain why some groups have undergone speciation while others have not in the Hawaiian environment. The question raised in these hypotheses is twofold: what is the extent of genetic variability at the gene-product level in speciating and nonspeciating organisms, and how does this relate to the specific hypothesis under consideration?

By the technique of electrophoresis, the extent of genetic variability in Hawaiian *Drosophila* has been explored. The results have been published or are in the process of being prepared (Rockwood, 1969; Rockwood et al., 1971; Steiner and Carson, TR 50; Steiner et al., TR 57). The point that is most apparent about these studies is that, in the Hawaiian *Drosophila,* there exists no paucity of genetic variability in the genus as a whole. Indeed, levels of variability are on a par with widespread continental species (Ayala et al., 1972; Berger,

*By W. W. M. Steiner

E.M., 1971; Johnson and Schaffer, 1973; Richmond, 1972), European species (Lakovaara et al., 1972), and other island species (Johnson, 1971).

One of these Hawaiian species that was investigated most from the genetic point of view has been *D. mimica*. A wealth of ecological (Heed, 1968; Montgomery, 1975; Richardson and Smouse, 1975; Kambysellis, TR 39), behavioral (Spieth, 1966; Richardson, 1974; Richardson and Johnston, 1975a, 1975b), and taxonomic (Yoon et al., 1972; Kaneshiro, 1969) data are available on this species. Its primary breeding niche appears to be fallen and fermenting fruit of *Sapindus saponaria* where it subdivides the niche with *D. imparisetae*, another endemic *Drosophila*. This tree, along with *D. mimica*, is found abundantly at only two places, at approximately 1,240 m elevation at Kīpuka Kī and Kīpuka Puaulu. These two mesic forests are about a mile apart and are described in Mueller-Dombois and Lamoureux (1967). A brief description of their climatic and biotic differences is given in table 13-1.

The life span of *D. mimica* in the natural situation is not known; however, it is thought to be quite long since individuals have been observed to live as long as three to four months under laboratory culture conditions. *Drosophila mimica* has a life cycle of three weeks, with individuals reaching sexual maturity at ten days of age (Carson et al., 1970).

Drosophila mimica

Rockwood (1969) and Rockwood et al. (1971) have investigated the genetics of several polymorphic gene loci in *D. mimica*. Considerable genetic

TABLE 13-1 *Comparison of Some Environmental Parameters between Kīpuka Kī and Kīpuka Puaulu*

Parameter	Kīpuka Kī	Kīpuka Puaulu
Canopy density	More open	Relatively closed
Evapotranspiration	Greater	Less
Cloud cover	Greater	Less
Rainfall	Less	Greater
Relative humidity	Lower	Higher
Temperature	Lower	Higher
Wind outside canopy	Greater	Less
Moisture content (lower soil profile)	Less	Greater
No. of endemic plant species	30	48
No. of exotic plant species	33	44

variation was observed at the acid phosphatase (ACPH), alkaline phosphatase (APH-3), esterase (EST-2), and octanol dehydrogenase (ODH) loci. At the ACPH locus, a trend toward cyclicity was observed in which the homozygote for a fast-migrating electrophoretic allele decreased in frequency from December 1967 to June 1968. The change was significant in Kīpuka Kī but not at Kīpuka Puaulu. The ODH and APH-3 loci demonstrated no changes in either allele or phenotype frequencies although the latter had four alleles maintained at frequencies above 10 percent. The EST-2 locus was sampled only once and displayed three alleles (Rockwood et al., 1971). Yoon et al. (1972) describe two chromosome inversions in this species, one extremely rare and the second occurring at low but unknown frequencies in nature. The relationships between enzyme and chromosomal diversity await investigation.

Steiner (1974, 1975, TR 52) and Steiner and Carson (TR 50) have extended the enzyme studies to include more gene-enzyme loci and to examine allele frequency changes over a twenty-month period in *D. mimica*. The study provided evidence indicating that *D. mimica* may be the most genetically variable Hawaiian *Drosophila* studied to date. The average heterozygosity over seven loci is 0.194. In addition, genetic variability was consistently higher in Kīpuka Kī than in Kīpuka Puaulu. The study also confirmed Rockwood's (1969) suggestion that the gene-frequency changes associated with the ACPH locus may be cyclical. The study was especially important because it also demonstrated changes in allele frequencies at three loci: isocitrate dehydrogenase (IDH), phosphoglucomutase (PGM), and EST-2 (found to have three more alleles than Rockwood et al. [1971] found). Steiner (1974) related these

TABLE 13-2 *Observed Correlations between Genetic Events at Three Enzyme Loci in* D. mimica *and Specific Weather Variables*

Locus	No. of Alleles at least 5% in frequency	Weather parameter[a]	Correlation[b]
EST-2	3	Rainfall	Frequency of fast-migrating allele increases with rainfall
IDH	2	Rain 1	Frequency of fast-migrating (common) allele increases with rainfall in this period
PGM	2	Rain 2	Frequency of slow-migrating (common) allele increases with rainfall in this period

[a] Rainfall = number of rain events in the month preceding collection; rain 1 = amount of rainfall in the two weeks preceding collection; rain 2 = amount of rainfall in the third and fourth week preceding collection.

[b] $P < 0.05$.

allele changes to different weather parameter shifts (table 13-2), indicating that the genetic changes might be directly following the environmental changes and serving some deeper adaptive function.

These findings suggested that some predictive value exists in the observations. For example, correlations drawn at the three loci in table 13-2 suggest that levels of heterozygotes at these loci might vary according to the fluctuation in the related weather parameter in nature. Steiner (1974) found that this was indeed the case; the frequency of heterozygotes was higher for the EST-2 locus during periods of more rainfall while it was lower at the PGM and IDH loci. A test for random association of the data was rejected ($P < 0.005$), indicating that genetic variability was tracking in a consistent pattern at these loci.

It is at this stage that the ecology of *D. mimica* becomes important. After larval development, this species, like many other Hawaiian *Drosophila*, pupates in the soil. In these developmental stages, *D. mimica* is most susceptible to environmental moisture. Both the amount of moisture (rain) and the number of rainfall events (rainfall) can affect development (table 13-2). Rain 1 is a short-term effect (one to two weeks) while rain 2 and rainfall are long term (three to four weeks). Thus genetic activity associated with the EST-2 and PGM loci may be considered long term; environmental moisture levels at prepupal stages while the fly is still feeding in the *Sapindus* fruit may be hypothesized as the key selective agent. On the other hand, these flies spend seven to ten days in the pupal stage. Thus amounts of rainfall affecting ground moisture may be the key selective feature acting on the IDH locus. It appears, then, that when conditions are dry, polymorphism is favored at the IDH and PGM loci, both of which are associated with energy production in the organism. On the other hand, moist conditions lead to homozygosity at these loci and increased heterozygosity at the EST-2 locus. It is tempting to suggest that some type of correlated genetic response exists, and Steiner (1974) has found some supportive evidence for this case.

Surprisingly no correlations were observed between rainfall and the cyclical ACPH locus. Rockwood (1969) earlier had hypothesized that rainfall made acid by volcano effluents might be a causal factor for the decrease in the fast allele (ACPH-1^3) frequency. Volcanic activity was constant during the period of this study, yet no correlations were observed between allele frequencies at this locus and the various moisture parameters. In addition, it is hard to think that acidity per se may be the selective factor at this locus since ground soil pH in both kīpukas is between 5.0 and 6.0 and the acidity of fermenting *Sapindus* fruit is below 5.0 (Steiner, personal observations). The acid rainfalls were reported by Smathers (see Kambysellis, TR 39) to be between pH 4.0 and 6.0, indicating that acid environments commonly encountered by developing larval forms of *D. mimica* are not different from rains made acid by volcanic activity. Nevertheless a possible explanation exists. Steiner (1974) found that a strong negative correlation ($P < 0.01$) exists between heterozygosity at the ACPH locus and that at the IDH locus. This indicates that during periods of wetness, the ACPH locus will display higher heterozygosity, while the IDH locus is approaching near-homozygous levels. Thus a possible genetic basis for the cyclicity at the ACPH locus may exist that is predicated on either linkage or some type of coadaptive process involving the whole genome. At the time of this report, no information exists for the extent of coadaptation or of linkage in this species, so the above explanation for cyclicity at the ACPH locus is speculative at best.

The studies on genetic variability in this representative endemic species, then, are essentially of three kinds. The first type involves a dynamic element in which the loci ACPH, IDH, PGM, and EST-2 participate by demonstrating short-term cyclical or directional changes. Another type of locus showed genetic changes, uncorrelated to meteorological parameters and demonstrating fluctuations almost random in nature (leucine aminopeptidase-1). Two other loci, hexokinase-3 (HK-3) and APH-3, demonstrated variability that was stable over the length of this study. Chi-square tests for independence between the collections proved negative, underscoring the stability in gene frequencies at these loci. Some changes in heterozygosity frequencies could be shown, however, that correlated positively with each other and with similar changes at the PGM locus. The lack of linkage information prevents further speculation since these types of correlations may also be due to nongenetic causes.

Desiccation Resistance: Variability in a Quantitative Adaptive Trait

In order to define the relationships between moisture in the environment and its influence on the genetic factors, additional physiological tests were conducted. Briefly, the same samples collected above were exposed to a zero percent relative humidity environment after giving them a routine standardization procedure. After obtaining an F_1 generation from females, males and females were individually desiccated, their relative survival time in minutes recorded, a thorax measurement for body size taken, and then electrophoresis conducted. This procedure enabled three-way correlations to be drawn among environmental variables, the quantitative trait termed desiccation resistance (DR), and the electrophoretic (genetic) data.

Several interesting and consistent associations were brought to light. The ability to resist desiccation was shown to vary with the season. In drier seasons flies were less susceptible to desiccation, and in moist (rainy) periods DR was greatly reduced (Steiner, 1974). Although the component due to physiological homeostasis was unknown, regression of offspring on parents' estimates of desiccation resistance indicated the latter to have a high genetic component as well. The trait was interpreted as having a high adaptive value.

Since strong correlations were found between allele frequencies at some loci and environmental moisture, similar analysis was done to define the relationships between allozyme phenotypes and DR. Steiner (1974) found that at each polymorphic locus, one phenotype was more or less resistant than the other phenotypes combined. In addition, it was noted that enzymes involved in known metabolic functions accounted for more of the variability in desiccation resistance than enzymes hypothesized to act on external metabolic substrates. This was interpreted to indicate that increased heterozygosity, if related to locomotor and physiological activity, could be advantageous in nature during especially dry field conditions. A model was produced that predicted that flies heterozygous at both the IDH and PGM loci should occur most often in dry or stressful conditions. The expectation was met.

D. engyochracea

Steiner (1974) conducted similar studies to those above on *D. engyochracea,* a member of the picture-wing species of the Hawaiian *Drosophila.* This species occurs sympatrically with *D. mimica,* breeding in the bark of fallen and rotting branches of *Sapindus saponaria.* Chromosomally monomorphic, the fly is several times larger than *D. mimica* and is most often found sitting on the boles or branches of the understory plants in the kīpukas, while *D. mimica* is most often found in or on the leaf litter on the ground.

Genetically *D. engyochracea* is not as variable as *D. mimica,* with only 30 percent of its sampled loci having two or more alleles. Heterozygosity levels are 12.8 percent in Kīpuka Kī and 11.6 percent in Kīpuka Puaulu, again on a par with observations made for other *Drosophila* species. Variability was observed at the EST-2, EST-3, LAP-2, octanol dehydrogenase-1 (ODH-1), aldehyde oxidase (ALDOX), and PGM loci.

Steiner (1974) found allele frequencies and heterozygosity levels at only the PGM locus related to environment in this species. Again the PGM allele frequencies were associated with the amount of rainfall occurring in the third and fourth week prior to collection (rain 2), just as in *D. mimica.* Here, EST-3, ODH-1, and LAP-2 heterozygosity levels correlated positively with each other and negatively with PGM and EST-2, while EST-2 correlated positively with PGM. As in *D. mimica,* the genetic events in this species theoretically could be related to a locus showing relationships to changes in a meteorological parameter. Once more the lack of knowledge concerning linkage prevented further consideration of the genetic relationships.

A similar series of desiccation experiments was also conducted with *D. engyochracea.* As in *D. mimica,* physiological shifts were observed that paralleled changes in the amount of moisture that the ecosystem of the fly was receiving. Relationships of the genetic factors to the physiological trait were similar to *D. mimica,* although not as clear-cut with respect to energy production (due primarily to the fact that the PGM locus was the only variable enzyme occurring associated with energy production). Such similar findings across both species occurring in different niches in the same ecosystem have important implications for genetic and physiological evolution and the evolution of adaptive responses in general in insular species.

The studies on genetic variability for *D. mimica* and *D. engyochracea* differ quite strongly from those for *D. silvestris*. *D. silvestris* normally is found in an environment that receives much more rainfall. The nature and pattern of the ecosystem is such that very wet conditions are the norm; moisture is retained in the rain forest areas of Hawai'i such that relative humidity is high most of the time. This is in sharp contrast to the isolated kīpuka forests. As a result, consistency in meteorological observations appears to result in consistency in genetic variability; no large changes are observed in *D. silvestris* gene frequencies over time. Findings of extensive genetic variability in all three species raise the problem of its mode of maintenance. Environmental differences leading to genetic adaptation appear important for *D. mimica* and *D. engyochracea,* and yet without further study they cannot be evoked for *D. silvestris*. Steiner and Carson (TR 50) have suggested a plausible explanation that involves both selection and neutrality theory. It is possible that mutations may arise and be carried along in a rare but largely neutral and drifting state. Rapid changes in environments that are normally encountered by these species may suddenly place a high selective premium on such alleles; they may reach high frequencies before environmental norms are reestablished. The original allele may then regain any previous advantage. Thus some of this genetic variation may serve as a genetic resource during periods of environmental fluctuation, with rare occasions of environmental disequilibrium preventing loss of drifting alleles. It might be expected that such action would result in higher levels of genetic variability than expected. Steiner and Carson (TR 50) have observed more alleles than theory predicts in *D. engyochracea,* which could be used as evidence to favor this hypothesis. These authors point out that the neutrality-selective fluctuation hypothesis would not only provide an efficient way of handling environmental overruns, but it may also serve as a source of variation that may contribute in some manner to the high rates of speciation observed in some endemic insular fauna.

One way that the environment can act to affect the organism is by shifting its energy requirements. Based on studies of *D. mimica* and *D. engyochracea,* Steiner (1974) has advanced the hypothesis that polymorphism in energy-producing systems enables the organism to adapt to environmental stress. The conclusion was reached that the dominant meteorological feature occurring in isolated, insular ecosystems may most likely affect a genetic response, although it may or may not be direct in its action. Such a response will be observed only after basic homeostatic buffering mechanisms have been overreached.

GENETIC VARIABILITY IN A MONTANE RAIN FOREST*

D. silvestris in Kīlauea Forest

Investigation of the genetic variability in a third endemic *Drosophila* species, *D. silvestris,* provides yet another pattern of intraspecific genetic variation. This species is one of the larger picture-winged flies endemic to the island of Hawai'i and typically occurs in montane rain forest areas between altitudes of 1,040 and 1,620 m. It is highly polymorphic chromosomally (Carson and Stalker, 1968; Craddock and Carson, 1975) and in this regard differs from *D. engyochracea,* which is chromosomally monomorphic. *D. mimica* has one or two low-frequency inversion polymorphisms. The differentiation within *D. silvestris* has been well studied, and all known populations of the species have been characterized both chromosomally and electrophoretically (Johnson and Carson, 1975; Craddock and Johnson, 1979). The *silvestris* population in the Kīlauea Forest Reserve is relatively large, although individuals are nowhere near as abundant as those of either *engyochracea* or *mimica*. The data from this population have been presented and analyzed in detail in Craddock and Johnson (TR 45), which provides the basis for the following discussion.

The Kīlauea forest population contains six of the seven autosomal paracentric inversion polymorphisms known within *D. silvestris*. These include one inverted sequence on chromosome 2, one on chromosome 3, and four on chromosome 4. The chromosomal heterozygosity of this population is thus relatively high but by no means extreme for the species. Chromosomal inversion polymorphisms of this kind have been classified as either flexible or rigid (Dobzhansky, 1962) depending on whether the alternative chromosome sequences show marked frequency variations in time and in space. The temporal frequency changes usually follow a cyclical pattern that corresponds to seasonal fluctuations in climate. Such flexible chromosomal polymorphisms, which have been studied in a number of mainland *Drosophila* species (see da Cunha, 1955, and Dobzhansky, 1970, for reviews), are presumed to promote the genetic adaptation of the population to its environment. It was unknown whether chromosomal polymorphisms behaved similarly in tropical environments, and the study of the Kīlauea forest population of *D. silvestris* attempted to provide some data on this question. Fluctuations in inversion frequencies have been observed previously in a neotropical species, *D. willistoni;* nevertheless the available data did not permit a decision on whether the observed variations were part of a seasonal and cyclic pattern. The correlation between chromosomal variation and genetic variation is another little-understood factor, so this was an additional point to be considered in the analysis of the genetic

*By E. M. Craddock and W. E. Johnson

composition and temporal variability of the *D. silvestris* population in Kīlauea forest.

The population was sampled five times during a fifteen-month interval between September 1971 and December 1972. All samples were scored for inversion frequencies. The first three were scored for both inversions and electrophoretic protein variants at twelve loci. Analysis of both kinds of genetic data indicated that successive samples from the population were homogeneous. In other words, the study failed to detect any significant temporal changes in the genetic composition of the population in either chromosomal or allozymic characters.

The conclusion that *Drosophila* populations in tropical habitats are genetically stable cannot be accepted unequivocally at this stage and indeed is not warranted by the data available from the Kīlauea forest population of *D. silvestris*. This study was necessarily limited in time and scope. Confirmation of genetic stability or, alternatively, detection of consistent cyclical changes in the frequencies of alleles and/or chromosomal sequences in this population would require a longer-term study, extending over a minimum of two or three years. Moreover, the population must be sampled at regular, brief intervals so that changes of short duration, if there are any, not be overlooked. (The sampling intervals in this particular study were ten, twenty-four, twenty-six, and four weeks, respectively.) Most importantly, in order for the results to be amenable to meaningful interpretation, the individual sample sizes must be sufficiently large to reflect significant frequency changes. Lack of adequate samples was a major deficiency of this study and the main cause of unreliability in the comparisons and conclusions derived from such comparisons. The inversion frequency data were based on samples of 70, 58, 96, 40, and 24 chromosomes respectively; the allozyme frequency data were based on 106, 152, and 346 genomes (sample sizes smaller for some loci). The chromosomal and electrophoretic data were obtained from the same field samples, and the discrepancy in sample sizes for the two kinds of genetic data is a consequence of the fact that field-collected males did not contribute to the chromosomal data, and further, when isolated in the laboratory as isofemale lines, only 60 to 70 percent of the field-collected females produced larval progeny for the chromosomal analysis. The value of the larger sample sizes for the allozymic analysis was reduced by the fact that several of the loci scored had multiple alleles, many of which were represented in the population at very low frequencies. The PGM-1 locus, for example, proved to have six alleles, and even in the largest sample, some genotypes appeared only once or not at all. Statistical comparisons of the inversion frequencies and allelic frequencies in the various samples did in fact reveal some minor variations. The common inverted sequence in chromosome 3 (3m) fluctuated in frequency in the first three samples from 30 to 41 to 27 percent (total family data). Similarly the 4t inversion fell from 89 percent in the initial sample to 73 percent at the end of the study (mean frequency of the last two samples, first larval data). At the ME-1 locus, the allelic frequencies remained stable in females but not in males of the various samples; the major allele showed consecutive male frequencies of 0.51, 0.72, and 0.72. These and correlated variations noted in heterozygosity values were mostly on the borderline of significance, and since the sample sizes were small and the variations inconsistent, such deviations were overlooked and attributed to sampling errors. Conceivably some of these minor variations may have been biologically real, but the present data are inadequate to permit us to make any such claims. Only more detailed and extensive studies can eliminate the doubt raised by our data and verify our provisional conclusion that temporal changes in the Kīlauea forest *silvestris* population are insignificant.

Leaving aside reservations about the significance of the observations, temporal stability in the genetic composition of this *Drosophila* population can be argued as a logical expectation on two counts. First, if it is true that flexible chromosomal polymorphisms are a consequence of variable fitnesses of different karyotypes in varying environments (Dobzhansky, 1962; Haldane and Jayakar, 1963), then if the environment is stable, presumably the several genotypes will be maintained in the population in a dynamic equilibrium at their optimum frequencies as dictated by their relative fitnesses in that particular environment. Accordingly, in a constant environment, the frequencies of the several chromosomal morphs would be expected to remain constant with time. Genic polymorphisms should behave similarly if such polymorphisms are selectively maintained in the same way as chromosomal polymorphisms. The climatic environment in Kīlauea forest shows remarkably little annual fluctuation. This is documented by continuous records of meteorological parameters at the IBP climatic station at 1,650 m in Kīlauea forest (very close to the *Drosophila* collecting site) available for the period from the inception of the IBP program in November 1970 through 1974 (Bridges and Carey, TR 22, TR 38, TR 59). Mean daily air temperature shows an annual fluctuation of about 10°C. (The actual minima and maxima for mean daily temperature were 9°C and 19°C for 1971, 9°C and 18°C for 1972, and 8°C and 16°C for 1973.) Much of this variation is short term rather than seasonal; there may be as much as 7°C variation within a particular month. The mean monthly air temperatures show 4°C annual fluctuation on the average (12–16°C in 1971, 11–16°C in 1972, and 11–14°C in 1973). The lowest temperatures tend to occur during the winter months of December to February, and the highest temperatures tend to occur during the summer between June and October. This pattern is not completely rigid, however; mean daily temperatures as low as 8°C have been recorded in June, and mean temperatures of 15°C are not uncommon between December and February. Thus, temperature fluctuations in Kīlauea forest are of rather small magnitude and are not consistently correlated with time of year.

The humidity (percentage saturation) in Kīlauea forest is consistently high throughout the year as a partial consequence of the relatively high annual rainfall in excess of 1,500 mm. Mean monthly humidities almost always exceed 90 percent, and almost every day the humidity reaches 100 percent or very close to maximum (97–100%). This pattern is consistent and independent of whether rain actually fell that day. Rainfall in Kīlauea forest is one of the more variable parameters, with the relative weekly amounts showing considerable variation, even though the overall distribution of rain throughout the year is more or less uniform. The humidities in this forest are more constant than the rainfall pattern would suggest, probably because of the very dense canopy that limits evaporation and also because of the daily mists that replenish the moisture content of the air but are not recorded as rain per se. It is important to note that the humidities (and, to a lesser extent, temperatures) actually experienced by adult flies in the forest are probably much more uniform than those given by the weather records because of the behavioral activities of the flies. Adults move freely about the forest only during periods of very high humidity, and once the humidity drops below certain thresholds, the flies take refuge in covered sites where the humidity is presumably higher than in the open environment. Larvae and pupae that inhabit rotting stems and the ground, respectively, would be subject to even less seasonal variation than occurs in the open air environment.

The *silvestris* population in Kīlauea forest thus can be presumed to experience a very stable environment little affected by the time of year, in marked contrast to populations of mainland *Drosophila* occupying temperate environments (for example *D. pseudoobscura*). Daily temperature and rainfall records covering five years are available for two sites in the San Jacinto Mountains in California, where the *pseudoobscura* populations show cyclic seasonal fluctuations in inversion frequency (Epling et al., 1953, 1957). The climate here is cold and wet in winter, with maximum daily temperatures frequently below 40°F (4.4°C), and summers are hot and dry, approaching drought conditions at times. The annual temperature differential is of the order of 50°F (approximately 28°C), and humidities show a similarly large differential, ranging from 100 percent during the winter rainfalls to approximately 40 percent during summer.

The *Drosophila* populations in these California environments undergo drastic changes in population density in the course of the seasonal cycle, and some of the inversion frequency changes parallel the population density changes more closely than they parallel actual changes in environmental parameters (Epling et al., 1953, 1957). Brncic (1972) noted a similar correlation between inversion frequencies and population density in *D. flavopilosa,* but he interpreted this as merely an outcome of independent responses by both to the weather conditions, and especially to variations in temperature. No data are available on density fluctuations in the Kīlauea forest population of *sylvestris,* but our collecting experience suggests that any such changes that occurred during the course of the study were not major. Nonetheless the possibility of density changes in this and other *silvestris* populations should be given serious consideration in any future studies.

The second reason for supposing that populations of *silvestris* should show negligible seasonal variation in their genetic composition relates to the life-cycle pattern of this species. Although the species reproduces throughout the year, the developmental cycle is very slow and the generation time long, with perhaps three or maybe at most four generations per year. This estimate is based on laboratory data, which probably underestimate the time taken in nature. In the field, larval development is probably slower as a consequence of lower nutritional levels than those provided in the laboratory. Furthermore the temperatures experienced, particularly by pupae in the soil, are for the most part substantially lower than the constant 17°C temperature at which cultures are maintained in the laboratory. In view of this extended generation time and the longevity of adults relative to the magnitude and duration of climatic fluctuations in the forest, it is unlikely that *silvestris* populations would be capable of undergoing any significant changes in response to their environment. This is an important factor, which can be appreciated by a consideration of the opposing situation of a population of a small and rapidly developing fly species that passes through many generations during one particular season and that can be modified genetically by selective forces acting for that period unidirectionally and at relatively constant pressure. Such species can track their environment to some degree, necessarily with a slight lag, and are more likely to display changes in their genetic composition that can be related to actual seasonal cycles. Such correlations have been assumed for observed frequency changes in chromosomal inversions, particularly within *D. pseudoobscura* (Dobzhansky, 1943, 1948; Epling et al., 1953, 1957; Crumpacker and Williams, 1974), and also within *D. persimilis* (Dobzhansky, 1956) and *D. flavopilosa* (Brncic, 1972) and some other species (cf. da Cunha, 1955); and, significantly, for allelic frequency changes in *D. melanogaster* (Berger, 1971) and in *D. pseudoobscura* and *D. persimilis* (Dobzhansky and Ayala, 1973).

A general hypothesis might be proposed that populations of large, long-lived species inhabiting tropical environments—for example, Hawaiian *Drosophila*—should show pronounced temporal stability in their genetic compositions. However, this can be discredited as a general rule by the data on other *Drosophila* species presented earlier in this chapter. It is not clear why the results for *silvestris* are at variance with those from the Kīpuka Kī and Kīpuka Puaulu populations of *D. engyochracea* and *D. mimica*. In each of those species, both of which have been comprehensively studied, there is evidence at several loci of significant seasonal changes in allelic variation, some of which show definite cyclicity. However, the total situation for these two species is only marginally different from that for *silvestris*. *D. engyochracea,* which is another large picture-winged species, must have a comparatively long life cycle, probably equivalent to that of *silvestris*. *D. mimica,* which is a considerably smaller fly, has a correspondingly shorter life-cycle length, although adults can live for several months. Field populations probably could pass through up to eight generations per year, and for this reason, *mimica* is considerably more likely than the two former species to display seasonal genetic changes should the environmental variation be sufficient to elicit such responses. Climatic data are available for both habitats from the records of

Smathers (see Kambysellis, TR 39) and for Kīpuka Kī in Bridges and Carey (TR 22, 38, 59). The temperature fluctuations are exactly comparable in magnitude to those observed in Kīlauea forest, but the humidities show greater fluctuations. Whereas the mean monthly humidities in Kīlauea forest are almost always in excess of 90 percent, those in Kīpuka Kī are consistently lower, for the most part below 90 percent, and notably more variable. Furthermore the humidity fluctuations in Kīpuka Kī and Kīpuka Puaulu show a definite seasonal pattern, with a gradual progressive decrease from January to July in parallel with the slight increase in mean temperature. Equally important is the more uneven distribution of rainfall in the kīpukas throughout the year; significant seasonal variability is observed, with the summer months May to August as the driest ones of the year. Total annual accumulation in the kīpukas is, moreover, substantially less than that in Kīlauea forest. The defoliation of *Sapindus* trees in these kīpukas causing seasonal changes in the canopy would also affect potential evaporation and humidity levels in a predictable manner. Superimposed on the gradual cyclical changes in the environments of the two kīpukas are abrupt fluctuations, particularly in relative humidity, which though important in the short term for the reproductive physiology of the flies (Kambysellis, TR 39), may or may not be important in terms of the genetic constitutions of the populations. Abrupt climatic fluctuations sometimes do occur in Kīlauea forest, but much more rarely than in the kīpukas, and such changes tend to be much less extreme and of much shorter duration.

The comparison of the environments experienced by *mimica* and *engyochracea* populations versus the environment of *silvestris* suggests that the most important difference is not in the magnitudes of the environmental fluctuations, which are actually comparatively small for both, but rather in the pattern of such changes. That in the kīpukas is somewhat more predictable in annual terms and shows a more pronounced seasonality, which may be an important criterion in determining the genetic responses of the *Drosophila* populations in these environments. The differences in sensitivity to environmental variations displayed by populations of both groups may also reflect some difference in the basic mechanisms for maintaining polymorphisms within these species. The selective basis of the chromosomal polymorphisms could very well be different from that responsible for maintenance of the allelic polymorphisms, but even so, the reasons for the differences in response by particular species are not yet understood.

Geographic Variation in *D. silvestris*

In marked contrast to the apparent lack of genetic variation within *silvestris* populations on a temporal basis is the observation of grossly significant differences between spatially isolated populations. The geographic interpopulation differences within *D. silvestris* are extreme (Craddock and Carson, 1975;

Craddock and Johnson, 1979); some pairs of populations show as much divergence as that between *silvestris* and its closely related and homosequential species *D. heteroneura*. Even populations that are geographically the closest have rather different genetic compositions. The population in 'Ōla'a forest, for example, the one geographically closest to the Kīlauea forest population, is not genetically the closest on either electrophoretic or chromosomal characters. It shows significant differences in the frequencies of at least four of the seven inversions, one of these being polymorphic in 'Ōla'a forest but fixed in the Kīlauea forest population. The inversion differences contribute much more to the diversity between populations than do the allozymic differences. The chromosomal polymorphisms in *silvestris* could thus be considered flexible in a geographic sense but rigid with regard to temporal variations.

The disparate character of local populations of *D. silvestris* is not unique among Hawaiian *Drosophila*. Thus sharp interpopulation differences in gene arrangement frequencies have been described for *D. crucigera* both within and between O'ahu and Kaua'i (Carson, 1966) and *D. grimshawi* from Maui, Moloka'i, and Lāna'i (Carson and Sato, 1969). In some of these cases, large gene frequency changes occur over very short linear distances. An extreme case has been described for *D. setosimentum* of the island of Hawai'i (Carson and Johnson, 1975). Although a number of populations in the high-altitude windward rain forest are closely similar to one another, the Puna populations (below 1,000 m altitude) show strong differences in gene arrangement frequencies. In addition, sharp differences occur in certain of the kīpuka populations of the Māwae area on the Saddle Road on the Mauna Loa lava flows. Kona populations of *D. setosimentum* are even more divergent from the main body of the species. This information is summarized in figure 13-1. Judgment as to systematic level of these divergent populations will be withheld pending further study. Nevertheless, it is clear that the isolations and ecological extremes of the island environments have had great impact on the genetic structure of the populations of the species that inhabit them.

GENETIC VARIATION WITHIN AND BETWEEN ENDEMIC HAWAIIAN SPECIES OF *DROSOPHILA**

Work on genetic variation in Hawaiian *Drosophila* that is associated with the island ecosystems research has stressed variation within certain selected species (*D. mimica, D. engyochracea,* and *D. silvestris*). The major value accruing from this work is that it makes it possible to relate genetic variability within a species to the specific environmental variables that species encounters in its daily life. These have been abundantly documented by IBP data. On the

*By H. L. Carson

other hand, the *Drosophila* project workers have devoted considerable attention to genetic distance between species. This is an important point, inasmuch as one of the major goals of the IBP program in Hawai'i was to identify the factors, genetic and environmental, that contribute to the extraordinarily high

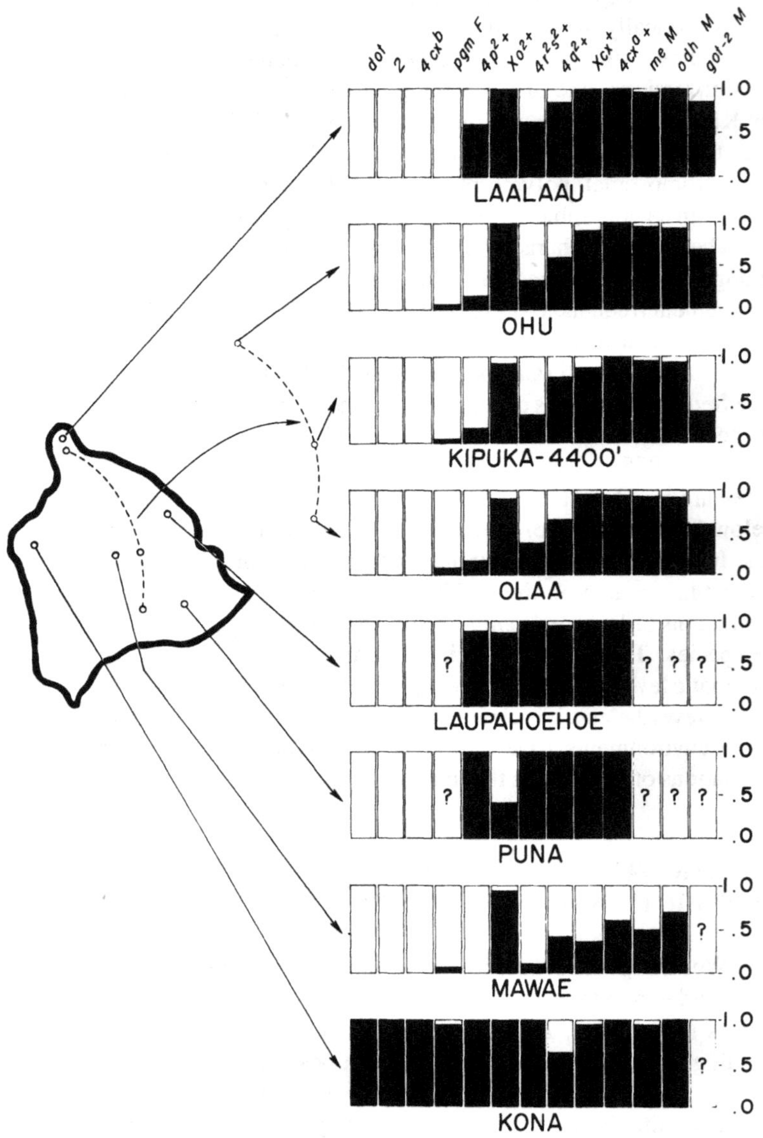

FIGURE 13-1. *Genetic variability between eight populations of* Drosophila setosimentum *from the island of Hawai'i. Each black column in each histogram represents the frequency of a chromosomal or allozyme variant. From Carson and Johnson (1975).*

rates of speciation in some groups of Hawaiian organisms relative to those groups that are less speciose.

Among the most actively speciating organisms in the islands are the picture-winged *Drosophila* (Carson et al., 1970). Recent studies (Craddock, 1974; Johnson et al., 1975; Carson et al., 1975b) have concentrated efforts to obtain genetic and ecological data on a small number of very closely related species. Because of the geological newness of the island, species confined to the island of Hawai'i may be judged to be themselves quite new. They thus serve as excellent objects for evolutionary study.

There is a great interpopulation variability, especially in chromosomes, of *Drosophila silvestris,* a species of the high-altitude montane forests of Hawai'i Island. Occurring sympatrically in a number of places is a very closely related species, *D. heteroneura*. Craddock (1974) and Ahearn and Val (1975) have succeeded in obtaining reciprocal F_1 laboratory hybrids between these species, despite the presence of considerable ethological isolation (Ahearn et al., 1974). Such a result has been obtained quite often in work on other *Drosophila,* but the remarkable circumstance in this case is that both sexes of both reciprocal hybrids are fully fertile. In nature, however, Craddock (1974) states that there is no evidence of hybridization. Detection of hybrids in this case is facilitated by several unusual conditions. The species differ strongly in color pattern of thorax and wings and in the shape of the head. *D. silvestris,* furthermore, is polymorphic for a number of inversions that *heteroneura* lacks; these serve as markers for the species. Although some evidence has turned up that natural hybridization occurs in one area (Kaneshiro and Val, 1977), the gene pools of the two species apparently remain intact in most sympatric areas. Accordingly, one must conclude that the ethological isolation between these species serves as a strong barrier to crossing in most natural situations.

Although there are a number of polymorphic inversion differences as well as morphological differences, *D. heteroneura* and *D. silvestris* are virtually identical in their allozyme patterns, as revealed by electrophoresis (Rogers' $S = 0.96$; Johnson et al., 1975). Rogers (1972), Nei (1972), and others have proposed that genetic data from two populations can be reduced and meaningfully compared by the use of a sensitive coefficient of similarity. This statistic takes into account both differences based on frequency changes between the alleles at a locus and differences due to differential fixation of alleles in the two populations.

In a recent landmark paper on the *Drosophila willistoni* group, Ayala et al. (1974) have proposed that these biochemical genetic differences can be used to suggest systematic levels of divergence. Thus intraspecific comparisons between pairs of local populations of the same species yield a similarity coefficient (Nei's I) of about 0.97, subspecies about 0.8, and full (nonsibling) species 0.35. The Hawaiian species, *D. heteroneura* and *silvestris,* which are virtually identical electrophoretically (Nei's $I = 0.94$), obviously do not conform to the pattern proposed for the continental *willistoni* group species (Sene and Carson, 1977).

Johnson and Carson (1975) and Carson et al. (1975b) have also studied populations of two other very closely related species from Hawai'i Island. These two species, *D. ochrobasis* and *D. setosimentum,* are rare at the IBP sites on the Mauna Loa transect and in Kīlauea forest, but samples may be obtained widely in rain forests from between 600 and 1,600 m elevation. Despite the fact that these two species hybridize at one location in the Ka'u district (Carson et al., 1975a), the gene pools of the two apparently remain virtually intact. Both carry extensive inversion and allozyme polymorphism. Despite some sharp genetic changes between kīpukas, for example, the widely distributed local highland populations of each species show rather high intraspecific similarity coefficients whether based on chromosome inversions or on allozymes. On the other hand, chromosomes are much more sensitive than allozymes in separating the species. For example, one population of *ochrobasis* from the Kohala range is electrophoretically so close to *setosimentum* (Nei's $I = 0.981$) as to be indistinguishable electrophoretically from the latter species. This recalls the situation in *D. silvestris* and *D. heteroneura.* Nevertheless, the distinctive cytological fixations and the distinctive wing patterns of the males serve as an unequivocal means of identifying the specimens. Thus it appears that allozyme differentiation has not kept pace with morphological, cytological, and behavioral evolution.

These cases suggest the following interpretation. Regardless of the pattern of immediate ancestry, these species are new, just as the island to which they are endemic is new. The genetic events that accompanied the speciation events in both pairs of species do not seem to have involved much change at the level of the structural genes as revealed by electrophoresis. Such lack of correspondence between electrophoresis and systematic data is not unknown in other organisms, for example, some fish (Turner, 1974; Avise and Ayala, 1976; see also Avise, 1975). Wilson et al. (1974) and King and Wilson (1975) have argued that evolution of structural genes has not kept pace with anatomical and behavioral evolution in the mammals. They cite the remarkable biochemical similarity of humans and chimpanzees and suggest that under some circumstances changes in regulatory genes are particularly important in the initial establishment of species differences. It appears that the biochemical-systematic criteria enunciated by Ayala et al. (1974) for the *willistoni* group of *Drosophila* do not apply to these pairs of very recently evolved species of Hawaiian *Drosophila.*

The species discussed here belong to the speciose picture-winged group of *Drosophila.* They are characterized by complex mating and territorial (lek) behavior. Spieth (1974) has argued that this condition enhances the rate of speciation in this subgroup of flies: "Lek behavior with its concomitant heightened sexual selection apparently results in isolated populations which occupy relatively and absolutely small geographical areas being able quickly to achieve reproductive isolation." Implicit in this formulation is the idea that the amount of genetic distance between such species should be small. Indeed

Richardson (1974) has proposed a population pattern that could lead to positive assortative mating and, ultimately, speciation in sympatric situations. Such neospecies would, of course, be very similar electrophoretically. Craddock (1974) has also proposed a sympatric speciation scheme for the evolution of *D. heteroneura* and *D. silvestris*. On the other hand, a number of other groups of Hawaiian drosophilids with fewer behavioral elaborations exist. The implication is that they should resemble the behaviorally more simple continental species in their structural gene similarities.

A final possibility for explaining the biochemical similarity of *Drosophila* species pairs on Hawai'i island relates to their newness and the high isolating potential of the island environments (Carson et al., 1975). Most of the picture-winged species of this island appear to have originated, by founder events, from ancestors stemming from the nearby island of Maui (Carson, 1970). Occupation of the island by the flies occurs by chance and follows the establishment of the plant communities on which the flies are dependent. The evidence indicates that many, if not all of them, stem from a few founders or possibly a single ancestral specimen. Formed in this way, the species must be far younger than the volcanoes and forests where they are found. A young species may not always immediately build, by selection, its allozyme variants into its newly acquired gene complexes. Under these circumstances, accumulation of structural protein differences might then become a function of time since isolation (Carson, 1976). Indeed many of these variants may be neutral or slightly deleterious (Ohta and Kimura, 1975).

GENETICS OF ISLAND POPULATIONS OF EXOTIC *DROSOPHILA**

Inversions in *Drosophila immigrans*

Natural populations of many *Drosophila* species carry large amounts of heterozygous genetic variability, which can be detected in terms of genic or chromosomal differences. Inversion polymorphisms are frequent in *Drosophila* species that are confined to a given geographic region, with marked quantitative and qualitative differences between both intraspecific and interspecific populations. The adaptive significance of the inversion systems in these species has been well studied and documented (Dobzhansky, 1970), but information on inversion polymorphism in cosmopolitan species of the genus is scarce and ineffectively collated (Carson, 1965).

In natural populations of *Drosophila immigrans* Sturtevant, a cosmopolitan species, three inversions are known to be universally distributed. All of

*By Y.K. Paik and K.C. Sung

these are located on the second chromosome. Despite the paucity of inversion polymorphism in this species, however, several workers have shown that considerable differences exist not only among the inversion types but also in heterozygous tendencies among geographic populations (Freire-Maia et al., 1953; Brncic, 1955; Gruber, 1958; Hirumi, 1961; Toyofuku, 1961; Paik, 1973). More recently, Richmond and Dobzhansky (1968) and Paik and Sung (1974) have shown different patterns of frequency distribution of these inversions among major islands of the Hawaiian archipelago. Such genetic differences among separate islands in the Hawaiian chain, as well as dissimilarities between the island populations and the continental ones, may be related to their respective distinctive biotic environments. But this view does not necessarily hold true for populations subject to harsh fluctuations in size, genetic drift, or founder effect (Mayr, 1963). Knowledge about this aspect of population dynamics in *Drosophila immigrans* is needed.

The present study was attempted primarily to see how the polymorphism of the species responds to changes in environmental conditions in a locality; observations were conducted simultaneously at three elevations on the Mauna Loa transect on the island of Hawai'i. Several weather variables were measured during the *Drosophila* collection period for correlation with the genetic data by the Hawaii IBP; the weather data are described by Bridges and Carey (TR 22, 38).

Four inverted gene arrangements in the second chromosome have been found in the collections of 1971 to 1973 on Mauna Loa; they occurred independently or in combinations with other arrangements. Following the conventional notations, the median inversion of the left arm will be designated as A, the subterminal one of the right arm as B, and the proximal one of the right as C in the tables and the text. The fourth variant is a complex inversion, probably new, which is located on the proximal part of the right arm; this will be designated as D. Frequency data were obtained for the three gene arrangements commonly found in three altitudinal populations between 1971 and 1973. The three populations, located respectively at 1,220, 1,550, and 2,040 m altitude on the Mauna Loa Strip Road, yielded information that is graphically presented in figure 13-2. (Details are found in Paik and Sung, TR 21, TR 65.)

The percentage frequencies of separate gene arrangements shown in the figure have been determined from the number of heterokaryotypes (single or multiple) as well as by those of homokaryotypes (single) identified in the larval samples. The D arrangements have been found sporadically during the three years at 1,220 and 1,550 m at a total frequency of 0.2 percent. They are too rare to be included in the figures.

The figure shows that in all collections arrangement A was present at the highest frequency, followed by either the B or C arrangement. The overall changes in the frequency of arrangement A in the period 1971 to 1973 are statistically significant, except the one noted for the 2,040 m population. The chi-square value of the 1,220 m populations is 17.0, with 9 degrees of freedom;

that of the 1,550 m is 16.2, with 6 degrees of freedom; and that of the 2,040 m is 1.9, with 3 degrees of freedom. The statistical insignificance obtained for the high-elevation population does not necessarily represent a true parameter of the population because the chi-square test is based upon a relatively small number of collections with small sample sizes. In contrast to arrangement A, no significant changes were seen for both B and C arrangements over the three-year period in most populations. However, as determined by the chi-square test, a significant change ($P = 0.03$) occurred in the B arrangement frequency at 1,550 m; but the statistical significance is due primarily to the unusually low occur-

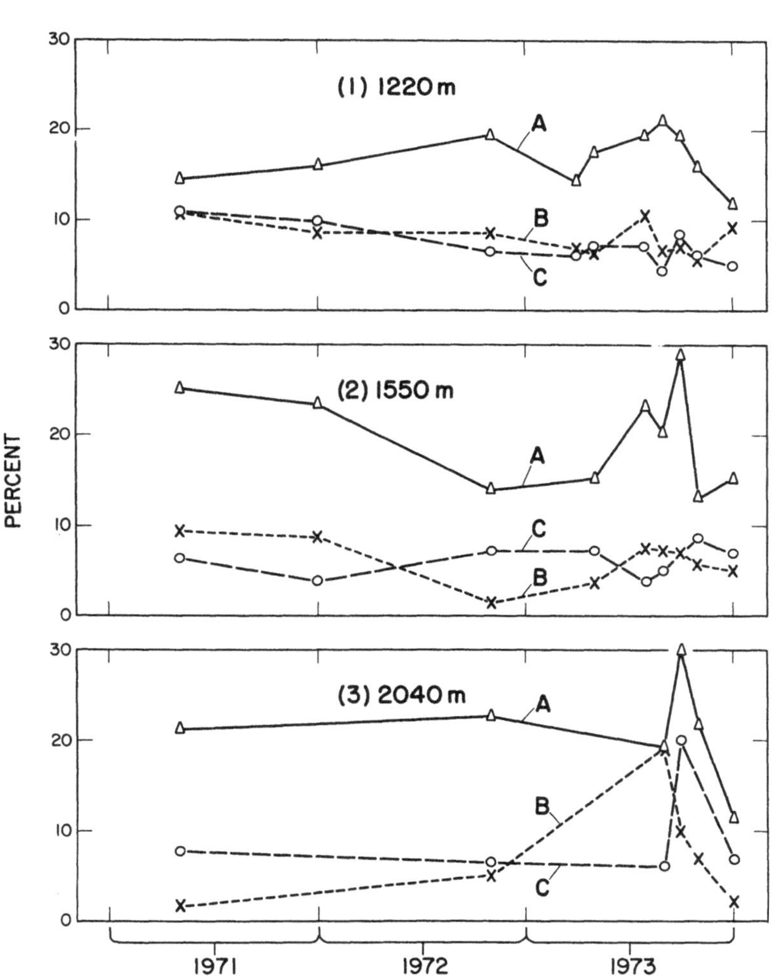

FIGURE 13-2. *Frequencies of arrangements A, B, and C, plotted against date of collection at 1,220 m (1), 1,550 m (2), and 2,040 m (3).*

rence of this arrangement (1.6 percent) in the October 1972 sample; this has amplified the chi-square value up to 14.1, with 6 degrees of freedom.

In 1971 the two monthly samples taken eight months apart at the same elevations were very similar in the relative frequencies of the three arrangements, although no comparison for this seasonal interval was made at 2,040 m. However, there were marked differences in comparing elevations for gene arrangement frequencies. For example, considering the April and December samples together, the A arrangement frequency at 1,220 m is significantly lower than those from the higher elevations ($P = 0.004$ by chi-square); a comparable P value for the C arrangement shows the borderline of significance (P is just below 0.05); for the B arrangement it is equal to 0.5 to 0.1.

In the October 1972 collection, the only sample taken that year, the frequencies of both A and B arrangements declined considerably at 1,550 m. The variations between elevations were tested for significance by the hypergeometric approximation. It was found that the difference in the B arrangement frequencies between 1,220 and 1,550 m is the only one that is statistically significant ($P = 0.01$). The year-to-year changes during the period 1971–1972 were the changes in arrangement A and B at 1,550 m; the frequencies of A and B dropped 10 and 7 percent, respectively, in 1972 from their average frequencies in 1971. These differences are significant when tested by the hypergeometric approximation: $P = 0.007$ for the A arrangement and $P = 0.003$ for the B.

An examination of the 1973 frequency data reveals that regular changes occurred in the frequencies of arrangement A in all populations. At 1,220 m, the frequency of arrangement A was relatively low in the spring period (March to April), increased considerably in the summer period (July to September), and returned to approximately its spring value during the winter period (October to December). The overall changes from March to December in 1973 are significant ($P = 0.03$ by chi-square), and the results suggest that the changes are of a seasonal trend. The same trend was also discernible at 1,550 m when the frequency data were grouped into the three seasonal intervals. The hypergeometric tests indicate that the P value is slightly below 0.05 for the comparison of the April frequency with the combined frequency of the July to September samples; in comparing the July to September frequency with that of October to December, it is 0.04. Visual inspection of the 2,040 m data is consistent with these findings in that the pooled frequency of arrangement A (23 percent) from the August and September samples is appreciably higher than the October to December frequency (14 percent). In a statistical sense, the difference is not significant ($P = 0.15$ by the hypergeometric), but the statistical insignificance is probably the result of the inadequate sample sizes at a time when large changes in frequency occurred. The monthly samples taken from this elevation in 1973 are probably too small to give an accurate estimate of the true proportions of gene arrangements in the natural population.

On the other hand, the frequency variations recorded for the B and C arrangements in the 1973 monthly collections were generally small and showed no regular patterns of change. The month-to-month variations have been tested for significance by the hypergeometric probability, but none of the variations is significant for any of the two arrangements or for any of the three populations. The data thus suggest that the frequencies of the B and C arrangements of the population were essentially constant.

In comparing the seasonally pooled frequencies of the three gene arrangements of the April to September 1973 samples from 1,220 m with those from 1,550 m, the hypergeometric tests have revealed that the value of *P* for each arrangement is always above 0.13; similarly for the between-elevation comparison of the pooled frequencies of the arrangements from the October to December samples, *P* for each arrangement is always greater than 0.40. The monthly samples taken from 2,040 m are too small to warrant their inclusion in this comparison. The present suggestion is that the stratification of gene-arrangement frequency by elevation was not present in 1973 for any gene arrangement.

Temporal Changes in Gene Arrangement Frequencies

Regarding the possible temporal changes in arrangement frequencies of the Mauna Loa populations of *D. immigrans,* the first two years of observations were not informative because only one or two samples were taken each year. In 1973, however, a significant change in general in the frequency for arrangement A occurred so that the frequency was low in the spring populations, high in the summer, and low again in the winter populations. The regular pattern of change is particularly clear from month to month at 1,220 m where the population appears to maintain a consistently high population size and drift effects are probably of little importance. More important is the fact that a similar pattern is also seen at the high elevations (above 1,500 m) where the species does not appear to thrive to the same extent that it does at 1,220 m and it passes through bottlenecks of low frequency. This suggests that drift effects might be overridden by selection pressures at the high elevations so that the A arrangement can maintain the regular pattern of change. For these reasons it appears that the left arm gene arrangement is genetically flexible and sensitive to the seasonal variations, but it is difficult to clarify whether the changes reflect a seasonal cycle or may have been a nonseasonal response to other environmental factors because the observations were limited to about a year. The flexible polymorphism of the A arrangement of *D. immigrans* seems to be comparable to the situations found in other polymorphic *Drosophila* species (Borisov, 1969; Dobzhansky, 1943, 1956; Dobzhansky and Ayala, 1973; Dubinin and Tiniakov, 1945). In the present connection, however, one should remember that a gene arrangement of a species displays seasonal changes in frequency in many but not all populations.

Unlike the A arrangement, the frequencies of arrangements B and C have remained essentially constant during the same collecting periods of 1973, suggesting that the right arm gene arrangements are genetically rigid and insensitive to the seasonal changes. The stability of these arrangements was generally consistent during the entire study period between 1971 and 1973 as well. This finding appears to present a similar polymorphic pattern to that found in *D. robusta* (Carson and Stalker, 1949; Carson, 1958), while the same species showed seasonal changes in the frequencies of some X-chromosome gene arrangements of a local population (Levitan, 1957). In *D. melanica,* no seasonal responses were found in gene arrangements of the X-chromosome, whereas autosomal gene arrangements behaved seasonally (Tonzetich and Ward, 1973).

Carson (1959) has proposed that rigid chromosomal polymorphisms result when the heterozygotes have high fitness and are selected for, but the homozygotes are selected against due to low fitness in most of the habitats of the species. According to Dobzhansky (1962), the flexible and rigid patterns of chromosomal polymorphisms are different adaptive means that are not mutually exclusive and would be utilized by a species through selection whenever the genetic opportunities arise. Interestingly the polymorphism of the Mauna Loa populations of *D. immigrans* shows the two adaptive means within a chromosome.

In addition, the significantly higher range of arrangement A (21–25 percent) found in 1971 at the elevations above 1,550 m compared to those (14–16 percent) observed at 1,220 m appears to be instructive. A precise reason for this difference cannot be ascertained at present, but the monthly observations of 1973 have already shown that the A arrangement has a selective response to the environmental changes. Moreover, the differences occurred when the populations were obviously at high numerical levels. Considering also the relative constancy in the frequencies of the two right arm gene arrangements in 1971 at all elevations, drift effect on the variations of arrangement A can be expected to have had little importance. Taking all the information together, one can suggest that microenvironmental differences may be the basic reason for the between-elevation variations in the frequency of arrangement A in 1971. The general trend for arrangement B throughout the entire period of study was one of relative constancy in frequency. The only marked departure was with the single sample taken in October 1972 at 1,550 m. This difference might be explained by the consideration that the sample represented the immediate offspring of the small numbers of flies that had survived through the summer of 1972.

Comparisons with Other Populations, Island and Continental

On a pooled basis of our three-year observations on Mauna Loa, arrangement A (17–20 percent) was most common, and B (5–7 percent) and C (6–8 percent) arrangements were present at similar, low frequencies. Based on karyotype frequencies, the frequency of heterozygous A ranged from 30 to 33 percent, that of the B from 10 to 13 percent, and that of the C from 10 to 15 percent. The average number of heterozygous inversions per larva was 0.58.

Richmond and Dobzhansky (1968) analyzed the heterozygous incidence of inversions of *D. immigrans* among populations from the island of Maui and found 19 to 33 percent A, 11 to 21 percent B, and C ranging from 7 to less than 1 percent, with the mean frequency of heterozygous inversions per larva of 0.35. Paik and Sung (1974) studied other Hawaiian islands for the chromosomal polymorphism and found that the distribution of three inversions among the islands shows significant differences. On O'ahu, inversion A was most common (19–28 percent in heterozygous state), followed by B (2–9 percent), and C was absent or rare (0–1.6 percent). On Kaua'i, the average frequency of heterozygous A was 10 percent, that of the B was 7 percent, and C was as high as 20 percent. The average frequency of heterozygous inversions per larva was 0.32 for O'ahu and 0.37 for Kaua'i.

In continental areas, Freire-Maia et al. (1953) analyzed the heterozygous incidence of inversions among the populations of *D. immigrans* from Brazil and found the presence of only A inversion, with a mean frequency of 20 percent. In tests of widely separated Chilean populations, Brncic (1955) found 7 percent heterozygous A, 13 percent heterozygous B, and only 0.3 percent heterozygous C, with the mean inversion heterozygosity of 0.20. Hirumi (1961) also discovered the three inversions to be prevalent among Japanese populations, with the mean frequency of the heterozygosity of 0.19; here inversion B was most common (12 percent in heterozygous state), followed by A (7 percent), while C was only 0.4 percent. Similar results were arrived at by Toyofuku (1961) in her study on other Japanese populations. Paik (1973) analyzed a Korean population for polymorphism and found that the results are quite similar to those from the Japanese populations. Gruber (1958) found Israeli populations to be strikingly different in that the A inversion was the only aberration present, occurring at a frequency of 34 percent in heterozygous condition.

This review clearly shows a wide diversity of the inversion frequencies between the continents, between the continents and islands, and even between the islands. The observation that the populations on the island of Hawai'i show a particularly greater amount of inversion polymorphism than do the populations on the other islands is instructive. An even more interesting fact is that the island populations on the whole display much higher proportions of heterozygous inversions per individual compared to the continental ones. This is probably an unusual situation, contrasting strongly with the findings from island populations of *D. willistoni* for comparisons between continents and islands (Dobzhansky, 1957; Ayala et al., 1971). Nevertheless Sperlich (1961) observed no reduction of the amount of inversion polymorphism in island populations of *D. subobscura* as compared to continental ones. Da Cunha, Burla, and Dobzhansky (1950), assuming that inversions have adaptive properties, postulated that the amount of chromosmal polymorphisms reflects richness of natural resources that the species can utilize in a locality. It is not clear at present, however, whether the hypothesis can explain our finding of an increase of the amount of chromosomal polymorphism in the island population of *D. immigrans;* more data on the ecology of this species are needed.

It has been postulated that the populations isolated on oceanic islands would show strong divergence in the outcome of natural selection processes (Mayr, 1954). The marked differences in the distribution pattern of inversions among the Hawaiian islands appear to be compatible with the above hypothesis. The patterns were also affected by random drift at least sometime in the past because the existing populations might be established through a long time period from small numbers of occasional migrants that might have survived the new, founding environments. However, the interisland variations observed may be defined not in terms of the origin of the differences but in terms of how they are maintained. In this connection, Sung (1974) has tested the three inversions of *D. immigrans* derived from two Hawaiian islands (Kaua'i and Hawai'i) in population cages kept in our laboratory and found that all inversions are maintained by heterotic balancing selection. He also found that the inversions (B inversion was not tested) are genetically coadaptive when they are combined with chromosomes of the same geographic origin. Considering also the frequency equilibria of the inversions found in the natural populations on Mauna Loa, balancing natural selection is the most likely process to account for the interisland differences we have observed.

Relation to Weather Variables

Several investigators have recorded correlations of specific weather variables such as temperature, rainfall, and moisture with the frequencies of particular gene arrangements in different species of *Drosophila* (da Cunha et al., 1959; Dobzhansky, 1952; Steiner, 1974; Strickberger and Wills, 1966), although no such correlation was found in other species (Krimbas, 1967). Particularly it has been well established that environmental dryness can affect insects and their gene pools in various ways.

Correlations between temperature, rainfall, and relative humidity and the frequencies of arrangement A were done for this study, but they showed no special relationship. However, a correlation was found with the saturation deficit, a measure of the evaporative power of the atmosphere. According to the weather data collected from 1971 to 1973 on the Mauna Loa transect, the saturation deficit changed cyclically every year, and the changes appeared to be synchronous in all elevations studied. Figure 13-3 shows regressions of monthly frequencies of arrangement A on the square root of the mean saturation deficit of a month prior to collection at two elevations. The specimen collections at 1,220 m have been matched with the weather data from 1,280 m and specimens from 1,550 m with the data from 1,646 m. In the lower-elevation relationship, two data points, one for April 1971 and another for October 1973, are missing due to the lack of pertinent weather data; for the higher elevation, two data points for 1971 are excluded. The data points for the lower site fit the regression line fairly well, and the regression is highly significant (F = 10.29,

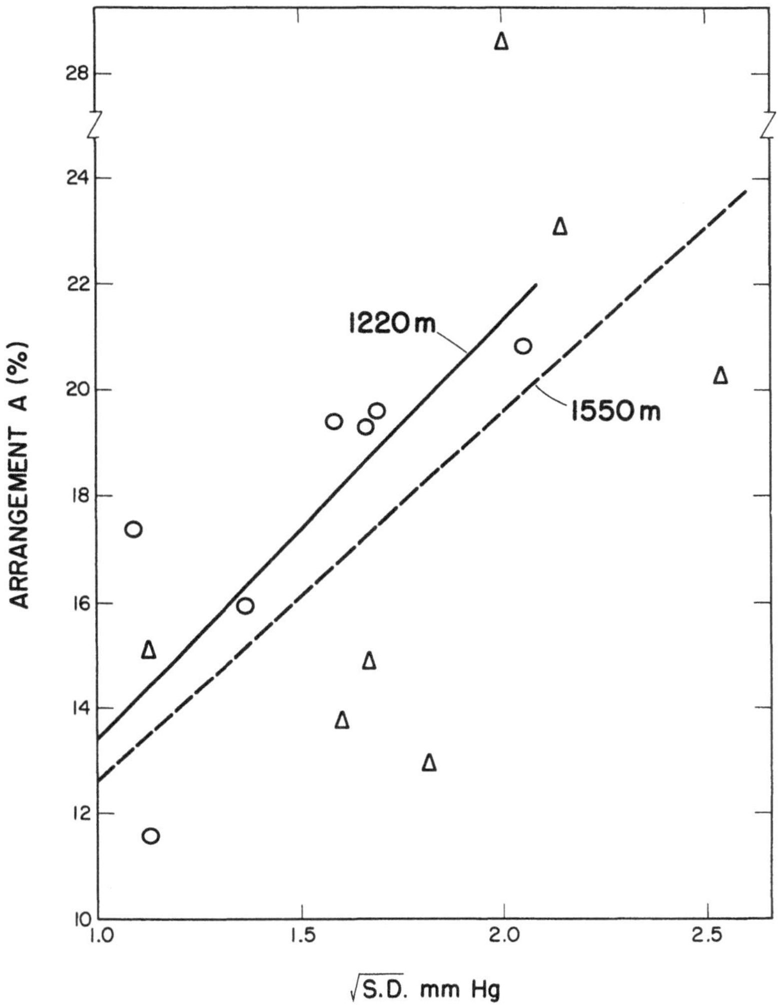

FIGURE 13-3 *Correlation between frequency of gene arrangement A and square root of the saturation deficit. The weather data from 1,280 m and gene arrangement samples from 1,220 m are plotted as circles. The weather data from 1,600 m and gene arrangement samples from 1,550 m are plotted as triangles.*

$df = 1$; $r = 0.80$). It should be remembered that large populations generally occur at this site. The higher site shows data points that deviate considerably from the regression line, and the slope of regression does not differ significantly from zero ($F = 2.07$, $df = 1$ and 5; $r = 0.54$). However, the lack of a close correlation at this site may not necessarily represent the reality of the situation that might occur there because the large amount of residual variance (144.79) is mainly attributed to the two decidedly small samples (the July and September 1973 samples). We suggest, therefore, that dry conditions may be an important contributor to the frequency variations observed, showing a selective advantage of arrangement A under dry conditions.

The primary reason for excluding the 1971 data points was the occurrence of unusual temperatures that year (table 13-3). The temperatures recorded for 1,550 m were markedly different from the subsequent two years regarding the minimum temperatures and the number of days that freezing, or near freezing, temperatures occurred. During 1971, there were numerous occasions when the daily minimum temperature approached or dropped below freezing. This was not the case in either of the years after 1971. In addition, the general pattern of temperatures, when comparing 1,220 and 1,550 m for the difference between the monthly mean temperatures, appears to be relatively heterogeneous in 1971 ($k = \bar{d}/SE_{\bar{d}} = 6.2$) and in 1972 ($k = 7.5$), while in 1973 it is relatively homogeneous ($k = 0.84$). The differences in the patterns of low temperatures may require an additional consideration of temperature in explaining the environmental correlations with the A arrangement frequency. At present, however, we do not have sufficient data to pursue this suggestion so the two data points in 1971 have been excluded from this anlaysis. There is a need for further investigation into the climatic factors in relation to the gene arrangement frequencies in natural populations of *D. immigrans*.

Protein Variability in *D. simulans* and *D. immigrans**

Electrophoretic studies of genic variability in introduced species conducted simultaneously with similar studies on endemic species may be informative, especially if background information from other geographic areas is available for the introduced species as well. Such studies may demonstrate relative levels of variability in species types and whether variability in introduced types has reached a stable equilibrium with respect to its new habitat.

Genetic variation is not lacking in endemic *Drosophila* species, even in those known to be less (probably much less) than 750,000 years old (*D. mimica* and *D. engyochracea*). For comparative purposes, *D. simulans* and *D. immigrans* were also investigated. Possible adaptive shifts in chromosome inversion

*By W. W. M. Steiner

TABLE 13-3 *Distribution of Daily Minimum Temperatures over Three Years at 1,550 m Site*

	1971 : °C					1972 : °C					1973 : °C				
Month	*≤0*	*1–4*	*5–8*	*9–12*	*≥13*	*≤0*	*1–4*	*5–8*	*9–12*	*≥13*	*≤0*	*1–4*	*5–8*	*9–12*	*≥13*
January	19.4	35.5	38.7	6.5			46.7	43.3	10.0			25.8	54.8	19.4	
February	30.8	30.8	34.6	3.8			31.0	48.3	20.7			48.1	33.3	18.5	
March	19.4	29.0	41.9	9.7			9.7	58.1	32.3				9.7	90.3	
April		13.3	60.0	26.7				26.7	63.3	10.0					
May		41.9	58.1					25.0	75.0			41.2	41.2	17.6	
June	3.3	33.3	56.7	6.7				10.0	86.7	3.3		10.0	50.0	40.0	
July		41.9	41.9	16.1				19.4	51.6	29.0		4.0	28.0	60.0	8.0
August			58.1	41.9				48.4	48.4	3.2		3.2	61.3	29.0	6.5
September			13.3	76.7	10.0			30.0	70.0		3.3	23.3	36.7	36.7	
October			41.9	48.4	9.7			45.2	54.8			3.2	32.3	58.1	6.5
November		6.7	50.0	40.0	3.3		13.3	46.7	40.0				36.7	63.3	
December		35.5	61.3	3.2			12.9	71.0	16.1			6.5	58.1	29.0	6.5
Year	5.9	22.3	46.5	23.4	1.9		9.5	39.5	47.1	3.9	0.3	13.7	40.4	43.1	2.6

Notes: The figures given in body of table are percentage of days that temperatures occurred in the various ranges during the month. For November and December 1970, the corresponding figures are 7.5, 40.7, 42.6, 9.2, and zero.

frequencies have already been demonstrated for *D. immigrans;* however, *D. simulans* has no inversions and thus may rely on its genic variability to a larger extent for adaptive purposes. The results of these studies have been presented by Steiner and Sung (TR 57) and Steiner et al. (1976).

These workers found that *D. simulans* and *D. immigrans* populations were variable at intermediate levels (7.3 percent of the genome heterozygous in *D. simulans,* and 11.5 percent heterozygous in *D. immigrans*). This compared to about 16.2 percent in Texas populations of *D. simulans* (Kojima et al., 1970) and no or almost none in other continental populations (Berger, 1970; O'Brien and MacIntyre, 1969) of this species. For *D. immigrans,* it was found that Korean populations demonstrated lower variablity (9.3 percent of the genome heterozygous).

Explanations were offered for the discrepancies between the Hawaiian and Nearctic populations. These included biases due to selection of the lines used in the Berger (1970) and O'Brien and MacIntyre (1969) studies (only the samples in the Kojima et al., 1970, study were drawn directly from nature). If the Kojima et al. study was selected as representative for *D. simulans,* then it is clear that variability is reduced in Hawaiian populations. This may be due to a founder effect or to selection pressures associated with adaptation to a new environment.

Heed (1968) presents evidence that *D. simulans* currently occupies three endemic niches while *D. immigrans* occupies seven. Variability is higher in *D. immigrans* (Hawai'i) than either *D. simulans* or *D. immigrans* (Korea). This appears to fit Levene's (1953) hypothesis concerning the effect of niche variability on genetic variation. The effect of adaptive processes on the genetic constitution may be real for *D. immigrans* since the chromosome shows within-island differences between Kīpuka Puaulu and higher elevations on the same transect. Heed's (1968) data indicate that *D. immigrans* may take longer to develop in the saponin-laden fruit of *Sapindus,* saponin being a naturally toxic compound in this fruit.

It is difficult to conclude that these cosmopolitan species are using their variablity to invade and adapt better to the Hawaiian ecosystems. Although they are as variable as endemic types, it is not certain whether this variability has arisen since introduction some time in the last two thousand years or is due to continual founding events. The new environment is not without some heterogeneity, and the invasion of endemic niches, which is interpreted as successful because of the size and extent of these Hawaiian populations, must include competition with endemic species. In addition, Steiner (1974) has demonstrated the effect that meteorological factors may have on genetic systems in Hawaiian ecosystems. Thus ample opportunity exists for unique types of selection to act on these species.

EVOLUTION OF GNATS OF THE GENUS *TELMATOGETON**

Species of the genus *Telmatogeton* that live in the intertidal zone have been described from the Pacific, Atlantic, and Indian oceans, but it is only in the Hawaiian Islands that they have been found in fresh water. Here these gnatlike flies complete their life cycle in rapidly moving streams and in the splash zone of waterfalls. In the definitive taxonomic work on *Telmatogeton* in the Hawaiian Islands, Wirth (1947) describes two marine and five freshwater species. The marine species are *Telmatogeton japonicus* (Tokunaga) collected from the island of Hawai'i and *T. pacificus* (Tokunaga) from the islands of Kaua'i, O'ahu, and Hawai'i. The freshwater species listed by Wirth are *T. torrenticola* (Terry) found on the islands of Hawai'i, Maui, and Moloka'i, *T. williamsi* (Wirth) in Wai'anae Mountains of O'ahu, *T. fluviatilis* (Wirth) in the Ko'olau Mountains of O'ahu, *T. hirtus* (Wirth) on Kaua'i, and *T. abnormis* (Terry) in Kaua'i and the Ko'olau Mountains of O'ahu.

The purpose of the study reported here and in TR 56 was to determine the evolutionary relationships between species of the Hawaiian *Telmatogeton* by chromosomal analysis. With the exception of *T. williamsi* and *T. abnormis* from O'ahu, all of the species described by Wirth (1947) were collected. Two new species were found—one from East Maui designated *Telmatogeton* n. sp. 1 and the other from Kaua'i designated *Telmatogeton* n. sp. 2 (D. Elmo Hardy, personal communication). Larvae were collected in the field and fixed in Carnoy's fixative. Tissue was stained with lacto-acetic orcein. Inversion differences between and within species were determined by examining the banding patterns in the giant polytene chromosomes of the salivary gland. All species except *T. pacificus* contain excellent polytene chromosomes. The polytene chromosomes of *T. pacificus* are poorly developed and are not expected to yield any banding information. Chromosome number of the various species was determined from an examination of meiotic and mitotic cells.

The polytene chromosomes consist of six long elements, designated as A, B, C, D, E, and F, and one extremely short and generally diffuse element, designated as G. In those species with seven chromosomes, each element is believed to be a separate chromosome and, in species with chromosome numbers less than seven elements, they join together to form specific chromosomes. In some species it is not possible to determine the elements that make up a chromosome since the elements tend to fall apart at the centromere. Element associations are known for *T. abnormis* and *T. hirtus* but are not known for *T. torrenticola* from Moloka'i, East Maui, and the island of Hawai'i.

For each of the long elements, a pattern of bands found in most of the freshwater species was established as the standard sequence. The only species

*By L. J. Newman

that contains the standard sequence for all of the elements is *T. torrenticola* collected from West Maui. Inversion changes relative to the standard band sequence were determined and designated by the element letter and a numerical subscript.

The proposed phylogenetic scheme as determined by analysis of unique fixed inversions and chromosome number is illustrated in figure 13-4. Firm associations between species resulting from inversion analysis are indicated by solid arrows, and tentative associations on less firm grounds are indicated by dashed arrows. (Details are given in TR 56 and in Newman, 1977.)

T. torrenticola from Moloka'i, East Maui, West Maui, and from the Kohala Mountains and Mauna Kea on the island of Hawai'i form a complex of allopatric sibling species. The chromosome numbers and fixed inversions relative to the standard sequence for the *T. torrenticola* species in areas east of West Maui are East Maui ($n = 4$) with inversion D_5, Kohala ($n = 5$) with inversions D_5 and F_4, and Mauna Kea ($n = 4$) with inversions D_5, F_4, and D_3. *T.*

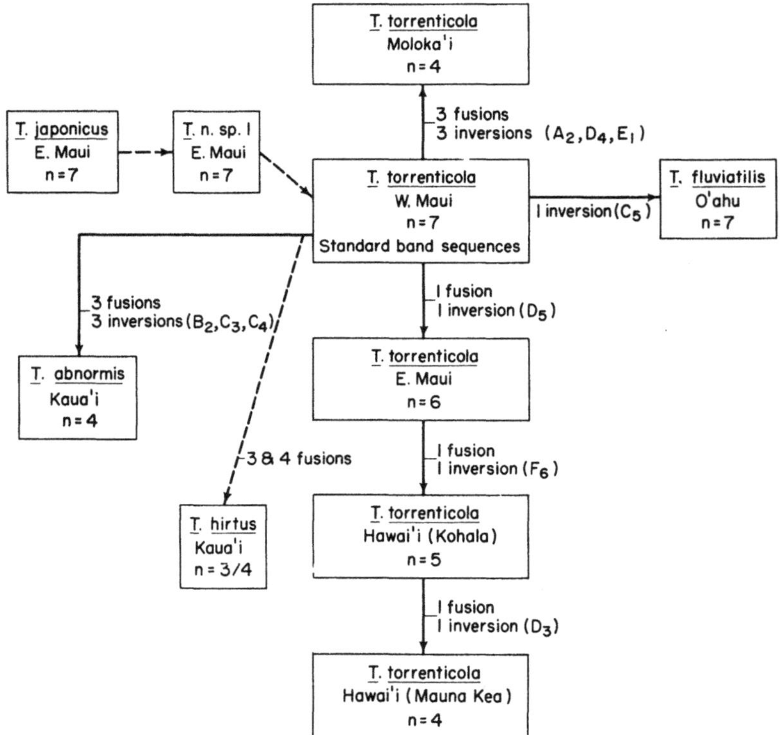

FIGURE 13-4. *Chromosomal relationships between nine freshwater species of* Telmatogeton *and their proposed marine ancestor,* T. japonicus. *The haploid chromosome numbers are given, along with specific inversions whereby the species differ.*

torrenticola-Moloka'i ($n = 4$) differs from the standard by inversions A_2, D_4, and E_1. It is proposed that evolution occurred in a stepwise fashion from West Maui to the east with fixation of one chromosomal fusion and one inversion for each step and from West Maui to Moloka'i with the fixation of three chromosomal fusions and three inversions. The intermediate steps in the evolution of the Moloka'i form have not been found.

The banding pattern of *T.* n. sp. 1 differs from the standard sequence by at least six inversions. Elements A, B, and E differ from standard by one inversion each and element F by three inversions. Elements C and D are so complex that it has not been possible to relate them to the standard sequence. A portion of element A is also complex. Larvae of *T.* n. sp. 2 were not collected.

T. fluviatilis and *T. abnormis* differ from the standard band sequence by one and three inversions, respectively. The polytene elements of *T. hirtus* are so complex that they cannot be related to the standard sequence. Nevertheless *T. abnormis* and *T. hirtus* are tentatively placed together in the phylogenetic scheme because both have fused elements A-B and C-D and they are the only *Telmatogeton* species with differentiated sex chromosomes. In both species element E is a sex chromosome.

A simple XY sex chromosome system was found in *T. abnormis*. The polytene elements are identical between the sexes except in males there is an unpaired region in the E element adjacent to the centromere. The unpaired segment of one homolog is identical to the banding sequence of the female and represents the X-chromosome. The other unpaired segment that represents the Y-chromosome is slightly longer than the X-chromosome and has a unique banding sequence. The sex chromosome difference is not obvious in meiotic and mitotic cells. *T. abnormis* differs from the standard banding sequence by one inversion and is thus thought to be related to *T. torrenticola*-West Maui.

T. hirtus has an XY_1Y_2 sex chromosome system that is obvious in meiotic, mitotic, and polytene chromosomes. Fused elements E and F represent the X-chromosome and unfused elements E and F represent the Y-chromosomes. Males ($2n = 7$) contain two pairs of autosomes, a fused E-F element, an E element, and an F element. Females ($2n = 6$) contain two pairs of autosomes and a pair of fused E-F elements. During meiosis in the testis, a sex chromosome trivalent disjoins, with the two Y-chromosomes going to one pole and the X-chromosome to the opposite pole. Portions of the Y-chromosomes are heterochromatic. The relationship of *T. hirtus* to the other species is uncertain since most of its polytene chromosomes are complex when compared to the standard sequence.

Inversion polymorphism was found in all freshwater species except *T. torrenticola*-East Maui, *T. abnormis,* and *T. fluviatilis*. *T. torrenticola*-Kohala, *T. torrenticola*-Mauna Kea, and *T. hirtus* were each polymorphic for one inversion, *T.* n. sp. 1 for two inversions, *T. torrenticola*-Moloka'i for three inversions, and *T. torrenticola*-West Maui for five inversions. Inversion frequency data fit closely to those expected according to the Hardy-Weinberg equilibrium (Newman, TR 56).

Wirth (1947) proposed that the marine species *T. japonicus* is the likely ancestor of the freshwater species, however, the chromosomal relationship between the freshwater and marine *Telmatogeton* is unclear since the polytene chromosomes of *T. pacificus* are poor and the collection of *T. japonicus* larvae yielded polytene chromosomes unsatisfactory for a complete band analysis. The primitive freshwater chromosome number may have been seven for it appears that chromosome fusions have reduced the chromosome number from seven to four in at least three lines: from *T. torrenticola*-West Maui east to Mauna Kea, west to Moloka'i, and west to Kaua'i (figure 13-4). If this is the case, *T. japonicus* with seven chromosomes is a more likely candidate as the ancestor of the freshwater species than *T. pacificus,* which has four chromosomes.

The species of freshwater *Telmatogeton* with a chromosome number of seven are *T. torrenticola*–West Maui, *T. fluviatilis,* and *Telmatogeton* n. sp. 1. There is conflicting evidence as to which of these species is the link with *T. japonicus*. *Telmatogeton* n. sp. 1 is a candidate since it shares a unique sequence of bands in one element with *T. japonicus*.

Another approach to the problem has been the analysis of the genus *Paraclunio,* which according to Wirth (1949) is phylogenetically very close to *Telmatogeton*. As with the *Telmatogeton* species the polytene chromosomes of *Paraclunio trilobatus* collected on the Oregon coast are organized into six large elements and one extremely short one. All of the *Telmatogeton* elements except for B are at least recognized in *P. trilobatus,* and elements A and E of *P. trilobatus* are homosequential to the *Telmatogeton* standard. Sequences shared by two genera would indicate that these sequences are primitive and were present in the common ancestor of the genera.

According to this line of reasoning, *T. torrenticola*-West Maui ($n = 7$), which is standard for all of the elements, and *T. fluviatilis* ($n = 7$), which differs from standard by one inversion in element C, are likely candidates for the original freshwater species. *T.* n. sp. 1 would not be considered since its element A differs from standard by one inversion. The proximal portion of A is complex, and element F differs from standard by one inversion.

In conclusion, a combination of metaphase and polytene chromosome analyses has led to the proposal of certain phylogenetic schemes for the evolutionary history of the present-day species. West Maui appears to have been the center of evolution for the freshwater forms, although their mode of dispersal is not known. All evidence points to a marine ancestor similar to the present-day *T. japonicus,* a fly that is widespread in the Pacific area. Several of the species carry considerable intraspecific polymorphism.

The absence of *Telmatogeton* from Mauna Loa, Kīlauea, and Hualālai volcanoes seems correlated with the absence of permanent streams in these geologically very recent mountains.

14

Microevolution in Insular Ecosystems

H. L. Carson

DO INSULAR SPECIES HAVE UNIQUE GENETIC PROPERTIES?

Biotas of oceanic islands have a number of extraordinary features. Zimmerman (1948) and more recently Carlquist (1974) have considered the extensive and convincing evidence that terrestrial areas of isolated archipelagos like Hawai'i are populated mainly by long-distance dispersal. A dispersal event introduces a large element of chance into the founding of a new population. Despite this unquestionably strong initial stochastic element, insular biotas manifest many bizarre adaptations. In fact, islands are noted for their novelties (Carson, 1974). From a modest founder, furthermore, a wealth of species may evolve, leading to the inference that founding events do not permanently restrict evolutionary potential.

The isolation, and to some extent the ecological simplicity, of many insular cases appears to be mirrored in certain continental situations. Thus, for example, Hedberg (1970) has given an account of evolution and speciation in the high mountain flora of Africa. The products of evolutionary episodes in these enclaves reflect many of the attributes of oceanic island archipelagos. The same properties of endemism, sharp geographic isolation, and progressive adaptive evolution are observed in the two situations. Carlquist (1974), furthermore, is at some pains to point out the insular nature of certain other continental situations, for example, the flora of southwestern Australia. These likenesses lead one to question whether there is indeed a distinctive pattern of evolution on islands or whether certain directive factors of evolution such as isolation are not more clearly displayed in insular situations.

The genetic data presented here provide no evidence that there are unique properties to the genetic variation systems of insular species. Thus the levels of genetic variation found within a series of endemic and introduced species of *Drosophila* are basically similar to their continental counterparts. This is manifested in a number of ways. First, the picture-winged *Drosophila* of Hawai'i show no dearth of naturally occurring chromosome variability in the form of paracentric inversions. As in some continental groups, certain species are replete with intraspecific chromosomal polymorphisms, whereas others carry only one or two such variants. Some among the best-studied species, have no such polymorphisms; they may be described as homokaryotypic. Such cases are also known in continental representatives of the genus (Wasserman, 1963).

The picture-winged *Drosophila* of Hawai'i have been studied intensively and represent a useful paradigm (Carson et al., 1970; Carson and Kaneshiro, 1976). At first glance they appear to be unusual in their remarkable similarity of chromosome number; all of the species so far examined have the same diploid chromosome number ($2n = 12$). Yet closer examination shows a number of alterations from microchromosomes to acrocentrics, and from acrocentrics to metacentrics. These have been shown (Clayton, 1969; Clayton and Wheeler, 1975; Baimai, 1975) to be due wholly to heterochromatin alterations, a type of chromosome change that is also abundantly found in continental faunas.

Although there has not been any reduction in chromosome number by arm fusion among the picture-winged *Drosophila,* such changes have occurred in the modified-mouthpart flies (Clayton, 1971; Yoon et al., 1972). These give every evidence of having resulted from the translocation and centromere-loss system so well known from the continents (White, 1973).

Structural protein variability, as detected by electrophoresis, does not seem to be very different between insular and continental species. Thus it has been estimated that an individual of *D. mimica* is heterozygous, on the average, at 19.4 percent of its electrophoretically detected loci. This average individual heterozygosity ($\bar{H}$) is slightly higher than that found for the neotropical species, *D. willistoni,* in which the estimate is 18.4 percent (Ayala et al., 1972). This latter species has a natural geographic distribution from extreme southern Brazil to northern Mexico and central Florida, a land area of about 12×10^6 km^2. This is about two million times larger than the area occupied by *D. mimica,* which is virtually restricted to two kīpukas on the island of Hawai'i totaling about 6 km^2 in area. The population size of *D. mimica*—in fact, of any of the Hawaiian species—must be very small relative to that of most continental species. Obviously intrapopulation heterozygosity level is not related to population size. *D. mimica,* nevertheless, forms exceedingly dense local populations under fruiting *Sapindus* trees. My qualitative judgment is that local populations of the two species are quite similar in density.

Perhaps the most important point is that island species, even some with quite small total populations, are capable of carrying as much genetic variability in a local population as are species with very large populations. The importance of this observation lies in the key importance of local populations in the evolutionary process. Such populations, called demes, represent the locus where active gene frequency changes occur. In a real sense, such populations are the growing point of evolutionary change. Although an island species like *mimica* may consist of only a few demes and a continental species like *willistoni* of hundreds of thousands, population structure at the local level is similar in the two.

Such considerations appear to go far to explain why insular populations appear to suffer no permanent disability when forced to go through a number of population bottlenecks as a growing archipelago like Hawai'i is occupied. It has been argued (see Carson 1971) that at the time of a founding event the new population may receive only a small fraction of the genetic variability of the donor population. This implies that, immediately following the founding event, the new population would be depauperate in genetic variability. The observed high variabilities found in species endemic to the island of Hawai'i indicate that the mutation process is capable of replenishing the supply of variability.

Not all of the island species have heterozygosities as high as that of *D. mimica*. Rockwood et al. (1971), Selander and Johnson (1973), Carson and Johnson (1975), and Steiner (1974, 1975) have recorded average individual heterozygosities for a number of species. They range from zero in *D. odontophallus* of Maui to a high of 39.8 percent in *D. reducta* of Hawai'i. Species with large amounts of chromosomal polymorphism are not necessarily high in allozyme polymorphism. For example the chromosomally very polymorphic species *D. setosimentum* has a moderate individual allozyme heterozygosity level ($\bar{H} = 0.069$). High-inversion polymorphism is not confined to *Drosophila* but is also found in the midge genus *Telmatogeton* (Newman, 1977).

Among the species studied are several cosmopolitan forms, such as *D. immigrans* and *D. simulans*. The data show that these also are not depauperate in genetic variability. *D. immigrans* carries the same three inversions that occur widely in continental populations, as well as several unique ones, and the allozymes of both species show variability that is not very different from levels reported from other parts of the world (Steiner et al., 1976).

We can conclude that insular populations, even in remote, isolated areas such as Hawai'i, do not have unusual genetic properties. Accordingly demes from islands should be able to respond to selection and track the environment just as vigorously as those from elsewhere in the world. This, indeed, has been experimentally verified in the IBP work by Steiner (1974) and by Paik and Sung (1974). Steiner showed that both *D. mimica* and *D. engyochracea* can track environmental variables related to desiccation resistance. Paik and Sung provide evidence that the genetic material of an introduced species is also sensitive to weather variables and responds differently on different islands and at different altitudes on the same island.

That these genetic systems are integrated and balanced is manifested by the population cage work of Sung (1974). When artificial populations are begun with genetic materials of *D. immigrans* collected from Kaua'i, the results differ from those obtained in experiments using comparable genetic materials from Hawai'i. When flies from the two sources are mixed and placed in a population cage, the results are erratic, suggesting that hybridization unbalances the original geographic gene coadaptations.

Insular and continental species differ in their gross population sizes. Continental species have many local populations, whereas the former have few.

Accordingly the total genetic variance carried in a continental species should be far greater than that found in insular ones. A continental deme will be able to draw variability from adjacent demes, given a capacity for gene flow between demes. Any deme is limited in the amount of genetic variability it can carry. Continental demes, on the other hand, have the advantage of being able to be enriched continually by gene flow. This probably represents an important difference between island and continental populations. The isolated nature of most island demes may be conducive to the evolution of restrictive specializations, whereas continental conditions are capable of giving rise to the genetic basis of a generalism, wherein the organism is homeostatic. This difference may underlie the observed failure among island organisms to evolve aggressive weedy organisms that have genotypes adapted to general purposes (Baker, 1965).

Not all island organisms operate with restricted demes. Certain nonspeciating organisms in the islands tend toward a continental type of population structure. Understanding of these genetically should provide valuable clues to the evolutionary origin of generalist tendencies in organisms.

ISOLATION IN ISLAND ECOSYSTEMS AND ITS EFFECT ON THE GENETIC PROPERTIES OF SPECIES

If there are no unique features of genetic systems in island organisms, then where else may one look for an explanation for the great diversity of species and adaptations that are encountered in insular biotas? There are many indications, in Hawai‘i at least, that a major contributing factor is isolation.

One can view isolation from a number of perspectives. First, there is the fact that the Hawaiian archipelago is, and perhaps always has been, the most isolated oceanic island group on the earth. Recent geological evidence appears to favor the view that the Hawaiian Islands have been forming sequentially as the Pacific tectonic plate moves northwestward over an immobile hot spot in the earth's molten mantle. If this is so, it would mean that islands so formed always have been isolated by great ocean distances from the nearest continent. Erosion appears to have reduced the older islands, which have drifted closer to the continents, to submerged seamounts. High islands were, and still are, the most isolated.

Such oceanic isolation can be broken only by chance long-distance dispersal. As an older high island moves away from the hot spot and a new island is formed, a second-order isolation is encountered. A new island thus appears near to, and yet partially isolated from, an older one. The new terrestrial life of this island is probably derived in most cases by dispersal of propagules from an adjacent, slightly older island. At each of these events, drastic effects on the genetics of the new species resulting from founders would be expected.

At a fourth level—within an island—further isolation over even shorter distances can occur. The island of Hawai‘i, for example, was formed by the sequential eruption of five volcanoes, each of which provides a new area for colonization that is, at least initially, disjunct with regard to the older volcano, even on the same island.

Isolation occurs at an even closer level, on a single volcano in several more ways. A volcano high enough to acquire cloud formation by intercepting the northeast trade winds supports a rain forest on the windward slopes but grades to almost desert conditions at lower levels, especially on leeward slopes. These conditions function as strong ecological, and thus genetic, isolating mechanisms over short distances.

Perhaps the most dramatic isolations of all, and indeed perhaps the most important for genetic effects, are those that are continually occurring as lava flowing from rifts builds the characteristic shield volcanoes of Hawai'i. Forests clothing the slopes are continually destroyed by new flows as the mountain builds. They are replaced by a continuing series of successional events. On a microscale, these changes mirror the larger history of the volcano, the island, and the archipelago.

Destruction by lava, however, is never complete. As it flows down the slope, pockets of forest remain intact as kīpukas, islands of vegetation in the broader sea of more recent lava flows. Kīpuka formation has a number of important genetic effects. First, there is a reduction in population size. This reduction operates by chance in that no organism can be biologically adapted for escape from such a conflagration as a lava flow produces. This means that the gene pool preserved as a deme in a kīpuka is favored by raw chance and not because of special biological properties. Under these circumstances, population size reduction and accompanying genetic drift may be responsible for the striking genetic dissimilarities between local conspecific populations. If the population is to survive at all, it must make do with what genes the vicissitudes of chance have preserved.

One effect of such events, then, will be a forced reorganization of the genetic systems within the remaining demes. Natural selection may be called upon to alter epistatic organizations at the level of the genetic system. Such alterations should be distinguished from those imposed by external environmental challenges, such as temperature, moisture, and competitors. Even these factors, however, tend to be somewhat unpredictable in the tropical high-island ecosystem. The strict seasonality of the temperate continents, for example, is replaced by haphazard, aperiodic episodes of meteorological extremes involving temperature, rainfall, and high wind.

Accordingly species that inhabit a moderately old archipelago such as Hawai'i have populations that both ancestrally and currently have been and are being rent by forces of chance to which no specific adaptational response is possible. Some of the manifestations of this have intrigued and puzzled biologists such as Darwin and Wallace, Gulick, and Crampton to the present day. Our genetic perspective in this book enables us to appreciate the resilience of the local deme in the face of all the bottlenecks and vicissitudes that it has faced in its past history. Vast continental populations are not needed to provide variability for the forging of new adaptations. Yet it is abundantly obvious that a species population, buffeted and attenuated by chance, often manifests some of the effects of the wheel of fortune.

SPECIATING VERSUS NONSPECIATING ISLAND ORGANISMS

Certain phylads of organisms of similar geological age contain many more contemporary species than others. This phenomenon is very striking in the Hawaiian biota. At one extreme are such plants as *Metrosideros collina,* the most abundant tree species in the islands. Specimens ascribed to the same species are known from islands throughout the Pacific area. In Hawai'i, the species shows extraordinary phenotypic variability (Corn and Hiesey, TR 18). *M. collina* grows in a great variety of ecological habitats from new open lava flows to wet fern forests with deep soil, exposed ridges, and even very wet bogs. The species in Hawai'i is extraordinarily variable phenotypically not only in the different habitats but it shows extensive polymorphism within populations.

Most taxonomists consider *M. collina* to be a single, highly polymorphic species; some recognize two subspecies, and a number of form names have been employed to refer to certain of the more conspicuous variants. Nevertheless this species appears to occupy these diverse environments by adjustments based on a highly open, freely dispersing, recombining gene pool. This is aided by the fact that the seeds are airborne. Attempts to apply analytical genetic methods to this species have not met with success so that how much of the overt variation can be ascribed to genetic causes is not known.

In strong contrast is the extensive speciation in the plant genus *Cyrtandra*. St. John (1966) recognizes 129 full species on the island of O'ahu alone. These are morphologically distinct entities, usually confined to single valleys. In this, the situation resembles the classic case of land snails of the family Achatinellidae (Cooke and Kondo, 1960). Neither of these extraordinary cases has been studied with contemporary genetic methods; accordingly it is again necessary to turn to Hawaiian *Drosophila* for clues to the genetic situation.

The very large number of Drosophilidae that have been described in Hawai'i has been referred to earlier. From this large number, one might conclude that these constitute a speciating rather than a nonspeciating group. When the situation is examined more carefully, however, such a simple characterization is misleading. In the first place, these flies do not show the very confined distributions manifested by *Achatinella* and *Cyrtandra*. Although the majority of the species appear to be confined to single islands or a single volcano, most are widely distributed within that island or volcano. Second, several genera and subgenera are recognized within the family; each of these may be divided into subgroupings.

In this case, groupings of species have been done primarily by very sensitive methods involving chromosomal sequencing of the giant chromosomes. This, coupled with laboratory hybridizations and detailed anatomical and morphological studies, has led to very realistic groupings of species. These methods are extensions of similar ones employed earlier for the study of continental species of *Drosophila* (Patterson and Stone, 1952).

Among Hawaiian drosophilids, the most striking and best studied are the so-called picture-winged flies. The common name refers to the fact that most of them have dark maculations on the wings. The most sensitive definition of the group, however, is afforded by the banding sequences of the giant chromosomes. Each of more than one hundred species can be grouped together on the basis of this chromosomal similarity. Indeed, several species from Kaua'i (such as *D. primaeva)* lack wing maculations but nevertheless clearly can be joined with the rest on chromosomal characters.

The picture-winged species are a good example of a species-rich group (about 120 species are known from the six major islands). Contrasting with these is a species-poor group of six species ascribed by Kaneshiro (1969) to a special subgenus, *Engyscaptomyza*. Like the picture-winged group (which has been ascribed to the subgenus *Drosophila*) this group is found from Kaua'i through the other islands to Hawai'i. Cytological studies (Yoon et al., 1975) indicate that they have speciated stepwise down the archipelago, from northwest to southeast, in much the same pattern as that shown by the picture-wing species. In so doing, however, they have remained species-poor, even though similar periods of geologic time appear to have been involved. These two groups, accordingly, may be used as examples of speciating and nonspeciating groups, a tendency that is seen so widely elsewhere in the Hawaiian biota.

ARE THE SPECIATING ORGANISMS THE DYNAMIC ELEMENTS IN ISLAND ECOSYSTEMS?

At first glance it might seem that those groups that show the greatest number of cladistic evolutionary events must be more active, from the evolutionary point of view, than those that form fewer species over the same period of geologic time. The question comes down to a comparison of cladogenesis, or splitting, with anagenesis, or evolution in a single phyletic line.

Avise and Ayala (1975) have studied electrophoretic variability in two groups of North American fishes of similar geological age. Minnows of the family Cyprinidae are highly speciose, whereas sunfish of the genus *Lepomis* are depauperate, with only eleven species. When pairwise comparisons between species in the species-rich and species-poor phylads are made, however, the genetic distances, measured by biochemical variants, are not noticeably different in the two groups. This suggests that the biochemical differences are accumulated as a function of time since divergence from a common ancestor rather than the number of intermediate cladogenetic events.

This point recalls an earlier account that stressed the very great electrophoretic similarity of several species of Hawaiian picture-winged *Drosophila,* suggesting that biochemical differences between them are not of the essence as far as speciation is concerned. The finding of Avise and Ayala that multiple cladistic events are not associated with an increased rate of genetic divergence appears to be a correlated phenomenon. I believe that this evidence fits the idea that the structural protein differences between species do not reflect their most important biological differences. One can conclude further that although behavioral differences may be important in some species of insects, such as the picture-winged *Drosophila,* these could hardly be instrumental in the evolution of speciose plant groups, such as *Cyrtandra*. Rather it would appear that the rapid speciations of Hawaiian organisms reflect the accumulation of gene differences of a regulatory nature.

Speciating organisms differ from nonspeciating ones in another quite different way, one, indeed, that may serve as a basis for a dynamic difference between the two kinds. When separate gene pools are evolved through cladogenesis, these serve as opportunities for one or more of these split products to move into new niches. In this manner, new gene pools are provided by the isolation and the stochastic effects of the interisland and intervolcano founder events. Such new populations may be freed from the tendency that the ancestral population might have had to become bogged down, evolutionarily speaking, in specialisms. In this manner cladogenesis provides an escape from specialism, a very important atttibute for an organism that is required to become integrated into a new ecosystem on a new island or volcano.

Anagenesis, on the other hand, when carried over considerable geological time, would be expected to lead to specialism, especially if the vagility of the organism is reduced and isolation is pronounced. Close specialism obviously restricts evolutionary potential. For example, the genus *Titanochaeta,* a group of Hawaiian scaptomyzoids (Drosophilidae), are extreme specialists and are nonspeciose (Hardy, 1965). All ten species apparently are restricted in that they can complete their life cycle only by ovipositing on the eggs of endemic spiders of the family Thomisidae.

On the other hand, certain individual species, especially of higher plants, appear to have achieved enormous populations that are saved from isolation and speciation by their powers of dispersal. Few crucial data are available for dispersal differences in speciose versus nonspeciose groups in *Drosophila*. Nevertheless, the great dispersal powers of *Metrosideros* (Corn, 1972), for example, provide the possibility for a genetically variable, polymorphic gene pool. Although basically anagenic, such species are able to escape specialization and restriction by virtue of the maintenance of widely dispersed and variable gene pools.

Metrosideros collina and *Acacia koa* are important pioneers in various Hawaiian ecosystems. This type of nonspeciating organism in a sense lays the groundwork for the development of the ecosystem. As nonspecialists, they are not required to play a subjugated, specialized role in the ecosystem but, conversely, provide a broad base. One of the outcomes of our exploration of genetic systems and genetic variability in Hawaiian organisms has been the identification of certain problems. As a project for the future, genetic study of such basic organisms is especially important.

ADAPTATION AND SPECIATION IN INSULAR ECOSYSTEMS

A biologist citing a case of adaptation or adaptive radiation usually is merely stating an observation of facts regarding existing character states of certain organisms. Accordingly no dynamic process is available for study. The adaptive states themselves are the outcome of evolutionary processes that are now essentially finished. Most attempts to infer the nature of the processes from direct observations on organisms have met with failure. The only recourse the evolutionary biologist has is to attempt to deal with the process of adaptation on a microscale at the level of the species population.

Adaptation at its basic level occurs when natural selection, operating through the environment, progressively changes the pool of genes of a species so that the species tracks environmental conditions. The process is difficult to observe at the population level, however, because the genetic systems of most organisms are likely, at any time level, to be at equilibrium with the environment. Thus there is a low probability that the investigator will be able to locate for study a case that is in some dynamic stage. The process nonetheless can be observed in certain instances. A case in point is that of *Drosophila mimica,* the gene pool of which tracks the somewhat unpredictable changes in moisture levels in the two kīpukas on Mauna Loa.

That the small isolated demes of *D. mimica* can so respond is a point of fundamental interest. Most continental populations are so large that when the deme is observed to respond in this way, one cannot exclude the possibility that its capability is causally related to gene flow into it from adjacent demes. The response of a closed deme to selection has been observed widely in artificial plant and animal breeding and in experiments with population cages of flies and other insects. Populations of caged insects, of course, are prime examples of closed demes. Dramatic evolutionary changes can be observed in such artificial populations.

Organisms foreign to the Hawaiian ecosystems occasionally are extraordinarily successful after colonization of the islands. Much of this success is ascribable to ecological factors; the invader frequently has already been selected elsewhere for certain competitive and aggressive qualities. Thus it brings its already evolved genetic capacity along with it. Nevertheless the unfamiliar ecosystem in which it finds itself may provide new opportunities for selective response. Just as a natural interisland founder can produce a population capable of extensive further evolution, so also can the populations of organisms newly introduced from continental ecosystems.

The process of adaptation is a phyletic, or anagenetic, process. It occurs in a series of populations that succeed one another in time. At no time are they split into isolated daughter populations as descent occurs. Speciation, on the other hand, is characterized by a splitting process, which may be referred to as cladogenesis. Of all the special characteristics of island evolution, none is more striking than the amount of cladogenesis as evidenced by the numbers of species encountered in some island organisms.

The classical formulation of speciation, enunciated by Mayr (1963), views it as basically a geographic process. In its purest form, the geographic speciation theory visualizes an ancestral species that spreads out gradually over a large area with somewhat diverse habitats. Selection to these different local environments results in differently adapted subspecies. Isolation of a subspecies may be followed by the appearance of genetic differences that will restrict the amount of gene exchange that can take place should the new split-off entity again come in contact with the parent species. Thus reproductive isolation, if it is complete, is the essence of the splitting process. It prevents the return to the former single gene pool.

This scheme is easy to visualize in continental situations and recently has been related to gradual accumulation of biochemical differences in speciating continental forms (Ayala et al., 1974). This scheme seems less applicable, however, to the geographic situation on oceanic archipelagos like Hawai'i. Here speciation occurs from volcano to volcano and island to island without the possibility of connecting or intervening populations. Accordingly the pattern of allopatric speciation seems better explained by evoking the founder principle. This proposes that a new species may originate allopatrically from one or very few chance migrants that somehow make their way to an open environment on a newly formed volcano or island (Carson, 1971). The founder theory is an extreme case of peripheral speciation, a process discussed in detail by Mayr (1954). Mayr proposed that the isolation of the peripheral population would, because it is no longer in contact with the main body of the species, result in a genetic revolution. His discussion stressed that the isolating events could cause a change in genetic environment, forcing the population to seek, by selection, a new balanced genetic state. This concept has been explored further in a recent paper (Carson, 1975). I have extended this idea to include cases where bottlenecking of population size is an accompaniment of the system of colonization.

The stressing of the founder effect does not preclude the possibility of the functioning of other modes of speciation. Indeed there is some evidence (Carson and Johnson, 1975) that some species on the very large island of Hawai'i form clines of genetic difference reminiscent of chromosomal races in continental species. In these, a more conventional mode of geographic speciation is possible. In several instances, furthermore, the data suggest that a scheme of sympatric speciation could be evoked for the appearance of certain species pairs (Craddock, 1974; Richardson, 1974). Despite these considerations, the bulk of the luxuriant proliferation of Hawaiian species seems to be related to the dispersive effects of isolation on the gene pools of island organisms.

In the conventional allopatric and sympatric speciation schemes, adaptation precedes the completion of reproductive isolation. The founder scheme, however, injects a large element of chance into the initial colonization. Indeed it has been suggested that in some cases isolation may result as a by-product of the perturbing effect of the disruption of genetic balances in the new population. In such cases, speciation might be said to be in a sense a catastrophic event and to precede a new adaptive stage to which all allopatric species must soon come. As certain insects speciate in Hawai'i, the frequent host shifts that are observed show that narrow specialism is not a rule and that new adaptive breakthroughs frequently occur.

GENERAL SUMMARY

In the foregoing discussion, the following points have been made:

1. Local populations of island species can be fully as polymorphic genetically as those of widespread continental species. Accordingly, such populations are highly competent for adaptive evolution.
2. Most island species have small total populations. Thus, even though the local populations may be rich in genetic variability, the total variability sequestered within widespread continental species is much larger.
3. Other than the above, island populations and species, both introduced and endemic, have no unique genetic properties. Heterozygosity levels are generally high.
4. In high-altitude oceanic archipelagos like Hawai'i, many factors promote isolation. The continuing series of isolations to which island populations are subjected have effects of great importance on the genetic structure and thus the biological future of island organisms.
5. The most striking mode of speciation is through one or a few inter-island or inter-volcano founders. Enforced isolation promotes speciation.
6. Some island groups (genera) are highly speciose whereas others, evolving over the same time-scale, do not show this tendency.
7. Although other factors are also involved, organisms with high vagility tend to be species-poor, again emphasizing the importance of isolation.
8. Speciose organisms are dynamic in that the chance nature of the founder mode of speciation permits them to enter a new niche rather than remain closely adapted.

9. Certain trees basic to the island ecosystem are non-speciating. They appear to have high vagility and a highly outcrossed and complex genetic system.
10. Hawai'i Island harbors some unique species which are newly formed in time. Allozymic genetic similarities between such species are very high suggesting that species can be formed with only minor alteration of the proteins encoded by structural genes. It is implied that regulatory genes are important in initial species differences.

VI
Concluding Survey

The broader research aims of the Hawaii IBP were to concentrate on aspects that are unique and different in island ecosystems as compared to continental ecosystems and to assist in solving regional problems related to wildland management and conservation of biological resources. Our approach was to study the biological organization in relatively undisturbed natural communities and ecosystems. For synthesis purposes we selected four areas of emphasis: the spatial distribution characteristics of island biota, their community structure and general niche differentiation, their annual-cyclic behavior, and their genetic variations and microevolutionary trends.

These areas have little overlap with the well-known island studies of Carlquist (1965, 1970, 1974), whose emphasis is primarily taxonomic-phylogenetic. They also overlap little with the theoretical studies in island biogeography of MacArthur and Wilson (1967) or with Fosberg's (1948, 1963, 1966) exploratory studies of island ecology. Our synthesized results cannot be implied as representing a complete coverage of the characteristics of island biota and their ecosystems. Unlike some other IBP projects, which built on substantial ecological background, our studies began from a very small information base. Therefore our research included several other facets that were not emphasized in the preceding parts but that provided necessary background information. One such research facet was directed specifically to adding knowledge relevant to the biological conservation of Hawaiian forest ecosystems. This area of emphasis will form the material for the final chapter. Chapter 15 will focus on the more unique and characteristic aspects that emerged from our four areas of synthesis.

15

Island Ecosystems: What Is Unique about Their Ecology?

D. Mueller-Dombois

ENVIRONMENTAL CONSTRAINTS

We may now ask whether we have found anything unique or different in the ecology of island ecosystems and their organisms. This is not an easy question to answer. It is clear that the biological evolution of our island ecosystems has been rather unique. Four factors stand out that contribute to the unique biological evolution of ecosystems here: the extreme isolation of the island group, the small size of the island habitats, the recency of the oceanic islands as a group, and their perturbation history in connection with volcanism.

The extreme isolation had a significant screening effect on what organism groups could get here and establish themselves successfully. This screening effect excluded many plants with large seeds or small seeds of short longevity. It also excluded among animals all terrestrial mammals (except the hoary bat), large reptiles, and primates, except humans.

The small size of the island habitats is the result of small island land masses jutting high out of the ocean. Thus we have distinct altitudinal segregations of habitats with their own temperature regimes. Furthermore these small land masses are segregated into windward (pluvio-tropical) and leeward (xerotropical) habitats with their own rainfall regimes. The habitats are further fractioned by great variations in substrate, ranging from recent volcanic flows to old, skeletonized, and nutrient-depauperated latosols (Oxisols and Ultisols). This island habitat mosaic brings about another factor of important ecological consequences: the very limited recurrence of similar habitats across the island chain. These narrow habitat dimensions strongly limit the population sizes of the island biota.

The recency of oceanic islands as a group is undoubtedly of evolutionary significance also. They originated in the late Tertiary, when the modern angiosperm flora had already evolved. In contrast, some of the continental tropical ecosystems evolved without major geological disturbances, forming a primary succession from seed fern forests to primitive gymnosperm and angiosperm forests to modern angiosperm forests. These continental ecosystems developed during a much greater evolutionary time span. The high tree species diversity of some continental tropical forests may be largely attributable to this.

Volcanism causes major geological disturbances. These perturbations are a significant part of island ecosystem development, from the highest mountain top to sea level. Volcanic perturbations are of many kinds and differing degrees and are erratic or unpredictable. They were once effective on each island and left their traces long after individual volcanoes became extinct. As such they had and still have a great effect on the evolution of the island biota.

There is little doubt, then, that the island biota evolved under unique environmental conditions. Much has been written also about their adaptive characteristics, which sometimes resulted in the development of rather bizarre island life forms. But has this also made the ecology of island ecosystems different?

The answer, as revealed from our studies, appears to be that ecological principles do not differ for island ecosystems. However, our studies have brought some new dimensions to island ecosystem ecology, which should add to both their scientific understanding and appropriate management.

DISTRIBUTIONAL CHARACTERISTICS OF ISLAND BIOTA

The spatial distribution analysis along the Mauna Loa mountain gradient confirmed Whittaker's (1970, 1975) individualistic species distribution model established for temperate mountains. It also confirmed the spatial association model of species distribution that is, in part, an affirmative answer to MacArthur's (1972) question on species and community patterns in the tropics. However, spatially associated species groups along environmental gradients are not to be considered unique for tropical areas, since such patterns have been demonstrated many times also in temperate environments. They are, like Whittaker's individualistic species distribution patterns, a universal phenomenon applicable to islands and continents, temperate and tropical environments alike. Therefore island biota form communities, which are just as easy or difficult to define by species distribution characteristics as are biological communities in continental mountains of temperate or tropical areas.

In addition we found a number of other distribution trends. Three of these were typical of wide-ranging species: bimodal, multimodal, and almost unrestricted. Two included narrow-ranging species, individuals or groups with decreasing tolerances toward one end of the gradient and species with high modal similarity but low range similarity. While wide-ranging species typically exhibit generalistic behavior, narrow-ranging species cannot be considered automatically as more specialized, especially in our example. The distribution trend of synchronized modality but low range similarity was particularly exemplified by a group of exotic plant species. Their distribution pattern indicated a common establishment center in a heavily pig-disturbed habitat, from which they spread into other habitats. Their range dissimilarity may indicate merely that they are still on the move and thus have not yet fully naturalized. The dynamic nature of the gradient environment in terms of human-induced habitat disturbances and the introduction of exotic organisms did not allow many generalizations about narrow-ranging species.

Nevertheless, a few clear relationships existed. One was the congeneric splitting of certain plant species along the Mauna Loa transect, such as *Styphelia douglasii* versus *S. tameiameiae* and *Vaccinium peleanum* versus *V. reticulatum,* which were represented with nearly nonoverlapping upper- and lower-elevation species populations, respectively. Similar congeneric spatial segregations were found in the soil microfungi. Another clear case of specialization associated with narrow-ranging distributions was represented by the host-specific cerambycid beetles.

The seven distribution trends established from our analysis and the existence of wide- and narrow-ranging species among the different organism groups studied are not characteristics unique to island ecosystems. What then did we find that can be said to be unique in the spatial organization of our island communities? Two points of relative uniqueness may be considered here.

Among our reference species there was a rather high proportion of wide-ranging species. These generalistic species were common among the endemics of several organism groups (plants, birds, canopy arthropods, sciarids, and the scaptomyzas among drosophilids). The high proportion of wide-ranging species was further brought about by a number of exotics, also from the same and additional organism groups (plants, birds, rodents, their ectoparasites, canopy-residing insect predators, certain soil arthropods, and drosophilids). This high proportion of generalists among our biota is perhaps characteristic also for geologically young areas on continents, and it may not apply so much to older volcanic islands. However, wherever the same endemic species are present on the older Hawaiian Islands, they occur there similarly as wide-ranging generalists. What appears to be different on older volcanic islands is the presence of a greater number of range-restricted endemics.

This aspect of island biota distribution was not tested through our analysis since we lacked the time and resources to do a comparative transect on an older island mountain. If we had been able to study a number of comparative transects, probably we could have demonstrated another biological distribution phenomenon characteristic of islands: that similar habitats are often occupied by quite dissimilar species assemblages. This implies that the proportion of the same species occurring in similar habitats is relatively small, while the proportion of species unique to the locality is relatively great in island ecosystems in comparison to similar continental ecosystems. If one considers the contribution of unequally invading exotics in island ecosystems, this is almost a truism. However, this relationship is expected to hold also for the endemic organisms and is here considered as an hypothesis. This pattern would correlate with the observation that there are many rare species in Hawaii whose high number is also responsible for the biological richness of the islands as a whole. The main reasons for this high dissimilarity pattern between similar habitats appear to be the concomitant factors of habitat isolation and loss of dispersability. Habitat isolation is not only significant between islands but also within islands. Habitat fragmentation in islands is caused initially by volcanism and later by unequal geomorphological weathering and erosion. Loss of dispersability has been explained as a factor for rareness in many island species (Carlquist 1966a, b).

Another aspect brought out by our gradient analysis relates to the distribution characteristics of endemic and exotic species among our organism groups. Tree-associated arthropods contained a high number of endemic species, while soil arthropods were determined to be comprised of mostly exotic species. The other soil organisms tested, the soil fungi and algae, were considered to be indigenous; they apparently got to the islands themselves without the aid of humans, but they did not speciate. This has an interesting application insofar as soil fungi and algae can be used as biological indicators of habitat similarity between island and continental habitats, probably on a global scale. It should also be reemphasized that the soil fungi in our analysis exhibited a high degree of gradient sensitivity, meaning that species responded significantly to habitat differences along our mountain transect. The origin determination of soil arthropods as exotics was only tentative. Sometime in the future, it may be shown from comparative gradient analyses in other island and continental ecosystems, provided they are relatively undisturbed, that some of the soil arthropods found along our transect are also indigenous rather than exotic in origin.

The soil substrate of island ecosystems is not so unique as is the plant substrate, which forms the life-support medium for the tree-associated arthropods. However, it is the dispersability of the respective organism group rather than the degree of uniqueness of the life-support medium that determines the origin characteristics of organism groups on islands. Because of their high degree of dispersability, soil fungi and algae can maintain a cosmopolitan gene flow that is prevented in island organisms with lesser dispersability. Another interesting point in this connection was the high endemism occurring in the fungus gnats (the sciarids). Their main food source is the indigenous fungi, yet island speciation is typical for this group of flies.

Apart from the differences in origin between organism groups with different roles in the ecosystem, we did not detect any significant pattern difference in the spatial distribution characteristics of endemic versus exotic species. Both exotic and endemic species of plants, birds, tree-associated arthropods, scaptomyza flies, and the others contain wide-ranging and narrow-ranging species. Therefore the successful exotic species do not differ in their spatial characteristics from the successful endemic species. However, there is a clear tendency of introduced species to be more concentrated in places disturbed by humans.

COMMUNITY STRUCTURE AND NICHE DIFFERENTIATION

Island ecosystems have the same gross-structural characteristics as found in continental communities with similar climatic regimes and soils. For example, montane tropical rain forests and lava tube ecosystems are also found on continents. At the species assemblage level, our island communities are almost totally unique with two exceptions; these relate to the indigenous soil organisms, which are also present elsewhere, and to the exotic species, which have been brought to Hawaii by humans. Of course, one cannot expect exotic invaders always to occupy habitats in the islands that are equivalent to their native environments. They may occupy much larger terrains or narrower terrains than in their native ecosystems depending largely on the degree of habitat disturbance by humans and the degree of resistance offered by the native island organisms and communities. Therefore, the exotics occasionally may display rather unique ecological relationships in island ecosystems.

Our community structure analysis focused on the general niche level, a functional ecological unit concept, intermediate between that of individual species and the total ecosystem. We identified general niches by species of closely similar function and structure or life-form types—synusiae in plants and guilds in animals. This provided us with a basis for recognizing the diversity of ecological roles assumed by the island organisms, as well as that of the exotic invaders in the two types of ecosystems studied.

In the montane rain forest we found a plant life-form spectrum as complete as in any continental tropical rain forest. There were no missing plant-life forms other than the heterotrophic vascular type. What was found to be unique, however, was a different balance in the representation of the plant life forms. The upper tree synusia was occupied by only two species, both of them relatively shade-intolerant pioneer species. In continental tropical montane rain forests, the species diversity is usually much greater in this synusia, and shade-tolerant climax species are more typical. Other synusiae were much richer in species. For example, the lower tree synusiae (meso- and microphanerophytes) contained fifteen indigenous species, all of them shade tolerant. The species diversity of these undergrowth synusiae in our island rain forest may approach that of continental tropical montane rain forests. This hypothesis is worth testing in future studies.

Another important and probably rather unique phenomenon in our island forest is the fact that all indigenous tree species, with the exception of the tree ferns and *Myoporum sandwicense,* are able to start growing as epiphytes. Also, most of the normally ground-rooted endemic plants of the herbaceous layer were found to grow as epiphytes. This capacity for occupying two spatial positions in the same forest was not displayed by any of the nineteen exotic plant species that had invaded the forest.

A significant invasion of exotic plant species was found in only one synusia, the creeping or mat-forming ground-cover plants. Seven new species had invaded this synusia, which also contained five endemic species. The invasion in this case was related to periodic disturbances of the forest floor by feral pigs. It was not related to a poor representation of endemic species in this ground-cover plant synusia.

The endemic animal community in our montane rain forest clearly has a reduced niche spectrum in comparison to continental tropical forests of this kind. This is because of the absence of terrestrial forest mammals among the endemic fauna. This reduced niche spectrum, however, cannot be interpreted as having resulted in a functional deficiency because the island forest ecosystem evolved into a self-perpetuating system without the interaction of such mammals.

Analysis of the faunal groups with high endemism in our sample forest—the tree-associated arthropods and forest birds—revealed similarly complete guild spectra as the life-form spectrum of the plant community. No important functional groups were missing among the native fauna.

The guild spectrum of arthropods on 'ōhi'a (*Metrosideros collina* subsp. *polymorpha*) trees was closely comparable to that on koa (*Acacia koa*) trees. Among the primary consumers, four phytophagous guilds and one anthophagous guild were recognized on each tree species. Among secondary consumers there were two entomophagous and two saprophagous guilds identified. Only one additional saprophagous guild was detected on koa. This was represented by a single species, a dead-wood consuming bostrichid beetle.

The endemic forest birds displayed a relatively diverse spectrum of guilds with only few species. Four different insectivore niches were occupied by one endemic species each, three of these by honeycreepers and one (the 'elepaio) by a member of the Muscicapidae family. The primarily nectar-seeking guild, in contrast, was represented by four species, three of them also by honeycreepers and one by an exotic bird, the Japanese white-eye (*Zosterops japonica*). The feeding behavior of this exotic bird was sufficiently different to consider it as occupying a separate niche, however. One can see a certain parallelism of the guild spectrum of the birds with the synusial spectrum of the plants insofar as some general niches in each spectrum show a very low species diversity relative to other niches. This unequal distribution of species diversity among the general niches, and particularly the low diversity in certain niches, is probably characteristic of island communities in general.

In terms of exotic species invasion, there are also some parallels between the plant and bird life-form spectra. The phanerophyte and epiphyte synusiae among the plants and the insectivore guilds among the birds are almost exclusively represented by endemic species. Among the endemic plants these two synusiae contain the species with greater spatial versatility because of their capacity to grow both on the ground and above the ground on fallen trees and in the crotches of branches. A high spatial versatility was found also among the endemic insectivores insofar as three of them were described as full-range canopy users. Significant invasion was found in the bird community in the frugivore and nectarivore guilds and in the plant community in the mat-forming ground-cover plant synusia. In both the bird and plant communities, these general niches are represented by ecological types of species that are spatially more restricted in the ecosystem.

Among the tree-associated arthropods, species diversity was greatest among the secondary consumers, particularly the insect predators. This was considered to be an island characteristic according to Janzen (1973), who believes that generalists at higher trophic levels become more easily established in islands than the more host-specific primary consumers. Both native and exotic species displayed greater diversity among the secondary as compared to the primary consumers. The natives among the secondary consumers may be represented more by indigenous nonendemic rather than endemic species. This difference was not clarified, but it would help to explain this phenomenon as a function of greater dispersability or colonizibility among the secondary consumers. No significant invasion of new general niches was observed among the tree-associated arthropods.

The terrestrial mammals, all of them exotics, now occupy a wide, general niche spectrum in our rain forest. The spectrum includes several species of rodents (rats and a mouse), two carnivores (feral cat and feral dog), and two omnivores (mongoose and feral pig). Their interaction was only generally established; much more work is needed to obtain a clear picture of their food relations and activity patterns. But some of the food resources are significantly provided by exotic organisms, such as soil fauna eaten by rats and pigs, and rats and mice eaten by the feral dog, the cat, and the mongoose. The roof rat in particular has become an important food source for the last three exotic mammals. In addition, a major food source of the feral pig and probably also for the rats are starchy cores of the native tree fern trunks. The tree ferns are particularly abundant among the endemic plants, and they regenerate well by both vegetative and sexual means wherever the pig density is not too high. The rodents were not fluctuating greatly in numbers, and a new equilibrium has become established. This equilibrium was regarded as slightly depressant on the existing bird populations, but it may also be considered depressant on the endemic plant species diversity, particularly in the undergrowth layers.

Analysis of the cave fauna also resulted in a general niche spectrum as diverse as can be expected from the limited resources available in the Hawaiian lava tubes. Except for accidentals (organisms straying into the lava tubes) and the primary food resources (tree roots and fungi), all organisms were arthropods. They included three types of root feeders, two types of saprophages (fungus feeders and scavengers on other microorganisms), three types of predators, and an omnivorous cricket. Exotic species had invaded all but two general niches, the omnivore and fungus-feeder niche. The greatest invasion of exotics in terms of species number and population quantities occurred among the accidentals and the scavengers. This again indicates a direct relationship between disturbance (here in form of new accidentals, such as rats, straying into the caves) associated with the addition of new resources and the invasion of additional exotic species.

We can now reevaluate the two more problematic niche hypotheses that were answered negatively in chapter 8. A new hypothesis stated there that greater native species diversity in a general niche is associated with a greater diversity of exotic species (a reversal of our initial hypothesis) can be answered with both yes and no. The affirmative answer can be given for the species relationships across certain trophic levels. For example, the secondary consumers among the tree-associated arthropods contain more exotics than the primary consumers perhaps because species at a lower trophic level create new resources for species at a higher trophic level. If exotic species invade niches at a lower trophic level, the likelihood for more exotic species invasion at the next higher trophic level appears to be increased. The negative answer can be given for species in the same trophic level. For example, among the birds a new guild was added in the form of three seedeating bird species. Apparently the seedeating niche was never exploited by any endemic species in this forest. This is not too surprising because of the dominance of spore-producing plants, tree ferns and because of the second dominant tree species, the ‘ōhi‘a tree, which has tiny airborne seeds that contain very little food material. We can conclude that endemic species diversity in a general niche of an organism group belonging to the same trophic level (primary producers, for example) is unrelated to exotic species invasion.

The second more problematic niche hypothesis relates to the question of empty niches. We determined empty niches only through exotic organisms with ecological roles not represented among the endemic organisms. This is perhaps a simplified view since general niches were not measured by resource parameters. Other than indicated by the mammals and seedeating birds, these empty niches were very few in the rain forest. A few other empty niches were identified along the Mauna Loa transect. Moreover, most of these were not really empty in the sense that a functional balance presumably had existed in these ecosystems before the invasion of exotic species with additional ecological roles. Instead these exotics can be seen as interfering with the community balance that was established without the influence of humans during the evolution of these island ecosystems. These invaders with new ecological roles exist in part on new resources introduced with other exotic organisms, but they also overlap in their resource requirements with those of endemic species. Therefore the new niches that are now filled by exotics with new functions cannot really be considered as having been empty before their arrival.

SEASONAL RESPONSES IN ISLAND BIOTA

The five organism groups studied for their year-round variation patterns—the trees, rodents, the psyllid on *Acacia koa,* the sciarids, and drosophilids—

all displayed some seasonal behavior. However, their seasonal behavior could not always be related to environmental seasonality.

The flowering phenology of the thirteen Hawaiian trees studied along the extended Mauna Loa transect appeared in three major patterns of peak flowering. Individual peak flowering of different species at almost any time of the year was characteristic for the montane rain forest. This pattern can also be described as successive peaking with broad overlaps in flowering periods of individual species. The overall outcome is an around-the-year flowering of the rain forest, whereby individual species maximize their flower production in approximately successive and variously overlapping periods. Synchronized peaking of different tree species was characteristic of the seasonal environments outside the rain forest. Several species displayed synchronized flowering peaks during the moister and slightly cooler season of the year, and another group of species had synchronized flowering peaks during the drier and slightly warmer season of the year. A third pattern was that of random-peaking displayed by a single species, *Metrosideros collina* subsp. *polymorpha*. Along the extended Mauna Loa transect, this species displayed flowering peaks at almost any time of the year except during the warmest months (August and September). Whether these peak-flowering periods (winter, spring, summer, fall) are fixed from year to year for the specific site populations of *Metrosideros collina* subsp. *polymorpha* for which these records were made requires further research. There seemed to be no obvious environmental relationship to this flowering pattern. Therefore further studies may be approached with the hypothesis that year-to-year patterns of peak flowering are random also for specific site populations of this species.

There is a possible evolutionary implication of this random peaking in *Metrosideros collina* subsp. *polymorpha*. This species is the main pioneer tree on new lava flows. Its success may be related to its ready seed availability at any time of the year. However, the relationship of peak flowering and seed availability has not yet been investigated.

Was there anything in the tree phenological observations that can be considered unique to island ecosystems? The three major patterns of maximum flowering as shown for the Hawaiian trees almost certainly are not unique. However, the very long flowering periods observed for most of the island trees tested may be unique. Many species flowered from ten to eleven months of the year, and some populations flowered year-round. The proportion of trees with such long flowering periods appears to be small in continental tropical forests, where species with short flowering periods are more prevalent. This may be related to a different pollination phenology, which in our island species appears to be more generalistic. *Metrosideros collina* subsp. *polymorpha,* for example, can be pollinated by birds, insects, or through selfing (TR 76; Carpenter, 1976). Thus there is an analogy in the generalistic flowering phenology and the broad spatial amplitude displayed by many of the dominant island tree species. These two generalistic behavior traits may be functionally related by the generalistic reproductive strategies, which are also characteristic for most of the dominant ecosystem-structure-forming island tree species.

The exotic rodents are perennial organisms. In that sense they are like the trees. The rodents only displayed some oscillations over a year's time in their populations and no drastic changes. These population oscillations gave little evidence for seasonality-related patterns. They could be related to changing or oscillating home ranges, which in turn may be traceable to food availability.

The psyllid on *Acacia koa* represented a good example of a successful exotic invader, in this case a phytophagous insect. Its annual-cyclic population development very closely matched the leaf-flushing phenology of its new island host species, *Acacia koa*. The exotic psyllid can be said to have been perfectly preadapted to its new ecosystem. Whether its adaptation was due to an opportunistic capacity for broad-range flexibility or to a coincidental matching of its inherited phasing pattern with that of its new host is a question worth pursuing.

The three endemic sciarids displayed an annual population pattern reflecting general seasonality. Their populations were more abundant during the first half of the calendar year, following the heavier rainfall period during the late fall and winter months. Since the sciarids are primarily fungus feeders, their annual population pattern may be related more closely to soil moisture and fungal activity than to rainfall.

The two exotic *Drosophila* species also displayed an annual population pattern reflecting general seasonality. Their populations were more abundant during the second half of the calendar year, following the reduction of rainfall during the summer.

There is probably nothing peculiar to island ecosystems in the seasonal behavior of the three ephemeral insect groups studied. They are, however, good examples of ecologically successful insects from the two major origin categories, endemic insects (the three sciarid species) and exotic insects (the psyllid and the two drosophilids, *D. immigrans* and *D. simulans)*.

MICROEVOLUTIONARY IMPLICATIONS

A special characteristic of island ecosystems is the high endemism among its plant and animal species. This high endemism has originated from a limited number of founder organisms that arrived in the islands through long-distance transfer. The successful establishment of populations from such founder organisms is an outstanding biological accomplishment. Following establishment, descendants of these founder populations underwent genetic changes, which resulted in the evolution of new species that now are unique to a particular island group, to individual islands, or even to a single island habitat. The nature of genetic variation and the processes that lead to adaptation and speciation in island populations formed the focus of attention of the microevolutionary study component in our project.

It was found that the genetic variation in island species is not different from that in continental species. As a rule, island species contain very large and similar amounts of genetic variation. This is remarkable because the population founders arriving in new island habitats (the original colonizers) can be visualized as carrying merely a small fragment of the genetic variability inherent in the species from which they are derived. However, as the founder organisms give rise to more and more offspring in the colonizing process, genetic variation apparently is replenished through the normal mutation process. As a result, the amplitude of genetic variation in the demes (local populations) of island species is soon restored to the same level as found in continental demes. The only difference is the number of demes, which in island species are usually much fewer than in continental species because island species are restricted by territory and the limited recurrence of suitable habitats.

Island species therefore track their environment just as vigorously as do species elsewhere. Experimental studies with small demes of endemic *Drosophila* species in field habitats have shown that they adapt genetically in allochronic populations in response to periodic changes in environmental moisture levels. Thus local population sizes are large and genetically diverse enough to permit microevolutionary response to environmental factors.

Only species with high colonizing capacity were able to become established successfully in island habitats. This applies to all indigenous founder organisms. Their colonizing capacity is not lost so rapidly in their offspring as has often been thought but appears to be maintained in many nonspeciating and speciating island organisms.

Nonspeciating island organisms are of two kinds. The ecologically successful ones are widespread. Among plants, they are the ecosystem builders, such as ʻōhiʻa (*Metrosideros collina* sp. *polymorpha*) and koa *(Acacia koa)*. The ecologically less successful ones are locally restricted, and characteristically they lack island relatives in form of congeneric species. This second group may never have developed a strong colonizing capacity, or their colonizing capacity may have been lost. These are the peculiar, often freakish or bizarre island organisms, but as a type they are rare rather than common. Speciating organisms are those represented by genera or families that have evolved into many island species. These include the majority of the endemic biota. Their speciation is largely due to geographically and ecologically imposed isolation and the stochastic effects this isolation may have on the genetic system.

The factor complex promoting speciation includes habitat isolation and limited dispersability. Habitat isolation appears at several levels in island ecosystems: first at the level of island group; second between islands of the group not merely as a function of distance but also by difference in island age and the concomitant difference in habitat characteristics; third within high islands due to contrasting windward and leeward climates and altitudinally through nonoverlapping, totally separated temperature regimes; fourth through differently aged volcanoes on the same island with their differently developed habitat mosaics; and fifth through the volcanic environment itself, which causes habitat isolation through intervening lava flows and island formation in form of kīpukas (islands of older volcanic ecosystems surrounded by newer ones). Speciating organisms typically have a lesser capacity for dispersal than do the ecologically successful nonspeciating island organisms, but their dispersal capacity is not totally lost. It is merely less predictive or due to rarer chance occurrences. There is no reason to assume an inevitable trend toward complete loss of dispersability in island organisms.

Accordingly these ecological factors promote genetic variance between the source and the newly founded population. They appear to lead to chance shifts in gene and chromosomal frequencies. The resulting new population may thus have a start toward cladogenesis as a direct outcome of the founder event, which may thus cause a temporary break in the adaptational tracking of the environment. If the new population is not swamped out by new arrivals (as is probably the case in nonspeciating organisms), this isolation may have a profound effect on the gene pool. As population size builds up, a new reorganized genetic base is formed. In this manner, such organisms escape from narrow specialism through continuing adaptation following the cladistic event. Accordingly new adaptive breakthroughs in speciating organisms are the rule rather than the exception. Therefore the majority of island biota, whether speciating or nonspeciating, have not lost the capacity to adapt to changing conditions.

ISLAND ECOSYSTEM STABILITY

An important original concern of our project was to address the question of island ecosystem stability. Island ecosystems have been considered to be much more fragile than their continental counterparts, primarily because of their high rate of extinction of original biota and their vulnerability to exotic species invasion.

The concept of ecosystem stability is very broad, and it has been defined in different ways (Preston, 1969; Orians, 1975). Most important in this concept is the capacity of an ecosystem to resist stresses without losing its fundamental biological structure. It was one of our synthesis objectives to develop means for comparing the fundamental biological structure among ecosystems. This we accomplished in part through diagramming species distribution complexes across several ecosystem types and through analyzing species with their ecological roles and portraying these in the form of life form spectra for different organism groups. The analysis of annual-cyclic behavior and the microevolutionary studies provided further guidance to understanding the biological structure in island ecosystems.

An analysis of the biological structure, however, is only a first step in the assessment of ecosystem stability. The important second step is to subject this biological structure to certain stresses. Such perturbation experiments can be of both theoretical and practical value, but they require more time and resources than were available for our project. To be of real use, such experiments require considerable resources and time; otherwise their results have little general value. We therefore had to resort to the method of comparative analysis, which was built into our project design to some extent.

In addition there are numerous circumstantial perturbation experiments performed in Hawaii. These may relate, for example, to volcanic eruptions with lava flows, ash-blanket deposits, flash floods, landslides, prolonged droughts, natural and human-caused fires, tropical storms, grazing and browsing damage, forest land clearing and conversion to pasture, and the general phenomenon of continued introduction of exotic species. Some of these can be observed in the form of perturbation gradients that are superimposed on the basic environmental gradients. From such observations one can extrapolate generally applicable information as to the possible stress responses expected in indigenous communities.

Initially we formulated four working hypotheses with regard to stability. We said that species diversity may be related to stability, that life form diversity may be related to stability, and that stability of biological structure may be related to climatic factors. Our fourth hypothesis was that island ecosystem stability may be related to the dominant life forms.

Much has been written (Woodwell and Smith, 1969; First International Congress of Ecology, 1974) on the idea that high species diversity can be used as an index of high ecosystem stability. The concept was based originally on the general observation that ecological succession, following a perturbation, will result eventually in a stable climax in which the biological community reaches its highest species diversity. The concept has been challenged. Some investigators found greater species diversity in seral or preclimax rather than climax stages. This departure is not difficult to understand. For example, a forest with canopy gaps may contain more species because of its shade-light mosaic in the subcanopy rather than either a uniformly open- or closed-canopy forest.

However, a climax community is stable only in the sense of constancy in time. This implies that the fundamental biological structure of a climax community, including its species composition, is considered to remain essentially the same over a long period of time. This concept says nothing about the capacity of a given climax community to withstand stress, which is the real meaning of stability. One of the major perturbations of concern to our project is the introduction of exotic species.

Our analysis indicates that species diversity is probably unrelated to stability. High species diversity is here understood in the conventional sense as a combination of high species number (species richness) with relatively even species quantities versus low species diversity, meaning few species with very uneven quantities. Stability is here understood in the sense of resistance against exotic species invasion in the same type of organism group (plants, birds, canopy arthropods, and so on). Guided by the general notion of a positive relationship between species diversity and stability, we set out to analyze species diversity (number of species and their quantities in terms of individuals or biomass) by grouping species in general niches (that is, according to their general roles in the ecosystem). In so doing we found that general niches filled with high endemic species diversity were often filled also with high exotic species diversity. Conversely a number of general niches filled with low endemic species diversity were not invaded by exotic species, but also certain niches filled with high endemic species diversity, such as the vascular epiphytes, showed no exotic species invasion.

Therefore, we cannot conclude as we originally speculated (Mueller-Dombois, 1975) that general niches filled with high endemic species diversity would be more saturated and thus would offer a greater resistance to exotic species invasion. Instead we found, for example, that exotic plant invasion was primarily related to ground disturbance by pigs and a lesser capacity of some endemic species to cope with such a disturbance. This applied particularly to the mat-forming ground-cover plants in the rain forest. However, a lesser capacity of endemic species to cope with disturbances can only be generalized with regard to a specific form of disturbance and with regard to a particular general niche. For example, most of the ground-rooted endemic herbaceous plants, with the exception of the shade-tolerant mat-forming ground-cover plants, also grew as epiphytes on the trees. This indicates a high degree of stability for these endemic species, which can survive even when the ground is heavily disturbed by feral pigs. A similar capacity to start growing above the ground on fallen, leaning, or standing trees was shown for all but two endemic woody plant species. This alternate reproductive strategy indicates a high degree of stability in the face of possible natural forest floor disturbances, such as inundation during flash-floods or ash blanketing following volcanic explosions.

Another form of exotic species invasion—the invasion of insect pests and diseases—may be thought of as holding a different relationship to species diversity in the organism groups attacked by such invaders. For example, if one of the two dominant tree species in our Hawaiian rain forest fell prey to an insect pest or disease, the stability of the ecosystem might be severely affected. Instead a higher diversity among canopy tree species may be considered as offering a greater stability, because another tree species may readily replace the tree that fell prey to the invading pest organism. However, since pest or disease organisms are often rather narrow in their requirements, the chances of their becoming successful invaders should also be less in communities that offer only low diversities in potential host species. Therefore, we can maintain our argument that species diversity is probably unrelated to stability.

This brings up the second initial hypothesis that life-form diversity may be related to stability. This idea implies a high number of life forms (plant synusiae or animal guilds) with different roles in the ecosystem versus a low number, indicating fragility. The original animal guilds in the Kīlauea rain forest were increased by the invasion of several new ones, which are represented by exotic terrestrial mammals. These introduced mammals with their new activity patterns have not yet upset the original balance of this rain forest entirely, but they certainly do not contribute to its stability. Therefore our second initial hypothesis on stability cannot be answered positively.

This may pose the question that if there had been endemic terrestrial mammals in the same guilds as now represented by the exotic mammals, would the native community be more resistant to the invasion of these exotic mammals? This question is not answerable because, as we have seen, the number of endemic species in a general niche provides no index for its resistance to the invasion of exotic species with similar resource requirements. One can, however, argue that the presence of endemic mammals with similar activity patterns would have resulted in the coevolution of the prey species. These prey species probably would have evolved defense mechanisms against overexploitation, and in this way a greater life form diversity associated with a more complex activity network would have created a greater biological stability in this island forest. However, the key relationship here is coevolution and ecosystem stability, not life form or species diversity and ecosystem stability.

Our third initial hypothesis that stability of biological structure in island ecosystems may be related to climatic factors was prompted by the general observation that the biota in rain forest terrains of the islands contain significantly more endemics than the biota in seasonal environments. For testing this hypothesis, one would need to analyze ecosystems for their proportion of endemic and exotic species in seasonal and rain forest environments, which have been similarly disturbed or undisturbed. We approximated this design by the Mauna Loa transect, which included seasonal as well as rain forest environments and by comparing these to the Kīlauea rain forest. The rain forest contained fewer exotics than did the transect, however, the rain forest on the transect in Hawaii Volcanoes National Park also harbored more exotic species than did the Kīlauea rain forest. Important exotic tree invaders in the park's rain forests were strawberry guava *(Psidium cattleianum)* and the fire tree *(Myrica faya)*. Both were absent in the Kīlauea forest. In addition, there were more exotic herbaceous plant species and many fewer endemic arthropod and forest bird species. The lesser number of endemic arthropods apparently was related to the effect of occasional sulfur dioxide steaming from the nearby Kīlauea volcano. This also depressed the number of exotic arthropods in the park's rain forest. The lesser number of endemic birds was probably in part related to the absence of *Acacia koa* as a major shelter tree in the park's rain forest. But other factors, such as the nearness of human habitation and the smaller spatial extent of the forest area, also contributed to the proportionately greater exotic species invasion in the park's rain forest as compared to the Kīlauea rain forest. Thus, disturbance factors are more important for exotic species invasion than climatic factors, although the latter may play an indirect role. Most exotic species are preadapted to warm lowland climates. Moreover, higher rainfall encourages faster growth rates, which means that greater competition is offered by the indigenous community to invaders not specifically adapted to rain forest climates.

Soil was an important factor along the Mauna Loa transect. Habitats with fine soil were occupied by a higher proportion of exotic plant species than habitats with rock outcrop soil, which were almost completely occupied by endemic plants. This pattern is related to two factors: the activity of feral pigs that only stir up the ground significantly when there are sufficient amounts of fine soil (with earthworms) present and also to the fact that the majority of exotic plants are adapted to grow in fine soil rather than on rock outcrop soil.

This brings out several points about the biological stability of island ecosystems, some of them well known. These ecosystems are particularly vulnerable to disturbances caused either directly or indirectly by humans because they did not coevolve with humans and human activities. Apart from the fact that humans are the original dispersal agents, invasion of exotic plant species in the majority of cases is attributable to physical disturbances. Only in rare cases do plants invade unaided by physical disturbances. Such unaided invasion is, however, very significant. Whether this type of invasion is damaging depends on the degree to which the indigenous community balance is disrupted.

Unaided invasion from the point of entry into an island requires first that the plant has a dispersal and reproductive mechanism, which is functional in the island environment, and, second, that the plant be adapted to overcome the resistance offered in the indigenous ecosystem. These two sets of circumstances do not come together very often, nor do they become equally or uniformly effective across all Hawaiian ecosystems.

The same two prerequisites for unaided invasion from the point of entry into indigenous communities apply to all other exotic organisms. Many of the introductions of new consumer organisms are successful because other exotic organisms have invaded before them. These provide new resources on which the new consumers can exist. Therefore strictly unaided invasions of exotic species into indigenous communities are rare. If humans had not entered the island environment as a new dispersal agent, island ecosystems could be considered biologically very stable.

This is an important insight to the question of island ecosystem stability, which has come from our studies of their biological organization. We have found that many island species are ecologically versatile organisms; they have not lost their pioneering characteristics that allowed them to become successful in the first place. They began their island existence completely unaided by humans. The successful ones grew into relatively large populations (like the successful exotics) and spread into a number of different habitats. They met with many rather unpredictable perturbations, such as volcanic explosions followed by lava flows, ash fallout, sulfur-dioxide steaming, fire, flash floods, tropical storms, and droughts. These disturbances provided a set of unequally distributed but recurring ecological catastrophies, which became part of the evolutionary history to which the surviving populations had to respond. Probably many populations that did not achieve a wide distribution were wiped out periodically. Those that achieved wider distributions were fractured or partly isolated by such perturbations. Some of these adapted cladogenetically by forming many related species (the speciating organisms); others adapted anagenetically by becoming polymorphic but less genetically separated (the successful nonspeciating organisms). Only a relatively small number of island species survived unperturbed in small local situations. These are the ones that are genetically isolated from the rest of the island species, which sometimes developed into bizarre life forms and which may easily be displaced by competition from exotics without further involvement by humans. However, these are not the typical island species, as has often been assumed.

There is no reason to believe that island species must necessarily evolve into an evolutionary dead end. If this were so, we would have to consider humans as the only important evolutionary stress factor; clearly they are not. The genetic variability and ecological versatility that we found to be characteristic for island species provide for an important index of island ecosystem stability. We may state as a new general hypothesis that these characteristics permit such species to form biologically stable communities in any volcanic island large enough to allow the biological refugia that are necessary during the perturbations typical of such island environments.

In island ecosystems, such as found in Hawai'i, humans recently (in the evolutionary sense) have become the dominant perturbation factor. Many of the ecosystem processes set in motion through their influences are now beyond their control and have become part of a new set of interactions. However, the dynamic balance in those island ecosystems that have not been converted to other uses so far have only been shifted and not destroyed, except in a few cases. Our findings indicate that the new stresses associated with humans, if carefully monitored and restrained, may not lead inevitably to the extinction of all original island biota. Instead there is hope of maintaining the indigenous character of island ecosystems through appropriate conservation management.

16

Understanding Hawaiian Forest Ecosystems: The Key to Biological Conservation

D. Mueller-Dombois

ECOLOGICAL IMPORTANCE OF KOA AND ʻŌHIʻA FORESTS IN HAWAIʻI

This final chapter focuses on the question of island forest maintenance under natural conditions. In this connection we will explore the fourth initial working hypothesis that island ecosystem stability is related to the behavior of the dominant life forms. Dominant life forms are here defined as those species or groups of species whose activity pattern has a profound stabilizing or destabilizing effect on the rest of the ecosystem. They are not necessarily always the most important organisms in terms of biomass. Such dominant life forms are found among both origin categories, native and exotic organisms. In contrast to the species-rich continental tropical lowland forests, it is usually not difficult in Hawaiian ecosystems to determine dominant life forms by just a few species.

Two such dominants in Hawaiian forest communities are the endemic trees ʻōhiʻa (*Metrosideros collina* subsp. *polymorpha*) and koa *(Acacia koa)*. They are the two most abundant native tree species that occur on all of the high Hawaiian Islands. They are usually also the taller-growing forest-canopy formers among the many other native Hawaiian tree species. Both species are of considerable interest because they provide the essential matrix of the more widely distributed indigenous and natural forest ecosystems in the Hawaiian Islands. These ʻōhiʻa and koa forests harbor much, if not most, of the remaining native Hawaiian fauna and flora. In addition they serve the important function of watershed cover on the windward slopes and on certain parts of the leeward slopes.

The characteristic distribution of these two tree species follows an interesting altitudinal pattern (figure 16-1). Going upslope on the windward sides of the high islands (Kauaʻi, Oʻahu, Molokaʻi, Lānaʻi, Maui, and Hawaiʻi), one first encounters cultivated or variously disturbed vegetation, where grass cover, shrub cover, and variously interrupted and patchy tree cover predominate. Here one often finds naturalized exotic trees, such as the java plum *(Eugenia cumini)* and guava *(Psidium guajava)*. Then one comes into a zone where koa trees occur and where they form stands with closed or open canopies. In this zone one can often find koa and 'ōhi'a in mixed stands with indigenous ferns growing beneath. Going higher and into wetter zones where the cloud layer prevails

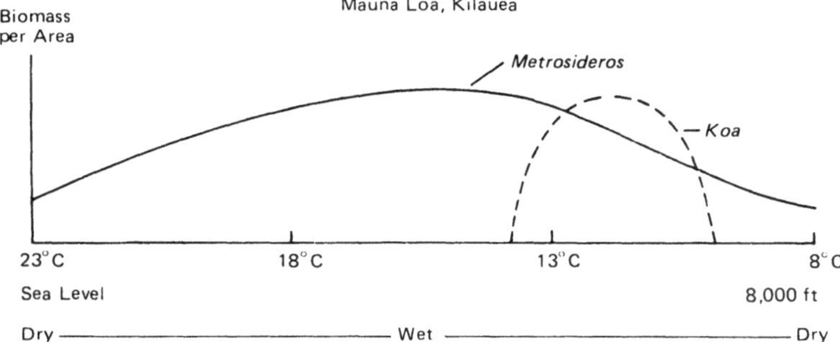

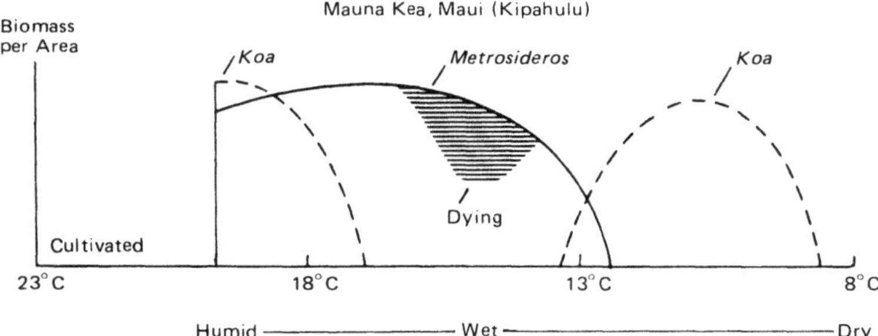

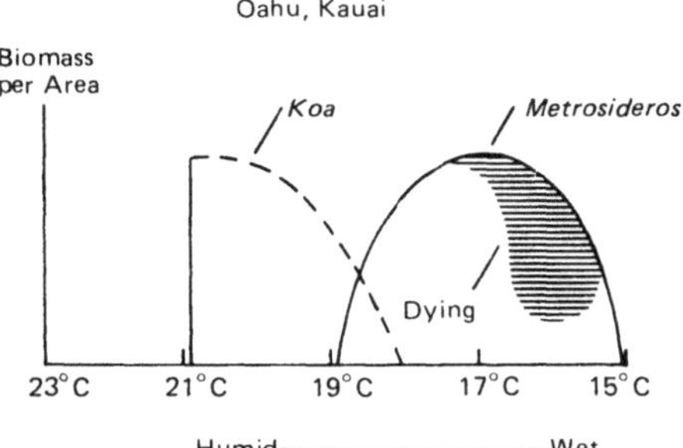

FIGURE 16-1. *General ecological amplitudes of the two most important native tree species* (Metrosideros collina *and* Acacia koa) *in the mountainous Hawaiian Islands. (From TR 19, Mueller-Dombois, 1975.)*

much of the time, koa drops out while ʻōhiʻa becomes the major canopy tree. ʻŌhiʻa dieback is also common in this zone. On Kauaʻi, Oʻahu, Molokaʻi, and Lānaʻi, ʻōhiʻa forests of low stature occur on the mountain tops and ridges. On the two very high islands, Maui and Hawaiʻi, koa reappears higher up at about 1,200 m altitude, first in mixture with ʻōhiʻa and then forming more or less pure koa stands with grassy undergrowth up to about 2,100 m altitude. From here upward, with some overlap in distribution, the endemic tree māmane *(Sophora chrysophylla)* becomes an important forest tree, which on Mauna Kea goes to 2,800 m altitude, there forming the tree-line ecosystem.

A more unique situation exists on Mauna Loa, the youngest high volcanic mountain. Here, ʻōhiʻa grows from sea level to the tree line at 2,500 m altitude, and koa is represented only in an upper elevational belt between 1,200 m and 2,100 m altitude.

From this distribution pattern, it is apparent that ʻōhiʻa has a very wide ecological amplitude. It grows from a warm tropical-dry climate through a wet mid-elevation rain forest belt into a cool subalpine climate, where the mean annual temperature is only 10°C. One of our IBP studies (Corn and Hiesey, TR 18, 1973) suggests that altitudinal races (or ecotypes) may exist in the ʻōhiʻa species complex.

Koa also has a wide ecological amplitude because it occurs in both a lower and upper elevational belt. There is a great probability that the upper- and lower-elevation koa also form different races. However, this question has not yet been resolved experimentally.

The maintenance strategies of these widely distributed indigenous forest builders are of special interest for two reasons. First, both koa and ʻōhiʻa are pioneer tree species, which are essentially shade intolerant. Thus the question arises from a purely scientific viewpoint as to how such tree species maintain themselves as the dominant forest-structure-forming species when they cannot grow successfully under a closed canopy. Second, the natural maintenance mechanisms and survival trends of these forest trees are important to understand from a practical viewpoint because the management of these koa and ʻōhiʻa forest ecosystems will largely determine the future of biological conservation in Hawaiʻi.

Conservation management requires information not only on the biological organization of indigenous communities and its evolutionary significance but also on how this organization works under natural conditions and how it stands up under stress. Thus the question of longer-term dynamics of Hawaiian ecosystems, particularly under the continuing stress of exotic species invasion, is of special relevance to biological conservation. This is a much broader question than could be answered under the short tenure of the Hawaiʻi IBP project. However, some effort was put into this broader question. In particular we were concerned with finding indexes that would reduce the complexity of biological information and yet would allow some predictions as to the relative stability of Hawaiian forest ecosystems. This led to the fourth working hypothesis that island ecosystem stability may be related to the activity patterns of the dominant life forms.

MAINTENANCE OF KOA IN THE UPPER MONTANE SEASONAL ENVIRONMENT

The upper montane seasonal environment occurs only on the two highest islands, Maui and Hawai'i. Here it represents on the windward slopes the area between the rain forest and the subalpine scrub and woodland. Much of this terrain is occupied by grassland and scattered shrub communities, which are associated with rock-outcrop soils and the grass-covered areas with fine soil deposits from volcanic ash (Mueller-Dombois, 1967).

There are a few remnant koa trees in this grassland on Mauna Kea, which are now dead or dying from old age. The same applies to the endemic māmane *(Sophora chrysophylla)* trees (Warner, 1972), which overlap with koa in this area and which form the backbone of the tree line ecosystem on Mauna Kea.

The reason that there is no replacement of young koa and māmane trees is simply browsing and grazing by introduced mammalian herbivores. Browsing of tree seedlings by feral sheep has been the major factor for loss of māmane forest in the upper elevations of this zone, and grazing by cattle is the main reason for the dying koa forest in the lower portion of this zone. Any young regeneration of koa that is coming up in this environment is eaten by introduced herbivores. This observation is not new, but important here for clarifying the causes of breakdown in island ecosystems.*

The great resilience of koa and the recovering capacity of the associated biological components (endemic grasses, forbs and shrubs, birds and tree-associated arthropods, and others) became apparent through the study of the mountain parkland ecosystem along the Mauna Loa transect. This ecosystem occurs in an equivalent basic environmental setting to the dying koa forest on Mauna Kea (Mueller-Dombois and Krajina, 1968). The parkland ecosystem is the result of recovery from degradation by cattle grazing after the National Park Service put a fence around this area in 1916. Since then koa has increased considerably in abundance and cover, and individual remnant trees have expanded vegetatively through outward growth from suckers (Mueller-Dombois, 1967).

Continued outward growth of suckers in the parkland ecosystem, however, is curtailed by feral goats and fire. We have studied the effect of goats in some detail with the help of exclosures (Spatz and Mueller-Dombois, TR 3, 1973). Released from goat-browsing pressure, koa suckers developed rapidly inside the exclosure, while outside the exclosure sucker growth was stunted. In

*The statement is not implied as a judgment on land-use decisions. Such decisions arise from socioeconomic considerations that are beyond the scope of this discussion.

the mid-1970s, the Park Service put a major effort into goat control, and now the koa colonies are expanding rapidly into the surrounding grassland matrix.

Fire is another factor that reduces koa colonies occasionally, but trees not killed by fire resume the process of sucker rejuvenation soon following the fire. Thus without cattle and goats, the koa is vigorously maintaining itself. Broad-area and intensive grass fires reduce koa, but if fires are not too frequent and extensive, koa will coexist with fire in the mountain parkland, there forming a substantial but patchy forest cover.

In contrast to the introduced grazers, fire is a natural perturbation factor in this ecosystem (Mueller-Dombois and Lamoureux, 1967), where it undoubtedly had an evolutionary significance in the development of the indigenous vegetation (Vogl, 1969). Koa seeds germinate readily after heat treatment (Judd, 1920), and it is possible that suckering is encouraged from the heat of grass fires just as in aspen *(Populus tremuloides)* stands of temperate forests in North America. A number of fire-resistant endemic bunch grasses and sedges occur in the grassy matrix surrounding and diffusing through the koa colonies. These have come back (Cuddihy, 1978) and are much more prevalent on Mauna Loa than on Mauna Kea where exotic temperate-zone grasses dominate.

The most prevalent endemic grass in the mountain parkland on Mauna Loa is *Deschampsia australis,* which, however, is becoming more and more displaced by the European grass *Holcus lanatus* (Spatz and Mueller-Dombois, TR 15, 1975). The reason is the widely occurring ground-digging activity of feral pigs. Like the feral goats, these are not controlled by any natural predator in the island environment.

Here we found that, in the displacement of *Deschampsia* by *Holcus* with the aid of pigs, the spatial factor is extremely important. The larger the pig-disturbed areas, the more firmly established becomes the exotic grass.

MAINTENANCE OF KOA AND ʻŌHIʻA IN THE MONTANE RAIN FOREST

The Kīlauea rain forest is representative for the upper-elevation mixed koa-ʻōhiʻa rain forest. This ecosystem occurs in an altitudinal belt from about 1,400 m to 2,000 m on the windward slopes of Mauna Kea and Mauna Loa.

This forest is particularly rich in native birds of the honeycreeper (Drepanididae) family. The forest structure is characterized by tall koa trees forming the overstory, ʻōhiʻa trees forming the mid-story, and tree ferns (*Cibotium* spp.) forming the understory. In addition there are many other native low-stature trees and shrubs. This forest has been called decadent by some foresters because individual tall, old koa trees or sometimes groups of overstory koa trees are blown down during Kona storms. Moreover, the current density of koa is obviously far below what the site could carry if this forest was managed for koa.

We studied the structure of this forest in detail through quantitative enumeration of koa and other tree individuals in all size classes. We found that there is replacement by koa regeneration from seedlings. Their quantity is not very great (about one hundred seedlings per ha), and their number is reduced in the sapling stage down to two to three individuals per ha. However, once young koa trees are established in a canopy gap, they grow rapidly. A koa tree can grow to a diameter of 40 cm and 15 m height in about thirty years. Thus, koa is a fast-growing tree in this forest, and the species is certainly not decadent. Nevertheless, its rate of replacement is low, and it appears to be maintained at an approximate one-to-one level. The effect of the relatively high feral pig population is probably an additional factor in the low-level replacement. Pigs have been found to feed on koa seedlings or to trample them into the ground. Koa seedlings established on root overturns appeared to have the best chance of survival.

We expect that every fifty to one hundred years there may be a hurricane-like storm, which then strongly decimates the overstory of koa, resulting in a wave regeneration of a new koa crop. Thus, koa is maintained in this forest through storm perturbations, which create canopy gaps through threshing down tall individuals and tree-groups.

ʻŌhiʻa appears to benefit from forest gap creation through koa in this ecosystem. We found that about one-quarter of the canopy gaps were filled with ʻōhiʻa reproduction in different stages of development (advanced regeneration, sapling groups, and pole-sized, near-mature tree groups).

Each gap group of ʻōhiʻa was rather uniform in size, indicating that seedlings were released after the opening had occurred. ʻŌhiʻa seedlings were abundant (about one per m^2) on the forest floor and almost always restricted to decaying logs. These in turn were rather randomly distributed through the forest. Thus, ʻōhiʻa seedlings appear to tolerate more shade, and they have an advantage over koa whenever a gap is formed. However, the growth rate of ʻōhiʻa saplings is certainly much less than that of koa saplings. Both koa and ʻōhiʻa compete with a third dominant plant component in this forest, the tree ferns. In certain gaps we found tree ferns to fill out the space almost totally, together with some other shade-tolerant native tree species. These include all others except the two tall-growing species, koa and ʻōhiʻa.

The fourth dominant life form that has entered this forest is the feral pig. Its major food source in addition to earthworms is the starchy core of tree ferns. The pig acts to thin down the tree fern density to some extent, and it also appears to interfere with koa reproduction. The pig has some effect on the interaction between the three indigenous and dominant plant competitors, but so far it has not yet upset their reproductive balance. This could easily happen, however, if the pig population increases in density. Unfortunately, we have not been able to establish any density values in relation to pig interaction in this forest.

Recently a South American vine *(Passiflora molissima),* locally called banana poka, has become dominant in small local sections of this forest ecosystem (but not yet in the Kīlauea forest). This vine, which covers the floor in the forest gaps, may cause smothering, particularly of the light-demanding koa reproduction. It now appears to offer an important threat to the maintenance of this important ecosystem.

Another factor in the maintenance of 'ōhi'a in this forest are several of the endemic honeycreepers, the 'apapane, 'amakihi, and i'iwi. Carpenter (TR 76, 1976) found that these endemic birds are essential for high levels of fruit set and outbreeding in 'ōhi'a. Thus 'ōhi'a seed supply, its viability, and genetic variability are dependent to some extent on the nectar-feeding activity of these birds. However, 'ōhi'a seed production is not totally dependent on these birds. Seeds can also be produced from selfing or through insect pollination. Carpenter suggests that insect pollination is the more usual mechanism in lower-elevation 'ōhi'a and particularly in the yellow-flowered form. This underscores again the generalistic attributes of 'ōhi'a.

There are a few other major rain forest ecosystems on the east flanks of Mauna Loa and Mauna Kea, which we did not study in detail under the IBP. For example, there is the lower-elevation type of koa-'ōhi'a-tree fern forest on Mauna Kea (not on Mauna Loa), which is structurally somewhat similar to our upper-elevation forest. The koa trees are not as tall in this forest; there are more koa seedlings and saplings throughout this forest, and endemic honeycreepers are very rare. It is not possible, without closer study, to extrapolate the same major maintenance mechanism (storm perturbations) to this forest.

The larger segment of rain forest in the mid-elevation belt lacks koa as an overstory tree. Throughout this territory are many areas covered with dead, standing 'ōhi'a trees (snags). Under the IBP we began to study one such snag-occupied area (a 2 km long belt-transect near 1,400 m altitude a few kilometers north of the Saddle Road between Mauna Loa and Mauna Kea). The snag forest occurred on poorly drained pāhoehoe lava with discontinuous shallow soil in the form of woody peat and ash. We found that the dieback was restricted to the canopy trees, almost entirely to 'ōhi'a, and that the undergrowth was vigorous and that there was abundant reproduction in the form of 'ōhi'a seedlings and saplings, mostly growing on decaying logs above the partly inundated forest floor. At that time it was widely believed that the dieback was caused by a disease (Mueller-Dombois, 1980).

MAINTENANCE OF INDIGENOUS WOODY VEGETATION IN THE COASTAL LOWLAND

The coastal lowland areas on all high Hawaiian islands contain very little original vegetation. This is understandable because most of the lowland areas have been urbanized or converted into agricultural lands. However, there is a considerable portion of wildland in coastal areas, which due to its rockiness and rugged terrain is not suitable for intensive agriculture. These areas have been used variously for grazing, and they had been burned over repeatedly by the Hawaiians long before the arrival of the Europeans. The Hawaiians apparently managed some of these areas for pili grass *(Heteropogon contortus)* production (Fosberg, 1972), which was used as a roof-thatching material. In addition some of these areas were tried out for agricultural crops and were later abandoned where too marginal for this kind of land use.

It is not surprising that these coastal lowland habitats are now stocked dominantly with weedy replacement vegetation of foreign origin. A very widespread scrub and low-stature forest cover is formed by koa-haole *(Leucaena leucocephala)*. Its establishment was actively supported by aerial seeding before World War II (McEldowney, 1930; Neil, 1965:411). Another important exotic dominant tree in the coastal lowlands is the mesquite *(Prosopis pallida)*, locally called kiawe. It forms forest groves in the drier parts of the coastal lowland and occurs often as scattered trees in the koa-haole scrub or in combination with introduced grasses. Egler (1947) has described this vegetation for O'ahu. Another account was given during the IBP (TR 12; Mueller-Dombois and Spatz, 1975).

Rainfall in the coastal lowlands is sufficient to support a woody vegetation anywhere, except perhaps at the northeast corner of the Island of Hawai'i (Kawaihae), where there is a true desert climate (figure 1-5). Thus the original vegetation in the coastal lowland, including the strongly seasonal leeward areas, was undoubtedly predominantly woody. Most areas may have had an open forest or woodland type of vegetation.

There are a few relict stands of indigenous lowland forests and woodlands still extant on Hawai'i, Maui, and O'ahu. During IBP we asked whether such relict stands can survive even if reasonably protected and whether it is possible to restore some of the original vegetation cover if the new stresses introduced by humans are removed.

We attempted to answer the first question through a detailed study (Wirawan, TR 34) of such a relict stand, which still occurs in the coastal lowland of northeast O'ahu, in the Mokulē'ia Forest Reserve. This quantitative analysis of an original dryland forest was particularly meaningful because the same forest had been analyzed in detail twenty years earlier by Hatheway (1952).

It was possible to relocate the same seven study plots used by Hatheway. The twenty-year reanalysis showed that all originally recorded endemic tree and shrub species (over thirty of them) were still present and actively reproducing. Only two species, the 'ōhi'a (*Metrosideros collina* subsp. *polymorpha*) and *Psychotria hathewayi,* did not show adequate recruitment for continued maintenance in this forest. The 'ōhi'a lacked seedlings and saplings under 2 m tall, a characteristic often found for this species also in rain forests under closed canopies. However, in dryland forests, which are advanced in primary succession, 'ōhi'a is a rare component, probably because there are other indigenous species better adapted to late seral and/or climax stages. *Psychotria hathewayi* had lost a considerable portion of the taller individuals due to mortality, and these were not yet replaced to the level as recorded by Hatheway in 1952. However, seedlings of *P. hathewayi* were present locally (more than ten per 100 m^2) indicating that the species was not yet becoming extinct in the area. Another endemic tree species, *Canthium odoratum,* was found to have its fruits and seeds badly infested and killed by the larvae of a moth, *Alucita (= Orneodes) objurgatella.* Zimmerman (1978) describes this moth as probably endemic and host specific to *Canthium odoratum.* The tree is still one of the most dominant, widely dispersed, and actively reproducing endemic woody plants in this relict dryland forest.

All nonnative woody plants (including six tree species) recorded by Hatheway were also still found in the same plots twenty years later. In the more mature closed forest plots, inside the Mokulē'ia Forest Reserve, there had been some quantitative changes among exotic and native woody species, but these did not indicate future shifts to greater dominance of exotics or natives. Both groups appeared to be in a dynamic balance. In the plots outside the forest reserve, the vegetation was seral, that is, not covered by mature forest.

Hatheway's analysis of these disturbed sites had shown that three native tree species, *Erythrina sandwicensis, Canthium odoratum,* and *Sapindus oahuensis,* participated in the succession following a major disturbance, probably from fire and intermittent grazing by cattle. Grazing still continues in this area. One plot on rock outcrop soil showed a shift toward dominance of the native *Sapindus,* which had displaced koa-haole under the now overtopping and expanding crown area of the *Sapindus* trees. In the two other plots on deep, fine soils, the exotic Christmas-berry tree *(Schinus terebinthifolius)* had gained in dominance, overtopping and displacing in part the native *Canthium. Erythrina* was losing ground because the thick mat-forming molasses grass *(Melinis minutiflora)* had moved in under these trees. The grass now was interfering with *Erythrina* seedling reproduction.

Another IBP study dealt particularly with the second question: is it possible to restore some of the original vegetation cover if the new stresses introduced by humans are removed? This study was done in the coastal lowland of Hawai'i Volcanoes National Park. Here feral goats, introduced to Hawai'i by Captain Cook in 1778 (Tomich, 1969), had reduced the original vegetation cover in a major section of their range to a sparse grassland covered primarily by the pantropical annual grass *Eragrostis tenella* (love grass). I had mapped this area in 1965 and suggested the construction of goat exclosures. One exclosure was built in 1968 by the Park Service and two others in 1972. These were monitored during the IBP (Mueller-Dombois and Spatz, TR 13, 1975; Mueller-Dombois, 1976).

Within two years, the annual grassland had changed rapidly in one exclosure by advancing through a stage with perennial bunch grasses, dominantly Natal red top *(Tricholaena rosea),* to one dominated by woody low shrubs (chamaephytes) and a vine, *Canavalia kauensis.* The vine covered eventually about one-half of the surface area of the 7 × 100 m exclosure. This vine turned out to be a new endemic species (St. John, 1972). However, few other endemic species participated in this succession.

The Park Service began a systematic goat-control program through fencing large areas and removing the goats. In the meantime the endemic vine climbed over the fence of the exclosure. It produced seeds that spread locally and germinated. Small colonies of the vine appeared in the now changing and densifying surrounding grassland.

This endemic vine did not appear in the other exclosures. Here, the succession also went through a bunch grass stage and a densification process with chamaephytes, principally the exotic *Indigofera suffruticosa* and the indigenous *Waltheria americana*. Another indigenous vine, a morning glory *(Ipomoea indica),* appeared in the neighborhood of the second exclosure, and this vine now also covers a broader area.

At present, the succession following goat removal is more or less arrested in the bunch-grass-chamaephyte-vine stage. No native trees have so far invaded although ‘ōhi‘a is in the area, and some other endemic trees, including *Canthium odoratum* and *Erythrina sandwicensis* (among others), are found at the extreme east end of the former goat-inhabited territory. The rainfall is such that trees will eventually invade. Currently the Park Service has to fight to keep koa-haole from overrunning certain parts of the former annual grassland.

The reason for this arrested succession is the absence of native trees in the area. Whatever endemic trees were in the area have been destroyed by the century-long goat occupation of this ecosystem. The goats probably fed on their seedlings, saplings, and bark and in this way effected an exhaustion of the supply of tree seeds. However, the situation has changed further from the original condition by the introduction and invasion of so many exotic grasses which now form a rather dense cover, which in addition to molasses grass includes now also the African savanna grass, *Hyparrhenia rufa*. For a further seminatural recovery process, it is necessary to find an indigenous lowland tree species that is capable of invading such a grass cover.

SOME CONCLUSIONS: EXOTIC ORGANISMS AND THE RESILIENCE OF INDIGENOUS BIOTA

The examples presented here have given some insight into the more obvious kinds of interferences between certain dominant exotic life forms and indigenous life forms that currently are occurring in specific island ecosystems. The examples covered a range of ecosystems from the grazing lands in the cool upper montane seasonal environments over the montane rain forest to the warm and alternately wet and dry seasonal lowlands. This range does not include all of the ecological zones in the Hawaiian Islands, but it does encompass some of the more outstanding ones, where indigenous trees, the original ecosystem builders and primary producers, display their resilience to new stresses introduced by humans and also to old stresses in the sense of island-specific perturbations. Such perturbations have been active since evolutionary time, and they are still active and will continue to be active in the future. They include such disturbances as fire, storms, volcanic explosions associated with ash-blanket deposits and lava flows, flash floods, landslides, tsunamis, severe droughts, and other physiological and mechanical interferences. These disturbance factors are not uniformly or randomly distributed in high islands; instead they operate in the context of specific ecosystems and their effects vary between ecosystems. The same can be said for the new stresses.

In the upper montane seasonal zone, cattle is a century old, but a new stress factor in terms of evolution, and it is surprising to find koa trees still existing in unprotected areas. The main reason for this appears to be the reproductive strategy of koa to send up vegetative shoots or suckers from its root system and from cable-like rhizomatous roots, which grow beneath the grass sod. Young koa suckers pushing through the grass sod are equipped with the feathery foliage typical of legumes. Above the grass layer, the feathery leaves are dropped and the leaf stems broaden into phyllodes. Early height growth is generally fast and then slows down somewhat after the feathery leaves are shed. Thus koa is ideally equipped to compete with grasses, and the endemic grasses that evolved in the upper montane seasonal environment, such as *Deschampsia australis,* are ideally equipped to cope with an occasional fire. There is little doubt that the evolutionary adaptation of indigenous organisms to fire in the upper montane seasonal environment is the major reason why the indigenous vegetation has not yet been totally destroyed by the introduced grazers. Both fire and grazing are primarily mechanical factors, which operate by reducing the leaf area of plants. If the leaf area is not totally destroyed, plants usually recover from such mechanical damage.

Fire and grazing operate very differently. Fire occurs only at certain intervals, whereas grazing operates more or less continuously where uncontrolled. Under natural conditions without humans, the fire frequency, perhaps related to lava flows, was very low. Thus there were long periods of recovery between fires. With grazing, this relationship was totally changed. The frequency of mechanical removal of foliage and other plant parts increased by several orders of magnitude. If grazers are allowed to persist long enough, they can totally destroy even mechanically resilient endemic plants.

In the upper-elevation montane rain forest, where koa forms the emergent tree layer, we found that natural maintenance of this ecosystem is related to storm perturbations. If this ecosystem were to be artificially sheltered from storms, koa would probably disappear. The reason for its disappearance would be the absence of large enough gaps for koa seedlings to grow toward the light. Its own undergrowth, particularly the tree ferns, would eliminate koa through light starvation. In this ecosystem koa exhibits an alternative reproductive strategy to the previously discussed one. It reproduces primarily from seed, not vegetatively from suckers. Koa seedlings here have a faster growth rate than the suckers (TR 17).

This example shows that island species compete with one another in the same community, but there is a harmonious coexistence among them in relation to the ecosystem-specific long-recurring (evolutionary) stresses. This harmonious coexistence can become seriously disrupted by certain dominant exotic life forms. The pig, for example, digs up the ground and promotes the invasion of exotic weeds. Many of the weeds, for example, the composite *Erechtites valerianaefolia,* are of little consequence because they are heliophytes (sun-loving plants), which have only a temporary position in such ground-disturbed places. The shade-adapted exotic mat-forming plants are of somewhat greater ecological importance because they assume a more permanent position. However, they will not, as a rule, disrupt the reproductive balance among the dominant ecosystem-structure forming plants.

An exotic plant that is capable of disrupting this balance is the vine *Passiflora mollissima* (the banana poka), which is adapted to take advantage of gaps formed in this ecosystem. By its rapid growth rate, it can form a dense foliage mass covering the low-growing vegetation in a forest gap. Moreover it can climb up on trees surrounding the gap to a height of 20 m or more, there forming thick, heavy curtains of foliage. Such an exotic plant life form clearly can cause the breakdown of the natural maintenance strategy that was established in this indigenous ecosystem during evolutionary time. The vine may also interact with the pig insofar as the pig is fond of the banana poka fruits. The pig probably can spread the plant through distributing its seed, and in turn the plant may promote the development of pig-concentration centers through attracting the pigs by its fruits.

With regard to the coastal lowland examples, we had asked two management-oriented questions. The first question—can such relict stands survive even if reasonably protected?—referred to the study of the relict dryland forest on O'ahu. We noted that all indigenous tree species were still present in the forest after their first analysis twenty years earlier and that all but two of them were actively reproducing and maintaining themselves. Further all exotic plant species were also still present and maintaining themselves. This seemingly stable pattern was true for the climax stand segment inside the Mokulē'ia Forest Reserve, which was fenced against cattle after an earlier history with some grazing influence through stray cattle. Stray cattle may still occur occasionally inside the reserve, but they had no effect on the reproductive balance of the study sites.

A different situation existed outside the reserve, where disturbance factors had resulted in seral (nonclimax stands) with participation of both indigenous and exotic plants. In one case, on sites having protection from grazing, the successional trend was leading toward the dominance of an indigenous tree, *Sapindus oahuensis,* and the competitive extinction of an exotic tree, *Leucaena leucocephala;* in another case, without protection from grazing, the trend was going toward the dominance of an exotic tree, *Schinus terebinthifolius,* and the competitive extinction of an indigenous tree, *Canthium odoratum*. In both cases, the reason could be identified as the respective greater height-growth capacity of the overtopping tree on that particular site. The extinction was only very local, in the sample plots. On a larger spatial scale, local extinction translated merely into some minor displacement; all four species still coexisted over the broader habitat. The overall successional outcome in terms of indigenous or exotic species dominance in the climax stage needs further study.

One important exotic plant life form had entered the seral dryland communities—the introduced grasses, particularly the thick mat-forming molasses grass *(Melinis minutiflora)*. It clearly interfered locally with the reproduction cycle of another endemic tree, *Erythrina sandwicensis*.

Where indigenous dryland stands are open and growing on fine soil rather than rock outcrop, the invasion of exotic grasses (and there are many other species; Kartawinata and Mueller-Dombois, 1972) will form a serious threat not only to the reproductive cycle of the indigenous woody plants but also because of the new fire danger associated with a dense dry-grass cover. Normal bunch-grasses, such as Natal redtop *(Tricholaena rosea)*, the pili grass *(Heteropogon contortus)*, and the broomsedge *(Andropogon virginicus)*, do not tend to form dense covers on rock outcrop soils. Indigenous dryland forests on rock outcrop soils therefore are better protected from fire than are those on fine soil.

Certain life forms among grasses, however, particularly the mat-forming molasses grass, can also form continuous covers on rock outcrop soils. If these relationships (grass-cover density in relation to soil and grass life form and tree crown closure to reduce grass density) are considered in the protection-management of indigenous dryland forests, there is a fair chance that such relict stands may survive.

The second question—is it possible to restore some of the original vegetation cover, if new stresses introduced by humans are removed?—related to the totally goat-degraded coastal lowland ecosystem in Hawaii Volcanoes National Park. Goat exclusion has shown that a partial recovery with indigenous plant life forms did occur. This is testimony to the great resilience of some indigenous life forms, in this case two vine species and an indigenous chamaephyte, *Waltheria americana*. However, this ecosystem was so far degraded that a spontaneous recovery with indigenous trees cannot be expected. An additional complication is the fire hazard, which has increased with the invasion and densification of exotic grasses. Further steps in recovery require more information and aids in conservation management.

Fire may have also been a natural stress factor in the coastal lowland of the Hawaiian Islands (Vogl, 1969). Wirawan's research (TR 34) on the O‘ahu dryland forest included the observation that almost all indigenous tree species reproduced from both seed and vegetative resprouting. To what extent indigenous tree species are killed by grass fires and to what degree their vegetative resprouting mechanisms operate in surviving trees are questions that should be addressed to answer more fully the two broader management-oriented questions.

From these examples, we can conclude that there is no evidence that the evolutionary stresses were less vigorous and that competition among indigenous island biota was any less harsh than among biota that evolved in continental ecosystems. However, when humans entered the island ecosystems, they introduced new stresses, some of which turned out to be very detrimental. The true nature of these destructive relationships does not seem to be found simply in the introduction of new life forms that were never part of the island ecosystems before humans arrived but in the introduction of life forms that find suitable conditions and for which there is no biological control. This applies to the introduced mammals as well as to the banana poka vine in the montane rain forest and to the introduced grasses in the coastal dryland ecosystem whose spread is favored by a greater fire frequency than existed under natural conditions without humans.

IS THE EXTINCTION OF INDIGENOUS ISLAND BIOTA INEVITABLE?

A number of authors starting with Darwin (1859), Hooker (1867), and Wallace (1880) have gone on record to proclaim that introduced biota inevitably will displace the indigenous biota on oceanic islands, primarily because the latter are not equipped to compete with the former. This general hypothesis was reemphasized in this century again by Degener (1930), Carlquist (1965), and others. From observing the same general relationships, other authors, such as Allan (1936), Egler (1942), Hatheway (1952), and Harris (1962), have concluded that the competitive capacity of a species is not determined by whether it originated on an island or continent. Egler and Hatheway, moreover, predicted from observations in the Hawaiian Islands that the indigenous biota eventually will succeed in the competitive struggle with aliens provided that humans do not interfere.

One explanation for opposite conclusions of this sort lies in the time scale that one wishes to apply. If one considers a geological time scale, all biota eventually will disappear on oceanic islands because of their eventual base-level erosion. Of course this extreme time scale was not meant by the cited authors. If we consider a time scale relating to processes of biological evolution, say tens of thousands of years, neither of the two general hypotheses appears to be acceptable because we have found that island biota track their environmental changes genetically just as vigorously as do continental biota. Thus it is not possible to freeze indigenous island biota and their communities on an evolutionary time scale. Neither can one expect continental biota and their communities to remain biologically stable on that time scale.

What was probably implied in the opposing viewpoints on indigenous versus alien biological interactions was a time scale relating to ecological succession. On this time scale, which involves decades in secondary successions (those starting from partial ecosystem destruction) or centuries in primary successions (those starting on new substrates, such as lava flows), predictions on the relationship are perhaps more realistic.

The insight gained from our IBP studies leads to yet a third viewpoint. Going back to the often-presented idea (Hubbell, 1968, 1972) that oceanic islands have a disharmonious biota, we can now state that this concept is valid only in the taxonomic sense and here relating only to the successful colonizers. It applies, for example, to the fact that coniferous trees (gymnosperms) or fig trees (Moraceae) never became part of the indigenous island biota because of their dispersal limitations. However, the idea of a disharmonious biota is not an ecologically and microevolutionary valid concept because indigenous island biota evolved into functional, self-perpetuating communities. From a few successful colonizers evolved many new species (the endemics), others remained unchanged (native species that occur naturally also outside Hawai'i). There is nothing disharmonious about island ecosystems that evolved in extreme isolation. Their biota are competing with one another, just as in continental communities, and their competition has resulted in a dynamic coexistence and a harmonious adaptation to their island habitats.

We have found that many important island species are not so easily displaced as one would expect if the first hypothesis was entirely valid. These island species offer, in fact, considerable resilience to the foreign invaders brought into their contact through humans. Our findings also do not support entirely the second hypothesis of Egler, Hatheway, and others, which ascribes to the indigenous biota the capacities of species in climax communities; that is, given enough time without interference by humans, such island species would displace the aliens. In fact, we found that many of the island species have retained their versatility and pioneering characteristics that were responsible for making them successful colonizers in the first place.

However, neither island species nor exotic species are ecologically homogeneous. Of the over forty-five hundred introduced plant species, fewer than four hundred (under 10 percent) have become adventive (St. John, 1973). Of those that have entered natural communities, most are pioneer species equipped to enter ecosystems that are partially disturbed; they are pioneers in secondary, not primary, successions. In primary succession, starting with new lava flows or ash-blanket deposits, for example, indigenous pioneers are much more important (Smathers and Mueller-Dombois, TR 10, 1974). Many of the exotic pioneer species active in secondary successions are only of temporary importance in Hawaiian ecosystems. Only a relatively small number of species (out of the 10 percent foreign plant invaders) have really exerted a dominant and destructive role in relatively undisturbed ecosystems. These species, like the banana poka vine in the upper montane rain forest, have entered into Hawaiian ecosystems not because there are no other endemic life forms of the same kind but because conditions were suitable and there was no effective natural (biological) control against its becoming disruptive.

The question of whether the extinction of indigenous island biota is inevitable cannot be answered easily. Our findings have shown that we cannot dismiss the indigenous biota as being weak competitors. Instead our findings put a much greater burden on humans. However, we cannot blame Captain Cook for introducing the goats or any other historical person involved in tipping the balance in our island ecosystems because there was no knowledge of the consequences of their actions. The burden now is on the managers and researchers who are dealing with island ecosystems that are either partly destroyed or whose balance is more or less displaced from the original. If there is a willingness to strive for biological conservation of island ecosystems wherever feasible in our socioeconomic framework, we may truthfully say that important elements of the indigenous island biota can be maintained in the form of functional ecosystems.

NEW QUESTIONS AND RESEARCH DIRECTIONS

Because of my own bias, the examples I have presented were restricted to those observed in the area of vegetation ecology. Focus on the primary producers as the dominant biological components of terrestrial island ecosystems has some validity in this context, but it gives only part of the story. It would be useful also to draw some conclusions with regard to exotic organisms and the resilience of indigenous biota for the other important island organism groups—the endemic birds, insects, snails, and the indigenous microorganisms. Such reviews would be important for a more complete picture of what humans can do or what is beyond our control regarding the future of biological conservation of island ecosystems.

Since our IBP research was directed primarily to the study of the biological organization in some better-preserved Hawaiian ecosystems, our results have only marginal information value to some other basic questions. Two other questions of high information value for biological conservation are: is there a way to predict which exotic organisms, once introduced to the islands, have the potential to become disruptive invaders? and, what makes some island ecosystems more stable, others less stable?

We did not address the first question directly, but we prepared the way to make this management-oriented question meaningful because there are two parts to this question. One part relates to the characteristics of the exotic organism itself and the other to the characteristics of the ecosystem, range of ecosystems, and their biological components that are threatened by such an exotic organism. We addressed the second part of this question—the island ecosystem and its biological content to be affected—by focusing on certain working hypotheses relating to ecosystem stability.

The fourth of these working hypotheses, that island ecosystem stability may be related to the activity pattern of the dominant life forms, provides, no doubt, for the best index among the four initial working hypotheses suggested on island ecosystem stability. It is superior to the two diversity-stability hypotheses, for which we found little evidence, and it carries more information value than the climatic factor-stability hypothesis. However, our fourth working hypothesis, while focusing on some important elements in the stability-fragility relationships of ecosystems, leaves out an important part relating to the nature of the disturbance or perturbation. Island ecosystem stability relationships cannot be fully understood without specifying the nature of the disturbance, particularly, whether the disturbance is natural or caused by humans. Thus a perturbation analysis has to be part of any useful index of ecosystem stability. Moreover, by focusing on the activity pattern of the dominant life forms, we have assumed that the associated life forms survive or collapse with the dominant life forms. There is probably some validity to this assumption, particularly in island ecosystems, where the consumers, such as the endemic birds, tree snails, *Drosophila* flies, and canopy arthropods can be said to have coevolved with the primary consumers. However, the assumption of coresponse to dominant life-form behavior may not be true for rare and endangered plant species, for example. The degree of correspondence of associated life forms with the activity pattern of the dominant life forms needs additional research if that proposed stability index is to become useful.

The second broader question raised—what makes some island ecosystems more stable, others less stable?—can now be addressed at least by drawing out some principle parameters requiring further research. These are size, nature of habitat, habitat disturbance, preadaptation of exotic invader and its furtherance by other organisms (indigenous or exotic), and reproductive strategy of indigenous ecosystem builders (dominant plants).

We found that the size or spatial extent is very important for the degree of stability of an indigenous ecosystem. Future research should address the question of critical size. This varies with the reproductive strategy of the dominant ecosystem builders and with many other variables—for example, the natural perturbation history, the home ranges of pollinators, and other factors. It would be tempting to say the smaller the area, the more fragile the ecosystem, but this would clearly be an oversimplification.

The nature of the habitat may indirectly be involved in the fragility-stability relations of an ecosystem. The habitat may, for example, act through screening out exotics that are not adapted to cope with the more extreme island habitats, such as lava flows, boggy soils, soils with aluminum toxicity, etc.

Physical habitat disturbance provides for a major avenue of exotic plant invaders. All exotic plant invasion processes studied during our IBP project could be related to direct physical disturbance of the habitat. For example, ploughing of the soil by feral pigs promoted advance of the European grass, *Holcus lanatus*. However, another type of physical disturbance, fire (when not too often repeated), does not appear to displace the endemic grass *Deschampsia australis* in the same habitat.

For exotic organisms to be adapted successfully for invasion, they must possess a long list of ecological properties. Most critical among these is a reproductive mechanism that is functional in the island environment. For example, a very important organism for facilitating invasion of fruit-bearing exotic plants in Hawai'i is the exotic passerine bird, the very versatile Japanese white-eye *(Zosterops japonica)*. After its introduction in 1929 (TR 29), this bird extended its range from sea level to tree line on all high Hawaiian Islands. According to Guest (TR 29), the Japanese white-eye is known to feed on the fruits of at least twenty exotic tree species, of which at least five (*Psidium cattleianum, P. guajava, Schinus terebinthifolius, Eugenia cumini,* and *Brassaia actinophylla*) are active invaders but mostly in physically disturbed habitats.

The reproductive strategy of indigenous ecosystem builders—for example, that of koa and 'ōhi'a—is as yet poorly understood. We began to focus attention on these properties during IBP. During that time the "'ōhi'a forest decline" (Petteys, Burgan, and Nelson, 1975) began to receive a lot of attention. The large but patchy areas of dead-standing 'ōhi'a trees occurring in the mid-elevation rain forest on Mauna Kea and Mauna Loa and the "'ōhi'a decline" discovered on other high islands was believed to be caused by a new disease, a rapidly spreading epidemic, to which the native 'ōhi'a rain forest was not adapted. After several years of intensive disease and insect-pest research, it turned out that the 'ōhi'a decline is not caused by a new biotic agent, not even by an indigenous one (Papp et al., 1979). The root pathogen *Phytophthora cinnamomi* was found in the rain forest soil and suspected to be the killer, but it showed no direct correlation with the so-called decline (Hwang, 1977). Through further detailed studies of the 'ōhi'a decline patterns and their relationship to soil and habitat types (Mueller-Dombois et al., 1977), it was found that the phenomenon is restricted only to canopy trees of 'ōhi'a, and in the more radical tree-to-tree dieback areas it is associated with abundant reproduction of 'ōhi'a seedlings and saplings. The phenomenon is not unlike the reproduction in some fire-killed jack pine *(Pinus banksiana)* stands in North Central America. Yet in the 'ōhi'a dieback and rejuvenation process, which clearly turned out to be a natural phenomenon (Mueller-Dombois, 1980), the trigger is not fire but another abiotic cause, which still has to be properly researched. We currently believe the dieback to be initiated by a climatic instability becoming effective through the soil moisture regime (therefore its patchiness) in combination with an ecophysiological readiness of mature 'ōhi'a stands to die in synchrony.

This example shows how important it is to revise our concept of island ecosystem stability. Biological communities in islands cannot be so easily dismissed as fragile. A tree dieback does not necessarily spell the decline and subsequent extinction of that species. It may merely indicate a very dynamic and somewhat unusual reproductive strategy. In this case, a dominant essentially shade-intolerant pioneer tree species has managed to maintain its dominance in ecological succession. In continental ecosystems of the same sort, the successional function of 'ōhi'a may be performed by different pioneer, seral, and climax species. In ‘ōhi‘a we may have instead successional races or ecotypes. This aspect also requires new research.

However, by focusing on the resilience of island biota, we should not overlook the equally possible reaction: the possibility of an invasion of a new pest or other locally destructive exotic organism, which can cause the disruption or breakdown of island ecosystems. Therefore, for the purpose of biological conservation on islands, where humans determine the doom or survival of indigenous ecosystems to a much greater degree than elsewhere, research efforts with high conservation value should be increased. This has happened to some extent concurrently with and following the Hawaii IBP project. Three federal agencies, the Fish and Wildlife Service, the U.S. Forest Service, and the National Park Service, have become engaged in the research of the rare and endangered species of Hawai‘i, particularly the endemic forest birds. Furthermore, the National Park Service, through a cooperative agreement with the University of Hawaii, has stepped up its management-oriented research. One important area of emphasis relates to studies of successful exotic invaders and to questions of their control. In addition, the National Science Foundation has sponsored another ecosystem-oriented research project on the montane rain forest with emphasis on a functional explanation of the dieback phenomenon, its relation to mineral nutrient dynamics, and its role in primary succession.

Research on the long-term dynamics of island ecosystems and their responses to old and new perturbations should be continued under the Man-and-the-Biosphere (MAB) Program of UNESCO. MAB project 7, “Ecology and Rational Use of Island Ecosystems,” has as yet received only token support by the U.S. government, which is administratively responsible for several important island groups in the Pacific outside Hawaii and in the Caribbean. Since it has become apparent through our IBP research that humans hold a much greater responsibility for the survival of island biota and communities than was thought so far, it appears to be appropriate to invest research funds into an MAB 7 component at least equivalent to that of France (Ricard et al., 1977) and a few other nations with responsibilities in the Pacific island region.

References

Ahearn, J.N., H.L. Carson, T. Dobzhansky, and K.Y. Kaneshiro. 1974. Ethological isolation among three species of the *plantibia* subgroup of Hawaiian *Drosophila*. *Proc. Nat. Acad. Sci.* **71:** 901–903.

Ahearn, J.N., and F.C. Val. 1975. Fertile interspecific hybrids of two sympatric Hawaiian *Drosophila*. *Genetics* **80:** s9.

Allan, H.H. 1936. Indigene versus alien in the New Zealand plant world. *Ecology* **17:** 187–193.

Alvin, Paulo de T. 1960. Moisture stress as a requirement for flowering of coffee. *Science* **132:** 354.

Apple, R.A. 1954. A history of land acquisition for Hawaii National Park to December 31, 1950. M.A. thesis, University of Hawaii, Honolulu. 158 pp.

Ashton, G.C. 1965. Serum tranferrins: a balanced polymorphism? *Genetics* **52:** 983–997.

Atkinson, I.A.E. 1969. Ecosystem development on some Hawaiian lava flows. Ph.D. diss., University of Hawaii, Honolulu. 191 pp.

Atkinson, I.A.E. 1970. Successional trends in the coastal and lowland forest of Mauna Loa and Kilauea volcanoes, Hawaii. *Pacific Sci.* **24:** 387–400.

Aubreville, A. 1938. *La foret coloniale: les forets d'Afrique occidentale française*. Ann. Acad. Sci. Colon., Paris. 9 pp.

Avise, J.C. 1975. Systematic value of electrophoretic data. *Syst. Zool.* **23:** 465–481.

Avise, J.C., and F.J. Ayala. 1976. Genetic differentiation in speciose versus depauperate phylads: evidence from the California minnows. *Evolution* **30:** 46–58.

Ayala, F.J., J.R. Powell, and T. Dobzhansky. 1971. Polymorphism in continental and island populations of *Drosophila willistoni*. *Proc. Nat. Acad. Sci.* **68:** 2480–2483.

Ayala, F.J., J.R. Powell, M.L. Tracey, C.A. Mourão, and S. Perez-Salas. 1972. Enzyme variability in the *Drosophila willistoni* group. IV. Genetic variation in natural populations of *Drosophila willistoni*. *Genetics* **70:** 113–130.

Ayala F.J., M.L. Tracey, D. Hedgecock, and R.C. Richmond. 1974. Genetic differentiation during the speciation process in *Drosophila*. *Evolution* **28:** 576–592.

Baimai, V. 1975. Heterochromatin and multiple inversions in a *Drosophila* chromosome. *Can. J. Genet. Cytol.* **17:** 15–20.

Baker, H. G. 1965. Characteristics and modes of origin of weeds, pp. 147–168. In H. G. Baker and G. L. Stebbins, eds., *The Genetics of Colonizing Species,* Academic Press, New York.

Baker, H. G. 1973. Evolutionary relationships between flowering plants and animals in American and African tropical forests, pp. 145–159. In B. J. Meggers, E. S. Ayensu, and W. D. Duckworth eds., *Tropical Forest Ecosystems in Africa and South America: A Comparative Review,* Smithsonian Inst. Press, Washington, D.C.

Baker, J. R., and I. Baker 1936. The seasons in a tropical rainforest (New Hebrides). Part 2. *Botany J. Linn. Soc.* **39:** 507–519.

Baldwin, D. D. 1887. Land shells of the Hawaiian Islands, p. 55–63. In T. G. Thrum, comp., *Hawaiian Almanac and Annual for 1887,* Press Publishing Co., Honolulu.

Baldwin, P. H. 1953. Annual cycle, environment, and evolution in the Hawaiian honeycreepers (Aves: Drepaniidae). *Univ. Calif. Publ. Zool.* **54:** 285–398.

Baldwin, P. H., and G. O. Fagerlund. 1943. The effect of cattle grazing on koa reproduction in Hawaii National Park. *Ecology* **24:** 118–122.

Baldwin, P. H., C. W. Schwartz, and E. R. Schwartz. 1952. Life history and economic status of the mongoose in Hawaii. *J. Mammal.* **33:** 335–356.

Barr, T. C., Jr. 1968. Cave ecology and the evolution of troglobites. *Evol. Biol.* **2:** 35–102.

Barr, T. C., and R. A. Kuehne. 1971. Ecological studies in the Mammoth Cave system of Kentucky. II. The ecosystem. *Ann. de Speleologie* **26:** 47–96.

Barron, G. L. 1968. *The Genera of Hyphomycetes from Soil,* Williams and Wilkins, Baltimore, 364 pp.

Bartram, E. B. 1933. *Manual of Hawaiian Mosses.* B. P. Bishop Museum Bull. No. 101, Honolulu, 275 pp.

Bartram, E. B. 1939, *Supplement to manual of Hawaiian mosses.* B. P. Bishop Museum Occas. Papers No. 15, Honolulu, pp. 93–108.

Beals, E. 1960. Forest bird communities in the Apostle Islands of Wisconsin. *Wilson Bull.* **72:** 156–181.

Beals, E. W. 1965. Ordination of some corticolous cryptogamic communities in southcentral Wisconsin. *Oikos* **16:** 1–8.

Beard, J. S. 1946. The Mora forests of Trinidad, British West Indies. *J. Ecol.* **33:** 173–192.

Beardsley, J. W. 1962. On accidental immigration and establishment of terrestrial arthropods in Hawaii during recent years. *Proc. Haw. Entomol. Soc.* **18:** 99–110.

Beardsley, J. W. 1971. Checklist of Homoptera for the Island of Hawaii. Unpublished.

Beardsley, J. W. 1975. Note on *Psylla uncatoides. Proc. Haw. Entomol. Soc.* **22:**4.

Becker, R.E. 1976. The phytosociological position of tree ferns *(Cibotium* spp.) in the montane rain forests on the Island of Hawaii, Ph.D. diss., University of Hawaii, Honolulu, 368 pp.

Behre, K., and G.H. Schwabe. 1970. Auf Surtsey Island im Sommer 1968. Nachgewiesene nicht marine Algen, pp. 31–100. In *Schr. Naturw. Ver. Schlesw.-Holstein,* Verlag Lipsius und Tischer, Kiel.

Bellinger, P.F., and K.A. Christiansen. 1974. The cavernicolous fauna of Hawaiian lava tubes. 5. Collembola. *Pacif. Insects* **16:** 31–40.

Berger, A.J. 1969a. The breeding season of the Hawaii Amakihi. *B.P. Bishop Museum Occas. Papers 24,* Honolulu, pp. 1–8.

Berger, A.J. 1969b. Discovery of the nest of the Hawaiian Thrush. *Living Bird,* **8:**243–250.

Berger, A.J. 1970. The eggs and young of the Palila, an endangered species. *Condor* **72:** 238–240.

Berger, A.J. 1972. *Hawaiian Birdlife.* University Press of Hawaii, Honolulu, 270 pp.

Berger, A.J. 1975. The warbling silverbill, a new nesting bird in Hawaii. *Pacific Sci.* **29:** 51–54.

Berger, A.J., C.R. Eddinberg, and S.C. Frings. 1969. The nest and eggs of the Anianiau. *Auk* **86:** 183–187.

Berger, E.M. 1970. A comparison of gene-enzyme variation between *Drosophila melanogaster* and *D. simulans. Genetics* **66:** 667–683.

Berger, E.M. 1971. A temporal study of allelic variation in natural and laboratory populations of *Drosophila melanogaster. Genetics* **67:** 121–136.

Berry, R.J. 1970. The natural history of the house mouse. *Field Studies* **3:** 219–262.

Bliss, C.I. 1970. *Statistics in Biology,* vol. 2, McGraw-Hill, New York, 639 pp.

Blumenstock, D.I. 1961. *Climates of the States: Hawaii.* U.S. Weather Bureau, Climatography of the United States No. 60–51, U.S. Govt. Printing Office, Washington, D.C., 20 pp.

Blumenstock, D.I., and S. Price. 1967. *Climates of the States: Hawaii.* Climatography of the United States No. 60–51, U.S. Env. Data Service. U.S. Govt. Printing Office, Washington, D.C., 27 pp.

Borisov, A.I. 1969. The adaptive significance of chromosomal polymorphism. *Genetika* **5:** 119–122.

Bousfield, E.L., and F.G. Howarth. 1976. The cavernicolous fauna of Hawaiian lava tubes, 8. Terrestrial Amphipoda (Talitridae), including a new genus and species with notes on its biology. *Pacif. Insects* **17:** 144–154.

Bratton, S.P. 1974. The effect of the European wild boar (*Sus scrofa*) on the high-elevation vernal flora in Great Smoky Mountains National Park. *Bull. Torrey Bot. Club* **101:** 198–206.

Bratton, S.P. 1975. The effect of the European wild boar, *Sus scrofa,* on grey beech forest in the Great Smoky Mountains. *Ecology* **56:** 1356–1366.

Bray, J.R., and J.T. Curtis. 1957. An ordination of the upland forest communities of southern Wisconsin. *Ecol. Monogr.* **27:** 325–349.

Briggs, T.S. 1973. Troglobitic harvestmen recently discovered in North American lava tubes (Travuniidae, Erebomastridae, Triaenonychidae: Opiliones). *J. Arachnology* **1:** 205–214.

Bristol, B.M. 1919. On the retention of vitality by algae from old stored soils. *New Phytolog.* **18:** 29.

Britten, E.J. 1962. Hawaii as a natural laboratory for research on climate and plant response. *Pacific Sci.* **16:** 160–169.

Brncic, D. 1955. Chromosomal variation in Chilean populations of *Drosophila immigrans*. *J. Heredity* **46:** 59–63.

Brncic, D. 1972. Seasonal fluctuations of the inversion polymorphism in *Drosophila flavopilosa* and the relationships with ecological factors. *Univ. Texas Publ.* **7213:** 103–116.

Brown, R.M., Jr. 1971. Studies of Hawaiian fresh-water and soil algae. I. The atmospheric dispersal of algae and fern spores across the Island of Oahu, Hawaii, pp. 175–188. In B.C. Parker and R.M. Brown, Jr., eds.*Contributions in Phycology*. Published by students of H.C. Bold, Allen Press, Inc., Lawrence, Kansas.

Brown, R.M., Jr., D.A. Larson, and H.C. Bold. 1964. Airborne algae: their abundance and heterogeneity. *Science* **143:** 583–585.

Burgan, R.E. 1970. Study plan to detect growth rings in koa. U.S. Forest Serv. Pacific S.W. Forest and Range Exp. Sta. Berkeley, Calif. (on file, U.S. F.S. Inst. of Pacific Island Forestry. Honolulu, Hawaii).

Burger, W.C. 1974. Flowering periodicity at four altitudinal levels in eastern Ethiopia. *Biotropica* **6**(1): 38–42.

Butcher, J.W., R. Snider, and R.J. Snider. 1971. Bioecology of edaphic *Colembola* and *Acarina*. *Ann. Rev. Entomol.* **16:** 249–288.

Carlquist, S. 1965. *Island Life: A Natural History of the Islands of the World,* Natural History Press, Garden City, N.Y. 451 pp.

Carlquist, S. 1966a. The biota of long-distance dispersal. II. Loss of dispersibility in Pacific Compositae. *Evolution* **20:** 433–455.

Carlquist, S. 1966b. The biota of long-distance dispersal. III. Loss of dispersibility in the Hawaiian flora. *Brittonia* **18:** 310–335.

Carlquist, S. 1970. *Hawaii: A Natural History,* Natural History Press. Garden City, N.Y. 463 pp.

Carlquist, S. 1974. *IslandBiology*. Columbia University Press, New York, 660 pp.

Carpenter, F.L. 1976. Plant-pollinator interactions in Hawaii: pollination energetics of *Metrosideros collina* (Myrtaceae). *Ecology* **57:** 1125–1144.

Carpenter, F.L., and R.E. MacMillen, 1976a. Energetic cost of feeding territories in an Hawaiian honeycreeper. *Oecologia* **26:** 213–223.

Carpenter, F.L., and R.E. MacMillen. 1976b. Threshold model of feeding territoriality and a test with a Hawaiian honeycreeper. *Science* **194:** 639–642.

Carson, H.L. 1958. The population genetics of *Drosophila robusta*. *Adv. Genetics* **9:** 1–40.

Carson, H.L. 1959. Genetic conditions which promote or retard the formation of species. *Cold Spring Harb. Symp. Quant. Biol.* **20:** 276–287.

Carson, H.L. 1965. Chromosomal morphism in geographically widespread species of *Drosophila,* pp. 503–531. In H.G. Baker and G.L. Stebbins eds., *The Genetics of Colonizing Species,* Academic Press, New York.

Carson, H.L. 1966. Chromosomal races of Oahu and Kauai, State of Hawaii. *Univ. Texas Publ.* **6615:** 405–412.

Carson H.L. 1970. Chromosome tracers of the origin of species. *Science* **168:** 1414–1418.

Carson, H.L. 1971. Speciation and the founder principle. *Univ. Missouri Stadler Symp.,* Columbia, Missouri, **3:** 51–70.

Carson, H.L. 1974. Three flies and three islands: parallel evolution in *Drosophila*. *Proc. Nat. Acad. Sci.* **71:** 3517–3521.

Carson, H.L. 1975. The genetics of speciation at the diploid level. *Amer. Nat.* **109:** 83–92.

Carson, H.L. 1976. Inference of the time of origin of some *Drosophila* species. *Nature* **259:** 395–396.

Carson, H.L., D.E. Hardy, H.T. Spieth, and W.S. Stone. 1970. The evolutionary biology of the Hawaiian Drosophilidae, pp. 437–543. In M.K. Hecht and W.C. Steere, eds., *Essays in Evolution and Genetics in Honor of Theodosius Dobzhansky,* Appleton-Century-Crofts, New York.

Carson, H.L., and W.E. Johnson. 1975. Genetic variation in Hawaiian *Drosophila*. I. Chromosome and allozyme polymorphism in *D. setosimentum* and *D. ochrobasis* from the Island of Hawaii. *Evolution* **29:** 11–23.

Carson, H.L., W.E. Johnson, P.S. Nair, and F.M. Sene. 1975a. Allozymic and chromosomal similarity in two *Drosophila* species. *Proc. Nat. Acad. Sci.* **72:** 4521–4525.

Carson, H.L., P.S. Nair, and F.M. Sene, 1975b. *Drosophila* hybrids in nature: proof of gene exchange between sympatric species. *Science* **189:** 806–807.

Carson, H.L., and K.Y. Kaneshiro. 1976. *Drosophila* of Hawaii: systematics and ecological genetics. *Ann. Rev. Ecol. Syst.* **7:** 311–345.

Carson, H.L., and J.E. Sato. 1969. Microevolution within three species of Hawaiian *Drosophila*. *Evolution* **23:** 493–501.

Carson, H.L., and H.D. Stalker. 1949. Seasonal variation in gene arrangement frequencies over a three year period in *Drosophila robusta* Sturtevant. *Evolution* **3:** 322–329.

Carson, H.L., and H.D. Stalker. 1968. Polytene chromosome relationships in Hawaiian species of *Drosophila*. II. The *D. planitibia* subgroup. *Univ. Texas Publ.* **6818:** 335–365.

Cattell, R.B. 1952. *Factor Analysis,* Harper and Bros., New York, 462 pp.

Ceska, A., and H. Roemer. 1971. A computer program for identifying species-relevé groups in vegetation studies. *Vegetatio* **23**(3/4): 255–277.

Chapman, J. D., and F. White. 1970. *The Evergreen Forests of Malawi*. Univ. Oxford, Commonwealth Forestry Inst., 190 pp.

Clark, G., A. Austring, and J.O. Juvik. 1975. The role of intercepted fog moisture in the water balance of forest ecosystems on windward and leeward Mauna Loa, Hawaii. Paper presented at Annual Meeting Assoc. Pacific Coast Geographers, Fresno, Calif., June 1975.

Clarke, F.E. 1949. Soil microorganisms and plant roots. *Advances Agron.* **1**:241–288.

Clayton, F.E. 1969. Variations in metaphase chromosomes of Hawaiian Drosophilidae. *Univ. Texas Publ.* **6918**:95–110.

Clayton, F.E. 1971. Additional karyotypes of Hawaiian Drosophilidae. *Univ. Texas Publ.* **7103**:171–181.

Clayton, F.E., and M.R. Wheeler. 1975. A catalog of *Drosophila* metaphase chromosome configurations, vol. 3, pp. 471–512. In R.C. King, ed., *Handbook of Genetics,* Plenum Press, New York.

Cline, M.G., ed. 1955. *Soil Survey of the Territory of Hawaii: Islands of Hawaii, Kauai, Lanai, Maui, Molokai, and Oahu*. U.S. Dept. Agri. in cooperation with Hawaii Agri. Exp. Sta. Soil Survey Ser. 1939, No. 25, 644 pp. + maps.

Cole, G. W., ed. 1976. *ELM:Version 2.O*. Range Science Dept. Sci. Ser. No. 20, Colorado State Univ. 662 pp.

Conant, S. 1977. The breeding biology of the Oahu 'Elepaio. *Wilson Bull.* **89**:193–210.

Cooke, C.M., and Y. Kondo. 1960. *Revision of Tornatellinidae and Achatinellidae (Gastropoda, Pulmonata)*. B.P. Bishop Museum Bull. No. 221, 303 pp.

Corn, C.A. 1972. Seed dispersal methods in Hawaiian *Metrosideros,* pp. 422–435. In J. A. Behnke, ed., *Challenging biological Problems: Directions toward Their Solution,* Oxford University Press, New York, 502 pp.

Corn, C.A., and W.M. Hiesey. 1973. Altitudinal variation in Hawaiian *Metrosideros. Amer. J. Bot.* **60**:991–1002.

Coster, C. 1923. Lauberneuerung und andere periodische Lebensprozesse in dem trockenen Monsun-Gebiet Ost-Javas. *Ann. Jard. Bot. Buitenzorg* **33**:117–189.

Coster, C. 1925. Die Fettumwandlung im Baumkörper in den Tropen. *Ann. Jard. Bot. Buitenzorg* **35**:75–104.

Coster, C. 1926. Periodische Blüteerscheinungen in den Tropen. *Ann. Jard. Bot. Buitenzorg* **35**:125–162.

Coster, C. 1927. Zur Anatomie und Physiologie der Zuwachszonen und Jahresringbildung in den Tropen. *Ann. Jard. Bot. Buitenzorg* **37**:49–160.

Coster, C. 1928. Zur Anatomie und Physiologie der Zuwachszonen und Jahresringbildung in den Tropen (cont.). *Ann. Jard. Bot. Buitenzorg* **38**:1–114.

Craddock, E.M. 1974. Reproductive relationships between homosequential species of Hawaiian *Drosophila. Evolution* **28**:593–606.

Craddock, E. M., and H. L. Carson. 1975. Chromosome variability in an endemic Hawaiian *Drosophila* species. *Genetics* **80:**s23.

Craddock. E.M. and W.E. Johnson. 1979. Genetic variation in Hawaiian *Drosophila*. V. Chromosomal and allozymic diversity in *Drosophila silvestris* and its homosequential species. *Evolution* **33:**137–155.

Creed, R. 1971. *Ecological Genetics and Evolution. Essays in Honour of E.B. Ford.* Appleton-Century-Crofts, New York, 391 pp.

Crumpacker, D. W., and J. S. Williams. 1974. Rigid and flexible chromosomal polymorphisms in neighboring populations of *Drosophila pseudoobscura*. *Evolution* **28:**57–66

Cuddihy, L. W. 1978. Effects of cattle grazing on the mountain parkland ecosystem, Mauna Loa, Hawaii. M.S. thesis, University of Hawaii, Honolulu, 198 pp.

Culver, D. C. 1970. Analysis of simple cave communities. I. Caves as islands. *Evolution* **24:**463–473.

da Cunha, A. B. 1955. Chromosomal polymorphism in the Diptera. *Adv. Genetics* **7:**93–138.

da Cunha, A. B., H. Burla, and T. Dobzhansky. 1950. Adaptive chromosomal polymorphisms in *Drosophila willistoni*. *Evolution* **4:**212–235.

da Cunha, A. B., T. Dobzhansky, D. Pavlovsky, and B. Spassky. 1959. Genetics of natural populations. XXVIII. Supplementary data on the chromosomal polymorphism in *Drosophila willistoni* in its relation to its environment. *Evolution* **13:**389–404.

Darwin, C. R. 1859. *On the Origin of the Species by Natural Selection, or the Preservation of Favoured races in the Struggle for Life,* John Murray, London, 251 pp.

Daubenmire, R.F. 1966. Vegetation: identification of typical communities. *Science* **151:**291–298.

Daubenmire, R. F. 1968. *Plant Communities: A Textbook of Plant Synecology,* Harper and Row, New York, 300 pp.

Daubenmire, R. F. 1972. Phenology and other characteristics of tropical semideciduous forest in northwestern Costa Rica. *J. Ecol.* **60:**147–170.

Davis, C. J., and K. Kawamura. 1970. Note on *Psylla uncatoides*. *Hawaii Coop. Economic Insect Rpt.,* July 31, 2 pp.

Davis, T. A. W., and P. W. Richards. 1933. The vegetation of Moraballi Creek, British Guiana: an ecological study of a limited area of tropical rain forest. Part I. *J. Ecol.* **21:**350–384.

Degener, O. 1930. *Plants of Hawaii National Park,* Edward Bros., Ann Arbor, Mich., 314 pp.

Degener, O. 1932–present. *Flora Hawaiiensis*. Books 1–6. Published by the author in form of leaflets for binding. Volcano, Hawaii.

Diamond, J. M., and J. W. Terborgh. 1967. Observations on bird distribution and feeding assemblages along the Rio Callaria, Department of Loreto, Peru. *Wilson Bull.* **79:**273–282.

Dobzhansky, T. 1933. Geographical variation in lady-beetles. *Amer. Nat.* **67:**97–126.

Dobzhansky, T. 1943. Genetics of natural populations. IX. Temporal changes in the composition of populations of *Drosophila pseudoobscura*. *Genetics* **28:**162–186.

Dobzhansky, T. 1948. Genetics of natural populations. XVI. Altitudinal and seasonal changes produced by natural selection in certain populations of *Drosophila pseudoobscura* and *Drosophila persimilis* in some locations in California. *Genetics* **33:**158–176.

Dobzhansky, T. 1952. Genetics of natural populations. XX. Changes induced by drought in *Drosophila pseudoobscura* and *D. persimilis*. *Evolution* **6:**234–243.

Dobzhansky, T. 1956. Genetics of natural populations. XXV. Genetic changes in populations of *Drosophila pseudoobscura* and *Drosophila persimilis* in some locations in California. *Evolution* **10:**82–92.

Dobzhansky, T. 1957. Genetics of natural populations. XXVI. Chromosomal variability in island and continental populations of *Drosophila willistoni* from Central America and the West Indies. *Evolution* **11:**280–293.

Dobzhansky, T. 1962. Rigid vs. flexible chromosomal polymorphisms in *Drosophila*. *Amer. Nat.* **96:**321–328.

Dobzhansky, T. 1970. *Genetics of the Evolutionary Process,* Columbia University Press, New York, 505 pp.

Dobzhansky, T., and F. J. Ayala. 1973. Temporal frequency changes of enzyme and chromosomal polymorphisms in natural populations of *Drosophila*. *Proc. Nat. Acad. Sci.* **70:**680–683.

Domsch, K. H., and W. Gams. 1972. *Fungi in Agricultural Soils,* John Wiley, New York, 290 pp.

Dorst, J. 1972. Parks and resources on islands. In Sir H. Elliot, ed., *Second World Conference on National Parks*. Natl. Parks Centennial Comm. Intl. Union for Conservation of Nature. Morges, Switzerland, 12 pp.

Doty, M. S. 1954. Floristic and ecological notes on Roroia. *Atoll Res. Bull.* **33:**1–41.

Doty, M. S. 1967. Contrast between the pioneer populating process on land and shore. *Bull. S. Calif. Acad. Sci.* **66:**175–194.

Doty, M. S., F. R. Fosberg, and D. Mueller-Dombois. 1969. Initial site studies for the International Biological Program in the tropical and far western Pacific. *Micronesica* **5**(2):283–293.

Doty, M. S., and D. Mueller-Dombois. 1966. *Atlas for Bioecology Studies in Hawaii Volcanoes National Park,* University of Hawaii, Hawaii Bot. Sci. Paper No. 2, 507 pp.

Dubinin, N. P., and G. G. Tiniakov. 1945. Seasonal cycles and the concentration of inversions in populations of *Drosophila funebris*. Amer. Nat. **79:**570–572.

Eddinger, C.R. 1970. A study of the breeding behavior of four species of Hawaiian honeycreepers (Drepanididae). Ph.D. diss., University of Hawaii, Honolulu, 212 pp.

Eggler, W. A. 1971. Quantitative studies of vegetation on sixteen young lava flows on the island of Hawaii. *Tropical Ecology* **12**(1):66–100.

Egler, F.E. 1939. Vegetation zones of Oahu, Hawaii. *Empire Forestry J.* **18:**44–57.

Egler, F.E. 1942. Indigene vs. alien in the development of arid Hawaiian vegetation. *Ecology* **23:**14–23.

Egler, F.E. 1947. Arid southeast Oahu vegetation, Hawaii. *Ecol. Monogr.* **17:**383–435.

Eisenberg, J.F., and M. Lockhart. 1972. *An Ecological Reconnaissance of Wilpattu National Park, Ceylon.* Smithsonian Contr. to Zool., No. 101, Smithsonian Inst. Press, Washington, D.C., 118 pp.

Eisenberg, J.F., C. Santiapillai, and M. Lockhart. 1970. The study of wildlife populations by indirect methods. *Ceylon J. Sci. (Biol. Sci.)* **8:**53–62.

Ellenberg, H. 1956. *Aufgaben und Methoden der Vegetationskunde,* Eugen Ulmer, Stuttgart, 136 pp.

Ellenberg, H., and D. Mueller-Dombois. 1967. A key to Raunkiaer plant life forms with revised subdivisions, *Ber. Geobot. Inst. ETH, Stiftg. Rübel, Zürich* **37:**56–73.

Elton, C. 1966. *Animal Ecology.* First publ. 1927, University of Washington Press, Seattle, 237 pp.

Elton, C.S. 1973. The structure of invertebrate populations inside neotropical rainforest. *J. Anim. Ecol.* **42:**55–104.

Emlen, J.T. 1971. Population densities of birds derived from transect counts. *Auk* **88:**323–341.

Epling, C., D.F., Mitchell, and R.H.T. Mattoni. 1953. On the role of inversions in wild populations of *Drosophila pseudoobscura. Evolution* **7:**342–365.

Epling, C., D.F. Mitchell, and R.H.T. Mattoni. 1957. The relation of an inversion system to recombination in wild populations. *Evolution* **11:**225–247.

Eskey, C.R. 1934. Epidemiological study of plague in the Hawaiian Islands. *Public Health Bull.* **213:**1–70.

Evans, L. 1971. Flower induction and the florigen concept. *Ann. Rev. Pl. Physiol.* **22:**365–394.

Fagerlund, G.O. 1947. *The Exotic Plants of Hawaii National Park,* Natural History Bull. No. 10, Hilo, Hawaii, 62 pp.

Fain, A. 1970. Diagnose de nouveaux Lobalgides et Listrophorides (Acarina:Sarcoptiformes). *Rev. Zool. Bot. Afr.* **81:**271–300.

Fennah, R.G. 1973a. The cavernicolous fauna of Hawaiian lava tubes. 4. Two new blind *Oliarus* (Fulgoroidea:Cixiidae). *Pacif. Insects* **15:**181–184.

Fennah, R.G. 1973b. Three new cavernicolous species of Fulgoroidea (Homoptera) from Mexico and western Australia. *Proc. Biol. Soc. Wash.* **86**(38):439–446.

Fichter, E. 1954. An ecological study of invertebrates of grassland and deciduous shrub savanna in eastern Nebraska. *Am. Midl. Nat.* **51:**321–439.

First International Congress of Ecology. 1974. *Unifying Concepts in Ecology: Report of the Plenary Sessions of the First International Congress of Ecology,* W.H. van Dobben and R.H. Lowe-McConnell, eds., Dr. W. Junk Publ., the Hague, 302 pp.

Foote, D.E. et al. 1972. *Soil Survey of the Islands of Kauai, Oahu, Maui, Molokai, and Lanai, State of Hawaii.* U.S. Soil Cons. Serv. in cooperation with Univ. Hawaii Agri. Exp. Sta., U.S. Govt. Printing Office, Washington, D.C., 232 pp.

Ford, E.B. 1931. *Mendelism and Evolution.* Methuen, London. 122 pp.

Ford, E.B. 1971. *Ecological Genetics,* 3d ed., Chapman and Hall, London. 442 pp.

Fosberg, F.R. 1948. Derivation of the flora of the Hawaiian Islands, vol. 1, pp. 107–110. In E.C. Zimmerman, ed., *Insects of Hawaii,* University of Hawaii Press, Honolulu.

Fosberg, F.R. 1961. *Guide to Excursion III: Tenth Pacific Science Congress.* Publ. jointly by Tenth Pac. Sci. Cong. and Univ. Hawaii, Honolulu. 207 pp.

Fosberg, F.R. 1963. The island ecosystem, pp. 1–6. In F.R. Fosberg, ed., *Man's Place in the Island Ecosystem: A Symposium.* Tenth Pacific Sci. Cong., Honolulu, Hawaii, 1961, B.P. Bishop Museum, Honolulu.

Fosberg, F.R. 1966. The oceanic volcanic island ecosystem, pp. 55–61. In R.I. Bowman, ed., *Gálapagos International Scientific Project,* University of California Press, Berkeley, 318 pp.

Fosberg, F.R. 1967. Opening remarks: Island Ecosystem Symposium, presented at XI Pacific Science Congress, Tokyo. *Micronesica* **3:**3–4.

Frankie, G.W., H.G. Baker, and P.A. Opler. 1974a. Tropical plant phenology: applications for studies in community ecology, pp. 287–296. In H. Lieth, ed., *Phenology and Seasonality Modeling,* Springer-Verlag, Berlin.

Frankie, G.W., H.G. Baker, and P.A. Opler, 1974b. Comparative phenological studies of trees in tropical wet and dry forests in the lowlands of Costa Rica. *J. Ecol.* **62:**881–919.

Freire-Maia, N., I.F. Zanardini, and A. Freire-Maia. 1953. Chromosome variation in *Drosophila immigrans. Dusenia* **4:**303–311.

Frenkel, R.E., and C.M. Harrison. 1974. An assessment of the usefulness of phytosociological and numerical classificatory methods for the community geographer. *J. Biogeog.* **1:**27–56.

Fretwell, S.D. 1972. *Populations in a Seasonal Environment,* Princeton University Press, Princeton, N.J., 217 pp.

Funasaki, G. 1968a. Notes and exhibitions. *Proc. Haw. Entomol. Soc.* **20:**8.

Funasaki, G. 1968b. Notes and exhibitions. *Proc. Haw. Entomol. Soc.* **20:**12.

Funasaki, G. 1968c. Notes and exhibitions. *Proc. Haw. Entomol. Soc.* **20:** 13–14.

Gagné, W. C. 1971. Notes and exhibitions. *Proc. Haw. Entomol. Soc.* **21:**25.

Gagné, W. C., and F. G. Howarth. 1975a. The cavernicolous fauna of Hawaiian lava tubes, 6. Mesoveliidae or water treaders (Heteroptera). *Pacif. Insects* **16:**399–413.

Gagné, W. C., and F. G. Howarth. 1975b. The cavernicolous fauna of Hawaiian lava tubes, 7. Emesinae or thread-legged bugs (Heteroptera: Reduviidae). *Pacif. Insects* **16:**415–426.

Gams, H. 1918. Prinzipienfragen der Vegetationsforschung. Ein Beitrag zur Begriffsklärung und Methodik der Biocoenologie. *Vierteljahrsschr. Naturforsch. Ges.* Zürich **63:**293–493.

Garrett, L. E., and F. H. Haramoto. 1967. A catalogue of Hawaiian Acarina. *Proc. Haw. Entomol. Soc.* **19:**381–414.

Gentry, A. H. 1974. Flowering phenology and diversity in tropical Bignoniaceae. *Biotropica* **6:**(1):64–68.

Gerdes, R. J. 1964. *History of the Feral Goat Control Program of Hawaii Volcanoes National Park,* Hawaii Natl. Park Headquarters, 17 pp. (mimeogr.).

Gerrish, G. 1978. The relationship of native and exotic plant species in two rain forest communities in the Koolau Mountains, Oahu, Hawaii. M.S. thesis, University of Hawaii, Honolulu, 158 pp.

Gertsch, W. J. 1973. The cavernicolous fauna of Hawaiian lava tubes, 3. Araneae (Spiders). *Pacif. Insects* **15:**163–180.

Goldstein, R. A., and D. F. Grigal, 1971. *Computer Program for the Ordination and Classification of Ecosystems.* Ecol. Sci. Div., Oak Ridge Natl. Lab., Tenn., Publ. No. 417, 125 pp.

Goodall, D. W. 1954. Vegetational classification and vegetational continua. *Angew. Pflanzensoziologie (Vienna), Festschrift Aichinger* **1:**168–182.

Graebner, P. 1895. Studien über die norddeutsche Heide. *Engler's Botan. Jahrb.* **21:**500.

Greig-Smith, P. 1964. *Quantitative Plant Ecology,* 2d ed., Butterworths, London, 256 pp.

Gressitt, J. L., and M. K. Gressitt. 1962. An improved malaise trap. *Pacif. Insects* **4:**87–90.

Gressitt, J. L., and C. M. Yoshimoto. 1964. Dispersal of animals in the Pacific, pp. 283–292. In J. L. Gressitt, ed., *Pacific Basin Biogeography: A Symposium.* B. P. Bishop Museum Press, Honolulu.

Giffin, J. G. 1972. *Statewide Pittman-Robertson Program,* Hawaii Div. of Fish and Game, Proj. No. W-15-1, 10 pp.

Gruber, F. 1958. Frequency of heterozygous inversions in a natural population of *Drosophila immigrans* from Israel. *Drosophila Info. Service* **32:**124.

Haas, G.E. 1969. Quantitative relationships between fleas and rodents in a Hawaiian cane field. *Pacific Sci.* **23:**70–82.

Haas, G.E., N. Wilson, and P.Q. Tomich. 1972. Ectoparasites of the Hawaiian Islands. I. Siphonaptera. *Contr. Am. Entomol. Inst.* **8:**1–76.

Haldane, J.B.S. 1957. The cost of natural selection. *J. Genetics* **55:**511–524.

Haldane, J.B.S., and S.D. Jayakar. 1963. Polymorphism due to selection of varying direction. *J. Genetics* **58:**237–242.

Hammer, M. 1972. Tahiti. Investigation on the oribatid fauna of Tahiti and on some oribatids found on the atoll Rangiroa. *Biol. Skr. Dan. Vid. Selsk.* **19:**1–65.

Hammer, M. 1973. Oribatids from Tongatapu and Eua, the Tonga Islands, and from Upolu, Western Samoa. *Biol. Skr. Dan. Vid. Selsk.* **20:**1–70.

Hardy, D.E. 1960. *Insects of Hawaii.* Vol. 10. *Diptera: Nematocera-Brachycera,* University of Hawaii Press, Honolulu, 368 pp.

Hardy, D.E. 1964. *Insects of Hawaii.* Vol. 11. *Diptera: Brachycera, Family Dolichopodidae; Cyclorrhapha, Series Aschiza; Families Lonchopteridae, Phoridae, Pipunculidae, and Syrphidae,* University of Hawaii Press, Honolulu, 458 pp.

Hardy, D.E. 1965. *Insects of Hawaii.* Vol. 12. *Diptera: Cyclorrhapha II, Series Schizophora, Section Acalypterae I, Family Drosophilidae,* University of Hawaii Press, Honolulu, 814 pp.

Hardy, D.E. 1974. Evolution in the Hawaiian Drosophilidae: introduction and background information, pp. 71–80. In M.J.D. White, ed., *Genetic Mechanisms of Speciation in Insects,* Australia and New Zealand Book Co., Sydney.

Harris, D.R. 1962. The invasion of oceanic islands by alien plants: an example from the leeward islands West Indies, pp. 67–82. *Trans. and Papers, Inst. of British Geographers, No. 31.*

Hatheway, W.G. 1952. Composition of certain native dry forests: Mokuleia, Oahu, T.H. *Ecol. Monogr.* **22:**153–168.

Hawaii Audubon Society. 1978. *Hawaii's Birds,* 2d ed., Hawaii Audubon Soc., Honolulu, 96 pp.

Hedberg, O. 1970. Evolution of the Afroalpine flora. *Biotropica* **2:**16–23.

Heed, W.B. 1968. Ecology of the Hawaiian Drosophilidae. *Studies in Genetics IV. Univ. Texas Publ.* **6818:**387–419.

Heed, W.B. 1971. Host plant specificity and speciation in Hawaiian *Drosophila. Taxon* **20:**115–121.

Heinrich, B., and P.H. Raven. 1972. Energetics and pollination ecology. *Science* **176:**597–602.

Hespenheide, H. 1966. The selection of seed size by finches. *Wilson Bull.* **78:**191–197.

Hillebrand, W. 1888. Die Vegetationsformationen der Sandwich Inseln. *Bot. Jahrb.* **9:**305–314.

Hillebrand, W. F. 1956. *Flora of the Hawaiian Islands: A Description of Their Phanerogams and Vascular Cryptogams,* Hafner Publ. Co., New York, 673 pp.

Hirumi, H. 1961. Studies on the chromosomal polymorphism in natural populations of *Drosophila immigrans* in Southern-Central districts of Japan. *Jap. J. Genetics* **36:**297–305.

Hoe, W. J. 1974. Annotated checklist of Hawaiian mosses. *Lyonia* **1:**1–45.

Holdridge, L. R. 1970. A procedure for representing the structure of tropical forest associations, pp. 147–150. In H. T. Odum, ed., *A Tropical Rain Forest: A Study of Irradiation Ecology at El Verde, Puerto Rico.* U.S. Atomic Energy comm., Tennessee.

Holdridge, L. R., W. C. Grenke, W. H. Hatheway, T. Liang, and J. A. Tosi, Jr. 1971. *Forest Environments in Tropical Life Zones: A Pilot Study,* Pergamon Press, New York, 747 pp.

Holthuis, L. B. 1974. Caridean shrimps found in land-locked saltwater pools at four Indo-west Pacific localities (Sinai Peninsula, Funafuti Atoll, Maui and Hawaii Islands), with the description of one new genus and four new species. *Zool. Verhandel.* **128:**1–48.

Holttum, R. E. 1931. On periodic leaf-change and flowering of trees in Singapore. *Garden's Bull.* (Singapore) **5:**173–206.

Hooker, J. D. 1867. Insular floras. *Gardeners' Chronicle and Agric. Gazette* **27:**6–7, 50–51, 75–76.

Horn, H. S. 1971. *The Adaptive Geometry of Trees.* Princeton Univ. Press, Princeton, N.J., 144 pp.

Howarth, F. G. 1972. Cavernicoles in lava tubes on the island of Hawaii. *Science* **175:**325–326.

Howarth, F. G. 1973. The cavernicolous fauna of Hawaiian lava tubes, 1. Introduction. *Pacif. Insects* **15:**139–151.

Hubbell, T. H. 1968. The biology of islands. *Proc. Nat. Acad. Sci.* **60:**22–32.

Hubbell, T. H. 1972. The biology of islands, pp. 385–395, in E. A. Kay, ed., *A Natural History of the Hawaiian Islands,* University of Hawaii Press, Honolulu.

Hubby, J. L., and R. C. Lewontin. 1966. A molecular approach to the study of genic heterozygosity in natural populations. 1. The number of alleles at different loci in *Drosophila pseudoobscura. Genetics* **54:**577–594.

Hwang, S. H. 1977. *Ohia Decline: The Role of Phytophthora cinnamomi,* CPSU Univ. Hawaii Bot. Dept. Tech. Rpt. 12, 71 pp.

Hyatt, A. 1912–1914. Genealogy and migrations of the Achatinellidae in the Hawaiian Islands, pp. 370–399. In H. A. Pilsbry and C. M. Cooke, eds., *Manual of Conchology: Structure and Systematics,* vol. 22, 2d. ser., Pulmonata, Conchological Dept. Acad. Natl. Sci., Philadelphia.

Jaccard, P. 1901. Ētude comparative de la distribution florale dans une portion des Alpes et du Jura. *Bull. Soc. Vaud. Sci. Nat.* **37:**547–579.

Jackson, M. T. 1966. Effects of microclimate on spring flowering phenology. *Ecology* **47**:407–415.

Jackson, W. B. 1962. Sex ratios. In T. I. Storer, ed., *Pacific Island Rat Ecology,* B. P. Bishop Museum Bull. 225.

Janzen, D. H. 1967. Synchronization of sexual reproduction of trees within the dry season in Central America. *Evolution* **21**:620–637.

Janzen, D. H. 1973. Sweep samples of tropical foliage insects: effects of seasons, vegetation types, elevation, time of day and insularity. *Amer. Nat.* **54**:687–708.

Janzen, D. H. 1977. Why are there so many species of insects? *Proc. XV Int. Cong. Entomol.* **1976**:84–94.

Janzen, D. H., M. Ateroff, M. Fariñas, S. Reyes, N. Rincon, A. Soler, P. Soriano, and M. Vera. 1976. Changes in the arthropod community along an elevational transect in the Venezuelan Andes. *Biotropica* **8**:193–203.

Johnson, D. H. 1962. Rodents and other Micronesian mammals collected, pp. 21–38. In T. I. Storer, ed., *Pacific Island Rat Ecology,* B. P. Bishop Museum Bull. 255.

Johnson, F. M. 1971. Isozyme polymorphisms in *Drosophila ananassae:* genetic diversity among island populations in the South Pacific. *Genetics* **68**:77–95.

Johnson, F. M., and H. E. Schaffer. 1973. Isozyme variability in species of the genus *Drosophila*. VIII. Genotype-environment relationships in populations of *D. melanogaster* in the eastern United States. *Biochem. Genetics* **10**:149–163.

Johnson, W. E., and H. L. Carson. 1975. Allozymic variation in *Drosophila silvestris*. *Genetics* **80**:s46.

Johnson, W. E., H. L. Carson, K. Y. Kaneshiro, W. W. M. Steiner, and M. M. Cooper. 1975. Genetic variation in Hawaiian *Drosophila*. II. Allozymic differentiation in the *D. planitibia* subgroup, pp. 563–584. In C. L. Markert, ed., *Isozymes,* vol. 4, *Genetics and Evolution,* Academic Press, New York, 965 pp.

Jones, A. E. 1943. Classification of lava surfaces. *Amer. Geophys. Union. Trans. 1943*. **1**:265–268.

Jones, E. W. 1950. Some aspects of natural regeneration in Benin rain forest. *Empire For. Rev.* **29**:108–125.

Joyce, C. R. 1967. Notes and exhibitions. *Proc. Haw. Entomol. Soc.* **19**:334.

Judd, C. S. 1920. The koa tree. *Hawaii Forestry and Agri.* **17**:30–35.

Julliet, J. A. 1963. A comparison of four types of traps used for capturing flying insects. *Can. J. Zool.* **41**:219–223.

Juvik, J. O., and P. C. Ekern. 1978. *A Climatology of Mountain Fog on Mauna Loa, Hawai'i Island,* Water Resources Research Center, Univ. of Hawaii Tech. Rpt. No. 118, 63 pp.

Kami, H. T. 1964. Foods of the mongoose in Hamakua District, Hawaii. *Zoonoses Research* **3**:165–170.

Kaneshiro, K. Y. 1969. The *Drosophila crassifemur* group of species in a new subgenus. *Univ. Texas Publ.* **6918:**79–83.

Kaneshiro, K. Y., and F. C. Val. 1977. Natural hybridization between a sympatric pair of Hawaiian *Drosophila*. *Amer. Nat.* **111:**897–902.

Karr, J. R., and F. C. James. 1975. Ecomorphological configurations and convergent evolution in species and communities, pp. 258–291. In M. L. Cody and J. M. Diamond, eds., *Ecology and Evolution of Communities,* Belknap Press of Harvard University Press, Cambridge, Mass.

Kartawinata, K., and D. Mueller-Dombois. 1972. Phytosociology and ecology of the natural dry-grass communities on Oahu, Hawaii. *Reinwardtia* (Bogor) **8:**369–494.

Kartman, L., and R. P. Lonergan. 1955. Observations on rats in an enzootic plague region in Hawaii. *Publ. Health Rpt.* **70:**585–593.

Katnelson, H. 1965. Nature and importance of the rhizosphere, pp. 187–209. In K. F. Baker and W. C. Snyder, eds., *Ecology of Soil-borne Plant Pathogens, Prelude to Biological Control,* University of California Press, Berkeley.

Kennedy, M. V. 1974. Survival and development of *Bradysia impatiens* (Diptera: Sciaridae) on fungal and nonfungal food sources. *Ann. Entomol. Soc. Amer.* **67:**745–749.

King, M. C., and A. C. Wilson. 1975. Evolution at two levels in humans and chimpanzees. *Science* **188:**107–116.

Knapp, R. 1965. *Die Vegetation von Nord- und Mittelamerika und der Hawaii-Inseln,* Fischer, Stuttgart, 373 pp.

Koelmeyer, K. O. 1959. The periodicity of leaf change and flowering in the principal forest communities of Ceylon. *Ceylon Forester* **4:**157–189.

Koelmeyer, K. O. 1960. The periodicity of leaf change and flowering in the principal forest communities of Ceylon. *Ceylon Forester* **4:**308–364.

Kojima, K., J. Gillespie, and Y. N. Tobari. 1970. A profile of *Drosophila* enzymes assayed by electrophoresis. I. Number of alleles, heterozygosities and linkage disequilibrium in glucose-metabolizing systems and some other enzymes. *Bioch. Genetics* **4:**627–637.

Koriba, K. 1958. On the periodicity of tree growth in the tropics, with reference to the mode of branching, leaf-fall, and the formation of the resting bud. *Gardens' Bull. (Singapore)* **17**(1):11–81.

Krajina, V. J. 1963. Biogeoclimatic zones of the Hawaiian Islands. *Newsletter of Hawaiian Bot. Soc.* **2:**93–98.

Krauss, B. 1978. *Honolulu Advertiser,* October 21, A-3.

Krimbas, C. 1967. The genetics of *Drosophila subobscura* populations. III. Inversion polymorphism and climatic factors. *Molec. and Gen. Genetics* **99:**133–150.

Kullback, S., M. Kupperman, and H. H. Ku. 1962. Tests for contingency tables and Markov chains. *Technometrics* **4:**573–608.

Lakovaara, S., S. Saura, and C. T. Falk. 1972. Genetic distance and evolutionary relationships in the *Drosophila obscura* group. *Evolution* **26:**177–184.

Lamb, S. H. 1938. *Wildlife problems in Hawaii National Park.* Class actions of the North Amer. Wildlife Conf. Wildlife Mgt. Inst., Washington, D.C. **3:**597–602.

Lanner, R.M. 1965. *Phenology of* Acacia koa *on Mauna Loa, Hawaii,* U.S. Forest Service Res. Note PSW-89, 10 pp.

Lanner, R.M. 1966. *The Phenology and Growth Habits of Pines in Hawaii,* Pacific S.W. Forest and Range Exp. Sta., U.S. Forest Service Res. Paper PSW-29, 25 pp.

Laurie, E. M. O. 1946. The reproduction of the house mouse *(Mus musculus)* living in different environments. *Proc. Royal Soc.* **133:**248–281.

Lee, K. W., and H. D. Bold. 1974. Phycological Studies. XII. *Characium* and some *Characium*-like Algae. *Univ. Texas. Publ.* **7403:**1–127.

Leleup, N. 1967. Existence d'une fauna cryptique relictuelle aux îles Galapagos. *Noticias de Galapagos* (1965) **5/6:**4–16.

Leleup, N. 1968. Introduction. Mission zoologique belge aux îles Galapagos et en Ecuador (N. et J. Leleup, 1964–65) Résultats scientifiques. *Koninkliji Museum voor Midden-Africa = Musée Royal de L'Afrique Centrale.* **1:**9–34.

Levene, H. 1953. Genetic equilibrium when more than one ecological niche is available. *Amer. Nat.* **87:**331–333.

Levitan, M. 1957. Natural selection for linked gene arrangements. *Anat. Rec.* **127:**430.

Lewontin, R. C. 1974. *The Genetic Basis of Evolutionary Change.* Columbia University Press, New York. 346 pp.

Lieth, H. 1974. *Purposes of a Phenology Book,* pp. 3–19. In H. Lieth, ed., *Phenology and Seasonality Modeling,* Springer-Verlag, New York, 444 pp.

Lippmaa, T. 1939. The unistratal concept of plant communities (the unions). *Amer. Midland Natur.* **21:**111–145.

MacArthur, R. H. 1972. *Geographical Ecology: Patterns in the Distribution of Species,* Harper and Row, New York, 269 pp.

MacArthur, R. H., and E. O. Wilson. 1967. *Theory of Island Biogeography.* Princeton University Press, Princeton, N.J., 203 pp.

McCammon, R. B. 1968. The dendrograph: a new tool for correlation. *Geol. Soc. Amer. Bull.* **79:**1663–1670.

McCammon, R. B., and G. Wenniger. 1970. The dendrograph. State Geol. Survey. *Univ. Kansas Computer Contr.* **48:**1–17.

McClure, H. E., B.-L. Lim, and S. E. Winn. 1967. Fauna of the dark cave, Batu Caves, Kuala Lumpur, Malaysia. *Pacif. Insects* **9:**399–428.

McColl, H. P. 1975. The invertebrate fauna of the litter surface of a *Nothofagus truncata* forest floor, and effect of microclimate on activity. *New Zealand J. Zool.* **2:**15–34.

Macdonald, G. A. 1945. Structure of aa flows. *Geol. Soc. Amer. Bull.* **56:**1179–1180.

Macdonald, G. A., and A. T. Abbott. 1970. *Volcanoes in the Sea: The Geology of Hawaii,* University of Hawaii Press, Honolulu, 441 pp.

McEldowney, G. A. 1930. Forestry on Oahu. *Hawaiian Planter's Record* **34:**267–287.

McGurk, Linda-Lee (nee Watson). 1974. Algae as food for *Onychiurus folsomii.* M.S. thesis, University of Hawaii, Honolulu, 66 pp.

Maciolek, J. A., and R. E. Brock. 1974. Aquatic survey of the Kona Coast ponds, Hawaii Island. *Sea Grant Advisory Rpt.* Apr. 1–73.

Magnusson, A. H. 1954. A catalogue of the Hawaiian lichens. *Arkiv för Botanik* (Stockholm) **3**(10):223–402.

Malaisse, F. P. 1974. Phenology of the Zambezian woodland area with emphasis on the Miombo ecosystem, pp. 269–286. In H. Lieth, ed., *Phenology and Seasonality Modeling,* Springer-Verlag, New York, 444 pp.

Mayr, E. 1954. Change of genetic environment and evolution, pp. 157–180. In Julian Huxley, A. C. Hardy and E. B. Ford, eds. *Evolution as a Process,* Allen and Unwin, London, 367 pp.

Mayr, E. 1963. *Animal Species and Evolution,* Harvard University Press, Cambridge, Mass., 797 pp.

Metcalf, C. L., W. P. Flint, and W. L. Metcalf. 1962. *Destructive and Useful Insects,* McGraw-Hill, New York, 1087 pp.

Miller, H. A. 1956. A phytogeographical study of Hawaiian Hepaticae. Ph.D. diss. Stanford University, 123 pp.

Miller, R. S. 1967. Pattern and process in competition, p. 1–7. In *Advances in Ecological Research,* vol. 4, Academic Press, New York.

Milstead, W. W. 1972. Toward a quantification of the ecological niche. *Am. Midl. Nat.* **87:**346–354.

Mitchell, C. J. 1964a. Ectoparasitic and commensal arthropods occurring on the rats of Manoa Valley, Oahu (Acarina, Anoplura, and Siphonaptera). *Proc. Haw. Entomol. Soc.* **18:**413–415.

Mitchell, C. J. 1964b. Population structure and dynamics of *Laelaps nuttalli* Hirst and *L. echidninus* Berlese (Acarina: Laelaptidae) on *Rattus rattus* and *R. exulans* in Hawaii. *J. Med. Entomol.* **1:**151–153.

Mitchell, R. W. 1969. A comparison of temperate and tropical cave communities. *Southwestern Nat.* **14:**73–88.

Montgomery, S. L. 1975. Comparative breeding site ecology and the adaptive radiation of picture-winged *Drosophila* (Diptera:Drosophilidae) in Hawaii. *Proc. Haw. Entomol. Soc.* **27:**65–102.

Mueller-Dombois, D. 1966a. Soils, pp. 93–142. In M. S. Doty and D. Mueller-Dombois, eds., *Atlas for Bioecology Studies in Hawaii Volcanoes National Park,* Hawaii Bot. Sci. Paper No. 2, University of Hawaii, Honolulu, 507 pp.

Mueller-Dombois, D. 1966b. The vegetation map and vegetation profiles, pp. 391–441. In M. S. Doty and D. Mueller-Dombois, eds., *Atlas for Bioecology Studies in Hawaii Volcanoes National Park,* Hawaii Bot. Sci. Paper No. 2, University of Hawaii, Honolulu, 507 pp.

Mueller-Dombois, D. 1967. Ecological relations in the alpine and subalpine vegetation on Mauna Loa, Hawaii. *J. Indian Bot. Soc.* **96**:403–411.

Mueller-Dombois, D. 1972. Crown distortion and elephant distribution in the woody vegetations of Ruhuna National Park, Ceylon. *Ecology* **53:** 208–226.

Mueller-Dombois, D. 1973. A non-adapted vegetation interferes with water removal in a tropical rain forest area in Hawaii. *Trop. Ecol.* **14:**1–18.

Mueller-Dombois, D. 1975. Some aspects of island ecosystem analysis, pp. 353–366. In F.B. Golley and E. Medina, eds., *Tropical Ecological Systems–Trends in Aquatic and Terrestrial Research.* Ecol. Studies Ser. vol. 11, Springer-Verlag, New York.

Mueller-Dombois, D. 1979. Succession following goat removal in Hawaii Volcanoes National Park, pp. 1149–1154. In R.M. Lim, ed., *Proc. First Conf. on Sci. Research in the Natl. Parks,* USDA Natl. Park Service Trans. and Proc. Ser. No. 5, vol. 2.

Mueller-Dombois, D. 1980. The ‘ōhi‘a dieback phenomenon in the Hawaiian rain forest, pp. 153–161. In J. Cairns, Jr., ed., *The Recovery Process in Damaged Ecosystems.* Ann Arbor Sci. Publ., Ann Arbor, Mich., 167 pp.

Mueller-Dombois, D., and H. Ellenberg. 1974. *Aims and Methods of Vegetation Ecology,* John Wiley, New York, 547 pp.

Mueller-Dombois, D., and F.R. Fosberg. 1974. *Vegetation Map of Hawaii Volcanoes National Park (at 1:52,000),* Cooperative Park Studies Unit, Univ. Hawaii Bot. Dept., Honolulu, Tech. Rpt. 4, 44 pp.

Mueller-Dombois, D., J.D. Jacobi, R.G. Cooray, and N. Balakrishnan. 1977. *‘Ōhi‘a Rain Forest Study: Ecological Investigations of the ‘Ōhi‘a Dieback Problem in Hawaii,* Coop. Park Studies Unit, Botany Dept., Univ. Hawaii, Honolulu, Tech. Rpt. No. 20, 117 pp.

Mueller-Dombois, D., and V.J. Krajina. 1968. Comparison of east-flank vegetations on Mauna Loa and Mauna Kea, Hawaii, pp. 508–520. In R. Misra and B. Gopal, eds., *Recent Advances in Tropical Ecology II,* 773 pp.

Mueller-Dombois, D., and C.H. Lamoureux. 1967. Soil-vegetation relationships in Hawaiian kīpukas. *Pacific Sci.* **21:**286–299.

Mueller-Dombois, D., and M. Perera. 1971. Ecological differentiation and soil fungal distribution in the montane grasslands of Ceylon. *Ceylon J. Sci. (Biol. Sci.)* **9:**1–41.

Mueller-Dombois, D., and G.A. Smathers. 1975. Sukzession nach einem Vulkanausbruch auf der Insel Hawaii, pp. 159–188. M.W. Schmidt, ed., *Sukzessionsforschung.* Ber. Intl. Symp. Intern. Vereinig. Vegetationskunde. J. Cramer FL-9094 Vaduz.

Mueller-Dombois, D., and G. Spatz. 1975a. Application of the relevé method to insular tropical vegetation for an environmental impact study. *Phytocoenologia* **2**(3/4):417–429.

Mueller-Dombois, D., and G. Spatz. 1975b. The influence of feral goats on the lowland vegetation in Hawaii Volcanoes National Park. *Phytocoenologia* **3**:1–29.

Nash, S. M., and W. C. Snyder. 1962. Quantitative estimations by plate counts of propagules of the bean root rot *Fusarium* in field soils. *Phytopathology* **52**:567–572.

Neal, M. C. 1965. *In Gardens of Hawaii,* B. P. Bishop Museum Special Publ. 50, Honolulu, 924 pp.

Nei, M. 1972. Interspecific gene differences and evolutionary time estimated from electrophoretic data on protein identity. *Amer. Nat.* **105**:385–398.

Nelson, R. E., and P. R. Wheeler. 1963. *Forest Resources of Hawaii*–1961, Forestry Div., Dept. of Lands and Nat. Resources, State of Hawaii in cooperation with Pacific S.W. Forest and Range Exp. Sta. USDA Forest Service, 48 pp.

Newell, C. L. 1968. A phytosociological study of the major vegetation types in Hawaii Volcanoes National Park, Hawaii. M.S. thesis. University of Hawaii, Honolulu, 191 pp.

Newman, L. J. 1977. Chromosomal evolution of the Hawaiian Telmatogeton. Chironomidae, Diptera. *Chromosoma* **64**:349–369.

Newsome, R. D., and R. L. Dix. 1968. The forests of the Cypress Hills, Alberta and Saskatchewan, Canada. *Amer. Midland Natur.* **80**:118–185.

Nichols, L., Jr. 1962. *Ecology of the Wild pig,* State of Hawaii, Division of Fish and Game, Honolulu. W-5-R-13, Job 46(13), 20 pp. (mimeo.).

Njoku, E. 1963. Seasonal periodicity in the growth and development of some forest trees. I. Observations on mature trees. *J. Ecol.* **51**:617–624.

O'Brien, S. J., and R. J. MacIntyre. 1969. An analysis of gene-enzyme variability in natural populations of *Drosophila melanogaster* and *D. simulans*. *Amer. Nat.* **103**:97–113.

Odum, E. P. 1971. *Fundamentals of Ecology,* 3d ed., W. B. Saunders, Philadelphia, 574 pp.

Ohta, T., and M. Kimura. 1975. Theoretical analysis of electrophoretically detectable polymorphisms: models of very slightly deleterious mutations. *Amer. Nat.* **109**:137–145.

Opler, P. A., G. W. Frankie, and H. G. Baker. 1976. Rainfall as a factor in the release, timing, and synchronization of anthesis by tropical trees and shrubs. *J. Biogeog.* **3**:231–236.

Orians, G. H. 1975. Diversity, stability and maturity in natural ecosystems, pp. 139–150. In W. H. van Dobben and R. H. Lowe-McConnell, eds., *Unifying Concepts in Ecology,* Dr. W. Junk Publ., the Hague.

Orloci, L. 1967. An agglomerative method for classification of plant communities. *J. Ecol.* **55**:193–205.

Orloci, L. 1975. *Multivariate Analysis in Vegetation Research,* Dr. W. Junk, B.V., the Hague, 276 pp., 2d ed. 1978, 451 pp.

Paik, Y.K. 1973. Frequency of heterozygous inversions in a Korean population of *Drosophila immigrans. Drosophila Info. Serv.* **50:**114.

Paik, Y.K., and K.C. Sung. 1974. Variation in chromosomal polymorphism in Hawaiian populations of *Drosophila immigrans. Jap. J. Genetics* **49:**159–169.

Papp, R.P., J.T. Kliejunas, R.S. Smith, Jr., and R.F. Scharpf. 1979. Association of *Plagithmysus bilineatus* (Coleoptera: Cerambycidae) and *Phytophthora cinnamomi* with the decline of ʻōhiʻa-lehua forests on the island of Hawaii. *Forest Science* **25:**187–196.

Parker, B.C., H.C. Bold, and T.R. Deason. 1961. Facultative heterotrophy in chlorococcacean algae. *Science* **133:**761–763.

Patterson, J.T., and W.S. Stone. 1952. *Evolution in the Genus* Drosophila, Macmillan, New York, 610 pp.

Paulian, R. and A. Grjebine. 1953. Une campagne speleologie dans la Reserve naturelle de Namoroka. Naturaliste Malgache V:19–28.

Pearsall, G. 1951. The phenology of ornamental plants in the Honolulu area. M.S. thesis, University of Hawaii, Honolulu, 159 pp.

Peck, S.B. 1973. A review of the invertebrate fauna of volcanic caves in the western United States. *Bull. Natl. Speleol. Soc.* **35**(4):99–107.

Peck, S.B. 1974. The invertebrate fauna of tropical American caves. Part II: Puerto Rico, an ecological and zoogeographic analysis. *Biotropica* **6:**14–31.

Peck, S.B. 1975. The invertebrate fauna of tropical American caves. Part III: Jamaica, an introduction. *Intl. J. Speleol.* **7:**303–326.

Perkins, R.C.L. 1899. *Hymenoptera aculeata.* In D. Sharp ed., *Fauna Hawaiiensis.* Vol. I. Cambridge Univ. Press, Cambridge, 600+ pp.

Peterson, J.B. 1935. Studies on the biology and taxonomy of soil algae. *Dansk Botan. Arkiv* **8:**180.

Peterson, D.W., and D.A. Swanson. 1974. Observed formation of lava tubes during 1970–71 at Kilauea Volcano, Hawaii. *Studies in Speleology* **2:**209–222.

Petteys, E.Q.P., R.E. Burgan, and R.E. Nelson. 1975. *Ohia Forest Decline: Its Spread and Severity in Hawaii,* USDA Forest Serv. Res. Paper PSW-105, Pacific S.W. Forest and Range Exp. Sta., Berkeley, Calif., 11 pp.

Phipps, J.B. 1971. Dendrogram topology. *Syst. Zool.* **20:**306–308.

Platt, T., and K.L. Denman. 1975. Spectral analysis in ecology. *Ann. Rev. Ecol. Syst.* **6:**189–210.

Poore, M.E.D. 1964. Integration in the plant community. *J. Ecol.* **52** (suppl.):213–226.

Poulson, T.L., 1975. Symposium on life histories of cave beetles: an introduction. *Intl. J. Speleol.* **7:**1–5.

Poulson, T.L., and W.B. White. 1969. The cave environment. *Science* **165:**971–981.

Preston, F.W. 1969. Diversity and stability in the biological world, pp. 1–12. In G.M. Woodwell and H.H. Smith, eds., *Diversity and Stability in Ecological Systems,* Brookhaven Symp. in Biol. No. 22, National Tech. Inform. Service, U.S. Dept. of Commerce, 264 pp.

Price, S. 1966. The climates of Oahu. *Bull. Pacific Orchid Soc. Hawaii* **24:** 9–21.

Pukui, M.K. and S.H. Elbert. 1975. *Hawaiian Dictionary.* University Press of Hawaii, Honolulu, 402 + 188 pp.

Radovsky, F.J., J.M. Tenorio, P.Q. Tomich, and J.D. Jacobi. 1979. Acari on murine rodents along an altitudinal transect on Mauna Loa, Hawaii, pp. 327–333. In *Proc. 4th Intl. Cong. Acarology,* Saalfelden, Austria, 1974.

Rasid, R. 1963. Phenology and floral ontogeny of the skunk tree, *Sterculia foetida* Linnaeus. M.S. thesis, University of Hawaii, Honolulu.

Raunkiaer, C. 1918. Recherches statistiques sur les formations végétales. *Det. Kgl. Danske Vidensk. Selsk. Biol. Medd.* **1:**1–80.

Raunkiaer, C. 1934. *The Life Forms of Plants and Statistical Plant Geography; Being the Collected Papers of C. Raunkiaer, Translated into English by H.G. Carter, A.G. Tansley, and Miss Fausboll.* Clarendon, Oxford, 632 pp.

Raunkiaer, C. 1937. *Plant Life Forms,* Clarendon, Oxford, 104 pp.

Reese, E.S. 1969. Behavioral adaptations of intertidal hermit crabs. *Amer. Zool.* **9:**343–355.

Ricard, M., G. Richard, B. Salvat, and J.L. Toffart. 1977. *Coral Reefs and Lagoon Research in French Polynesia, 123 Publications with abstracts.* Revue Algologique Fascicule Hors Sene No. 1 (Muséum National D'Histoire Naturelle Antenne Tahiti, Papeete, Tahiti), 44 pp.

Richards, P.W. 1939. Ecological studies on the rain forest of Southern Nigeria. I. The structure and floristic composition of the primary forest. *J. Ecol.* **27:**1–61.

Richards, P.W. 1952. *The Tropical Rain Forest,* Cambridge University Press, 450 pp.

Richardson, R.H. 1974. Effects of dispersal, habitat selection and competition on a speciation pattern of *Drosophila* endemic to Hawaii, pp. 140–164. In M.J.D. White, ed., *Genetic Mechanisms of Speciation in Insects.* Australia and New Zealand Book Co., Sidney, 170 pp.

Richardson, R.H., and J.S. Johnston. 1975a. Behavioral components of dispersal in *Drosophila mimica. Oecologia* **20:**287–299.

Richardson, R.H., and J.S. Johnston. 1975b. Ecological specialization of Hawaiian *Drosophila.* I. Habitat selection in Kīpuka Puaulu. *Oecologia* **21:**193–204.

Richardson, R. H., and P. E. Smouse. 1975. Ecological specialization of Hawaiian *Drosophila*. II. The community matrix, ecological complementation and phyletic species packing. *Oecologia* **22:**1–13.

Richmond, R. C. 1972. Enzyme variability in the *Drosophila willistoni* group. III. Amounts of variability in the superspecies *D. paulistorum*. *Genetics* **70:**87–112.

Richmond, R. C., and T. Dobzhansky. 1968. Chromosomal polymorphism in populations of *Drosophila immigrans* on the island of Maui, Hawaii. *Univ. Texas Publ.* **6818:**381–386.

Richmond, T. de A., and D. Mueller-Dombois. 1972. Coastline ecosystems on Oahu, Hawaii. *Vegetatio* **25**(5/6):367–400.

Ripperton, J. C., and E. Y. Hosaka. 1942. Vegetation zones of Hawaii. *Hawaii Agri. Exp. Sta. Bull.* **89:**1–60.

Robbins, R. G. 1959. The use of the profile diagram in rain forest ecology. *J. Biol. Sci.* **2:**53–63.

Rock, J. F. 1913. *The Indigenous Trees of the Hawaiian Islands,* 1st ed., privately publ., Honolulu, Hi., 518 pp. Reprinted 1974, Charles E. Tuttle Co., Tokyo, Japan, 548 pp.

Rockwood, E. S. 1969. Enzyme variation in natural populations of *Drosophila mimica*. *Univ. Texas Publ.* **6918:**111–132.

Rockwood, E. S., C. G. Kanapi, M. R. Wheeler, and W. Stone. 1971. Allozyme changes during the evolution of Hawaiian *Drosophila*. *Univ. Texas Publ.* **7103:**193–212.

Rogers, J. S. 1972. Measures of genetic similarity and genetic distance. *Univ. Texas Publ.* **7213:**145–153.

Root, R. B. 1967. The niche exploitation pattern of the blue gray gnatcatcher. *Ecol. Monogr.* **37:**317–319, 331–349.

Sailer, R. E. 1978. Our immigrant insect fauna. *Bull. Entomol. Soc. Amer.* **24:**3–11.

St. John, H. 1966. Monograph of *Cyrtandra* (Gesneriaceae) on Oahu, Hawaiian Islands. *B. P. Bishop Museum Bull.* **229:**1–465.

St. John, H. 1972. *Canavalia kauensis* (Leguminosae), a new species from the Island of Hawaii. Hawaiian Plant Studies 39. *Pacific Sci.* **26:**409–414.

St. John, H. 1973. *List and Summary of the Flowering Plants in the Hawaiian Islands,* Pacific Tropical Bot. Gard. Mem. No. 1, Lawai, Hawaii, 519 pp.

Sato, H. H., W. Ikeda, R. Paeth, R. Smythe, and M. Takehiro, Jr. 1973. *Soil Survey of the Island of Hawaii, State of Hawaii,* USDA Soil Cons. Serv. in cooperation with Univ. Hawaii Agri. Exp. Sta., 115 pp. + 195 map sheets.

Schiller, E. L. 1956. Ecology and health of *Rattus* at Nome, Alaska. *J. Mammal.* **37:**181–188.

Schultz, G. A. 1973. The cavernicolous fauna of Hawaiian lava tubes, 2. Two new genera and species of blind isopod crustaceans (Oniscoidea: Philosciidae). *Pacif. Insects* **15:**153–162.

Scowcroft, P. G., and R. E. Nelson. 1976. *Disturbance during Logging Stimulates Regeneration of Koa*. USDA Forest Serv. Res. Note PSW-306, Pacific S.W. Forest and Range Exp. Sta., Berkeley, Calif., 7 pp.

Selander, R. K., and W. E. Johnson. 1973. Genetic variation among vertebrate species. *Ann. Rev. Ecol. Syst.* **4:**75–91.

Sene, F. M., and H. L. Carson. 1977. Genetic variation in Hawaiian *Drosophila*. IV. Allozymic similarity between *D. silvestris* and *D. heteroneura* from the island of Hawaii. *Genetics* **86:**187–198.

Smathers, G. A., and D. Mueller-Dombois. 1974. *Invasion and Recovery of Vegetation after a Volcanic Eruption in Hawaii,* National Park Service Scientific Monogr. Ser. No. 5, 129 pp.

Smithies, O. 1959. An improved procedure for starch gel electrophoresis: further variations in the serum proteins of normal individuals. *Bioch. J.* **71:**585–587.

Sneath, P. H. A., and R. R. Sokal. 1973. *Numerical Taxonomy: The Principles and Practice of Numerical Classification,* W. H. Freeman, San Francisco, 573 pp.

Snow, D. W. 1965. A possible selective factor in the evolution of fruiting seasons in tropical forests. *Oikos* **15**(2):274–281.

Sokal, R. R., and P. H. A. Sneath. 1963. *Principles of Numerical Taxonomy.* W. H. Freeman, San Francisco, 359 pp.

Southwood, T. R. E. 1960. The abundance of the Hawaiian trees and the number of their associated insect species. *Proc. Haw. Entomol. Soc.* **17:** 299–303.

Spatz, G. 1970. Pflanzengesellschaften, Leistungen und Leistungspotential von Allgäuer Alpweiden in Abhängigkeit von Standort und Bewirtschaftung. Dissertation, Techn. Univ. Munich, 160 pp.

Spatz, G., and D. Mueller-Dombois. 1973. The influence of feral goats on koa tree reproduction in Hawaii Volcanoes National Park. *Ecology* **54:**870–876.

Spatz, G., and D. Mueller-Dombois. 1975. Succession pattern after pig-digging in grassland communities on Mauna Loa, Hawaii. *Phytocoenologia* **3:**346–373.

Spatz, G., and J. Siegmund. 1973. Eine Methode zur tabellarischen Ordination, Klassifikation und ökologischen Auswertung von pflanzensoziologischen Bestandsaufnahmen. *Vegetatio* **28:**1–17.

Spencer, W. P. 1932. The vermilion mutant of *Drosophila hydei* breeding in nature. *Amer. Nat.* **66:**474–479.

Sperlich, D. 1961. Untersuchungen über den chromosomalen Polymorphismus einer Population von *Drosophila subobscura* auf den Liparischen Inseln. *Z. Vererbungsl.* **92:**74–84.

Spieth, H. T. 1966. Courtship behavior of endemic Hawaiian *Drosophila*. *Univ. Texas Publ.* **6615:**245–313.

Spieth, H. T. 1974. Mating behavior and evolution of the Hawaiian *Drosophila,* pp. 94–101. In M. J. D. White, ed., *Genetic Mechanisms of Speciation in Insects*. Australia and New Zealand Book Co., Sydney, Australia.

State of Hawaii. 1970. *An Inventory of Basic Water Resources Data, Island of Hawaii,* Rpt. R34, Dept. of Land and Nat. Res., Div. of Water and Land Development, Honolulu, Hawaii.

Stearns, H. T., and G. A. Macdonald. 1946. *Geology and Ground-water Resources of the Island of Hawaii,* Hawaii Div. of Hydrography Bull. 9, 363 pp.

Steffan, W. A. 1973. Studies of *Ctenosciara hawaiiensis* (Hardy) (Diptera: Sciaridae). *Pacif. Insects* **15:**85–94.

Steffan, W. A. 1974. Laboratory studies and ecological notes on Hawaiian Sciaridae (Diptera). *Pacif. Insects* **16:**41–50.

Steiner, W. W. M. 1974. Enzyme polymorphism and desiccation resistance in two species of Hawaiian *Drosophila*. Ph.D. diss., University of Hawaii, Honolulu, 301 pp.

Steiner, W. W. M. 1975. Enzyme variability in exotic and endemic *Drosophila* in Hawaii. *Genetics* **80:**s78.

Steiner, W. W. M., W. E. Johnson, and H. L. Carson, 1973. Molecular differentiation in *Drosophila grimshawi*. *Drosophila Info. Service* **50:** 100–101.

Steiner, W. W. M., K. C. Sung, and Y. K. Paik. 1976. Electrophoretic variability in island populations of *Drosophila simulans* and *Drosophila immigrans*. *Biochem. Genetics* **14:**495–506.

Stiles, F. G. 1975. Ecology, flowering phenology and hummingbird pollination of some Costa Rica *Heliconia* species. *Ecology* **56:**285–310.

Stiles, F. G. 1976. On taste preferences, color preferences, and flower choice in hummingbirds. *Condor* **78:**10–26.

Stiles, F. G. 1977. Coadapted competitors: the flowering seasons of hummingbird-pollinated plants in a tropical forest. *Science* **198:**1177–1178.

Stiles, F. G. 1978. Temporal organization of flowering among the hummingbird foodplants of a tropical wet forest. *Biotropica* **10:**194–210.

Stoner, M. F. 1967. Diet food media for the culture of phytopathogenic fungi and bacteria. *Phytopathology* **57:**447.

Stoner, M. F. 1974. Ecology of *Fusarium* species in noncultivated soils of Hawaii. *Proc. Amer. Phytopathol. Soc.* **1:**102.

Stoner, M. F. 1976. Reference fungus method for ecological comparisons of soils. *Proc. Amer. Phytopathol. Soc.* **3:**276–277.

Strickberger, M. W., and C. J. Wills. 1966. Monthly frequency changes of *Drosophila pseudoobscura* third chromosome gene arrangements in a California locality. *Evolution* **20:**592–602.

Strong, D.R., E.D. McCoy, and J.R. Rey. 1977. Time and number of herbivore species: the pests of sugar cane. *Ecology* **58:**167–175.

Sukachev, V.N. 1928. Principles of classification of the spruce communities of European Russia. *J. Ecol.* **16:**1–18.

Sung, K.C. 1974. Behavior of inversion polymorphisms in artificial populations of *Drosophila immigrans* Sturtevant. Ph.D. diss., University of Hawaii, Honolulu, 106 pp.

Sweeny, B.M. 1969. *Rhythmic Phenomena in Plants,* Academic Press, New York, 147 pp.

Swezey, O. 1954. *Forest Entomology in Hawaii.* B.P. Bishop Museum Special Publ. No. 44, 266 pp.

Synave, H. 1954. Un cixiid troglobie decouvert dans les galeries souterraines du systeme de Namoroka (Hemiptera-Homoptera). *Le Naturaliste Malgache* **2**(1953):175–179.

Taliaferro, W.J. 1959. *Rainfall of the Hawaiian Islands,* Hawaii Water Auth., 394 pp.

Tenorio, J.M., and M.L. Goff. In press *Ectoparasites of Hawaiian Rodents (Siphonoptera, Anoplura and Acari),* Spec. Publ. Dept. Entomol. B.P. Bishop Museum, Honolulu.

Terborgh, J. 1971. Distribution on environmental gradients: theory and preliminary interpretation of distributional patterns in the avifauna of the Cordillera Vilcabamba, Peru. *Ecology* **52:**23–40.

Terborgh, J., and J.M. Diamond. 1970. Niche overlap in feeding assemblages of New Guinea birds. *Wilson Bull.* **82:**29–52.

Terborgh, J., and J.S. Weske. 1975. The role of competition in the distribution of Andean birds. *Ecology* **56:**562–576.

Thornton, R.H. 1960. Growth of fungi in some forest and grassland soils, pp. 84–91. In D. Parkinson and J.S. Waid, eds., *The Ecology of Soil Fungi,* Liverpool University Press, Liverpool.

Tomich, P.Q. 1968. Coat color in wild populations of the roof rat in Hawaii. *J. Mammol.* **49:**74–82.

Tomich, P.Q. 1969a. *Mammals in Hawaii: A Synopsis and Notational Bibliography,* B.P. Bishop Museum Spec. Publ. 57, 238 pp.

Tomich, P.Q. 1969b. Movement patterns of the mongoose in Hawaii. *J. Wildlife Management* **33:**576–584.

Tomich, P.Q. 1970. Movement patterns of field rodents in Hawaii. *Pacific Sci.* **24:**195–234.

Tonzetich, J., and C.L. Ward. 1973. Adaptive chromosomal polymorphism in *Drosophila melanica. Evolution* **27:**486–494.

Torii, H. 1960. A consideration of the distribution of some troglobionts of Japanese caves. (I). *Jap. J. Zool.* **12:**555–584.

Toyofuku, Y. 1961. Chromosomal polymorphism found in natural populations of *Drosophila immigrans. Jap. J. Genetics* **36:**32–37.

Treub, M. 1888. Notice sur la nouvelle flore de Krakatau. *Ann. du Jard. Bot. de Buitenzorg* **7:**211.

Troll, C. 1959. Die tropischen Gebirge. *Bonner Geogr. Abhandlungen.* Heft 25. Univ. Bonn, W. Germany.

Tschetverikov, S. S. 1926. On certain aspects of the evolutionary process from the standpoint of genetics. (Translated from Russian 1959.) *Proc. Amer. Phil. Soc.* **105:**167–195.

Turesson, G. 1923. The scope and impact of genecology. *Hereditas* **4:** 171–176.

Turner, B. J. 1974. Genetic divergence of Death Valley pupfish species: biochemical versus morphological evidence. *Evolution* **28:**281–294.

Uehara, G. 1973. Soils, pp. 39–41. In R. W. Armstrong, ed., *Atlas of Hawaii,* University of Hawaii Press, Honolulu.

Ueno, S.-I. 1971. The fauna of the lava caves around Mt. Fuji-san. I. Introductory and historical notes. *Bull. Natl. Sci. Mus. Tokyo* **14:**201–218.

UNESCO Expert Panel on MAB Project 7. 1973. *Ecology and Rational Use of Island Ecosystems,* UNESCO Man and the Biosphere Program, Final Rpt., MAB Ser. No. 11, 80 pp.

U.S. Department of Agriculture. 1975. *Soil Taxonomy: A Basic System of Soil Classification for Making and Interpreting Soil Surveys,* U.S. Dept. Agri. Handbook 436, U.S. Govt. Printing Office, Washington, D.C., 754 pp.

Vandel, A. 1965. *Biospeleology. The Biology of Cavernicolous Animals,* trans. B. E. Freeman, Pergamon Press, Oxford, 524 pp.

van Riper, C., III. 1976. The influence of food supplementation upon the reproductive strategy and movement in the Hawaii 'Amakihi *(Loxops virens),* pp. 227–228. In C. W. Smith, ed., *Proc. First Conf. Nat. Sci. in Hawaii,* CPSU/UH, Dept. of Botany, University of Hawaii, Honolulu.

van Riper, C., III. 1978. The breeding ecology of the Amakihi *(Loxops virens)* and the Palila *(Psittorostra bailleui)* on Mauna Kea, Hawaii. Ph.D. diss., University of Hawaii, Honolulu, 165 pp.

Vogl, R.J. 1969. The role of fire in the evolution of the Hawaiian flora and vegetation, pp. 5–60. In *Proc. Ann. Tall Timbers Fire Ecol. Conf.,* April 10–11, 1969, Tallahassee, Florida.

Waggoner, P. E. 1974. Using models of seasonality, pp. 401–405. In H. Lieth, ed., *Phenology and Seasonality Modeling,* Springer-Verlag, New York, 444 pp.

Wallwork, J. A. 1970. *Ecology of Soil Animals,* McGraw-Hill, London, 283 pp.

Walter, H. 1955. Die Klimadiagramme als Mittel zur Beurteilung der Klimaverhältnisse für ökologische, vegetationskundliche und landwirtschaftliche Zwecke. *Ber. Deutsch. Bot. Ges.* **68:**331–344.

Walter, H. 1957. Wie kann man den Klimatypus anschaulich darstellen? Die Umschau in Wissenschaft und Technik. *Heft* **24:**751–753.

Walter, H. 1971. *Ecology of Tropical and Subtropical Vegetation,* trans. D. Mueller-Dombois, Oliver and Boyd, Edinburgh, 539 pp.

Walter, H. 1973. *Vegetation of the Earth,* trans. Joy Wieser, Springer-Verlag, New York, 237 pp.

Walter, H., E. Harnickell, and D. Mueller-Dombois. 1975. *Climate-Diagram Maps of the Individual Continents and the Ecological Climatic Regions of the Earth,* Springer-Verlag, New York, 36 pp. + maps.

Walter, H., and H. Lieth. 1960–1967. *Klimadiagramm-Weltatlas,* VEB Fischer Verlag, Jena.

Warner, R.E. 1968. The role of introduced diseases in the extinction of endemic Hawaiian avifauna. *Condor* **70:**101–120.

Warner, R.E. 1972. A forest dies on Mauna Kea, pp. 39–44. In F. R. Fosberg, *Guide to Excursion III,* Tenth Pac. Sci. Congr. Univ. Hawaii and Hawaii Bot. Gard. Foundation, Honolulu.

Wasserman, M. 1963. Cytology and phylogeny of *Drosophila. Amer. Nat.* **97:**333–352.

Watson, J.S. 1956. The present distribution of *Rattus exulans* Peale in New Zealand. *New Zealand J. Sci. Tech.* **37:**560–570.

Wheeler, M.R., and N. Hamilton. 1972. Catalog of *Drosophila* species names, 1959–1971. *Univ. Texas Publ.* **7213:**257–268.

Wheeler, M.R., and L. Wheeler. 1972. Notes on some introduced *Drosophila* in Hawaii. *Drosophila Info. Service* **48:**77.

White, M.J.D. 1973. *Animal Cytology and Evolution,* 3rd ed., Univ. Press, Cambridge, 961 pp.

Whitesell, C.D. 1964. *Silvicultural characteristics of koa (Acacia koa Gray),* U.S. Forest Serv. Pacific S.W. Forest and Range Exp. Sta. Res. Paper PSW-16, Berkeley, Calif., 12 pp.

Whitmore, T.C. 1966. The social status of *Agathis* in a rain forest in Melanesia. *J. Ecol.* **54:**285–301.

Whitmore, T.C. 1975. *Tropical Rain Forests of the Far East,* Clarendon Press, Oxford, 282 pp.

Whittaker, R.H. 1956. Vegetation of the Great Smoky Mountains. *Ecol. Monogr.* **26:**1–80.

Whittaker, R.H. 1960. Vegetation of the Siskiyou Mountains, Oregon and California. *Ecol. Monogr.* **30:**279–338.

Whittaker, R.H. 1967. Gradient analysis of vegetation. *Biol. Rev.* **42:**207–264.

Whittaker, R.H. 1969. Evolution of diversity in plant communities, pp. 178–196. In G.M. Woodwell and H.H. Smith, eds., *Diversity and Stability in Ecological Systems.* Brookhaven Symp. in Biol. No. 22. National Tech. Inform. Service, U.S. Dept. of Commerce, 264 pp.

Whittaker, R.H. 1970. *Communities and Ecosystems,* Macmillan, Collier-Macmillan Ltd., London, 162 pp.

Whittaker, R.H. 1975. *Communities and Ecosystems,* 2d ed., Macmillan, New York, 385 pp.

Wick, H.L. 1970. *Lignin Staining—a Limited Success in Identifying Koa Growth Rings,* U.S. Forest Serv. Res. Note PSW-205, Pacific S.W. Forest and Range Exp. Sta., Berkeley, Calif., 3 pp.

Willson, M.F., J.R. Karr, and R.R. Roth. 1975. Ecological aspects of avian bill-size variation. *Wilson Bull.* **87**:32–44.

Wilson, A.C., V.M. Sarich, and L.R. Maxon. 1974. The importance of gene rearrangement in evolution: evidence from studies on rates of chromosomal, protein and anatomical evolution. *Proc. Nat. Acad. Sci.* **71**:3028-3030.

Wirawan, N. 1978. Vegetation and soil-water regimes in a tropical rain forest valley on Oahu, Hawaiian Islands. Ph.D. diss. University of Hawaii, Honolulu, 420 pp.

Wirth, W. 1947. A revision of the genus *Telmatogeton* Schiner, with descriptions of three new Hawaiian species (Diptera: Tendipedidae). *Proc. Haw. Entomol. Soc.* **13**:143–191.

Wirth, W.W. 1949. A revision of the clunionine midges with descriptions of a new genus and four new species (Tendipedidae). *Univ. Calif. Publ. Entomol.* **8**:151–182.

Wium-Anderson, S., and B. Christensen. 1978. Seasonal growth of mangrove trees in southern Thailand. II. Phenology of *Bruguiera cylindrica, Ceriops tagal, Lumnitzera littorea* and *Avicennia marina. Aquatic Bot.* **5**:383–390.

Woodside, D.H. 1958. *Food Habits Study of Game Birds,* Wildlife Management Research, Hawaii State Div. of Fish and Game, 11 pp. (Job completion report, mimeo.)

Woodwell, G.M. and H.H. Smith. 1969. *Diversity and Stability in Ecological Systems.* Brookhaven Symp. in Biol. No. 22, National Tech. Inform. Service, U.S. Dept. of Commerce, 264 pp.

Yoon, J.S., K. Resch, and M.R. Wheeler. 1972. Cytogenetic relationships in Hawaiian species of *Drosophila.* II. The *D. mimica* subgroup of the "modified mouthparts" species group. *Univ. Texas Publ.* **7213**:201–212.

Yoon, J.S., K. Resch, M.R. Wheeler, and R.H. Richardson. 1975. Evolution in Hawaiian Drosophilidae: chromosomal phylogeny of the *Drosophila crassifemur* complex. *Evolution* **29**:249–256.

Zimmerman, E.C. 1948a. *Insects of Hawaii,* vol. 1, *Introduction,* University of Hawaii Press, Honolulu, 206 pp.

Zimmerman, E.C. 1948b. *Insects of Hawaii,* vol. 2, *Apterygota to Thysanoptera,* University of Hawaii Press, Honolulu, 475 pp.

Zimmerman, E.C. 1948c. *Insects of Hawaii,* vol. 3, *Heteroptera,* University of Hawaii Press, Honolulu, 255 pp.

Zimmerman, E.C. 1948d. *Insects of Hawaii,* vol. 4, *Homoptera: Auchenorhyncha,* University of Hawaii Press, Honolulu, 268 pp.

Zimmerman, E. C. 1948e. *Insects of Hawaii*, vol. 5, *Homoptera: Sternorhyncha*, University of Hawaii Press, Honolulu, 464 pp.
Zimmerman, E. C. 1957. *Insects of Hawaii*, vol. 6, *Ephemeroptera-Neuroptera-Trichoptera and Suppl. to Vol. I to V*, University of Hawaii Press, Honolulu, 209 pp.
Zimmerman, E. C. 1958a. *Insects of Hawaii*, vol. 7, *Macrolepidoptera*, University of Hawaii Press, Honolulu, 542 pp.
Zimmerman, E. C. 1958b. *Insects of Hawaii*, vol. 8, *Lepidoptera: Pyraloidea*, University of Hawaii Press, Honolulu, 456 pp.

List of Technical Reports

1. *Hawaii Terrestrial Biology Subprogram. First Progress Report and Second-Year Budget.* D. Mueller-Dombois, ed., December 1970, 144 pp.
2. *Island Ecosystems Stability and Evolution Subprogram. Second Progress Report and Third-Year Budget.* D. Mueller-Dombois, ed., January 1972, 290 pp.
3. *The Influence of Feral Goats on Koa* (Acacia koa Gray) *Reproduction in Hawaii Volcanoes National Park.* G. Spatz and D. Mueller-Dombois, February 1972, 16 pp.
4. *A Non-adapted Vegetation Interferes with Soil Water Removal in a Tropical Rain Forest Area in Hawaii.* D. Mueller-Dombois, March 1972, 25 pp.
5. *Seasonal Occurrence and Host-Lists of Hawaiian Cerambycidae.* J.L. Gressitt and C.J. Davis, April 1972, 34 pp.
6. *Seed Dispersal Methods in Hawaiian* Metrosideros. Carolyn Corn, August 1972, 19 pp.
7. *Ecological Studies of* Ctenosciara hawaiiensis *(Hardy) (Diptera: Sciaridae).* W.A. Steffan, August 1972, 7 pp.
8. *Birds of Hawaii Volcanoes National Park.* A.J. Berger, August 1972, 49 pp.
9. *Bioenergetics of Hawaiian Honeycreepers: the Amakihi* (Loxops virens) *and the Anianiau* (L. parva). R.E. MacMillen, August 1972, 14 pp.
10. *Invasion and Recovery of Vegetation after a Volcanic Eruption in Hawaii.* G.A. Smathers and D. Mueller-Dombois, September 1972, 172 pp.
11. *Birds in the Kilauea Forest Reserve. A Progress Report.* A.J. Berger, September 1972, 22 pp.
12. *Ecogeographical Variations of Chromosomal Polymorphism in Hawaiian Populations of* Drosophila immigrans. Y.K. Paik and K.C. Sung, February 1973, 25 pp.
13. *The Influence of Feral Goats on the Lowland Vegetation in Hawaii Volcanoes National Park.* D. Mueller-Dombois and G. Spatz, October 1972, 46 pp.
14. *The Influence of SO_2 Fuming on the Vegetation Surrounding the Kahe Power Plant on Oahu, Hawaii.* D. Mueller-Dombois and G. Spatz, October 1972, 12 pp.
15. *Succession Patterns after Pig Digging in Grassland Communities on Mauna Loa, Hawaii.* G. Spatz and D. Mueller-Dombois, November 1972, 44 pp.
16. *Ecological Studies on Hawaiian Lava Tubes.* F.G. Howarth, December 1972, 20 pp.

17. *Some Findings on Vegetative and Sexual Reproduction of Koa*. Günter O. Spatz, February 1973, 45 pp.
18. *Altitudinal Ecotypes in Hawaiian* Metrosideros. Carolyn Corn and William Hiesey, February 1973, 19 pp.
19. Some aspects of Island Ecosystems Analysis. Dieter Mueller-Dombois, February 1973, 26 pp.
20. *Flightless Dolichopodidae (Diptera) in Hawaii*. D. Elmo Hardy and Mercedes D. Delfinado, February 1973, 8 pp.
21. *Third Progress Report and Budget Proposal for FY 74 and FY 75*. D. Mueller-Dombois and K. Bridges, eds., March 1973, 153 pp.
22. *Supplement 1. The Climate of the IBP Sites on Mauna Loa, Hawaii*. Kent W. Bridges and G. Virginia Carey, April 1973, 141 pp.
23. *The Bioecology of* Psylla uncatoides *in the Hawaii Volcanoes National Park and the* Acacia koaia *Sanctuary*. John R. Leeper and J. W. Beardsley, April 1973, 13 pp.
24. *Phenology and Growth of Hawaiian Plants: A Preliminary Report*. Charles H. Lamoureux, June 1973, 62 pp.
25. *Laboratory Studies of Hawaiian Sciaridae (Diptera)*. Wallace A. Steffan, June 1973, 17 pp.
26. *Natural Area System Development for the Pacific Region, a Concept and Symposium*. Dieter Mueller-Dombois, June 1973, 55 pp.
27. *The Growth and Phenology of* Metrosideros *in Hawaii*. John R. Porter, August 1973, 62 pp.
28. *EZPLOT: A Computer Program Which Allows Easy Use of a Line Plotter*. Kent W. Bridges, August 1973, 39 pp.
29. *A Reproductive Biology and Natural History of the Japanese White-Eye* (Zosterops japonica japonica) *in Urban Oahu*. Sandra J. Guest, September 1973, 95 pp.
30. *Techniques for Electrophoresis of Hawaiian* Drosophila. W. W. M. Steiner and W. E. Johnson, November 1973, 21 pp.
31. *A Mathematical Approach to Defining Spatially Recurring Species Groups in a Montane Rain Forest on Mauna Loa, Hawaii*. Jean E. Maka, December 1973, 112 pp.
32. *The Interception of Fog and Cloud Water on Windward Mauna Loa, Hawaii*. James O. Juvik and Douglas J. Perreira, December 1973, 11 pp.
33. *Interactions between Hawaiian Honeycreepers and* Metrosideros collina *on the Island of Hawaii*. F. Lynn Carpenter and Richard E. MacMillen, December 1973, 23 pp.
34. *Floristic and Structural Development of Native Dry Forest Stands at Mokuleia, N.W. Oahu*. Nengah Wirawan, January 1974, 49 pp.
35. *Genecological Studies of Hawaiian Ferns: Reproductive Biology of Pioneer and Non-pioneer Species on the Island of Hawaii*. Robert M. Lloyd, February 1974, 29 pp.
36. *Fourth Progress Report and Budget Proposal for FY 1975*. D. Mueller-Dombois and K. Bridges, eds., March 1974, 44 pp.

37. *A Survey of Internal Parasites of Birds on the Western Slopes of Diamond Head, Oahu, Hawaii 1972–1973*. H. Eddie Smith and Sandra J. Guest, April 1974, 18 pp.
38. *Climate Data for the IBP Sites on Mauna Loa, Hawaii*. Kent W. Bridges and G. Virginia Carey, May, 1974, 97 pp.
39. *Effects of Microclimatic Changes on Oogenesis of* Drosophila mimica. Michael P. Kambysellis, May 1974, 58 pp.
40. *The Cavernicolous Fauna of Hawaiian Lava Tubes, Part VI. Mesoveliidae or Water Treaders (Heteroptera)*. Wayne C. Gagné and Francis G. Howarth, May 1974, 22 pp.
41. *Shade Adaptation of the Hawaiian Tree-fern* (Cibotium glaucum *(Sm.) H. & A.*). D. J. C. Friend, June 1974, 39 pp.
42. *The Roles of Fungi in Hawaiian Island Ecosystems. I. Fungal Communities Associated with Leaf Surfaces of Three Endemic Vascular Plants in Kilauea Forest Reserve and Hawaii Volcanoes National Park, Hawaii*. Gladys E. Baker, Paul H. Dunn, and William A. Sakai, July 1974, 46 pp.
43. *The Cavernicolous Fauna of Hawaiian Lava Tubes, Part VII. Emesinae or Thread-legged Bugs (Heteroptera: Redvuiidae)*. Wayne C. Gagné and Francis G. Howarth, July 1974, 18 pp.
44. *Stand Structure of a Montane Rain Forest on Mauna Loa, Hawaii*. Ranjit G. Cooray, August 1974, 98 pp.
45. *Genetic Variability in the Kilauea Forest Population of* Drosophila silvestris. E. M. Craddock and W. E. Johnson, September 1974, 39 pp.
46. *Linnet Breeding Biology on Hawaii*. Charles van Riper III, October 1974, 19 pp.
47. *The Nesting Biology of the House Finch,* Carpodacus mexicanus frontalis *(Say), in Honolulu, Hawaii*. Lawrence T. Hirai, November 1974, 105 pp.
48. *A Vegetational Description of the IBP Small Mammal Trapline Transects—Mauna Loa Transect*. James D. Jacobi, November 1974, 19 pp.
49. *Vegetation Types: A Consideration of Available Methods and Their Suitability for Various Purposes*. Dieter Mueller-Dombois and Heinz Ellenberg, November 1974, 47 pp.
50. *Genetic Structure and Variability in Two Species of Endemic Hawaiian* Drosophila. William W. M. Steiner and Hampton L. Carson, December 1974, 66 pp.
51. *Composition and Phenology of the Dry Forest of Mauna Kea, Hawaii, as Related to the Annual Cycle of the Amakihi* (Loxops virens) *and Palila* (Psittirostra bailleui). Charles van Riper III, January 1975, 38 pp.
52. *Environment-enzyme Polymorphism Relationships in two Hawaiian* Drosophila *Species*. W. W. M. Steiner, January 1975, 28 pp.
53. *A Review of the Hawaiian Coccinellidae*. John R. Leeper, February 1975, 54 pp.
54. *Integrated Island Ecosystems Ecology in Hawaii—Introductory Survey. Part I of Proposed Synthesis Volume for US/IBP Series*. Dieter Mueller-Dombois, February 1975, 46 pp.

55. *Soil Algal Relationships to* Onychiurus folsomi, *a Minute Arthropod.* Linda-Lee McGurk, March 1975, 66 pp.
56. *Cytogenetics of the Hawaiian* Telmatogeton *(Diptera).* Lester J. Newman, March 1975, 23 pp.
57. *Electrophoretic Variability in Island Populations of* Drosophila simulans *and* Drosophila immigrans. William W. M. Steiner, Ki Chang Sung, and Y. K. Paik, March 1975, 20 pp.
58. *Acari on Murine Rodents along an Altitudinal Transect on Mauna Loa, Hawaii.* Frank J. Radovsky, JoAnn M. Tenorio, P. Quentin Tomich, and James D. Jacobi, April 1975, 11 pp.
59. *Climate Data for the IBP Sites on Mauna Loa, Hawaii.* Kent W. Bridges and G. Virginia Carey, April 1975, 90 pp.
60. *Oxygen Consumption, Evaporative Water Loss and Body Temperature in the Sooty Tern,* Sterna fuscata. Richard E. MacMillen, G. Causey Whittow, Ernest A. Christopher, and Roy J. Ebisu, April 1975, 15 pp.
61. *Threshold Model of Feeding Territoriality: A Test with an Hawaiian Honeycreeper.* F. L. Carpenter and R. E. MacMillen, April 1975, 11 pp.
62. *Parasites of the Hawaii Amakihi* (Loxops virens virens). Charles van Riper III, April 1975, 25 pp.
63. *Pollination Energetics and Foraging Strategies in a* Metrosideros-*Honeycreeper association.* F. Lynn Carpenter and Richard E. MacMillen. May 1975, 8 pp.
64. *Seasonal Abundances of the Mamane Moth, Its Nuclear Polyhedrosis Virus, and Its Parasites.* Michael Conant, May 1975, 34 pp.
65. *Temporal Pattern of Gene Arrangement Frequency in Altitudinal Populations of* Drosophila immigrans *on Mauna Loa, Hawaii.* Y. K. Paik and K. C. Sung, May 1975, 14 pp.
66. *Integrated Island Ecosystem Ecology in Hawaii. Spatial Distribution of Island Biota, Introduction. Part II, Chapter 6 of Proposed Synthesis Volume for US/IBP Series.* Dieter Mueller-Dombois and Kent W. Bridges, June 1975, 52 pp.
67. *User Oriented Statistical Analysis Programs: A Brief Guide.* Kent W. Bridges, July 1975, 37 pp.
68. *Systematic Patterns of Foraging for Nectar by Amakihi* (Loxops virens). Alan C. Kamil, July 1975, 17 pp.
69. *The Island Ecosystems Data Bank.* Kent W. Bridges and G. Virginia Carey, August 1975, 15 pp.
70. *Climate Data for the IBP Sites on Mauna Loa, Hawaii.* Kent W. Bridges and G. Virginia Carey, August 1975, 55 pp.
71. *Evolution of the Endemic Hawaiian Cerambycid-beetles.* J. L. Gressitt, August 1975, 46 pp.
72. *Index to Technical Reports 1–66.* Winifred Y. Yamashiro, August 1975, 50 pp.
73. *The Use of Sheep Wool in Nest Construction by Hawaiian Birds.* Charles van Riper III, September 1975, 11 pp.

74. *Spatial Distribution of Bird Species on the East Flank of Mauna Loa.* Sheila Conant, October 1975, 98 pp.
75. *Ecology of Fungi in Wildland Soils along the Mauna Loa Transect.* Martin F. Stoner, Darleen K. Stoner, and Gladys E. Baker, November 1975, 102 pp.
76. *Plant-pollinator Interactions in Hawaii: Pollination Energetics of* Metrosideros collina *(Myrtaceae).* F. Lynn Carpenter, April 1976, 62 pp.
77. *Canopy-associated Arthropods in* Acacia koa *and* Metrosideros *Tree Communities along the Mauna Loa Transect.* Wayne C. Gagné, June 1976, 32 pp.

Taxonomic Index to Organisms

Author Index

Geographic Name Index

Hawaiian entries are followed by island abbreviations: Ha. = Hawai'i, Ka. = Kaua'i, Ma. = Maui, Oa. = O'ahu.

Subject Index

9 781932 846270

www.ingramcontent.com/pod-product-compliance
Lightning Source LLC
Chambersburg PA
CBHW072037020426
42334CB00017B/1306

* 9 7 8 1 9 3 2 8 4 6 2 7 0 *